Proceedings of the International Conference

Geotechnical Engineering

for Disaster Mitigation and Rehabilitation

Proceedings of the International Conference

Geotechnical Engineering for Disaster Mitigation and Rehabilitation

Singapore, 12–13 December 2005

Editors

CHU Jian
Nanyang Technological University, Singapore

PHOON Kok Kwang
National University of Singapore, Singapore

YONG Kwet Yew
National University of Singapore, Singapore

World Scientific

NEW JERSEY • LONDON • SINGAPORE • BEIJING • SHANGHAI • HONG KONG • TAIPEI • CHENNAI

Published by

World Scientific Publishing Co. Pte. Ltd.

5 Toh Tuck Link, Singapore 596224

USA office: 27 Warren Street, Suite 401-402, Hackensack, NJ 07601

UK office: 57 Shelton Street, Covent Garden, London WC2H 9HE

British Library Cataloguing-in-Publication Data
A catalogue record for this book is available from the British Library.

GEOTECHNICAL ENGINEERING FOR DISASTER MITIGATION AND REHABILITATION
Proceedings of the 1st International Conference

ISBN 981-256-469-1

Printed in Singapore by B & JO Enterprise

Acknowledgements

The Editors gratefully acknowledge the following people and organisations for their contributions and support:

- Executive members of the Joint Working Group on Geotechnical Engineering for Disaster Mitigation and Rehabilitation (JWG-DMR) for their dedication and support to the organisation of this conference;
- Ground Technology Association for co-hosting the conference;
- Members of the International Advisory Committee, in particular Dr Za-Chieh Moh and Professor Balasubramaniam, for their support and advice;
- Keynote lecturers and invited plenary lecturers by country
- Special session organisers and panelists for: (1) Extreme Events in Geotechnical Hazards in Asia, sponsored by ACT3 on Geotechnology for Natural Hazards, (2) Seismic Zoning and Geotechnical Data Base, sponsored by ACT10 on Urban Geo-Informatics, and (3) Dealing with future tsunami risks in Thailand, sponsored by the Norwegian Geotechnical Institute (NGI), Coordinating Committee for Geoscience Programmes in East and Southeast Asia (CCOP), and Department of Mineral Resources (DMR), Thailand.
- Gold sponsors — Surbana Consultants Pte Ltd and CPG Consultants Pte Ltd, Silver sponsor — Keller Foundations (SE Asia) Pte Ltd, for their generous sponsorship;
- Lee Foundation for their support;
- Conference Secretariat, Integrated Meetings Specialist Pte Ltd, in particular Mr Ivan Boo and Ms Lee Siew Hon, for their help;
- Support staff and students to the conference, in particular, Ms Meng Guanghui, Dr Darius Wanatowski, and Mr Subhadeep Banerjee.
- World Scientific Publishing Co Pte Ltd's publishing team, in particular Mr Rhaimie Wahap.

CONFERENCE ORGANISERS

Joint Working Group on Geotechnical Engineering for Disaster Mitigation and Rehabilitation (JWG-DMR)
&
Ground Technology Association

SUPPORTING ORGANISATIONS

Australian Geomechanics Society
Chinese Geotechnical Society
Indian Geotechnical Society
Indonesian Society for Geotechnical Engineering
Japanese Geotechnical Society
Korean Geotechnical Society
Southeast Asian Geotechnical Society
Sri Lankan Geotechnical Society
ASCE Geo-Institute
International Consortium on Landslides
ATC3 on Geotechnics on Natural Hazards
Institution of Engineers Singapore (Geotechnical & Transportation Technical Committee)
Institution of Engineers Malaysia (Geotechnical Engineering Technical Division)
National University of Singapore
Nanyang Technological University

ORGANISING COMMITTEE

Conference Chair — Prof K.Y. Yong
Organising Committee Chairs — Dr J. Chu & Dr K.K. Phoon

The Joint Working Group Executive Members

Prof D.T. Bergado, Thailand
Prof B. Indraratna, Australia
Prof S.R. Kim, South Korea
Dr R. Lo, Australia
Dr H.G.P.A. Ratnaweera, Sri Lanka
Dr W. Sengara, Indonesia
Dr S.P.R. Wardani, Indonesia
Dr C.T. Chin, Taiwan
Prof M.M. Kim, South Korea
Prof H.-L. Liu, China
Prof H. Ohta, Japan
Prof N. Seneviratne, Sri Lanka
Ir. Y.C. Tan, Malaysia
Prof J.-M. Zhang, China

INTERNATIONAL ADVISORY COMMITTEE

Dr Z.-C. Moh, Taiwan (Chair)
Prof R. Fell, Australia
Prof K. Ishihara, Japan
Dr M. Jaksa, Australia
Prof S.I. Kim, South Korea
Prof M.R. Madhav, India
Dr F. Nadim, Norway
Prof D. Santoso, Indonesia
Prof F. Tatusoka, Japan
Dr S. Yodmani, Thailand
Prof Z.-M. Zhang, China
Prof A.S. Balasubramaniam, Australia
Dr B.M. Hutapea, Indonesia
Prof A. Lerner-Lam, USA
Prof M. Kamon, Japan
Dr S. Lacasse, Norway
Maj. Gen. S.N. Mukherjee, India
Dr T.A. Ooi, Malaysia
Prof K. Sassa, Japan
Dr W.H. Ting, Malaysia
Prof Yudhbir, India
Eng. K.L.S. Sahabandu, Sri Lanka

Preface

The devastating tsunami disaster that occurred on 26 December 2004 in the South and Southeast Asia has compelled a group of geotechnical engineers and researchers from several countries around the region to come forward to form a Joint Working Group on Geotechnical Engineering for Disaster Mitigation and Rehabilitation (JWG-DMR) in the hope to contribute to the course of disaster mitigation and rehabilitation through their profession. The JWG-DMR is hosted by the Southeast Asian Geotechnical Society and is supported by the Australian Geotechnical Society, the Chinese Geotechnical Society, the Indian Geotechnical Society, the Indonesian Society for Geotechnical Engineering, the Japanese Geotechnical Society, the Korean Geotechnical Society and the Sri Lankan Geotechnical Society. This international conference on Geotechnical Disaster Mitigation and Rehabilitation (GEDMR05), held at the Furama Riverfront Hotel in Singapore from 12-13 December 2005, is one of the activities organised by the JWG-DMR. The GEDMR05 conference is organised jointly with the Ground Technology Association, Singapore, and supported by 15 national societies or organisations.

Seventy-four papers from 17 countries are included in this Proceeding. The papers were selected from more than 100 abstract submissions after a rigorous review process. This Proceedings contains two keynote lectures written by two international renowned experts and 13 Invited Plenary Lectures by Country prepared by leading geotechnical practitioners or researchers from 10 countries which reflect the state-of-the-practice on the topics discussed in this conference. The other 59 papers cover a range of topics including disasters related to earthquake, landslide and tsunami, soil dynamics, risk assessment and management, slopes, disaster mitigation and rehabilitation and others. Some papers presented at the special sessions held during the Conference are also included.

It is hoped that this Proceedings will be a useful source of reference to geotechnical engineers and professionals in other disaster related fields.

Editors
J Chu, KK Phoon and KY Yong

Contents

Landslides and Slopes

Tsunamis

Risk Assessment and Management

Others

KEYNOTE LECTURES

International Conference on Geotechnical Engineering for Disaster Mitigation & Rehabilitation
Chu, Phoon & Yong (eds) © 2005 World Scientific Publishing Company ISBN 981-256-469-1

EXTREME EVENTS IN GEOHAZARDS IN ASIA

TAKAJI KOKUSHO
Chairman of ATC3, ISSMGE.
Faculty of Science and Engineering, Chuo University, Bunkyo-ku, Kasuga 1-13-27
Tokyo, 112-8551, Japan

The 2004 Indian Ocean Tsunami has given us an opportunity to think about extreme events in natural disasters. Data analyses on disasters in 20th century revealed that extreme natural disasters as catastrophic as the Tsunami had actually occurred once in every 16 years in Asia. The reason the Tsunami shocked us so much is that normal people did not expect such an extreme event to really occur in that region. The recurrence time was too long for local population to transmit the disaster experiences to their descendents. It is the most importance task for geo-specialists to imagine what extreme events in geohazards have a probability to occur in Asia and how much we should be prepared for them. Typical examples of extreme events of geohazards in Asia were summarized by reviewing past experiences. Some basic research topics to understand the mechanisms and propose effective countermeasures against extreme geohazards are discussed.

ATC3 ACTIVITIES

Before discussing on extreme events in geohazards, recent activities of ATC3 will be briefly mentioned. The Asian Technical Committee (ATC3), "Geotechnology for Natural Hazards" was established in 1993 to deal with the geotechnical topics concerning earthquake or rain-induced natural hazards. Since then, ATC3 has always been very active in holding workshops in various places and occasions as well as in publishing manuals and proceedings.

Major recent activities include the followings.

- Workshop in Dalian, China, in 1997 on seismic zonation and rain-induced slope failure with international participants from China, Korea and Japan.
- Workshop in Seoul, Korea, in the occasion of the 1999 ISSMGE Asian Regional Conference on seismic zonation and rain-induced slope failure with international participants from Korea, China, Taiwan, Hong Kong, Vietnam and Japan.
- Publication of Zonation Manual for Rain-Induced Slope Failure (Asian Technical Committee No.3, 1997).
- Joint Workshop with the Taiwan Geotechnical Society in Taipei on Geotechnical Hazards during the 1999 Chi-Chi Earthquake.
- Workshop in Hong Kong in the occasion of 2001 SEA Geotechnical Conference on rain-induced slope failure with participants from many Asian countries.

- Two Technical Sessions on "Diversity and Common features of Recent Geotechnical Natural Hazards in Asia" and "Potential Geotechnical Natural Hazards in Asia" in 12 ARC in Singapore in 2003.
- A session on Geotechnical Hazards in the Areas of Collapsible soils during the International Conference on Continental Earthquakes in Beijing, China, in 2004.
- Taiwan-Japan Joint Workshop on Geotechnical Hazards from Large Earthquakes and Heavy Rainfalls, in Taipei, Nov. 2004.
- International Symposium on Geotechnical Aspects of Natural and Man-made Disasters, in Astana, Kazakhstan, June, 2005 with many participants from several central Asian countries, Russia, Japan, etc.

Since most of the natural hazards are closely related with geotechnical engineering, it is an important task for geotechnical engineers to exchange most advanced knowledge and information on geotechnical hazards in the Asian Region to mitigate the threats. ATC3 will be playing a greater role as a forum among Asian geotechnical engineers and researchers for natural disaster mitigation in Asia. Some of the future activities will be;

- Exchange of technical knowledge and information on geotechnical natural hazards in the Asian Region.
- Reconnaissance and investigation of geotechnical natural hazards occurred in the Asian Region in recent years.
- Development or improvement of design methodologies of structures for mitigating geotechnical natural hazards in the Asian Region.
- Promote technical publications on geotechnical natural hazards in the Asian Region.

NATURAL DISASTERS IN ASIA

The natural disasters are classified into several types as shown in Table-1 according to the Center for Research on the Epidemiology of Disasters (CRED, http://www.cred.be), among which several disasters in bold letters in the table are closely connected with geo-science and geo-engineering fields (named hereafter as "geodisasters"); earthquakes, floods, slides, volcanoes, waves/surges and wind storms. Fig.1 shows the regional distribution of damage by the geodisasters from 1990 to 2005, in terms of the number of events, the human fatality, the number of affected people and the economic loss. Obviously, Asia overwhelms the other regions of the world in all the aspects of the damage.

Table-1 Classification of Natural Disasters by CRED (Bold are geodisasters discussed here.)

Drought
Earthquake
Epidemic
Famine
Extrene Temprature
Flood
Insect Infestation
Slide
Volcano
Wave / Surge
Wild Fires
Wind Storm

Fig.2 illustrates the number of occurrence of the major geodisasters in Asia between 1990 to 2005 based

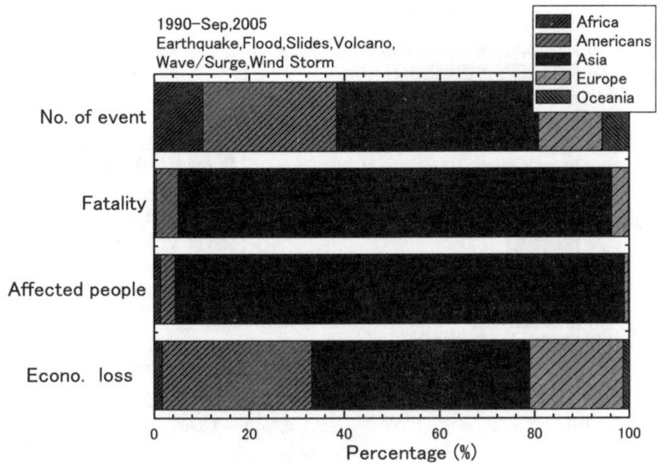

Fig.1 Regional percentage of geodisasters in recent 15 years (1990-2005) based on CRED data

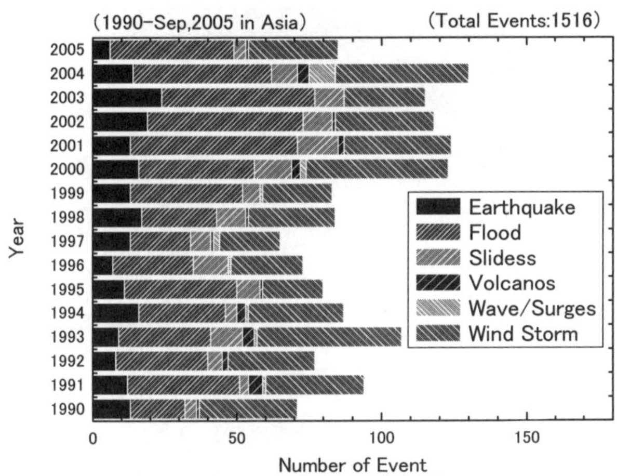

Fig.2 Bar chart on annual number ofgeodisaster events in Asia in recent 15 years (1990-2005) based on CRED data

on the newest statistics provided by CRED. It is remarkable that the number of events tends to increase recently, particularly in these five years, 2000-2004. Fig.2 also reveals that floods occurred most frequently among other natural disasters.

On the other hand, Fig.3 shows the annual loss of human lives by the disasters in the same years. This data does not include the most recent Pakistan earthquake which

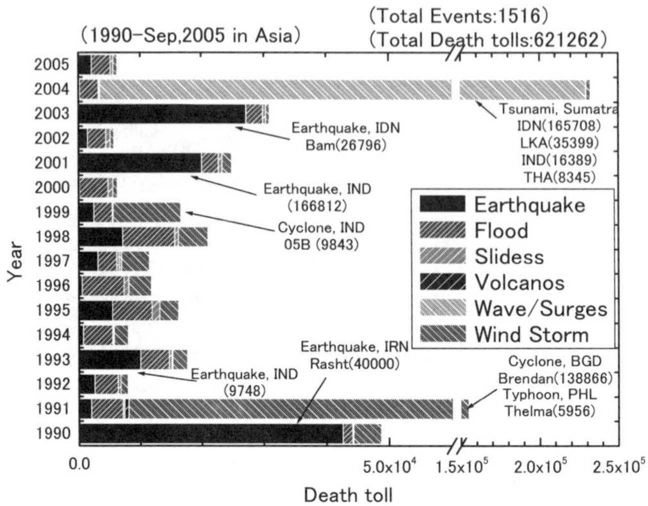

Fig.3 Bar chart on annual number of death toll in Asia in recent 15 years (1990-2005) based on CRED data

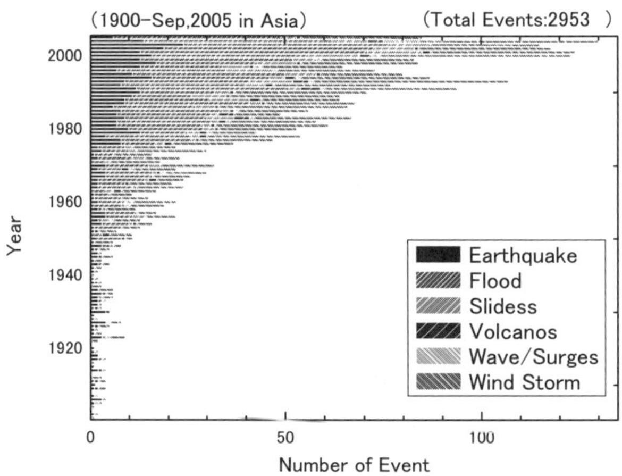

Fig.4 Bar chart on annual number ofgeodisaster events in Asia in the last 105 years (1900-2005) based on CRED data

occurred October 8, 2005. The largest tragedy was the Indian Ocean tsunami in 2004 and the second one the Cyclone flooding in Bangladesh in 1991. Earthquakes claimed a number of casualties other than tsunami in many countries in Asia, particularly in Iran. Thus, water disaster by wind storms or tsunamis and earthquakes are the two major calamities in the past 15 years in terms of fatality. It may also be pointed out for the landslides from Figs.2 and 3 that the fatality is not so high in comparison with the number of events. Totally 620 thousands lives were lost in these 15 years by the geodisasters; annually 40,000 lives on average in Asian region.

Fig.4 shows the annual occurrence of natural disasters in Asia in the 20th century plus the recent five years visualized from the data by CRED, which shows a drastic increase after the World War II. Although there are ups and downs up to now, it looks as almost monotonic increase. CRED database consists of disasters of more than 10 deaths and more than 100 casualties from the beginning of 20th century. It may be too naïve to believe that the chart accurately reflects the reality rather than to suspect it highly influenced by the development of global communication systems and supporting technologies. More and more information on local societies, even a smaller disaster, has been able to be heard in a long distance recently. However, it may also be probable that the number of disasters is actually increasing not so rapidly as in Fig.4 but more gradually because of population growth, higher development activities and rapid urbanizations throughout Asia. Global warming might be deeply involved recently in the increase through stronger tropical storms, heavier rains, etc.

Fig.5 shows annual fatalities during the same period of the 20th century plus 5 years. In contrast to the number of events in Fig.4, the bars are more dispersed with bigger events occurring from time to time throughout the century. It may be because big events could be recorded in documents or reported by news media better than smaller events even in the old times. However, care should be taken because data on older events tend to have greater uncertainties than newer ones. For instance, the death toll in the 1920 earthquake (M8.5) in Ningxia Province in China is 235 x 10^3 by China Earthquake Agency (2004) versus 180 x 10^3 by CRED, and that of the 1927 earthquake (M8.0) in

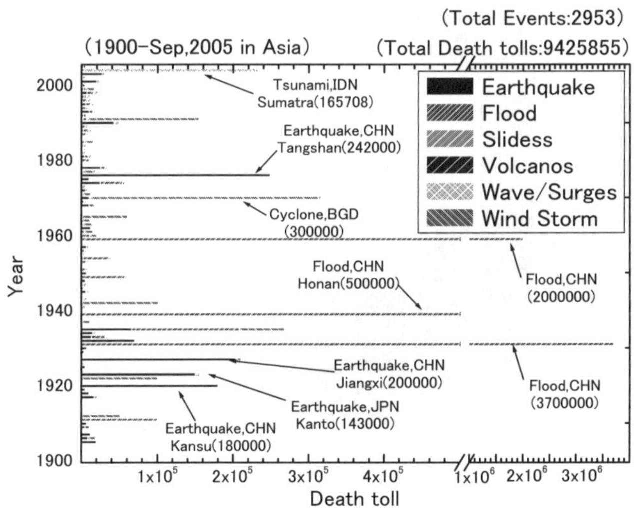

Fig.5 Bar chart on annual number of death toll by geodisasters in Asia in the last 105 years (1900-2005) based on CRED data

Gansu Province in China is 40 x 10^3 by CEA, while it is 200 x 10^3 by CRED. In contrast, for the 1976 Tangshan earthquake (M7.8) again in China, the death toll of 242 x 10^3 coincides in the two databases.

WHAT ARE EXTREME EVENTS IN NATURAL HAZARDS IN ASIA?

When the Indian Ocean Tsunami occurred in December last year, the first impression of almost every people was "Can such a catastrophic event which is really beyond one's expectation ever happen?". Such catastrophic natural hazards may be named as "extreme events", because they are supposed to occur very scarcely but far exceeds more frequent ones in devastation of human lives and social and economical impacts.

The 2004 Indian Ocean Tsunami really seems to be a typical extreme event in natural hazards claiming more than 240 thousands lives in many countries in Asia and Africa and also greatly affected social and economic aspects in the regions. It has given us an opportunity to think about potential extreme events so that we will be able to be prepared for such catastrophic disasters even if their probability of occurrence may not be so high.

In principle, extreme events should far exceed smaller events in their energies or magnitudes. For instance, in earthquakes, the formulation; $\log N = A - bM$ is well known to be applicable to previous events. This indicates that the larger the magnitude M of an event, the smaller number of occurrence N may be expected in a given time interval with constants A and b depending on regional characteristics. Similar relationships may well be anticipated to hold for different types of natural disasters.

Fig.6 depicts a relationship between the death toll by a single event and the number of

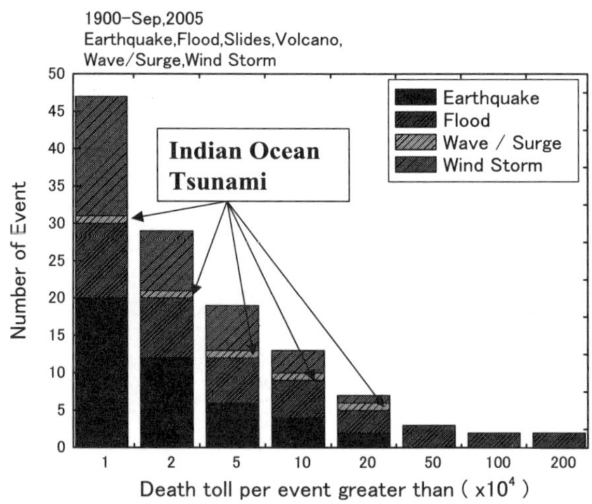

Fig.6 Death toll per event versus number of corresponding events occurred in Asia during 1900-2005

occurrence of the events exceeding that fatality in the 20th century plus 5 years based on the CRED database. It indicates that events of more than 50 x 10^3 deaths have happened 20 times (once in every 5 years on average), those of more than 100 x 10^3 deaths 13 times (every 8 years) and those of more than 200 x 10^3 deaths 7 times (every 16 years). Among the worst events are three floods in China, with the death toll exceeding 500 x 10^3 or even 2~3 million.

There is no clear boundary of the number of fatalities or casualties beyond which the term "extreme events" seems suitable. However, one may well agree that such a big disaster as the 2004 Indian Ocean tsunami in which more than 200 x 10^3 lives were lost is really the extreme event. This size of disaster have actually occurred not so infrequently but rather frequently (once in every 16 years) in the 20th century, indicating that even the extreme events have occurred rather frequently in Asia. The reason we were shocked is that among other disasters it was a single huge tsunami disaster which normal people did not expect to occur in that region. A similar extreme tsunami may have happened once in every hundred years in the same region. This long recurrence time was too long for local population to be well informed of and prepared for.

Thus, even such statistics as shown in Fig.5 may overlook some of really hazardous extreme events which may have happened more scarcely in rural areas, which were seldom recorded in historical documents or archives. In this context, it is the most importance task for geo-specialists to imagine what extreme events in geohazards have a probability to occur in Asia and how much we should be prepared for them.

One of the important aspects in considering the extreme events in future is that human-activities are expanding more than ever before in relation with natural environments and thereby increasing the potential of extreme geo-hazards and also creating new types of complex hazards particularly in rapidly developing areas.

Because of recent population growth and rapid economic development in Asia, mega-cities are expanding on deltaic soft ground along coasts and big rivers vulnerable to various natural hazards. In city areas, new types of facilities which have little experience of severe natural disasters are abundant; subways, skyscrapers, elevated or underwater highways, etc. These new comers may not be well prepared for infrequent natural hazards such as big earthquakes, severe tropical storms, floods, etc. and may lead to future extreme events.

There may be another type of emerging man-made hazards; Oil or gas pumping has caused and is still causing ground subsidence below sea-level in many natural resource rich areas (e.g. Aitaliyev and Baimakhan 2005). Groundwater dewatering for drinking, irrigation or industrialization poses similar ground subsidence problems throughout Asia. The subsided areas are more threatened than before by flooding because dikes protecting them from neighboring water are becoming vulnerable due to subsidence. Tropical storms may trigger severe flooding like in New Orleans which had suffered ground subsidence for a long time. The metropolitan Tokyo has a similar problem, where "the zero-meter low-land area" in the downtown Tokyo occupies about 20 % in terms of area and population of the city, with the lowest place -4 m below the average sea-level. Once a strong near-field earthquake attacks Tokyo and if dikes or other protecting facilities collapse, it may lead to a complex disaster combining earthquake and flood.

Furthermore, mines of coals or metals have produced a number of huge tailing piles or tailing dams in many parts of Asia, which may cause devastating slope failures unless they are carefully protected from heavy rainfalls or seismic effects (e.g. Aleshin and Torgoev 2005).

EXAMPLES OF EXTREME EVENTS IN GEODISASTERS IN ASIA

Asia occupies a wide area on the globe with widely different geological, geographical and climate conditions, which highly affect geodisasters that have occurred and will potentially occur in future. It may not be so easy to properly predict future extreme events in natural disasters in all over the Asia. However, typical examples of extreme events of geohazards in Asia may be briefly summarized by reviewing past experiences as follows.

Floods and wild storms

Flood is actually one of the most damaging natural disasters in Asia. Even in Japan, where flat alluvial plains are not so wide, Isewan-Typhoon in 1959 and its associated surge caused heavy floods in the middle part of Japan, killing more than 5 thousands people.

Much more catastrophic floods associated with heavy rainfall events often occurred on a recurring basis in the middle and lower reaches of the major rivers in China. During the 20th century, major flooding disasters occurred mainly along the Yangtze River Valley. Among them, the flood during July-August 1931 was the most severe, with over 51 million people affected, 3.7 million people perished from this greatest disaster of the century due to disease, starvation or drowning. (http://www. noaanews.noaa.gov/stories /334b.htm). It may well be expected that the hazards will be decreasing drastically because the river system is rapidly improving by construction projects of dams and river dikes currently undertaken in this part of China.

Among wild storm-prone areas in Asia, the Bay of Bengal is particularly vulnerable to cyclone disasters. Two catastrophic events occurred there in the 20th century. The Bhola cyclone in November 1970 with winds in excess of 190 km/h, combined with an exceptionally high tide of 5 to 6 m, drove a tidal surge into low lying, densely populated regions in East Pakistan (now Bangladesh). The result was widespread flooding, with many people being drowned in their sleep. The official death toll was put at 500,000, with 100,000 people missing, making it the deadliest tropical cyclone on record (http://en.wikipedia.org/wiki/1991_Bamgladesh _cyclone). The greatest loss of life and destruction occurred on the low lying islands of the Ganges Delta south of Dhaka. The city of Chittagong was also badly affected.

The next comparable disaster in the area was the 1991 Bangladesh cyclone with winds around 250 km/h, pressure measured at 898 millibars, and a 6 m storm surge. On April

1991 the cyclone struck the Chittagong region of Bangladesh. At least 138,000 people were killed by the storm, with the majority of deaths in the Chittagong area. Most deaths were from drowning, with the highest mortality among children and the elderly. Although cyclone shelters had been built after the 1970 Bhola cyclone, few people knew where to go for shelter with just a few hours of warning. The storm caused an estimated 1.5 billion dollars in damage. The high velocity wind and the storm surge devastated the coastline. Although a concrete levee was in place near the mouth of the Karnaphuli River, it was washed away by the storm surge. The cyclone uprooted a 100-ton crane and smashed it on the Karnaphuli River Bridge, breaking it into two sections. A large number of boats and smaller ships ran aground.

Thus, really catastrophic events claiming more than 100 thousands human lives had occurred from time to time in coastal low-land areas by huge waves or surges by tropical storms. Recently, the Indian Ocean Tsunami did impose similar damage on coastal areas in many countries. One of the countermeasures to such a catastrophe may be to construct earth-structures with high resistivity against wave action or scour. Such earth-structures may take forms of seawalls, dikes or mounds for temporary refuge, which should best fit to environments and histories of local societies. They should be constructed easily and inexpensively by local people by local soils and still have to keep the resistivity for long time ahead. Based on the experience during the 1991 Bangladesh cyclone, in which refuge facilities were not well known by general public, some special considerations will be needed so that such earth-structures should be respected and taken care of by local people as a memorial of past extreme disaster, and the disaster experiences should be transferred to many generations to go.

Earthquakes and tsunamis

Among the earthquake disasters, the collapse of houses or buildings by seismic shaking is said to be responsible for three fourths of the total casualties in history (Coburn A. and Spence R. 2002). However, seismically induced landslide and tsunami have also been responsible for a great number of casualties and can lead to extreme events in future.

It is clear from the global distribution of historical large earthquakes that almost all countries in Asia cannot be exempt from earthquake hazards. The east and south rim of the Asian continent is surrounded by near-shore earthquake belts where ocean plates are subducting beneath the continental crust. The 1923 great Kanto earthquake which claimed more than 140 thousands lives in and around Tokyo is one of them. In addition, a number of huge earthquakes occur along the inland plate boundaries like the one near Himalaya mountains or within the continental crust. Earthquakes in the offshore subduction zones sometimes caused devastating tsunamis such as the 2004 Indian Ocean tsunami, the 1896 Meiji Sanriku Tsunami in Japan claiming 22thousand lives and the 1960 Chili-Tsunami in the Pacific Ocean.

Inland earthquakes of large magnitudes are also devastating because they often occurred in shallow depths beneath big cities. The 1976 Tangshan earthquake (M7.8) in China which claimed 242 thousands lives are one of the typical extreme events by near ield earthquakes. More recently inland earthquakes occurred in Rasht, Iran (1990, M7.4), Kobe, Japan (1995, M7.2), Turkey (1999, M7.4) in Gujarat, India (2001, M7.9) and in Bam, Iran (2003, M6.6) claiming large number of human lives. The most recent one is the Pakistan earthquake (M7.6), which occurred on October 8, 2005, at the north border near India. The focal depth was very shallow, approximately 10 km, and caused countless building damage and slope failures, killing more than 75 thousands people.

Inland shallow earthquakes also cause ground displacements due to fault movements. According to experiences in Japan, the fault ruptures are considered to appear at the ground surface if the earthquake Magnitude is larger than around 6.5-7.0. A potential hazard due to fault-induced ground ruptures or displacements has been recognized through damage experiences of recent earthquakes; such as the 1999 Chi-Chi earthquake (M7.7) in Taiwan, the 1999 Kocaeli earthquake (M7.4) in Turkey, etc. Though the fault displacement is normally smaller than several meters, it may sometime become almost 10 meters like in the Taiwan earthquake, which may trigger extreme hazards by damaging very important structures such as high dams, high-speed railways, etc.

Seismically induced landslides

In mountainous areas prone to landslides by earthquakes and heavy rainfalls, a greater number of people are living now. Local soils; loess, decomposed granite, volcanic ashes make landslides very site-specific throughout Asia. In loess plateau areas spanning from China to the central Asia, near field earthquakes caused huge landslides from time to time.

Among them, the Haiyuan 8.5 earthquake in 1920 induced dense landslides in loess deposits of about 4,000 km^2 area, which buried numerous villages and caused 234,000 people died (Wang et al. 1998). The loess is not local soil but widespread in Central Asia and also in Central North America. It keeps relatively high strength normally but becomes collapsible by increasing water content. The loess soils mostly consist of silts with low plasticity with their plasticity index about 10 or less. If the water content is higher than the plastic limit, the excess pore-water pressure obviously increases by seismic loading, leading to partial or complete liquefaction.

Among a number of huge landslides, the Shibeiyuan landslide is believed to have occurred due to liquefaction. During the earthquake, the rising of ground water table is thought to have saturated late Pleistocene loess in the depth of 10-16 m. Under the effect of earthquake, the soil body of gentle slope of 2.5% slid forward by 1.5 km (Wang et al. 1996). On the other hand, an earthquake, the Magnitude of only 5.5 and the maximum acceleration 0.15 G caused a series of devastating landslides accompanied by a large scale mud flow. The reasons of such events are thought to be the collapsible nature of highly porous loessal silt which had been wetted by irrigation water. The silt-sized soil constituting the loess was of low plasticity and could easily slump and flow through a distance as long as 2.0 km (Ishihara et al. 1990). Similar destructive land slides of loessal slopes occurred during the 1949 Chait earthquake in USSR and also repeatedly by several other earthquakes in the same area. In every case, heavy rain had wetted the slopes before the earthquakes presumably decreasing the strength.

On the other hand, the most of failed slopes during the 1920 Haiyuan earthquake seem to have been dry. Seed (1968) speculated that pore air pressure if not pore water pressure caused the flow-type failure of the loess slopes. The mechanism of post-liquefaction flow in such a long distance as more than 1 km is not exactly known yet. Similar devastating flow-type slope failures of the unsaturated condition have also been observed in Taiwan during the 1999 Chi-Chi earthquake and the 2001 El Salvador earthquake (Earthquake Engineering Committee, 2001). More research efforts are certainly needed to clarify the exact mechanism of the long distance flow of unsaturated slopes.

Countermeasures by soil improvement or any other countermeasures to such devastating disasters seem almost impossible within reasonable cost. Establishing reliable hazard maps considering such long-term geotechnical hazards and urban planning based on it may be most crucial to avoid such calamities.

Huge debris flows caused by moraine lake failures

Due to global warming, glacial lakes are retreating, expanding vast moraine lakes behind in high altitude regions of central Asia. Threats of gigantic debris flows triggered by breaching are increasing as future extreme events. The following is a valuable case history on remedial measures for this type of hazards carried out by Kazakhstan people (Kazakhstan Geotechnical Association 2004).

The city of Almaty, the former capital of Kazakhstan, with its population of more than 1.5 million is situated at the foot of Tyan-Shan Mountains. Two rivers are running from the mountains through the city, which are considered as the most hazardous in terms of debris flow not only in Kazakhstan but also in the nearby region. More than 10 breach-hazardous lakes, 2 large mud notches and more than 30 extremely dangerous colluvial deposits were threatening the city. The disastrous debris flow in 1921 along the river Malaya Almantika brought 3 million cubic meters of rock and mud mixtures destroying houses and killing 20% of the population. Similar debris flow occurred quite often, 90 % of which happened during the summer time due to moraine lakes breaching.

To protect the city from debris flows, a project started in the late 1960s, including constructions of retention dams, consolidations of riverbeds, developing monitoring and alerting posts, etc. The first large dam was a rock-fill dam (height; 150 m, the crest length; 800 m) constructed by a blasting method. A large volume of the dam material was laid by pin-point blasting of slopes on both sides of the river. This construction method using totally 9400 tons of explosives contributed to substantial reduction of the construction period. The dam is equipped with no impermeable core zone, but with water intake units, draining water with a rate of up to 30 m^3/s from any elevation. The second dam, constructed along the other river, Bolshaya Almatinka, was a reinforced concrete cellular dam (height: 40 m, length: 400m), the inside of which was filled with river sediments.

In July 1973, a debris flow caused by the breach of a moraine lake at the upper reach of Malaya Almantika occurred, with its average speed of 10 m/s, flow rate 5200 m^3/s, average density 2.3 t/m^3 and the total entrapped volume was 5.5 million m^3. The crest of the dam, which was initially 110 m high, was just about over-flowed and the dam was raised to the present height of 150 m after this event. In 1977, another catastrophic debris flow occurred along the other river, with the total volume of 6 million m^3. The both catastrophic debris flows were stopped successfully by the retention dams which had been completed in the timely manner. In order to drain water separated from debris flow, sedimentation ponds with the capacity of more than 150 thousands m^3 were constructed and a 20 km long water channel was excavated through the city of Almaty. The number of larger debris flows is decreasing in recent years because of retention dams in the upper streams, consolidations of riverbeds and controlling water drainage of the moraine lakes. Thus, the city of Almaty has succeeded to control catastrophic huge debris flows which otherwise could be one of the extreme events.

Volcanic hazards

Though really extreme disasters did not occur in the 20th century, Asia has a large inventory of active volcanoes and their eruptions are highly hazardous in a long run. Japan suffered a lot of disasters by volcanic eruptions in the long history, such as Mt. Asama in 1783, Mt. Unzen in 1792 and 1991, Mt. Bandai in 1888, etc. killing many people. Among them, Mt. Unzen in 1792, at the end of the eruption, suddenly collapsed at its slope and 440 million m^3 debris rushed into seawater, generating huge tsunami and killing 15 thousands people (Ito, 2002). Let us briefly review representative big events which occurred relatively recently in Indonesia and Philippine in 19th and 20th centuries (http://en.wikipedia.org/wiki/Volcanic_eruption).

In 1815, the volcano of Tambora on one of the Indonesian islands suffered the most violent eruption in modern times. Approximately 92,000 people were killed because of pyroclastic flows or subsequent starvation and disease. 1815 became known as "the year without summer" because of the extreme weather conditions following the eruption. The explosion ejected an estimated 100 cubic km of ash, leaving a caldera 7 km across making the original mountain height of 4200 m to 2851 m after the explosion.

Krakatoa is a volcano near the Indonesian island of Rakata in the Sunda Strait. It has erupted repeatedly, massively and with disastrous consequences throughout recorded history. The best known 1883 eruption ejected more than 25 cubic kilometres of ash. 36,417 people died and uncountable people were injured mostly in tsunami which followed the biggest explosion. The eruption destroyed two-thirds of the pre-existing island. Prior to Krakatoa's massive eruption of 1883, the volcano lay dormant for two centuries. The volcano is still active, with its recent eruptive episode having begun in 1994.

Mount Pinatubo is an active volcano located on the island of Luzon in the Philippines. Before 1991, the mountain was inconspicuous and heavily eroded. It was covered in dense forest which supported a population of several thousand indigenous people. The volcano's most recent eruption in June 1991 came after 500 years of dormancy, and produced one of the largest and most violent eruptions of the 20th century. Successful predictions of the onset of the climactic eruption led to the evacuation of tens of thousands of people from the surrounding areas, saving many lives, but the surrounding areas were severely damaged by pyroclastic flows, ash deposits, and later, lahars caused by rainwater remobilizing earlier volcanic deposits and thousands of houses were destroyed. Several important river systems have their sources on Pinatubo, but the eruption filled many valleys with deep pyroclastic deposits. Since 1991, the rivers have been clogged with sediments, and the valleys have seen frequent lahars, suggesting that the river systems will take many years to recover from the eruption. The eruption ejected about ten cubic kilometers of material, some ten times larger than the 1980 eruption of Mount St. Helens. Though about 300 people were killed, the evacuation in the days preceding the eruption certainly saved tens of thousands of lives as a great success for eruption prediction.

One of the important lessons from the above cases is that most violent eruptions did occur after long dormancy of more than a few hundred years. The Mount Fuji located only 100km from Tokyo is dormant for about 200 years. Recently, more and more attention is paid by local governments and specialists for the potential hazards in the near future. Compared to earthquakes or tsunamis, it seems nearly possible to predict volcanic eruptions and save human lives by evacuations such as the success story in Pinatubo. However, material damage mainly by volcanic ash or subsequent mudflow disasters is the subject in which geo-engineers should deeply be involved.

BASIC RESEARCH NEEDS TO PREPARE FOR EXTREME EVENTS IN GEOTECHNICAL ENGINEERING

What can geo-scientists or geo-engineers do to mitigate extreme geo-hazards in future? Most people feel that it is the problem for government officials or social planners rather than for geotechnical researchers. However, the basic research will also be contributing in a long run in order to understand their basic mechanisms and to propose effective preparations to countermeasure potential extreme events in future. The followings are some of basic research subjects from the present author's perspective that geotechnical engineering should focus upon;

Geotechnical conditions affecting building damage during earthquakes

Some of the recent destructive earthquakes, such as the 1995 Kobe earthquake and the 1999 Kocaeli earthquake, indicate that the soil conditions are very influential to the degree of building damage. It is sometimes pointed out that, during really strong shaking in recent near-field earthquakes, building damage tends to decrease in soft soil sites than stiff soil sites. This trend is particularly pronounced in liquefied ground. It is probably because the liquefied soil serves as a base isolator or a part of the seismic energy is dissipated within the liquefied layer. The vertical array record in the Port Island in Kobe clearly demonstrated the base isolation effect (Kokusho et al. 1996).

It has been generally believed in Japan and probably in many other countries that earthquake damage concentrates in soft soil sites more than in stiff soil sites. However, during the Kobe earthquake, RC buildings and civil engineering structures suffered heavier shaking damage in inland competent soil sites than in soft and liquefiable reclaimed ground (Matsui and Oda, 1996). Though the damage concentration in the narrow inland belt may have reflected the fault mechanism including the directivity effect and the basin edge effect (Irikura 1998), the near-surface soil condition seems to have a great effect on the drastic decrease of damage along the coast. A similar damage distribution was again observed in Adapazari during the Kocaeli earthquake, in which the structural damage of buildings were relatively light in liquefied areas in the southern part of the city (The Japanese Geotechnical Society 1999).

In near-field earthquakes of shallow focal depths, the seismic wave energy comes up to the surface without great attenuation, causing severe shaking damage even in relatively competent soil sites. Though the resonance effect in structures and surface soil layers also plays an important role in structural damage, it may be said in general that liquefiable or soft soil layers tend to absorb a large part of incident seismic energy by the hysteretic soil damping under induced large strain, leading to lighter structural damage by inertia force than competent soil layers (Kokusho and Motoyama 2002). Of course, this benefit is in many cases a trade-off of geotechnical damage such as ground subsidence, foundation failure, etc.

Many of recent earthquakes which claimed more than thousands or tens of thousands human lives were very shallow earthquakes. For example, during the 2003 Bam earthquake in Iran, more than 25 thousands people were killed by the collapse of non-engineered adobe houses. Though the magnitude (M_w) was only 6.6, the focal depth was only 10 km. The soil condition there was rather stiff with no possibility of liquefaction or strong nonlinearity. It would not be so difficult to imagine that the adobe houses would not have been so severely destroyed if the surface soil were much softer and absorbed a large part of the incident wave energy. The effect of soil conditions on structural damage is one of the important research topics to prepare for future shallow destructive earthquakes.

Another potential geotechnical earthquake hazards may be extraordinary seismic amplification of long period seismic motions which potentially causes severe damage of tall buildings or sky scrapers recently constructed in big cities. It is well-known that, during the 1985 Mexican earthquake, the Mexico city, more than 300 km far from the epicenter, suffered the most severe damage because of thick soft clay deposits underlying the city. The soil, originated from lake deposits, was very peculiar in its exceptionally high water content, low S-wave velocity and ultra low damping characteristics even under large shear strain. The soil deposit in a basin, on which the city rested, amplified the long period seismic vibration tremendously, leading to the collapse of many tall buildings by the resonance effect.

Though the damage was much smaller than in Mexico, a similar phenomenon did occur in Ahmedabad, 230 km from the epicenter, during the 2001 Gujarat earthquake in west India. Though, the soil there might not be so special as the Mexico city, buildings of about 10 stories were selectively damaged presumably by the resonance effect of a thick soil deposit underneath. Not only vertically propagating SH waves but also horizontally traveling surface waves are likely to have been involved in such seismic damage in a long distance.

Some of the big cities in Asia are located on deep alluvial plains or basins such as Tokyo, Shanghai, Jakarta, Bangkok, etc. The soils are soft near the surface and deeply deposited. Although epicenters may not be so near, tall buildings including skyscrapers may be heavily damaged in resonance with long-period motions stimulated by surface waves unless those structures are properly designed against such seismic resonance. Suspension bridges and other long period structures need special considerations. Oil tanks covered by floating roofs may suffer structure damage and fire due to oil sloshing excited by the long-period motions as actually observed recently in Tomakomai City in Hokkaido, Japan, during the 2003 Tokachi-oki earthquake (M8.0). As a fundamental step, geotechnical and geophysical surveys from surface soils down to deep rock base are really necessary to evaluate the future potential of such events and to prepare for the worst.

Mechanism and evaluation in flow-type slope failures

Slope failures or landslides have occurred much more often by heavy rainfalls than by earthquakes. Though the post shaking mechanism is very similar to each other, seismically triggered landslides are much larger in scale and tend to be more devastating. The earthquake-induced slides are often followed by a series of sustained slope failures by subsequent rainfalls for a long period as observed in the Tateyama caldera after the 1858 Hietsu earthquake (M7.1) in Japan or in mountainous areas in Taiwan after the 1999 Chi-Chi earthquake (M7.7). On the other hand, heavy rainfalls prior to earthquakes often result in heavier earthquake-induced damage due to greater sliding soil mass and longer run-out distance. Sometimes, irrigation water tends to play a role of heavy rainfall in arid areas in making the slide much more devastating like the case in Tajik reported by Ishihara et al. (1990). Thus, earthquake-induced slope hazards should properly be considered under potential effects of rainfall or other wetting conditions.

The safety against slope failure is normally computed by conventional slip surface analyses, though it cannot evaluate post-failure deformation. Slope failures become more hazardous if they develop into flow-type failures. At present, few method including the Newmark-type method can reasonably evaluate the run-out distance once flow-type failure occurs, though it is one of the most significant parameters in establishing hazard maps due to slope failures.

Normally water is deeply involved in the process of flow-type failures, such as debris flows or mud flows. However, there are quite a few case histories in which unsaturated soil or dry rock debris was involved in flow slides. An empirical relationship is proposed in which the larger the volume of flow, the further it arrives with smaller internal energy dissipation (Hsu 1975). The mechanism of such granular flows remains a lot of mystery still, though there are several mechanical models and analytical tools proposed from a viewpoint of the hydraulic mechanics. On the other hand, approaches from the granular mechanics are scarce, although debris flow is sometimes composed of dense gravelly soils of the density as high as 2.3 t/m^3. The effects of soil density, particle gradation, fines content, crushability, excess pore-pressure, etc. on the run-out distance, the flow velocity, the impact pressure and so on are really important from the perspective of hazard mitigation.

As experienced in Almaty, Kazakhstan, huge debris flows are expected to occur increasingly from China to central Asia because of retreating glaciers and expanding moraine lakes in high altitude regions due to global warming. Most of these areas may not so well prepared to protect villages and towns from devastating debris flows. In order to predict the hazards more reliably, the understanding of the basic mechanisms from geotechnical viewpoint is essential.

Mechanism and evaluation of fault-induced ground displacement

Significant potential hazard due to fault-induced ground ruptures or displacements has increasingly been recognized through damage experiences of recent earthquakes. During the 1999 Chi-Chi earthquake in Taiwan, among various civil engineering structures and buildings affected by the fault movement, a gravity dam was ruptured by a thrust fault of maximum 9 m displacement. Though the dam did not fail completely and flood in the downstream was avoided, the event told us that the fault-induced ground displacement can be a great threat to the social safety. During the 1999 Kocaeli earthquake in Turkey, too, a strike slip fault and its associated fault groups displaced rail roads, highways, bridges, pipelines, etc. and also inundated vast areas along the fault by subsidence of a few meter, submerging houses and other facilities along a coast. For the oil-pipeline in Alaska, the effect of the fault displacement has been already considered in its siting and construction. For other important structures, too, the fault displacement is increasingly considered as an essential factor in seismic design.

One of the basic research subjects in geotechnical engineering for predicting the hazard is how to evaluate the ground deformations above earthquake faults. The shearing mechanism of overlying layers by different types of fault movements should be investigated to evaluate how the shear zone develops toward the ground surface. The soil structure interactions between surface structures and foundation ground or between buried structures and ambient soils under the influence of fault movement should also be investigated along with accumulating database from case histories in the past.

Mechanism and evaluation of submarine slides

One of geotechnical hazards on which geotechnical engineering has not focused so much is submarine slides. They are classified into two types. The first is the slides which occurred at the coasts and affected coastal areas in deltaic lands or man-made fills. For example, Valdez and Seward, port cities in Alaska, USA, suffered a great loss of human lives and properties by large scale submarine slides involving coastal areas during 1964 Alaskan earthquake (Coulter et al., 1966; Lemke et al., 1967). The inclination of the sea bed was steep near the beach, but actually 5° or less on average considering the long span of the slip surface of more than 2 km offshore, which is considerably much less than the internal friction angle of the soil. The analogous submarine failure recently occurred during the 1999 Kocaeli earthquake along the southern coast of the Izmit bay in Turkey (The Japanese Geotechnical Society, 2000). The slope of the sea-bed was about 10° consisting of coarse sand and gravel. The near-shore slides can be very hazardous because it sometimes involves coastal lands where cities or important facilities are located.

The second type is the submarine slides which occur offshore in the sea bottom. The slope is normally gentler than a few percent or even less than 1% (Hampton and Lee 1996). A typical example of this type occurred 60 km off the California Coast during a 1980 medium magnitude earthquake (Field et al. 1982). In this case, 2 by 20 km sea floor consisting of sand and mud of only 0.25 degrees slipped and became completely flat with some evidences of liquefaction such as sand boils on the sea bottom of 30 to 70 m depth. The flow failure in liquefied soil deposits seems to have been involved also in submarine slides triggered seismically.

There are possibly many coastal zones with high seismicity and relatively steep sea floor consisting of loose sand or gravel in Asia where submarine slide is likely to occur. Basic research is needed to understand its basic mechanism. As one of possible mechanisms to explain slides of low-gradient slopes, void redistribution effect may be considered (Kokusho 2003), which had been introduced in a committee report in US (National Research Council, 1985) and also discussed by Seed (1987) by using a special term, water interlayer. In this view, fine soil sublayers sandwiched in sand deposits are considered to play a key role in flow failure. This may be able to explain, as Seed (1987) already pointed out, why the steady-state strength of uniform sand leads significantly higher values of residual strength than those estimated by the case studies in the field.

The formation of water interlayers or water films between liquefied sand and overlying lower-permeability seams has been observed under level ground conditions in a number of model tests (e.g. Scott and Zuckerman 1972). For sloping ground conditions subject to lateral spreading, Kokusho et al. (1998) and Kokusho (2003) demonstrated, based on model tests, that the water film or void-redistribution effect plays an important role in post-earthquake large lateral flow in liquefied ground. The stress-strain response of sand to the inflow of pore water from adjoining zones under sloping ground conditions was investigated through soil element tests by Boulanger and Truman (1996).

Thus, liquefaction may be highly responsible in earthquake-induced submarine slides, particularly in near-shore sites where the seabed is composed of sand or gravel. However, in off-shore slides far from coasts, sea floors are normally composed of cohesive fine soils, in which liquefaction seems to hardly occur. Off-shore submarine slides are sometimes really huge in scale and suspected as potential causes of tsunami other than fault dislocations. Needless to say, much more research is needed to understand the mechanism correctly. Thus, a concern on submarine slides from geotechnical point of view is necessary as one of potential future extreme events in natural disasters in Asia.

CONCLUDING REMARKS

In this article, after the activities of ATC3 was briefly introduced, a concept of extreme events of natural hazards in Asia was first introduced by reviewing past natural hazards based on CRED data. It revealed that extreme natural disasters as catastrophic as the Indian Ocean Tsunami had actually occurred once in every 16 years in Asia. The reason the Tsunami shocked us so much is that normal people did not expect such extreme event to occur in that region because the recurrence time was too long for local population to transmit the disaster experiences to their descendants. Hence, it is the most importance task for geo-specialists to imagine what extreme events in geo-hazards have a probability to occur in Asia and how much we should be prepared for them.

Then, some examples of extreme events in Asia closely related to geotechnical engineering were addressed. Furthermore, some topics of basic geotechnical research needed to prepare for the extreme events are discussed together with future research needs.

It is needless to say that social planning, supporting budgets and social preparedness are really essential to mitigate extreme geo-hazards, effectively. However, the basic research in geotechnical engineering will also be contributing in a long run in order to understand the basic mechanisms and to propose effective countermeasures of potential extreme events in future.

ACKNOWLEDGMENTS

Tomohiro Ishizawa, PhD student in Science and Engineering Graduate School of Chuo University, Japan, is gratefully acknowledged because of his great contributions in analyzing the natural disasters in Asia based on the CRED data.

REFERENCES

Aitaiyev, S. M. ad Baimakhan, R. B. (2005): The possible negative phenomena of post-Tengiz: Pilot-model, *Proc. of the International Geotechnical Symposium on Geotechnical Aspects of Nastural and Man-made Disasters,* Astana, Kazakhstan, 123-128.

Aleshin, Y. G. and Torgoev, I. A. (2005). Geomechanical and geoechological resk sinergetics of uranium miningd in Kyrgyzstan, *Proc. of the International Geotechnical Symposium on Geotechnical Aspects of Nastural and Man-made Disasters,* Astana, Kazakhstan, 129-133.

Asian Technical Committee No.3 (1997). Manual for zonation on areas susceptible to rain-induced slope failure, Japanese Geotechnical Society.

Boulanger, R. W., and Truman, S. P. (1996). .Void redistribution in sand under post-earthquake loading.. *Canadian Geotech. J.*, Vol 33. pp 829-833.

China Earthquake Administration, (2004). "Earthquake disaster mitigation in China", CD publication distributed during International Conference on Continental Earthquake, Beijing, China.

Coburn A. and Spence R. (2002). "Earthquake protection", John Wiley & Sons, 2nd Edition.

Coulter, H. W., and Migliaccio, R. R. (1966). "Effects of the Earthquake of March 27, 1964 at Valdez, Alaska", *Geological Survey Professional Paper* 542-C, U. S. Department of the Interior, 36.

Earthquake Engineering Committee. (2001). "JSCE Investigation Report of Earthquake; 1. The January 13, 2001 Earthquake occurred off Coast of El Salvador, 2. The 2001 Kutch Earthquake, Gujarat State, India", Japan Society for Civil Engineers.

Field, M. E., Gardner, J. V., Jennings, A. E. and Edwards, B. D. (1982). "Earthquake-induced sediment failures on a 0.25° slope, Klamath River delta, California", *Geology*, V.10, 542-546.

Hsu, K. J. (1975). "Catastrophic debris streams (Sturzstroms) generated by rockfalls", *Geological Society of America Bulletin*, V.86, Doc. No. 50117, 129-140.

Hampton, M. A. and Lee, H. J. (1996). "Submarine landslides", *Reviews of Geophysics*, 34, pp.33-59.

Ishihara, K. (1990). "Liquefaction-induced flow slide in the collapsible loess deposit in Soviet Tajik", *Soils and Foundations*, Vol.30, No.4, 73-89.

Irikura, K. (1998). "Seismic motions, Report on Hanshin-Awaji Earthquake Disaster", *Report on Hanshin-Awaji Earthquake Disaster*, (in Japanese), pp.261-280.

Ito, K. (2002). "Japanese history on earthquakes and volcanic eruptions", *Iwanami Sinsho*, Iwanami Publishing Co. (in Japanese).

Kazakhstan Geotechnical Association (2004). "Mud protection system of the City of Almaty", Handouts for Field Trip during the Conference for Geotechnical Problems on Construction of Large-Scale and Unique Projects. Almaty, Kazakhstan.

Kokusho,T., Matsumoto,M. and Sato,K. (1996). "Nonlinear seismic properties back-calculated from strong motions during Hyogoken-Nambu EQ", *Proc. World Conference on earthquake Engineering* (Acapulco), CD publication.

Kokusho,T., Watanabe,K. and Sawano,T. (1998). "Effect of water film on lateral flow failure of liquefied sand", *Proc. 11th European Conf. on Earthquake Eng.* (Paris), CD publication, ECEE/T2/kokeow.pdf.

Kokusho, T. and Motoyama, R. (2002). "Energy dissipation in surface layer due to vertically propagating SH wave", *Journal of Geotechnical and Geoenvironmental Engineering*, ASCE, Vol.128, No.4, 309-318.

Kokusho, T.(2003). "Current state of research on flow failure considering void redistribution in liquefied deposits", *Soil Dynamics and Earthquake Engineering*, Elsevier, Vol.23 , 585-603.

Lemke, R. W. (1967). "Effects of the Earthquake of March 27, 1964 at Seward, Alaska", *Geological Survey Professional Paper* 542-E, US Depart of Interior.

Matsui, T. and Oda, K. (1996). "Foundation damage of structures", *Special Issue of Soils and Foundations*, pp.189-200.

National Research Council (1985). "Liquefaction of soils during earthquakes", Committee of Earthquake Engineering, Commission of Engineering and Technical Systems, National Academy Press, Washington, D.C.

Scott, R. F. and Zuckerman K. A. (1972). "Sandblows and liquefaction", *Proc. The Great Alaska Earthquake of 1964-Engineering Publication 1606*, National Academy of Sciences, Washinton, D.C., 179-189.

Seed, H.B. (1968). "Landslides during earthquakes due to soil liquefaction", Proc. ASCE, Vol.94, No. SM5, 1055-1122.

Seed,H.B. (1987). "Design problems in soil liquefaction", *Journal of Geotechnical Engineering*, ASCE, Vol.113, No.8, 827-845.

Technical Committee No.4. (1999). "Manual for zonation on seismic geotechnical hazards (Revised version)", the Japanese Geotechnical Society.

The Japanese Geotechnical Society. (1999). "Investigations of the 1999 Kocaeli earthquake", *Reports on the Investigations of the 1999 Kocaeli Earthquake in Turkey and the 1999 Chi Chi Earthquake in Taiwan.*

Wang, L., Zhang, Z. Lin, L. and Wang J. (1996). "Laboratory study on loess liquefaction", *Proc. 11th World Conference on Earthquake Engineering*, Paper No.1681. CD publication.

Wang, L., Zhang, X., Lin, X. and Yuan Z. (1998). "The recent progresses in loess dynamics and its application in prediction of earthquake-induced disasters in China", *Proc. 11th European Conference on Earthquake Engineering*. CD publication.

International Conference on Geotechnical Engineering for Disaster Mitigation & Rehabilitation
Chu, Phoon & Yong (eds) © 2005 World Scientific Publishing Company ISBN 981-256-469-1

NATURAL DISASTERS IN TAIWAN

ZA-CHIEH MOH
*MAA Group Consulting Engineers, 11th Floor, No. 3, Tunhwa South Road, Section 1
Taipei 105, Taiwan*

DANIEL T. C. YAO
*MAA Group Consulting Engineers, 11th Floor, No. 3, Tunhwa South Road, Section 1
Taipei 105, Taiwan*

Taiwan is an island that lies on the western edge of the Pacific-rim earthquake belt, which is an extremely active tectonic region with seismic activities among the highest in the world. The boundary of the Eurasian and Philippine Sea Plates is near the island of Taiwan. Earthquake occurs frequently, and a recent catastrophic event, the Chi-Chi earthquake, is known to the world. Taiwan also is located in the subtropical climate region with high average temperature and precipitation. May and August are periods influenced by continuous seasonal rains that occasionally come with an extraordinary intensity. From summer to fall, Taiwan is affected by typhoons. Some extreme events would accumulate significant rainfalls within a relatively short period of time and result in disastrous consequences such as floods. Given the topography of Taiwan, which is covered by slopelands and high mountains over two third of the island, landslides and debris flows often were triggered after heavy rainfalls. The remaining quarter of the land is plains, but they have been densely inhabited and developed. Due to excess extraction of groundwater for fish farming, significant portion of the plains have suffered severe ground subsidence over the past decades. As a result, Taiwan has to cope with various hazardous situations caused by the Mother Nature. Most significant natural disasters in this island include earthquakes, floods, landslides, debris flows, and land subsidence. The government and people in Taiwan have long being aware of the disastrous consequence of these natural hazards, and continuous efforts have been placed on prevention and mitigation of these natural disasters. This paper discusses these natural disasters that confront the people in Taiwan. Studies and investigations carried out to understand the characteristics of these natural hazards are discussed, and development of programs and approaches in preventing and mitigating natural disasters in Taiwan are also addressed.

INTRODUCTION

Since the end of the 20th century, natural disasters started to take heavier tolls than ever on human lives and properties around the world owing to the unprecedented global climate variation and accelerating urbanization in mankind history. Among all kinds of natural disasters in the world, the earthquakes, followed by floods and fierce winds (typhoons, hurricanes, tornados, etc.), are the most catastrophic and have taken the most monetary loss in urban areas. The damages caused by earthquakes, floods, and fierce winds have totaled almost 90% of the monetary loss from all the natural disasters worldwide in the past decade (Guy Carpenter & Company, 2000). In addition, the death tolls of earthquakes ranked the highest and were the major part of human live loss in natural disasters in the world history. Due to the continuous increase of live and monetary losses from natural disasters, prevention and mitigation of natural disasters became an urgent issue for governments and people in areas with high natural hazards potentials.

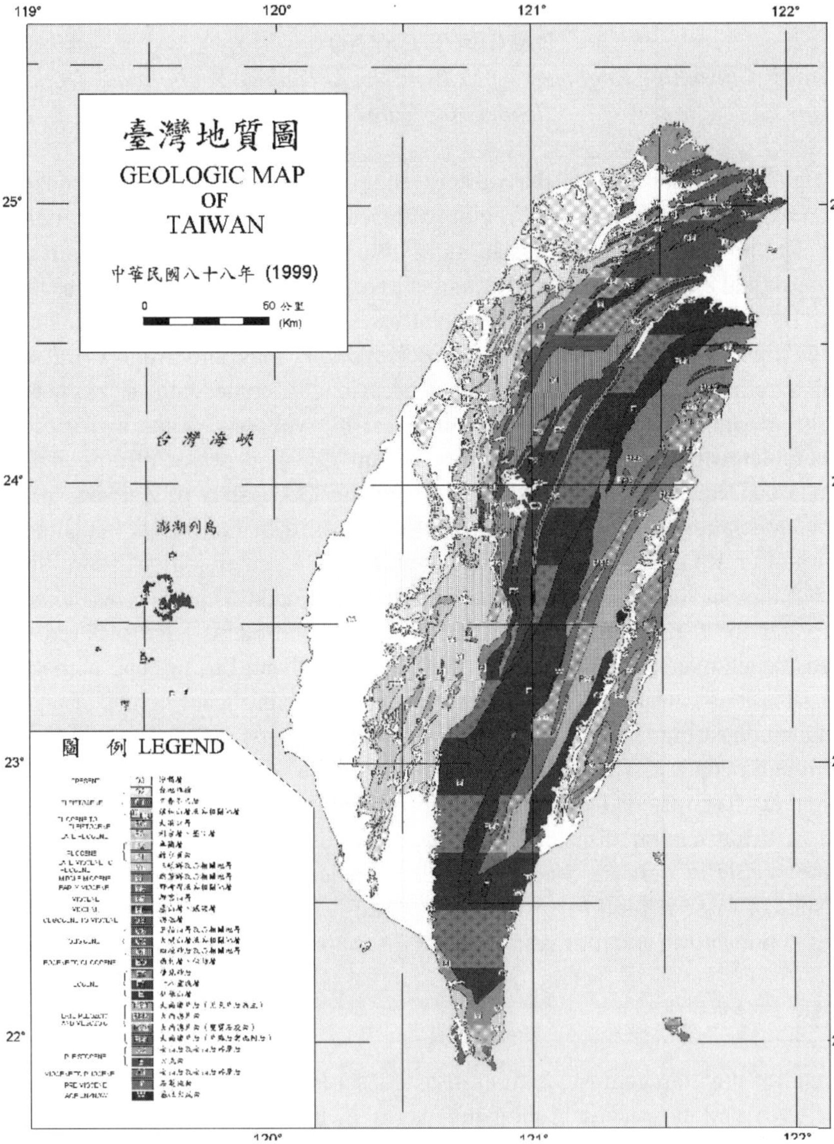

Figure 1. Geological map of Taiwan.

Taiwan is located along the west rim of Pacific Ocean with Tropic of Cancer across the middle. It is at all times under the threats of earthquakes, typhoons, and torrential rains. In 1999, the Chi-Chi earthquake declared a death toll of 2,434, 12,029 wounded, and a total loss of about 11.4 billion US dollars. In 2001, the Toraji typhoon claimed 214 lives including the missing. Later in September of the same year, the flood caused by the extremely heavy precipitation (425mm/day, record high in Taipei City) brought by the Nari typhoon cast a severe hit on northern Taiwan with a monetary loss estimated to 5.7 billion US dollars (Chang, 2002). The monetary loss in the event of Chi-Chi earthquake alone was approximately 3% of the Taiwan GDP in 1999 and human life loss was approximately one-ten thousandth of the total Taiwan population.

Along with the earthquakes and typhoons, landslides and debris flows in Taiwan seem to have higher occurrence rates after the Chi-Chi earthquake (Chin *et al.*, 2005). The loss of lives and properties from debris flows was the highest in 2001 with a death toll of 129. Damages from landslides and debris flows on landscapes, water resource systems, transportation and lifeline systems are much more often reported during heavy precipitation.

In light of the significant increase in damages caused by natural disasters island-wide and around the world, many studies initiated by government and private sectors have been carried out in various aspects for the prevention and mitigation of natural disasters in Taiwan. Some resolutions have been implemented. As the effort for prevention and mitigation of natural disasters continues, it is of supreme importance to understand the characteristics of these natural hazards and their interaction with local environment. The dynamic interaction between unpredictable natural events of earthquakes, typhoons, precipitation and changing local environment by continuous urbanization as well as civilization has significant impact on the outcomes of natural hazards.

NATURAL ENVIRONMENT OF TAIWAN

Tectonic formations, geology, and topography

The island of Taiwan is located approximately 150 km off the southeast coast of China mainland on the convergent boundary of the Eurasian and the Philippine Sea tectonic Plates. It is spindle-shaped, with the longitudinal axis extending roughly north-south for a length of 385 km and an east-west width of 143 km (Moh and Ou, 1979). The island is composed of geosynclinal deposition of Tertiary sediments of the most recent era (the Cenozoic) to a thickness of more than 10,000 meters on a metamorphic base formation (Figure 1). As the result of northwestward movement of the Philippine Sea Plate, the tectonic interactions between the Philippine Sea Plate and the Eurasian Plate are extremely complex in the vicinity of Taiwan. There has been apparent northward subduction of the Philippine Sea Plate beneath the Eurasian Plate and an eastward subduction of the Eurasian Plate underneath the Philippine Sea Plate (Figure 2). This active tectonic interaction and collision have produced the Central Mountain Range in Taiwan (Penglai Orogeny) and the prevailing structural pattern of rock formations on the island aligning in long narrow belts of elongated arcs with their convexities facing the west. All the major geological structures, including the faults and fold axes, align fairly well with this structure throughout the whole island and roughly parallel to the longitudinal axis of the island. The Penglai Orogeny has been continuously elevating the surface of Taiwan. In term of geological time, Taiwan is still a young and developing land (Lin and Chou, 1978).

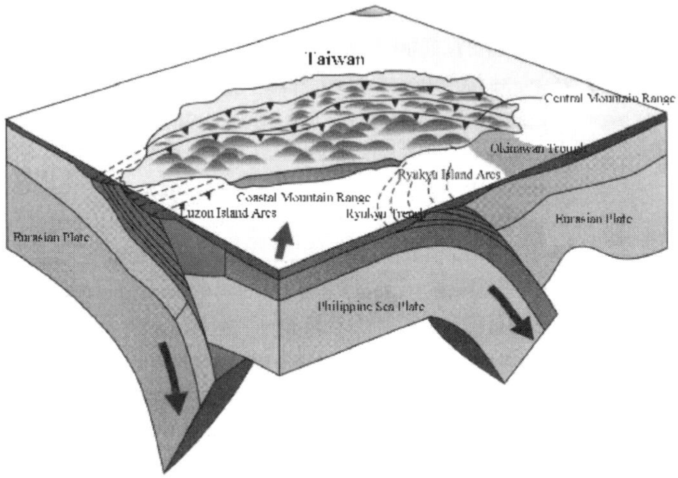

Figure 2. Tectonic activity in vicinity of Taiwan.

The topography and geological condition of Taiwan are closely related to the above tectonic activity. The rock formations are often fractured. In the island of approximately 35,960 sq. km, there are more than 200 peaks of over 3000 m and occupy approximately 1.3% of the island surface (Table 1). The plains with elevations below 100 m take about 31.3% of the island surface and lie mostly on the western side of island, which leaves approximately two-third of the island covered by slopeland merging from mountains to plains (Figure 3). Considering the dimensions of Taiwan, the heights of mountains, and the area of slopeland, the slopes in slopeland area are relatively steep and pose high threats to development in these areas.

Table 1. Topographical distribution of Taiwan (from Lin and Chou, 1978).

Elevations (m) (From/To)	0/100	100/500	500/1000	1000/2000	2000/3000	3000+
Area (%)	31.3	23.5	13.7	19.7	10.5	1.3

Central Mountain Range generally serves as the water divide of river system on the island of Taiwan. Conforming to the topography, the rivers are generally short with steep slopes and possess high scouring capacity in the upstream due to high gradients (Figure 4). Moreover, the flow rates are highly seasonal and fluctuated with the precipitation.

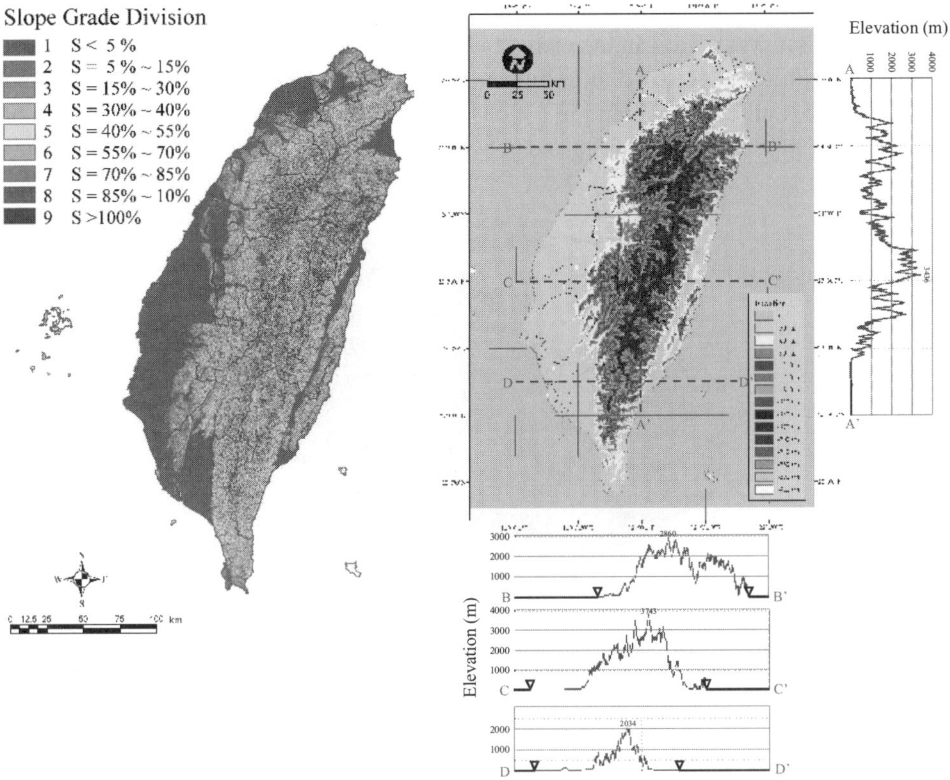

Figure 3. Topography of Taiwan.

River Name	Elevation (m) (origin-end)	River Length (km)	Average Slope (%)
Danshuei River	3529-0	158.7	0.82
Fongshan River	1320-0	45.5	0.44
Toucian River	1913-0	63.0	0.53
Jhonggang River	2616-0	54.0	0.67
Houlong River	2616-0	58.3	0.63
Da-an River	3488-0	95.8	1.33
Dajia River	3740-0	124.2	1.67
Wu River	2596-0	119.1	1.09
Jhuoshuei River	3220-0	186.6	0.53
Beigang River	516-0	82.0	1.69
Puzih River	1421-0	75.9	1.89
Bajhang River	1940-0	80.9	2.38
Jishuei River	550-0	65.0	0.85
Zengwun River	2609-0	138.5	0.50
Yanshuei River	350-0	41.3	0.34
Erren River	460-0	61.2	0.13
Agongdian River	300-0	38.0	0.14
Gaoping River	3330-0	171.0	0.67
Donggang River	1166-0	44.0	0.20
Sihchong River	1062-0	31.9	1.69
Lanyang River	3536-0	73.0	1.82
Heping River	3536-0	48.2	2.70
Hualien River	1200-0	57.3	0.35
Siouguluan River	2360-0	81.2	2.94
Beinan River	3293-0	84.4	0.61

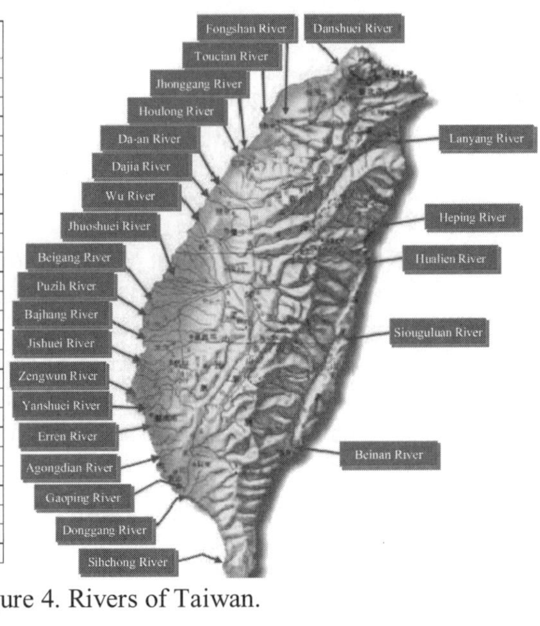

Figure 4. Rivers of Taiwan.

As the result of tectonic interaction and collision of the Eurasian and Philippine Sea Plates, the seismicity in Taiwan is among the highest in world (Figure 5). More than 200 earthquakes with tremors sensible to human (usually with Level 2 intensity and above) each year in Taiwan were recorded in the past century. In the last 15 years, more than 30 earthquakes of Ritcher's scale of 5.0 and above were recorded (Figure 6).

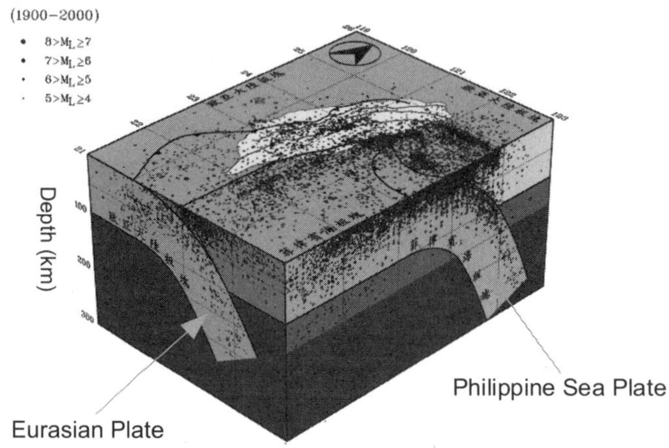

Figure 5. Earthquake hypocenter distribution in vicinity of Taiwan (after Lee, 1991).

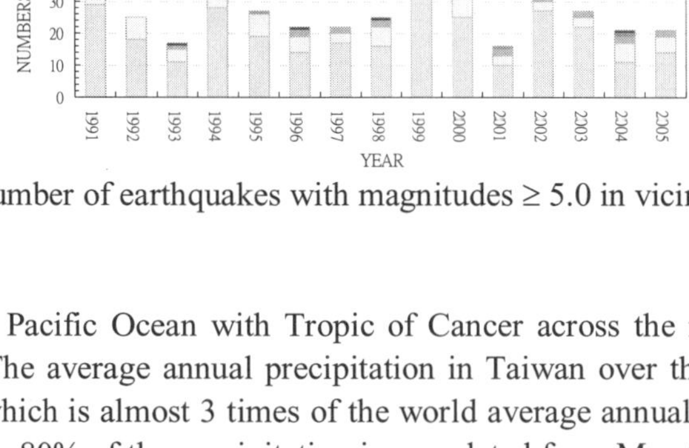

Figure 6. Number of earthquakes with magnitudes ≥ 5.0 in vicinity of Taiwan.

Climatic features

Located in western Pacific Ocean with Tropic of Cancer across the middle, Taiwan is warm and humid. The average annual precipitation in Taiwan over the past 30 years is around 2,400mm, which is almost 3 times of the world average annual precipitation, 800 mm. Approximately 80% of the precipitation is cumulated from May to October (Figure 7). With the topographical and geological conditions of Taiwan, approximately 71% of the precipitation becomes surface runoff, 24% evaporation, and 5% infiltration. Only about 15% of the precipitation can be retained in surface water bodies. The amount of fresh water stored in the island is around 20% of annual precipitation including 5% of infiltration into groundwater.

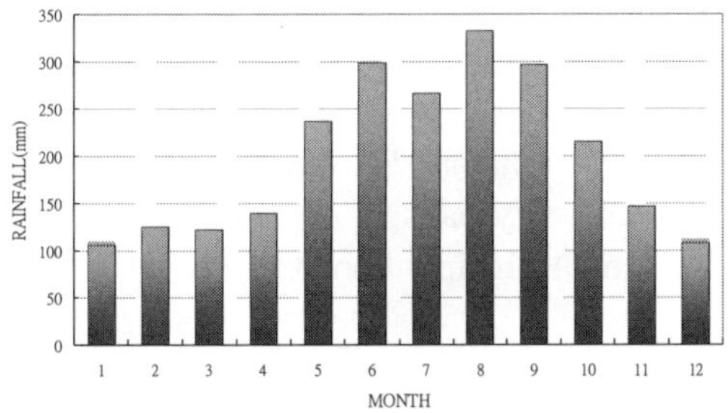

Figure 7. Average monthly precipitation in Taiwan.

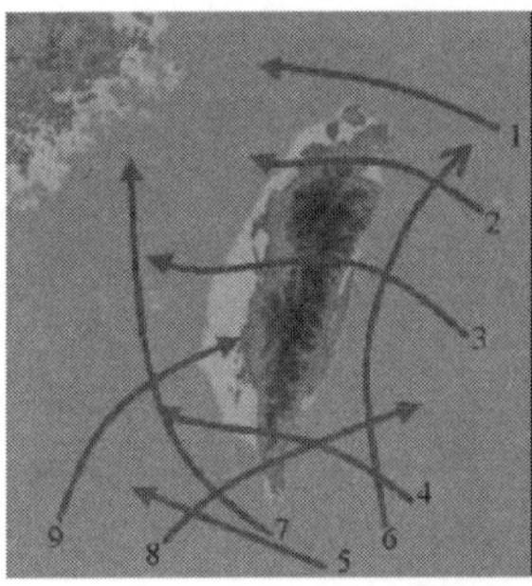

Tracks	Visits	%
1	50	13.4
2	46	12.3
3	40	10.7
4	35	9.3
5	76	20.3
6	53	14.2
7	30	8
8	16	4.3
9	28	7.5
Total	374	100
From 1897 to 1996		

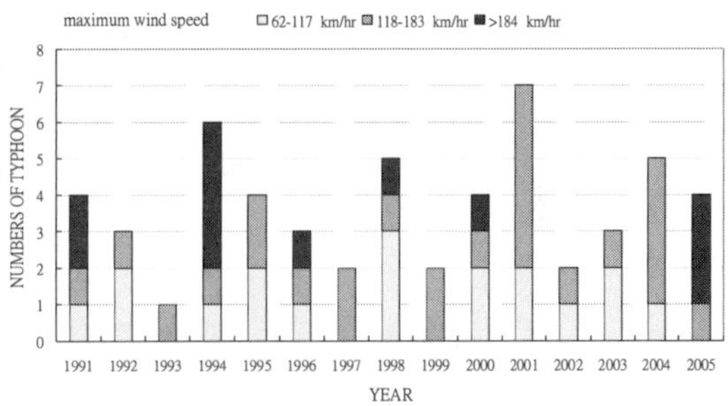

Figure 8. Major tracks and statistics of typhoons visiting Taiwan.

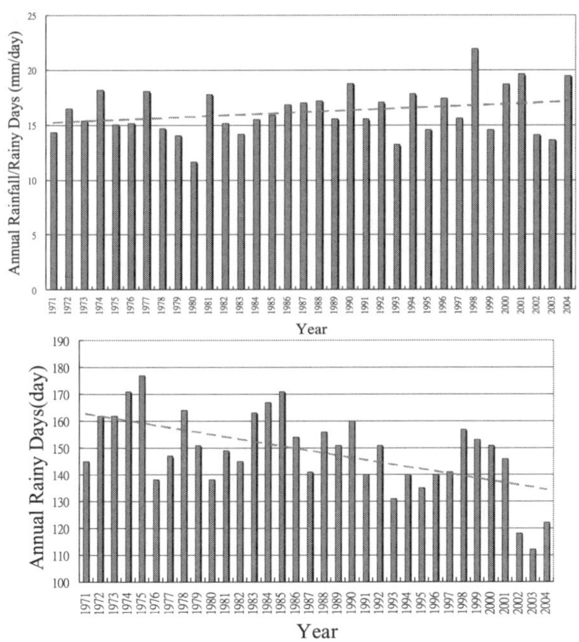

Figure 9. Precipitation pattern change in Taiwan.

In the past, Taiwan was subjected to the attack of typhoons mostly from May to October at an average rate of 3.7 times a year with dominant routes in northwestward direction (Figure 8). Heavy precipitation is usually associated with typhoons visiting Taiwan and constitutes one of many major parts of precipitation in addition to monsoon in May and June.

As a result of global climate variation, Taiwan has also been experiencing a gradual climate change. The number of rainy days is decreasing with higher rainfall intensity (Figure 9) even though the annual precipitation remains fairly constant in the past few years.

NATURAL DISASTERS IN TAIWAN

Earthquakes

The tectonic setting and dynamics of the Eurasian and Philippine Sea Plates are the major triggering mechanism of seismic activities in vicinity of Taiwan. Although occurring often, most of the seismic activities caused very minimal damage on the island. However, the catastrophic ones from time to time rock the island and induce significant loss of lives and properties.

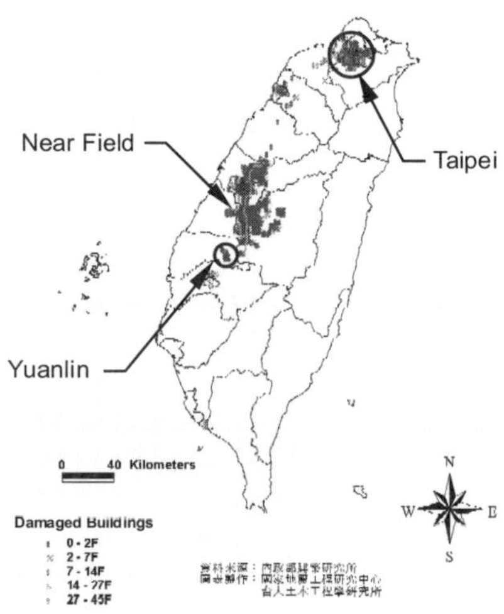

Figure 10. Most damaged areas in the Chi Chi earthquake.

In the Chi-Chi earthquake, the total loss in Taiwan was roughly 11.4 billion US dollars including an asset loss of more than 8 billion dollars and indirect loss of approximately 3 billion dollars. Three areas, respectively the vicinity of Chelungpu fault, the densely populated Taipei City and County area, and Yuanlin area, were identified later in study with major building and infrastructure damages (Figure 10) in the earthquake. In the area within a 6 km distance from Chelungpu and Shuangtung Faults, the damages were caused primarily by high seismic forces (a recorded horizontal peak acceleration of 989 gals in east-west direction at Station TCU084, and a vertical peak acceleration of 716 gals at Station CHY080) and large relative displacement at surface (average fault dislocations of 1.5 m horizontally and 3 m vertically). Most of the structures near the fault were severely damaged even for buildings designed in compliance with modern design codes. The bridges in central Taiwan suffered severe damages from collapsed spans, cracked piers, and distressed components (Figure 11). The damages on dams, lifelines and critical facilities, and power systems were also reported. In Hsinchu Industrial Park along, an estimated loss of 400 million US dollars was reported and most of which was incurred in semiconductor and silicon wafer production facilities.

In Yuanlin Town and Changhwa County, soil liquefaction seriously affected an area of approximately 60 sq. km in the Chi Chi earthquake (Moh *et al.*, 2002). Sand boils were found at locations at which the clay covers are either too thin or totally missing. At some locations where the clay covers are thick, the settlements of buildings up to 1 m or so as a result of ground subsidence were observed. The buildings in these areas suffered severe damages from soil liquefaction. Soil liquefaction also caused damages on levees and resulted in lateral spreading at several locations along Koniaokeng Creek near Chelungpu Fault. Moreover, over the reclaimed land of Taichung Port liquefaction, damaged 4 of its 45 docking wharves (Figure12).

Figure 11. Bridge damages in the Chi Chi earthquake.

Figure 12. Liquefaction induced damages in the Chi Chi earthquake.

Located approximately 150 km north of the epicenter, Taipei area suffered damages more than anticipated in the Chi Chi earthquake. The result of study indicates that the geological condition with thick soft and poorly cemented young sediment deposits could have adverse site response effect in bowl shape base formation of Taipei basin and be rationally argued to contribute to this consequence (Tsai, 1987). The adverse site response effects in Taipei basin include amplification of earthquake magnitudes and extended period of maximum design spectral response acceleration (Chen and Chen, 1995). In 1986, two earthquakes with magnitudes of 6.2 and 6.8 and epicenters approximately 15 and 10 km east of Taiwan respectively have caused damages on buildings in Taipei predominantly of 10 to 16 stories (estimated predominant structural periods between 1.0 and 1.6 sec). In light of the above findings, the seismic design code revised in 1989 included the site response effect and extended the predominant period with revised design response spectra for Taipei area. In July 2005 the design response spectrum was further modified with higher design seismic force (Figure 13).

Landslides were also reported during the Chi Chi earthquake. Almost all of the slope failures occurred east of Chelungpu Fault on the hanging wall with shallow slips in residual soils of depths between 1 m to 5 m. At Tsao Lin, the debris from a massive landslide slammed into the valley of Ching Shui River, blocked the river, and formed a reservoir behind (Moh *et al.*, 2002).

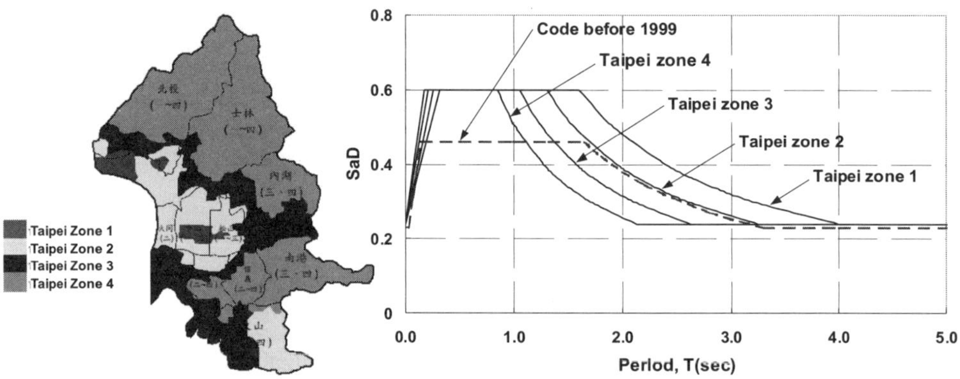

Figure 13. Design response spectrum for Taipei area.

Landslides and debris flows

Landslides and debris flows are two major natural disasters that threaten the people and environment in Taiwan, and the later has caused greater disasters in this island over the past few years. The geological settings, topographical features, and climatic conditions of Taiwan provide the inherent factors for potential cause of these disasters. In addition, the population congregation, excessive land development, and rapid economy growth over past decades contributed more potential adverse effects to the environmental condition. The precipitation statistics of Taiwan indicate the tendency of decreasing total raining days per year but increasing precipitation intensity per rainy day. The change in raining pattern suggests that the intensity of a raining event is becoming higher, which can be easily related to the increased possibility of landslide and debris flow hazards.

Table 2. Losses from recent debris flow disasters in Taiwan (from Chin *et al.*, 2005)

Time	Typhoon Event	Location (Township/County)	People Missing and Dead	Property Loss
6/23/1990	OFELIA	Sioulin / Hualien	35	35 houses damaged
7/31/1996	HERB	Sinyi / Nantou	2	21 houses and roadways damaged
7/31/1996	HERB	Shueili / Nantou	8	17 houses damaged
10/29/2000	XANGSANE	Rueifang / Taipei	8	20+ houses, roadways, and bridges damaged
7/28/2001	TORAJI	Fonglin / Hualien	6	3 houses and a wastewater treatment plant damaged
7/28/2001	TORAJI	Guangfu / Hualien	43	150+ houses damaged
7/28/2001	TORAJI	Sinyi / Nantou	16	67 houses, roadways, and bridges damaged
7/28/2001	TORAJI	Shueili / Nantou	1	21 houses and roadways damaged
7/28/2001	TORAJI	Shueili / Nantou	17	55 houses damaged and all public infrastructures fully destroyed
7/28/2001	TORAJI	Sinyi / Nantou	46	20+ houses, roadways, and bridges damaged
6/30/2004	MINDULLE	Heping / Taichung	3	30 houses and roadways damaged
6/30/2004	MINDULLE	Heping / Taichung	1	40 houses damaged
6/30/2004	MINDULLE	Dongshih / Taichung	1	Roadways damaged
6/30/2004	MINDULLE	Puli / Nantou	2	17 houses damaged
6/30/2004	MINDULLE	Hezuo / Nantou	1	6 houses damaged
8/23/2004	AERE	Wufong / Hsinchu	28	20+ houses damaged
9/11/2004	AERE	Jianshih / Hsinchu	4	1 house damaged

In 1999, the Chi-Chi earthquake triggered over 20,000 landslides island-wide (Chin *et. al.*, 2005). The deposited landslide masses of critically stable conditions became the source materials for debris flows during or after intensive precipitation. As a result, approximately 250 hazardous debris flow events have been reported during or after the earthquake. This hazardous threat of landslide and debris flow is aggravated with a higher occurrence rate and greater intensity after the Chi-Chi earthquake, and debris flows were easily triggered than ever during an event of heavy rain or typhoon (Table 2).

Typhoons and floods

Taiwan is located in the area with primary passages of typhoons originated in northwest Pacific Ocean. The average number of visitation of typhoons to the island is approximately 3.7 per year, which is roughly 15% of the typhoons originated in the region of northwest Pacific Ocean and South China Sea. When visiting Taiwan (Figure 8), the typhoons with tracks through either northern or southern Taiwan (Tracks 1, 2, 4 and 9) tend to pose higher threats to the island.

The damages by typhoons in Taiwan are primarily incurred by the strong winds, storm surges in the coastal area, and heavy precipitation. In addition to strong winds, the foehn and salty winds associated with typhoons have caused significant damages on agricultural produces and plants. The ground subsidence induced by excessive groundwater extraction has worsened the harm of storm surges to coastal areas, and the drainage of areas with seawater encroachment takes longer to complete. The precipitation further can easily trigger the landslides and debris flows. Moreover, the concentrated precipitation can cause extremely high flow rates in water channels and serious scouring problems on riverbeds and bridge foundations.

Nevertheless, the floods generated by heavy precipitation associated with typhoons have been the predominant cause of damage in Taiwan. In 2000 and 2001, the Xangsane, Troaji, and Nari typhoons caused severe damages on agriculture and transportation systems, and triggered landslides and debris flows island-wide. The flood in Taipei caused by the Nari typhoon severely damaged the MRT system at Taipei Station and it took several months for restoration to complete.

Ground subsidence

Ground subsidence can be triggered by many factors such as seismic activities, volcano eruptions, dissolution of limestone, etc. and exploitation of underground resources (groundwater, oil, coal, etc.) as well as underground space excavation. In Taiwan, the primary reason for regional ground subsidence is excessive exploitation of groundwater. Until now, more than 11% of the land in Taiwan has subsided, which was increased from 7.4% in 2001 and mostly along west coast of the island (Figure 14).

In 1960s, the ground subsidence in Taipei City was recognized as a result of excessive usage of groundwater for industry and domestic consumption. A surface settlement of more than 2 m was recorded near the center of city (Figure 15). Regulatory restriction on groundwater pumping activity was later reinforced within the city limit and had successfully stopped the subsidence in 1980s. With the gradually elevated groundwater levels, rebound of ground surface was observed from the elastic portion of soils due to decreased effective overburden stresses (Figure 15).

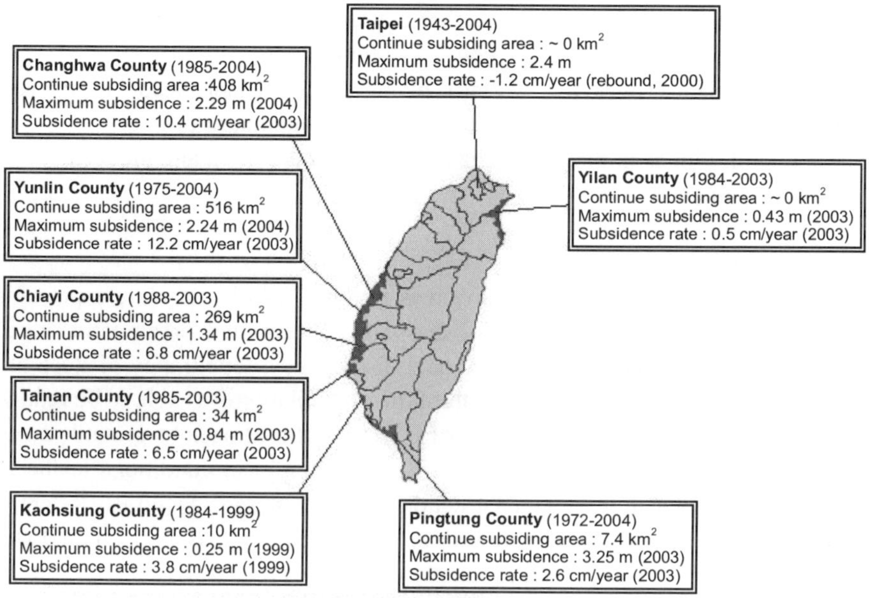

Figure 14. Ground subsidence in Taiwan.

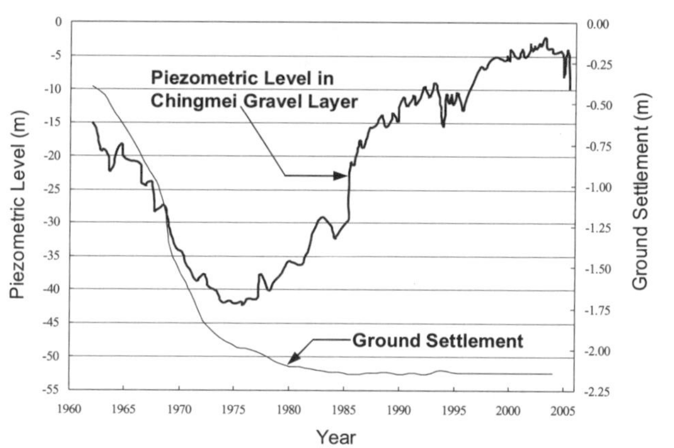

Figure 15. Ground subsidence in Taipei City.

However, continuous subsidence with extending territory still occurs along the southwest coast of Taiwan, where the land has long been used for agricultural purposes and fishery farms. Due to limited precipitation, relatively level topography, predominately granular alluvial formation, and unstable flow rates of rivers in the western plains of Taiwan, the surface water resources are rare and insufficient to provide steady supply of fresh water for consumption. Groundwater became the only economical and steady water supply in this area. Extensive pumping of groundwater for usage in fishery farm and agricultural purposes further aggravates ground subsidence and causes the intrusion of salt water into the groundwater systems.

Ground subsidence over the western plains of the island leads to irreversible consequences of infrastructure settlements and regional drainage difficulties in addition to the loss of land. Floods in these areas have occurred more often and the retention times become longer than ever.

Water deficiency

The average annual precipitation of Taiwan is roughly 2,400 mm/year, approximately three times of the world average. However, with concentrated precipitation over the period from May to October and with regional discrepancy, topography of steep slopes, and rivers of limited lengths, only one-sixth of the precipitation was able to be retained as surface water for consumption in addition to 5% of infiltration into groundwater. The surface water features and geological as well as topographical conditions in the island provide very limited locations for construction of water storage facilities. Retention of fresh water has long been a challenging task in Taiwan. In the densely populated western plains of Taiwan, sufficient and steady supply of fresh water for agricultural, industrial, and domestic consumptions has been sustained by continuous exploitation of groundwater, which is the main cause of ground subsidence in this area.

After the Chi-Chi earthquake, significantly increased events of landslides and debris flows, mostly in areas of watersheds, have stripped, to certain degrees, the slopeland and damaged hydraulic facilities. Surface erosion during heavy precipitation is aggravated on non-vegetated and steep slopes, which leads to large quantities of sediments in reservoirs, and generates high turbidity in surface water and causes much difficulty in water treatment processes.

In the past, the precipitation incurred from visitation of typhoons has been a major replenishing source of water storage in the island. However, in recent years, the high turbidity of surface water after typhoons significantly reduced the treatment capacity of water treatment plants and created water supply shortage in certain areas.

NATURAL DISASTERS MITIGATION

Disaster warning and monitoring systems

The issues regarding natural hazards started to receive more attention in Taiwan since the disastrous events of the Chi Chi earthquake in 1999 and the floods in 2000 and 2001. The issues regarding natural hazard forecast, prevention, and mitigation have been intensively discussed and studied by government agencies and private sectors. The government has taken initiatives to integrate loosely kept information in various different government agencies and to enhance capabilities in tasks of hazard warning and monitoring.

The Central Weather Bureau (CWB), Ministry of Transportation and Communications, is the primary government agency responsible for collecting and interpretation of climatic and seismic information in Taiwan. As part of the island-wide climatic monitoring system, it implements a total of 362 automated rain gauges island-wide (Figure 16) to collect real-time precipitation data, which is now incorporated with geographical information system (GIS) to provide early warnings on floods and debris flows. The CWB also operates one of the most advanced systems in the world on seismic monitoring in Taiwan area. The system includes 75 short period seismometers, 680 strong motion seismometers, 12 broadband seismometer stations, and 68 strong motion seismometers in major cities on buildings and bridges (Figure 17). With this system, the CWB made crucial information of the Chi Chi earthquake available within 102 seconds from the moment when earthquake occurred. The information included location of hypocenter, magnitude, and island-wide intensities.

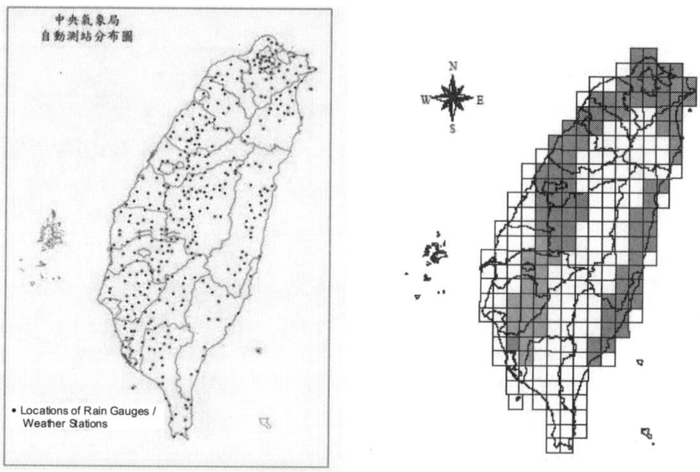

Figure 16. Locations of the CWB automated rain gauges and areas under natural hazard study.

Also in the Chi Chi earthquake period, the groundwater monitoring system (Figure 18) implemented and maintained by the Water Resources Agency (WRA), Ministry of Economic Affairs, supplemented invaluable information on groundwater variation in elevations and quality before, during and after earthquakes. The system, consisting of 492 monitoring wells mostly in the western plains of Taiwan, is set up to provide continuous monitoring data on groundwater levels and quality, and additional reference for ground subsidence in these areas.

Risk management and mitigation strategy formulation

To enhance the disaster response capability as part of the disaster mitigation resolution, the government enacted a Disaster Prevention and Response Act (DPRA) in 2000 (Bylaws in 2001) soon after the Chi Chi earthquake. The National Disasters Prevention and Protection Commission of the Executive Yuan was also established to cope with the natural disaster related issues in the central government. DPRA is basically structured on four stages of disaster management, i.e., mitigation, preparedness, response, and recovery. With respect to different types of natural disasters, different government agencies are in charge of the matters regarding various stages of disaster management according to DPRA (Table 3).

Table 3. Administrative agencies for natural disasters in Taiwan (DPRA).

Agency	Types of Natural Disasters
Ministry of the Interior	earthquakes and strong winds
Ministry of Economic Affairs	floods and droughts
Council of Agriculture, Executive Yuan	landslides and debris flows

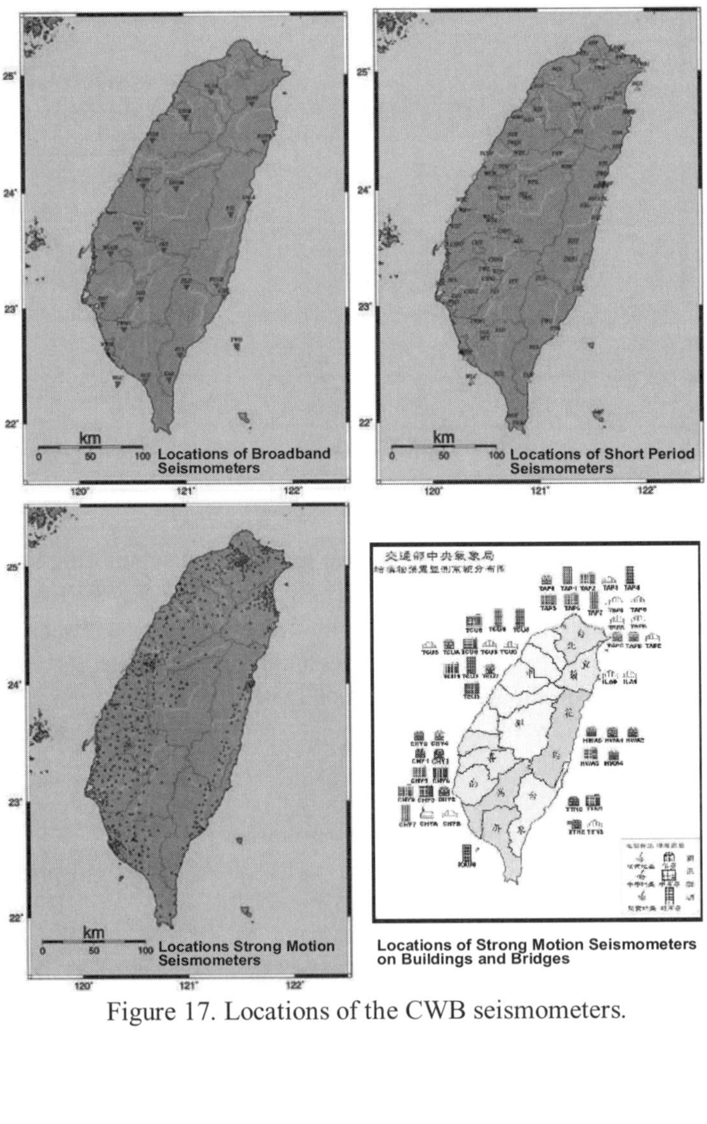

Figure 17. Locations of the CWB seismometers.

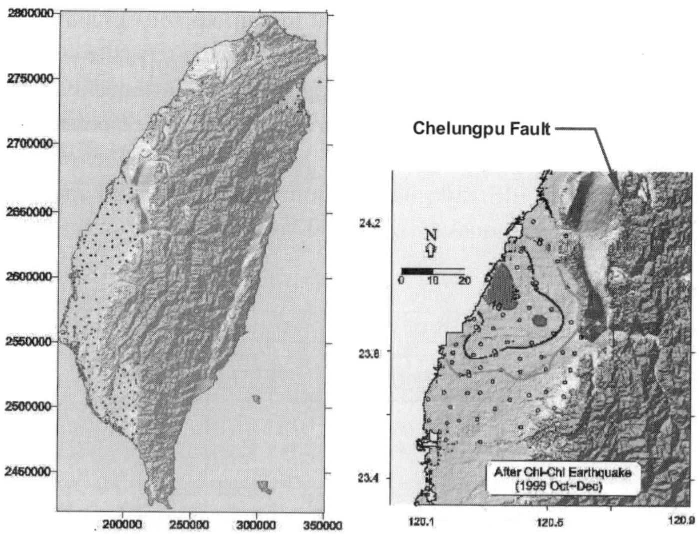

Figure 18. Distribution of the WRA groundwater monitoring wells.

The Taipei City is the most populated area in Taiwan and had suffered serious damages in various events of natural disasters. In response to the urgent need for disaster mitigation in the city, the Taipei Disaster Prevention and Rescue Committee was established under the city government to develop and promote hazard mitigation work on earthquakes, typhoons and inundations, and landslides as well as debris flows (Figure 19).

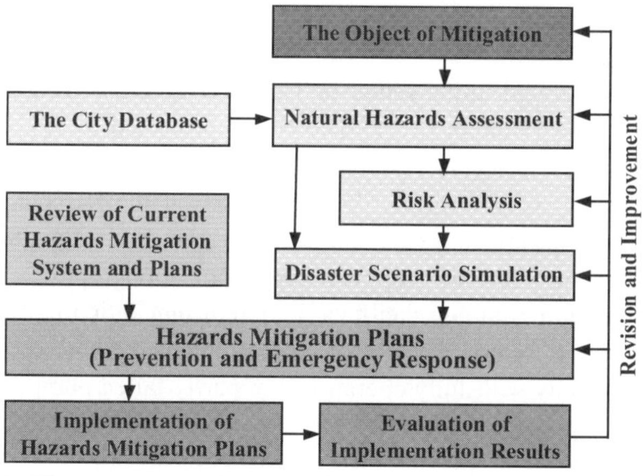

Figure 19. The conceptual framework of mitigation program of Taipei City.

The earthquake risk assessment model using HAZ-Taiwan and TELES (Taiwan Earthquake Loss Estimation System) based upon the framework of HAZUS (developed by the Federal Emergency Management Agency, FEMA, U.S.A.) is adopted by the Taipei City government to provide necessary information in forming the earthquake mitigation plans. Integrated with geological GIS information, demographic data, structure and building information, lifelines and critical facilities in the city, the assessment model can provide estimation on direct physical damages due to earthquakes and associated social losses (Figure 20). The city emergency preparedness and response plans are then developed in combination with transportation networks. With the city geotechnical boring data and geological GIS information, liquefaction potential maps can also be developed for reference of earthquake hazard and land use as part of mitigation plans in the city.

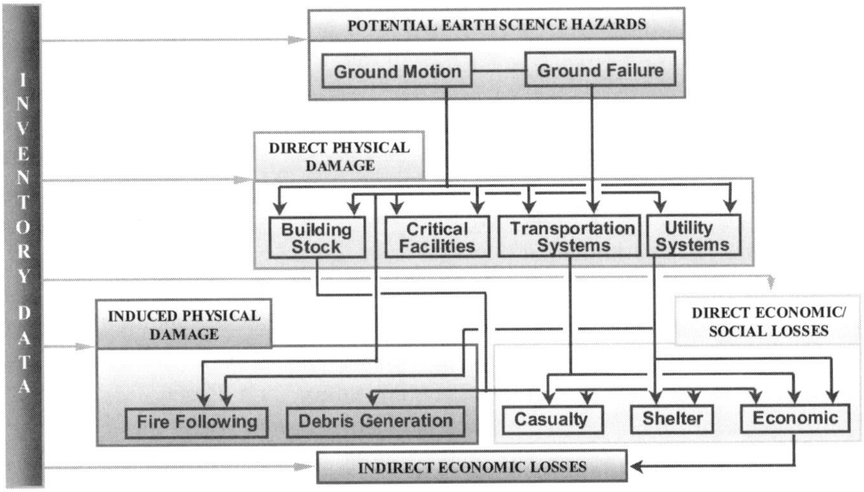

Figure 20. Structure of HAZ-Taiwan/TELES.

The Taipei City is located in the center of Taipei basin and surrounded by Tanshuei River, Keelung River and Hsintein Creek. The city is protected by levees and pumping facilities along these rivers against floods during heavy precipitations. The flood mitigation plan of Taipei City is primarily constituted on the basis of the inundation potential maps. These maps are constructed in consideration of precipitation with various intensity and elapsed times, land development conditions, topography, drainage capacity from the sewer system and pumping facilities. Different inundation hazard scenarios are formulated with different return periods of rainfall intensity and elapsed times and presumed drainage capacity with fully operational or partly failed pumping facilities. The inundation potential maps with inundated areas and depths of flood can be generated for mitigation plans and decision making process of evacuation (Figure 21). With the transportation GIS information, the routes of evacuation and shelter information can also be supplemented.

To form the landslide/debris flow mitigation plan, the Taipei City government built a comprehensive geological and natural environment database for the slopes in the area. The database includes 49 areas of potential debris flow rivers, situations of the existing slope management by the administration, documentation of historical landslides, the relationship between precipitation and landslide potential, and geological, hydrological, as well as geomorphic information of potential landslide areas. In addition to the rain gauges installed by the CWB, many other rain gauges were established by the Taipei City government to collect real-time precipitation data as a crucial index for issuing early warning of debris flows. In addition to the warning issued during precipitation in areas with high risk of landslides and debris flows, long-term resolutions of slope management and relocation of villages along the potential paths of debris flows were implemented as part of the city mitigation plans.

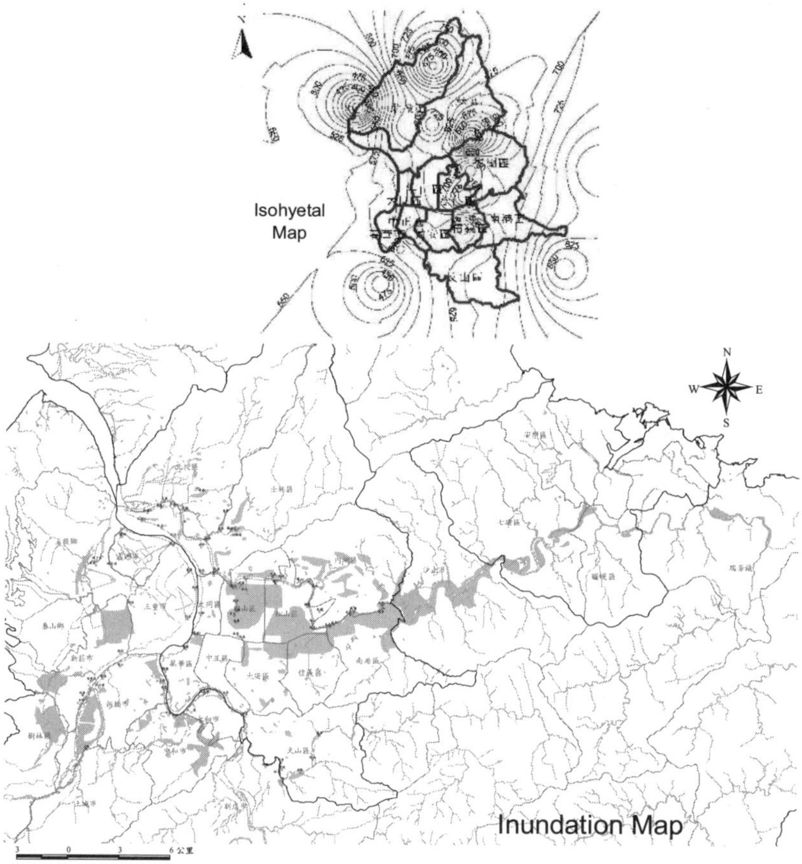

Figure 21. Isohyetal map and inundation map of the Nari typhoon.

Latest technical development

The CWB is conducting the Climate Variation and Severe Weather Monitoring and Forecasting Development Project, which is a multi-year project to establish monitoring and warning capabilities for climatic disaster prevention, to develop techniques of short-term climate analysis and prediction, and to strengthen the capabilities of very short-term severe weather analysis and forecasting. The capability of the CWB in typhoon monitoring and prediction, precipitation monitoring and forecasting, and marine climate forecast can be significantly improved. With the Third-Phase Strong-Motion Observing Project to develop an earthquake real-time warning system, the CWB will be able to reduce the earthquake rapid reporting time from the currently 60 seconds to 30 seconds by 2009 and even provide an early warning of major earthquakes in matter of seconds to organizations and facilities in pre-established system.

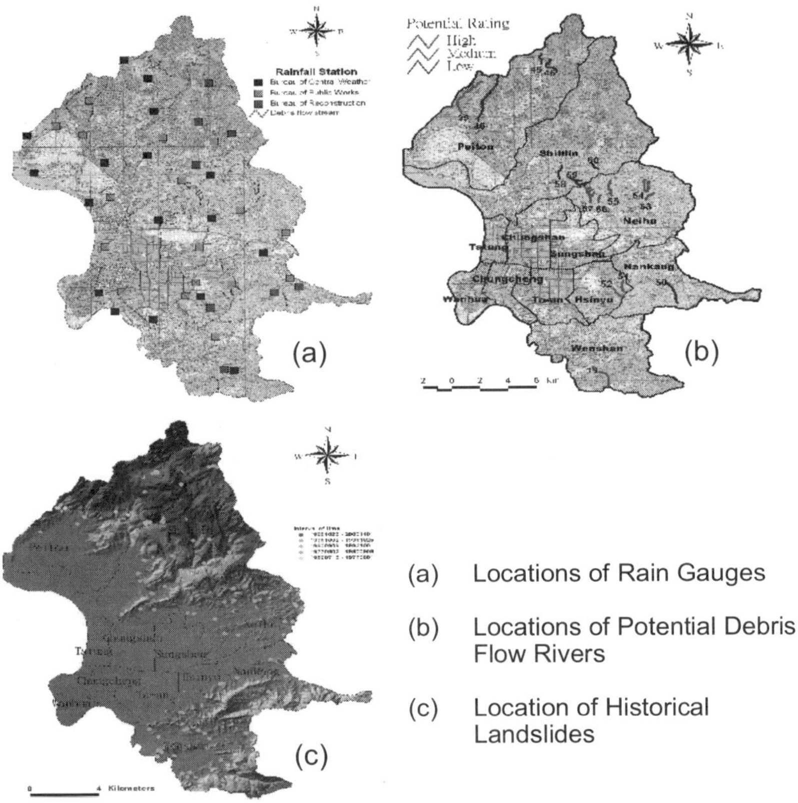

Figure 22. The city database of Taipei.

After the Chi Chi earthquake, the building seismic design codes in Taiwan were revised immediately in 2000 according to recorded seismometer data and field observations. Two seismic zones were adopted for seismic design of buildings and infrastructure in Taiwan (Figure 23). However, based upon further analysis of seismic data and new findings in earthquake, the latest revision of building seismic design codes was issued in July 2005. The major changes include larger seismic design loads in most areas, additional requirement on structure ductility under maximum considered earthquake with return period of 2,500 years, supplement of near fault effects and special consideration on site response characteristics of soft deposits, addition of provisions on seismic retrofit, isolation, and damper design guidance.

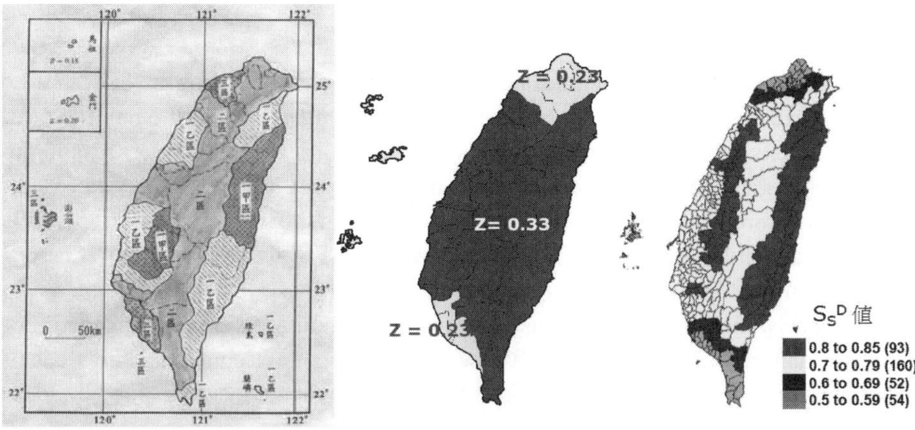

Figure 23. Seismic zoning change in design codes of Taiwan.

Modeling techniques of debris flows is also advanced in predicting the flow paths. A numerical solution using rheological model with two-layer system consisting of a plug region and the bottom boundary layer has provided better simulation on debris flow propagation (Liu and Huang, 2005). This advancement enables a physics based model to estimate the extent of influence of a given debris flow event. The result can be implemented to provide useful information for slopeland management and necessary information for urgent evacuation in debris flow areas.

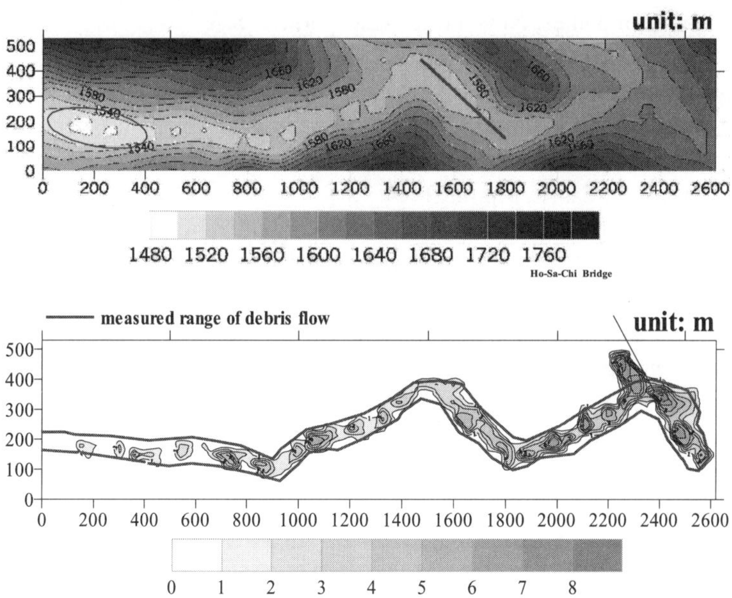

Figure 24. Result of numerical modeling of debris flow path (from Chin *et al.*, 2005).

CLOSING REMARKS

Typhoons, earthquakes, and precipitations are natural events occurring in Taiwan all the time. It is, however, the interaction between those natural events and the natural as well as man-made environments that determines the outcomes and consequences of these events. Even though the occurrences of these natural events are inevitable, the extent of preparedness beforehand and action taken afterwards certainly decide the cost incurred at each event. Taiwan has encountered many different types of natural hazards. The unique natural environment brings out the complexity of natural hazards. Systematic resolutions in many aspects to mitigate the disasters caused by natural events are needed. Even though the natural hazards may evolve with changing environments, knowledge of fundamental issues can always be exercised to provide valuable insights to the trends and resolutions to these disastrous events.

ACKNOWLEDGEMENT

The writers wish to acknowledge the assistance and advices provided by their colleagues in the MAA Group, particularly Dr. Chung-Tien Chin, Dr. Richard N. Huang, Dr. Jie-Ru Chen, Mr. Chi-Hung Chao, and Mr. Cheng-Chen Wu.

REFERENCES

Guy Carpenter & Company (2000). *"Natural Hazard Review of the Year 1999"*, Marsh and McLennan Companies, New York, New York, U.S.A.

Lin, C.H. and Chou, R.D. (1978). "*Taiwan Geology*", 1st edition, Maw Chang Book Co., Taipei, Taiwan.

Liu, K.F. and Huang, M.T. (2005). "Debris flow disaster mapping using numerical simulation", *Proc. International Conference on Environmental Fluid Mechanics*, Guwahati, Assam, India. (in printing)

Chang, C. C. (2002). "On the establishment of risk mitigation system for natural disasters and risk management", *Proc. Conference on Weather Analysis and Forecasting*, Taipei, Taiwan, 13-18.

Chen, C.H. and Chen, L.Y. (1995) "Ground period of Taipei Basin", *Earthquake Geotechnical Engineering Proceeding of IS-Tokyo*, Tokyo, 543-548.

Chin, C.T., Liu, K.F., and Chen, J.R. (2005). "Landslide and debris flow in Taiwan", *Proc. 3th International Conference on Geotechnical Engineering combined with 9th Yearly Meeting of the Indonesian Society for Geotechnical Engineering*, Semarang, Indonesia, 59-68.

Lee, C.T. (1991). "Earthquakes and disaster prevention", Website: *http://140.115.123.30/earth/edu98/edu98.htm*. (in Chinese)

Moh, Z.C. and Ou, C.D. (1979). "Building foundation practice in Taiwan", *Proc. 6th Asian Regional Conference on Soil Mechanics and Foundation Engineering*, Singapore, 205-209.

Moh, Z.C., Huang, R.N., Ueng, T.S., and Lin, M.L. (2002). "921 Chi Chi earthquake of Taiwan", *Proc. 17th Australasian Conf. on the Mechanics of Structures and Materials*, Gold Coast, Australia.

Ou, C.D. (2004). "Sustainable development and natural Hazard mitigation in Taipei City", *Proc. The 3rd Civil Engineering Conference in Asian Region*, Special Forum 2, Seoul, Korea, 51-60.

Tsai, I.C. (1987). "Review on normalized design coefficient, C, of Building Seismic Design Code from the strong motion seismic data at Taipei from earthquake on November 15, 1986", *Journal of Structural Engineering*, Vol. 2, No. 4, 13-20. (in Chinese).

SPECIAL PLENARY LECTURES BY COUNTRY

International Conference on Geotechnical Engineering for Disaster Mitigation & Rehabilitation
Chu, Phoon & Yong (eds) © 2005 World Scientific Publishing Company ISBN 981-256-469-1

TSUNAMI DEVASTATIONS AND RECONSTRUCTION WITH GEOSYNTHETICS

D. T. BERGADO and H. M. ABUEL-NAGA

School of Engineering and Technology, Asian Institute of Technology, Bangkok, 12120, Thailand

Great attention is directed to rebuild livelihoods and rehabilitate coastal communities affected by the Tsunami in the Indian Ocean in South Asia. It takes years of effort of different engineering disciplines to recover from recent devastations caused by the Tsunami. Geosynthetics can play important and vital roles in the protection, mitigation and rehabilitation efforts in affected coastal areas. The use of geosynthetics has advantages such as speed of construction, flexibility and durability, use of local soil materials rather than imported quarry product, and its cost effectiveness. Geosynthetics can be applied for reinforcement, filtration, drainage, protection, lining, and containment. More important, geosynthetics can be used effectively for erosion protection and for strengthening the earth embankments to resist failure during the occurrence of earthquakes associated with Tsunami. This paper demonstrates the various geosynthetic applications and the related engineering solutions to mitigate such Tsunami devastations.

INTRODUCTION

The Asian Tsunami of 26 December 2004, which struck the Indian Ocean Basin, affected hundreds of thousand of people in countries including Thailand, Sri Lanka, Indonesia and India. Its death toll has risen to over 260,000 victims. Many survivors had their lives disrupted since coastal tourism, fisheries, and agriculture have been seriously affected. Housing and public infrastructures have been destroyed Warnitchai (2005). There is urgent need to restore, rehabilitate and repair the damages of the affected people and the area.

Geosynthetics can play important and vital roles in the protection, mitigation and rehabilitation efforts in affected coastal areas. Geosynthetics have been used in hydraulic and geotechnical engineering for about the past two or three decades. Their use is well established for the purposes of material separation, filters, and reinforcement. In addition, all kind of bags are made now from synthetic fabric. The use of geosynthetics has advantages such as speed of construction, flexibility and durability, use of local soil materials rather than imported quarry product, and its cost effectiveness. Therefore, it is strongly recommended to use the geosynthetics engineering applications for restoring and rehabilitation of the recent devastations caused by the Tsunami. In the following, the Tsunami devastation effects in Thailand are presented briefly. Then, an overview for the many applications of geosynthetics and geosystems engineering that can be used for reconstructions of Tsunami devastations are highlighted and discussed.

TSUNAMI AND DEVASTATIONS IN THAILAND

The Tsunami wave height distributions in Thailand are shown in Fig. 1. The wave heights were greater at flat shorelines and shallow seawater depths. Consequently, beach resorts with deeper seawater depths and steeper shorelines only experienced slight damages (Warnitchai, 2005). Moreover, wide vegetated areas protect the shorelines and obstruct drifting debris that have destructive power when combined with water flow (Fig. 2).

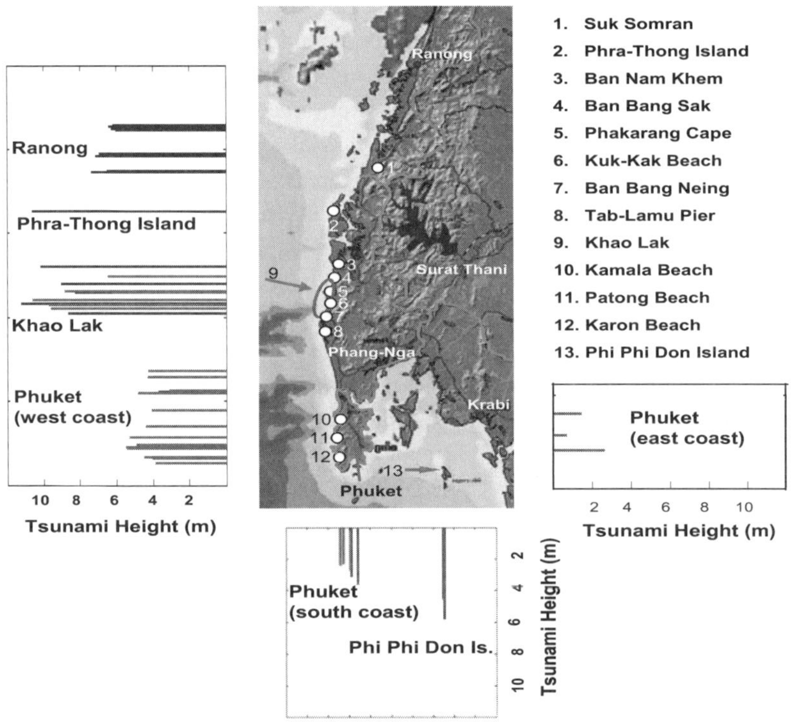

Figure 1. Tsunami waveheight distribution in Thailand (Warnitchai, 2005).

In the aftermath of the large-scale disaster, destructions related to coastal areas and waterways as well as infrastructures and buildings have been identified. Figure 3 shows the airphoto before and after tsunami at Khao Lak, Phanga, Thailand. The erosion and scouring in the coastlines and waterways can be observed. More details of coastal erosion at Khao Lak are shown in Fig. 4. Small and weak buildings directly open to the coastline were completely destroyed while large and strong buildings remain standing. The foundations of buildings are damaged by scouring and erosion. Erosion damage also occurs on seawalls and earth structures.

Figure 2. Mangrove forest along shorelines (Warnitchai, 2005).

Figure 3. Coastal erosion due to tsunami in Khao Lak, Phanga, Thailand (IKONOS Image, Space Imaging/CRISP-Singapore).

Figure 4. Coastal erosion due to tsunami in Khao Lak, Phanga, Thailand.

Ground elevation is a key factor. Even a small hill of 2 m height has saved houses on it. In fact, the natural sand dune deposits at Karon Beach in Phuket, Thailand has reduced the destructive effects of Tsunami. There were selective damages of the beaches in Phuket depending on the morphology, topography and depth of seawater of the coastal areas.

OVERVIEW OF GEOSYNTHETICS ENGINEERING

Geosynthetics are man-made synthetic materials utilized in geotechnical engineering applications. The earliest user of geosynthetics was Karl Terzaghi. In the late 1950s, Terzaghi made use of filter fabrics (today, geotextiles) as flexible forms. They were filled with a cement grout, thereby making closure between steel sheetpiling and rock abutments at the Mission Dam (now Terzaghi Dam) in British Columbia, Canada. During this same project, Terzaghi used pond liners (today, geomembranes) to keep an upstream clay seepage-control liner from desiccating.

There are various types of geosynthetics such as geotextiles (GT), geogrids (GG), geomembranes (GM), geonets (GN), geocomposites (GC), geosynthetic clay liners (GCL), geopipes (GP), geofoam (GF), geotubes, geocells, etc. Geosynthetics can be used for reinforcement, filtration, drainage, protection, lining and containment. The functions and geosynthetic types are tabulated in Table 1. The description of the varies geosynthetics types were given by Koerner and Soong (1995) and Bergado et al. (2005) as follows:

Geotextiles usually consisted of polypropylene (PP) or polyester (PET) polymers are made similar to standard textile manufacturing in either woven and nonwoven types. The yarns of woven geotextiles are arranged perpendicular to each other and may consist of slit film, monofilament or multifilament. The nonwoven geotextiles can be either made by mechanical needle punching or by heat bonding. Geotextiles are thin fabrics with open and porous structure. The mechanical and hydraulic properties vary widely depending on the weight per unit area. Geotextiles are primarily used for separation, filtration and reinforcements and secondarily for protection and drainage.

Geogrids are unitized woven yarns or bonded straps with either directional or unidirectional strength characteristics. This type of geosynthetics are mainly used for reinforcement applications such as reinforced vertical walls and steep slopes as well as reinforced foundations.

Geonets are usually made from high density polyethelene with parallel ribs as an integral unit. The primary function is in-plane drainage in either biplanar and triplanar directions.

Geomembranes are impermeable geosynthetics which are mainly used as barrier lining and containment of liquids and gases. It mainly consists of high density (HDPE) or low density (LDPE) polyethelene or PVC polymers. This product is required by regulations for waste containment as well as new applications in hydraulic and private developments.

Geosynthetic clay liners (GCL) consist of bentonite sandwiched between geotextiles. The bentonite infill is usually reinforced by needle-punching or stitching. Many other variations exist such as bentonite product bonded to geomembrane. This product is mainly function as barrier lining or containment as competition or in combination with geomembranes.

Table 1. Function vs. Geosynthetic Type.

Type of geosynthetics	Separation	Reinforcement	Filtration	Drainage	Containment
geotextile	√	√	√	√	
geogrid		√			
geonet				√	
geomembrane					√
geosynthetic clay liners					√
geopipes				√	
Geofoam	√				
geocomposites	√	√	√	√	√

Geopipes (GP) are really buried polymer pipe for drainage purposes. The high density polyethelene (HDPE) and polyvinyl chloride (PVC) are most common. The geopipes can be smooth walled or corrugated.

Geofoam (GF) consists of expanded polyesterene (EPS) in block form. This product is mainly used as lightweight fill on soft ground. It can effectively relieve lateral pressure on walls.

TSUNAMI RECONSTRUCTION WITH GEOSYNTHETICS

Geosynthetics can play important and vital role in the protection, mitigation, rehabilitation and reconstruction of affected coastal areas. Traditional construction techniques utilizing rock, concrete and steel for erosion protection are increasingly challenged by alternatives using geosynthetics for revetments, scour protection, groynes, berms, artificial reefs, reclamation, slope, protection, etc. The use of geosynthetics has advantages such as speed of construction, flexibility, durability, use of local materials rather than imported quarry products and its cost effectiveness. Koerner (2000) shows the emerging and future developments of selected geosynthetics applications. Moreover, Pilarczyk (2000) gave comprehensive review for the use of geosynthetics and geosystems in hydraulic and coastal engineering. In the following, some of geosynthetics engineering applications that can be used to restore the Tsunami destruction are presented.

Geotubes and geobags are made of geotextiles filled with dredged sand/silt/clay slurry using a hopper method with density of 1.6 tons per cubic meter and height of 1.8 to 2.0 m. Geotubes and geobags can be used as core for artificial sand dunes for shoreline protection. Heerten et al. (2000) reported that, based on case history in Island of Sylt,

Germany, the geotextile "sand cushions" shows superior effectiveness in protecting cliff area against erosion as shown in Fig 5. Geotubes and geobags can be also utilized for construction of shoreline protection and containment dykes to create artificial islands (Fig. 6). Restall et al (2002) gave a comprehensive review for the historical development and the types of geotextile containers used for marine structures in Australia and conclude that geotextiles containers structures can be successfully used to solve conventional coastal problems and non-conventional challenges.

Geocells are cellular confinement system for erosion control (Fig. 7). The geocells can be stacked together for slope erosion protection. These geocells are usually infilled with gravel, sand, silt or clay. Grasses and shrubs can grow at the infills. The geocells are made of ultrasonically-welded polyethelene sheets.

Gabions and mattresses are rock-filled baskets made of twisted hexagonal wires for erosion control. The wires can be coated by either zinc or PVC. Gabions have thicker dimensions than mattresses (Fig. 8). Gabions, geocells, and mattresses can be applied as revetments.

Figure 5. Defended cliff area with geotextile "sand cushions" on the Island of Sylt, Germany (Heerten et al. 2000).

a)

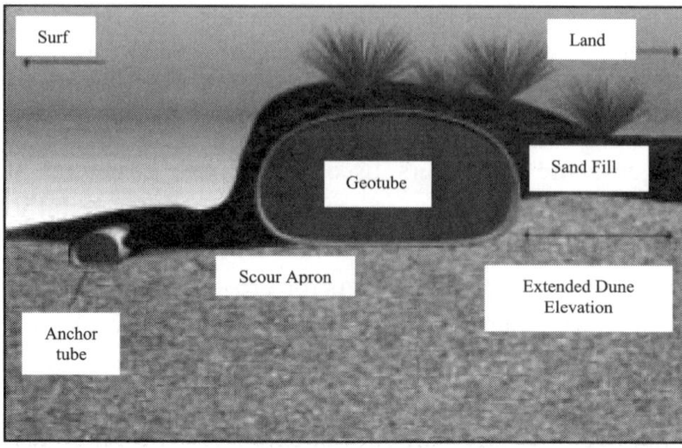

b)

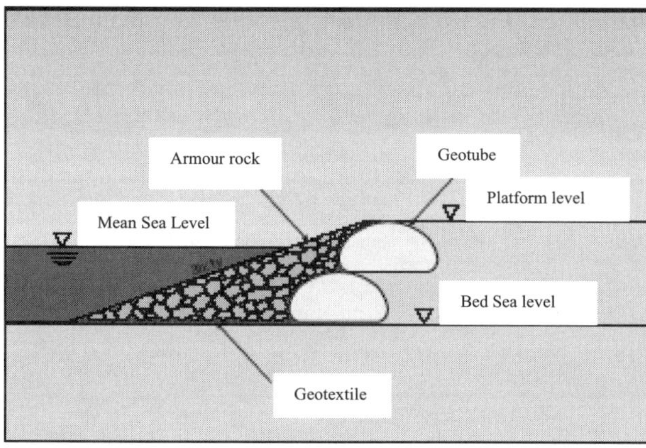

Figure 6. Geotube for a) shoreline protection; b) reclamation dykes to form artificial islands.

Revetments consisting of concrete facing, gabions, and mattresses or rock armour are effective for erosion control and scouring prevention (Fig. 9). Geotextiles underneath revetments serve as separators and filters.

Dykes and breakwaters protect harbors and recreation areas. Geotubes and geobags can be incorporated as components of breakwaters. Geotextiles at the base serve as separators, filters and membrane reinforcements (Fig. 10).

Geotextiles and geogrids can be applied as filters, wall reinforcements, slope reinforcements, erosion control and drainage functions in road and railroad embankments especially at approaches to bridges as shown in Fig. 11 (Bergado et al. 1996). In these structures, drainage can be provided by geocomposites and geopipes.

Geosynthetics are also utilized in geoenvironmental applications such as landfill liners, landfill covers, vertical cut off barriers, etc. In geoenvironmental applications, as shown in Fig. 12, almost all types of geosynthetics are utilized such as geomembranes and geosynthetic clay liners for liners, geotextiles for separation and protection, geonets, geocomposites and geopipes for drainage, etc.

For private development, geosynthetics have been used in sport fields, artificial lakes, golf courses, airfields, parks, playground, etc. For agriculture and aquaculture applications, geotextiles and geomembranes have been used as lining of shrimp farms, waste containment, fish baskets, and others.

a)

Geocell is three-dimensional cell made of HDPE strips ultrasonically welded together to contain soil

b)

Figure 7. a) Geocell cellular confinement system; b) The geocell are tufted above water and infilled with stone below waters

Figure 8. Gabions and mattresses are stone-filled box containers.

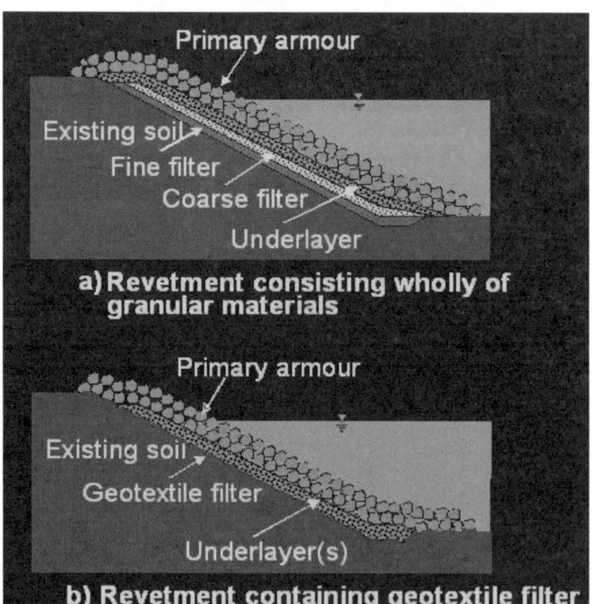

Figure 9. Revetments can have concrete facing gabions or rock armour.

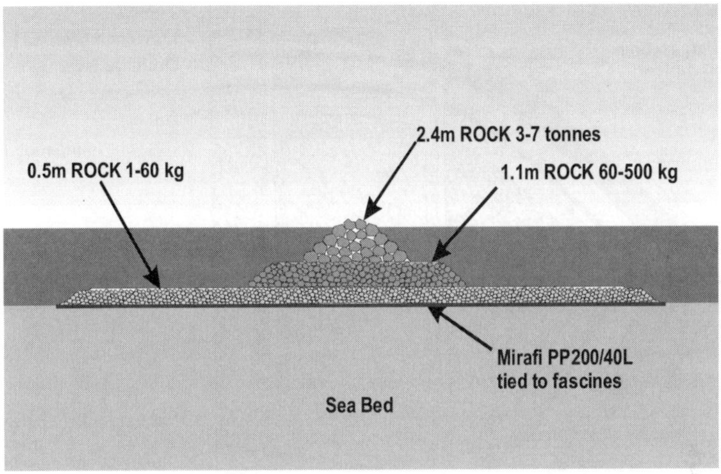

Figure 10. Dykes and breakwaters with geotextiles.

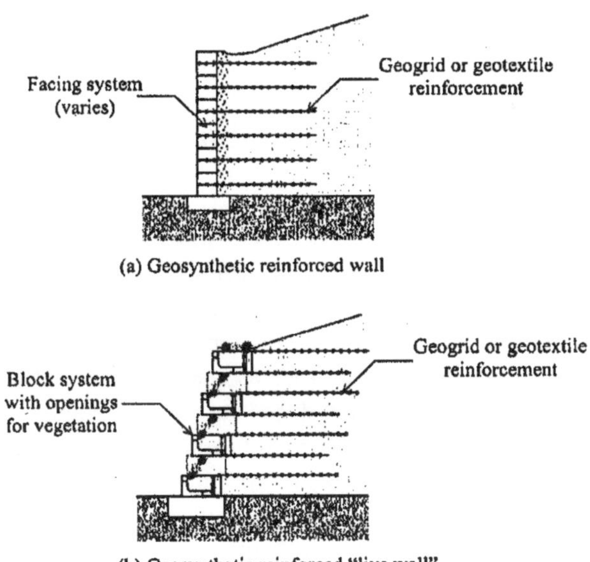

Figure 11. Geosynthetics use for reinforced wall.

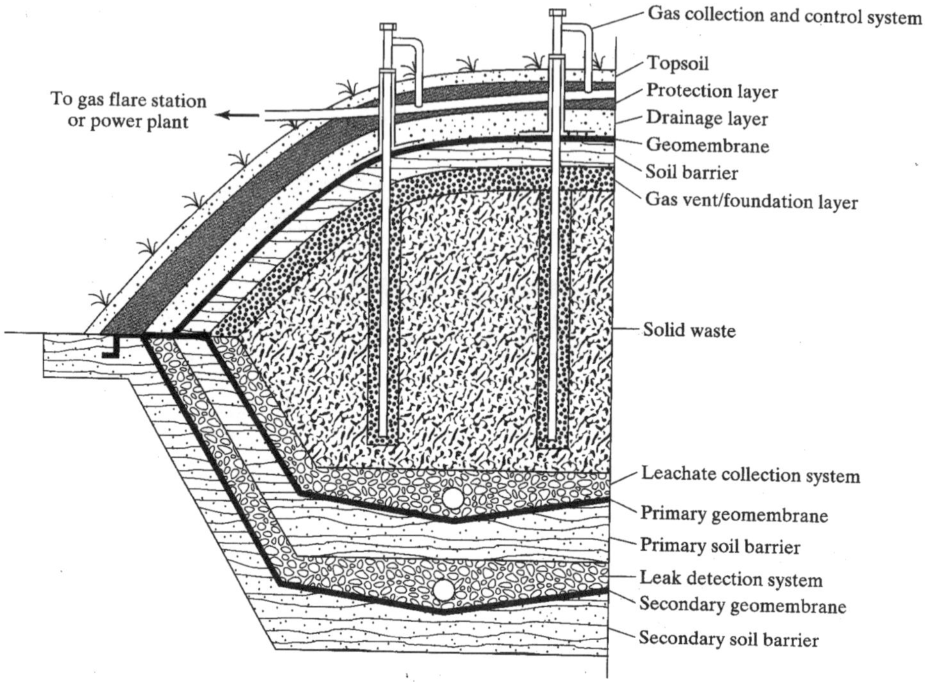

Figure 12. Geosynthetics application in geoenvironmental field.

PROPOSALS FOR MITIGATION AND REHABILIZATION OF COASTAL AREAS

Abednego (2005) presented the proposals by the Indonesian Engineering Association (IEA) for the contruction of buffer zone and canal (Fig. 13) as well as "escape mountain" and buildings (Fig. 14). The buffer zone serves to dissipate the impact of strong waves generated by Tsunami. Consequently, the buffer zones are located close to the seashore. These zones may consist of natural barrier such as mangrove forest. Man-made high road embankments and artificial elevated sand dunes can also function as buffer zones. The road embankments should have at least 2.0 m high and 6.0 m wide that can also function as coastal road. The road embankment should be reinforced and erosion resistant through the incorporation of geosynthetics (Fig. 15). The artificial sand dunes can be constructed using geotubes or geobags which are also made of geosynthetics.

Geosynthetic can be also incorporated in the design of the "escape mountain". The escape mountain or hill is constructed similar to a pyramid with geosynthetic reinforcements and erosion protection. The escape mountain or hill should be from 3.0 m to 5.0 m high with 30.0 m by 15.0 m rectangular area at the top in order to accommodate at least 3,000 people. Unlike the escape tower, the escape mountain or hill allows the access of people in all four sides (Fig. 16).

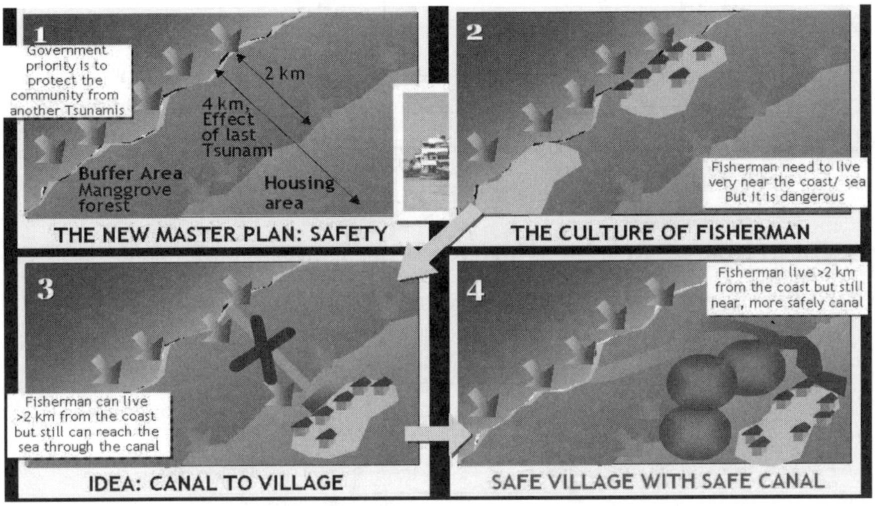

Figure 13. Proposal for buffer zone and channel (Abednego, 2005).

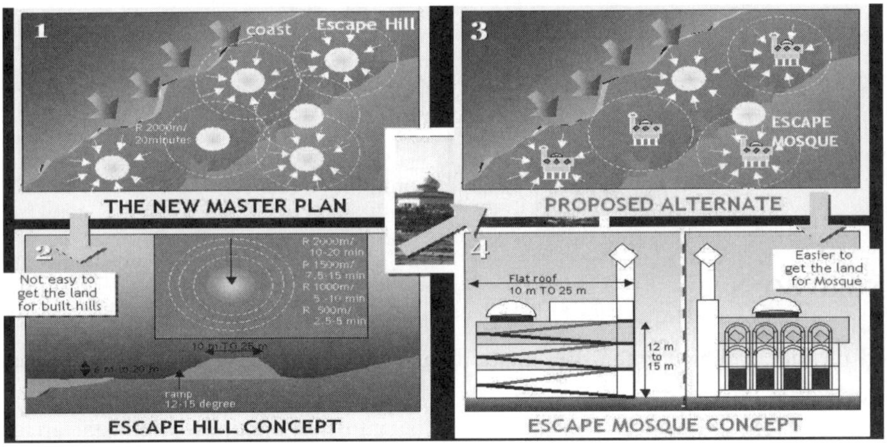

Figure 14. Proposal for escape building or hill (Abednego, 2005).

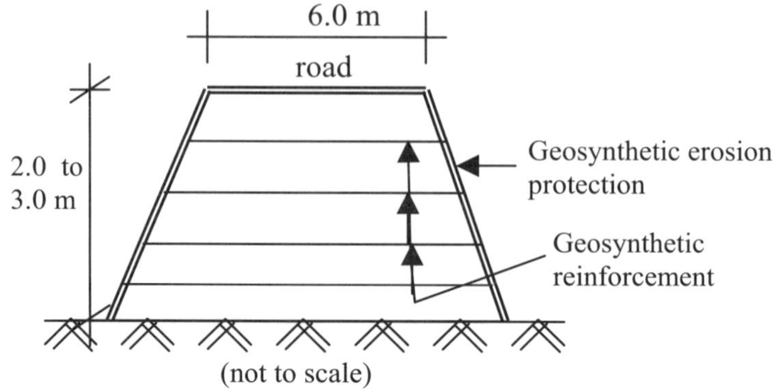

Figure 15. Reinforced high road embankment in buffer zone

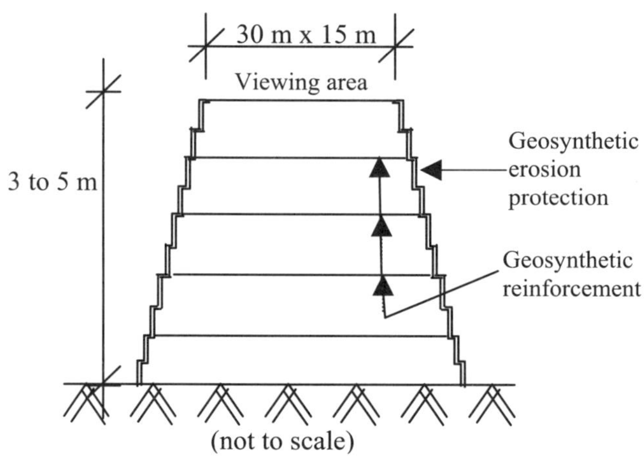

Figure 16. Escape mountain concept

CONCLUSIONS

Based on the aforementioned presentations and discussions, the following conclusions can be made:

1) Erosion and scouring resulted from tsunami devastations of coastal areas and waterways.
2) Geosynthetics can be incorporated in the reconstructions of infrastructures to be erosion resistant and for flexibility and durability.
3) Geosynthetics can be utilized as filters, reinforcements, drainage, containment and separators.

4) Geosynthetics are lightweight and easy to install with low handling and overall costs.
5) Geosynthetics can be utilized effectively for the construction of "escape mountain or hill" near the coast and earthen berms at buffer zones.

REFERENCES

Abednego, L.G. (2005). "The contribution of Indonesian Engineers Association to Aceh Province after the Earthquake and Tsunami", Proc. of Scientific Forum on Tsunami, its Impact and Recovery, 6 to 7 June 2005, AIT Conference Center, Bangkok, Thailand.

Bergado, D.T., Sadlier, M., Lawson, C. and Piyaboon, S. (2005). "Tsunami reconstruction with geosynthetics in coastal areas and waterways", Proc. 3rd International Conference on Geotechnical Engineering, Semarang , Indonesia.

Bergado, D.T., Anderson, L.R., Miura, N. and Balasubramaniam, A.S. (1996). Soft Ground Improvement in Lowland and other Environment, ASCE press, USA.

Heerten, G., Jackson, L.A, Restall, S.J and Stelljes, K. (2000). "Environmental benefits of sand filled geotextile structures for coastal applications", Proceeding of Geo-Eng 2000, Melbourne, Australia.

Koerner, R.M. and Soong, T.Y. (1995). "Use of geosynthetics in infrastructure remediation", *Journal of Infrastructure System,* ASCE, Vol. 1, No. 1, 66-75.

Koerner, R.M. (2000). "Emerging and future developments of selected geosynthetics application", *Journal of Geotechnical and Geoenvironmental Engineering*, ASCE, Vol. 125, No. 4, 293-306.

Pilarczyk, K. (2000). "Geosynthetics and Geosystems in Hydraulic and Coastal Engineering", A. A. Balkema Publishers, Rotterdam, The Netherlands.

Restall, S.J., Jackson, L.A. Heerten, G. and Hornsey, W.P. (2002). "Case studies showing the growth and development of geotextile sand containers: an Australian perspective", *Geotextile & Geomembranes*, Vol. 20, 321-342.

Warnitchai, P. (2005). "Lessons learned from the 26 December 2004 Tsunami Disaster in Thailand", Proc. of Scientific Forum on Tsunami, its Impact and Recovery, 6 to 7 June 2005 AIT Conference Center, Bangkok, Thailand.

International Conference on Geotechnical Engineering for Disaster Mitigation & Rehabilitation
Chu, Phoon & Yong (eds) © 2005 World Scientific Publishing Company ISBN 981-256-469-1

TSUNAMI AFFECTED COASTAL SOIL DISTURBANCE AND IMPLICATIONS ON RECONSTRUCTION WITH SPECIAL REFERENCE TO LOW-COST DWELLINGS AND RAIL TRACKS

BUDDHIMA INDRARATNA
Professor of Civil Engineering, University of Wollongong, Wollongong City, NSW 2522, Australia

MOHAMED A. SHAHIN
Research Fellow, School of Civil, Mining and Environmental Engineering, University of Wollongong, Wollongong City, NSW 2522, Australia

CHOLACHAT RUJIKIATKAMJORN
Research Associate, School of Civil, Mining and Environmental Engineering, University of Wollongong, Wollongong City, NSW 2522, Australia

HADI KHABBAZ
Jeffery and Katauskas, Sydney, NSW, Australia

On 26 December 2004, the largest earthquake for more than 4 decades (magnitude 9.0) occurred between the Australian and Eurasian plates in the Indian Ocean (along the overly stressed Sunda trench) to the west of Aceh Province (Northern Sumatra). The quake triggered a series of waves that increased in height rapidly close to the shore (tsunami) spreading thousands of kilometers across the Bay of Bengal. In the Eastern and Southern coastal belt of Sri Lanka, the floodwaters reached almost 1 km inland, causing unprecedented damage to infrastructure and over 4000 fatalities. Widespread destruction included several kilometres of rail tracks, dislocating the track elements from the remolded surface soil. Near the beach town of Hikkaduwa, the ferocity of the waves was evident with the total destruction of tracks within 100 m from the shore, and overturning a crowded intercity train. Based on visual examination and CPT tests conducted several weeks later at the site of the train disaster, it was observed that the sandy topsoil was turbulently blended with transported marine sediments including organic fines. Under excessive hydraulic gradients, the geotechnical properties of surface soils up to a meter or more have been significantly altered. At some locations near the surface, the void ratios have almost doubled once the waves receded and the soil re-deposited. In this paper, the relevant aspects of dwellings reconstruction and rail tracks on devastated coastal soils are elucidated. Guidelines for reconstructing robust cost-effective foundations are discussed, based on the first author's own experience. Revised ballast grading and enhanced track conditions are considered, including the benefits of increasing track confining pressure and the essential need for soft formation stabilization.

The use of geocomposites (i.e. bonded geogrid-geotextile layers) in achieving reduced track settlement, increased resilient modulus and decreased ballast degradation, is described.

INTRODUCTION

A tsunami travels from source area (usually earthquake epicenters) as a series of concentric waves. In the deep sea, these waves can travel at speeds of 500 to 800 km/h from the epicenter, but approaching the shore, the waves decrease in speed to 20-30 km/h while increasing the height as the kinetic energy transforms to potential energy. A wave that is only a meter in height in the deep ocean can grow to a few tens of meters at the shoreline (ITIC 2000). Destruction caused by tsunamis is the direct result of three factors: impact, inundation, and erosion. Both the incoming and receding waves can lead to a significant erosion of coastal sandy soils as well as inducing piping under excessive hydraulic gradients, undermining bridge piers, loss of foundation bearing capacity and confinement, apart from the obvious damage to structures upon wave impact.

The Boxing Day tsunami in December 2004 devastated several South and Southeast Asian countries including Indonesia (Northern Sumatra), Thailand (Phuket), Sri Lanka (Eastern and Southern Provinces), South India, Andaman and Maldives islands. The epicenter of this earthquake was measured as 9 on the Richter scale. As a consequence of the vertical 'throw' of the large discontinuity plane thus formed, high velocity ripples (more than 700 km/h) transformed to tsunami as the shallow depths (shorelines) were approached. While in most countries, the tidal waves directly impacted on the coastal belt, in Sri Lanka, while the eastern coast was destroyed by direct wave impact (waves up to 13 m), its southern coast was severely damaged by the turbulent and rebound waves with considerable angular momentum with heights still reaching close to 10 m. The damage caused by these turbulent waves were so catastrophic that hardly any houses and commercial buildings within a 100-150 m proximity to the shore were spared, and even the well built rail tracks and highways that have lasted for many decades were totally destroyed. The coastal stretch near the famous tourist beach town of Hikkaduwa was so badly hit. More than 1000 passengers lost their lives and many others injured, when the tsunami engulfed a crowded intercity train, derailing, toppling and submerging its carriages, making this train disaster the worst rail accident ever to be recorded. The lifting and piping of the formation soil beneath the track caused rapid demounting of the track, lifting it up with the tidal waters and displacing it several meters.

Site reconnaissance at the location of this train disaster was conducted under the guidance of the first author with the assistance provided by University of Moratuwa and Engineering Laboratory Services Ltd. The field observations indicated areas of significantly disturbed surface soils (at some locations up to 1 m), now a heterogeneously mixed medium composed of the original topsoil blended with very fine beach sands, silts and organic sediments transported by the waves and construction materials (debris) turbulently mixed *in situ*. Once the waves receded, the 'piping' sands, eroded top soils, transported fine sediments (including organics) and debris have settled at a much higher porosity compared to the originally compacted sandy soil that have existed for hundreds of years.

Sri Lanka has experienced a number of tsunamis given the past historical periods dating back over 2000 years, the last tsunami reported in June 1941 as a result of an earthquake of magnitude 8.1 near Andaman islands (Wattegama 2005). This paper will elucidate the need for rebuilding dwellings and rail tracks that are more resistant to future tsunamis as well as more resistant to possible earthquake tremors that Sri Lanka is now prone to. Given the recent history of earthquake epicenters and plate tectonics in the vicinity of the Sunda trench, Sri Lanka cannot be ruled out as earthquake free any longer. Especially where the coastal fine sands are prone to liquefaction, the roads and rail tracks require much more robust design implementation instead of conventional methods.

PRELIMINARY SITE INVESTIGATIONS

A trial open pit at an affected site indicated blended surface sands with no distinct layering, and clean relatively undisturbed sand was found at a depth exceeding 450 mm. The particle size distributions (Figure 1) indicated only a more obvious well-graded nature of the sand closer to the surface, in the areas where the beach sand was known to be very uniform before the tsunami. The uniformity coefficient has changed form 1.6 to 4.6 in this particular location as a result of mixing. A standard cone penetrometer test (CPT) and a cone penetrometer test with pore pressure measurement (CPTU) were conducted to re-examine the soil profile up to 10 m deep, on 2 different locations near the train disaster site. Figures 2 and 3 illustrate measured parameters obtained from the CPT and CPTU, respectively. The friction ratios determined for the shallow depths (less than 1 m) indicate metastable sands and/or mixed soils with increased sensitivity. It is also shown that soil layer up to 1 m was completely remolded by flooding. The surface compaction is essentially required to improve the ground condition. At greater depths exceeding 2 m, the stable cohesionless sand deposits (unaffected by the tsunami) could be established from the CPT and CPTU profiles (Figures 2 and 3). The piezocone tests also indicated the increased pore pressures at various depths due to the presence of organic (peat) sediments and clayey sand/silt deposits that now carry increased moisture content due to the infiltration of water through the relatively pervious top sand layers (Figure 3). The hydrostatic pore pressure increases linearly and starts at the ground surface. The presence of peat layer is clearly identified by the suddenly increased friction ratio.

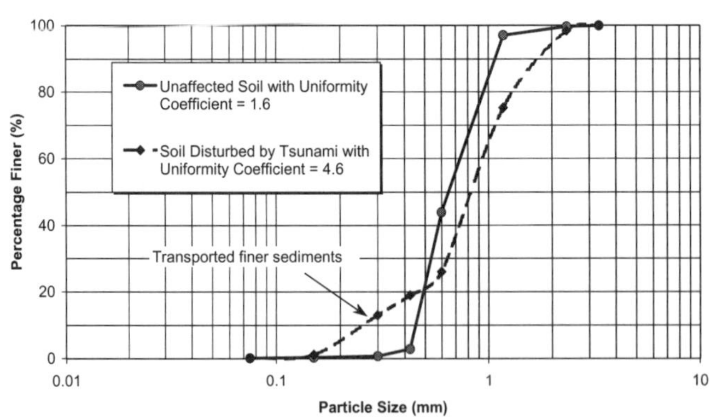

Figure 1. Particle size distribution curves of the tsunami affected and unaffected soils

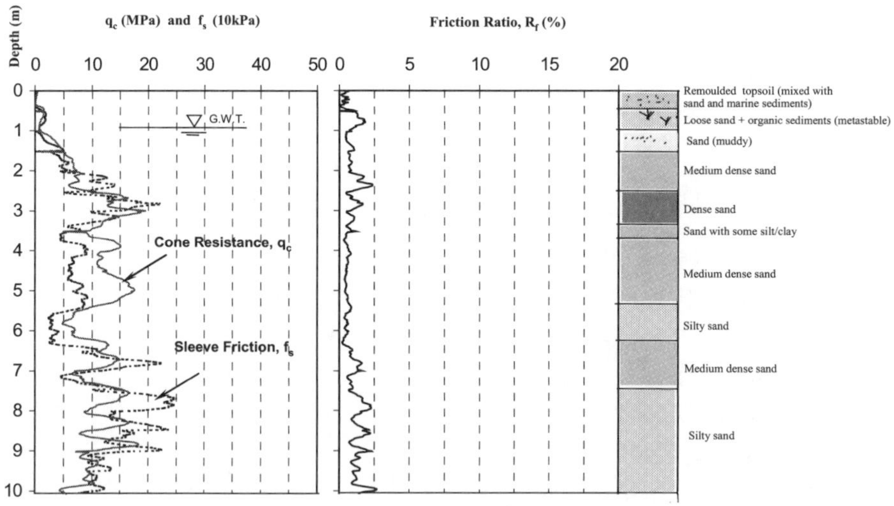

Figure 2. Cone penetration test (CPT) results of soil layers after tsunami occurrence (Site 1)

The soil properties of a soil sample in tsunami-affected area showed a void ratio of about 0.7 at 600 mm depth, which is much higher than the initial pre-tsunami void ratio estimated to be less than 0.5 on the compacted surface. Assuming this range of void ratio and an associated saturated unit weight of about 21 kN/m^3, the critical hydraulic gradient, i_{cr}, can be estimated from:

$$i_{cr} = (\gamma_s / \gamma_w) - 1 \approx 1.14 \tag{1}$$

where, γ_s= unit weight of saturated soil and γ_w = unit weight of water. The above value of i_{cr} implies that a rapid flood with a height greater than 1 m may induce piping of the shallow surface soils to a substantial depth.

RECONSTRUCTION GUIDELINES FOR HOUSING OF THE TSUNAMI AFFECTED AREAS

Rehabilitation of tsunami victims is of paramount importance in future sustainable development plans in Sri Lanka. In this section, some guidelines for reconstructing low-cost dwellings in the affected coastal areas are proposed, in terms of: (a) locations and alignment of houses and (b) appropriate design of foundations.

Locations and alignment of houses

1. Dwellings should be constructed at a 'safe distance' of at least 100-200 m away from the beach depending on the area (e.g. presence or absence of mangroves, sand dunes, etc). The Eastern part of Sri Lanka is more prone to direct waves from the Indian ocean, whereas the Southern part of the island may receive rebound waves.

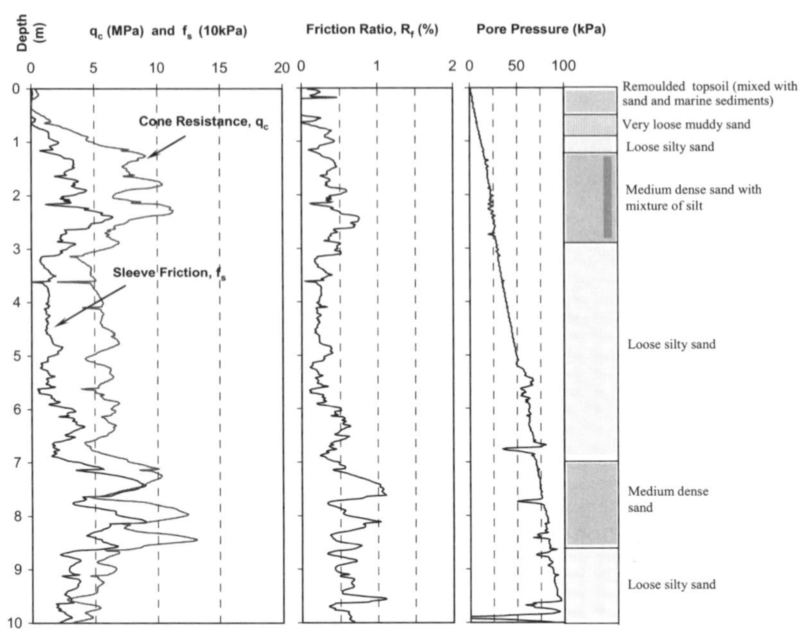

Figure 3. Cone penetration test with pore pressure (CPTU) results of soil layers after tsunami occurrence (Site 2)

2. Housing in areas behind densely grown palm trees (coconut and palmyrah) may be encouraged (Figure 4). These palms have fibrous root systems which facilitate the intake of groundwater from soil as deep as 2-3 meters. They provide high suction to hold together large volumes of fine soils that can otherwise be disturbed or remolded by extreme hydraulic gradients. Moreover, these strong roots provide excellent natural reinforcement for sandy coastal soils, increasing the apparent bearing capacity considerably.

3. For land at almost sea level, typical dwellings should be constructed on a raised earth platform in the order of 0.6-1 m in height, to provide protection against flooding and foundation scouring (Figure 5).

4. Dwellings may be constructed such that the strong masonry side faces the direction of waves (Figure 6). Mangroves must be encouraged as much as possible within the 'safe distance'. A group of houses should be arranged in a parallel pattern rather than random to create channels for the waves to pass through easily, with minimum damage.

Recommended foundation design for housing units

1. At sites affected by tsunami, proper soil investigations based on Cone Penetration Test and/or Standard Penetration Test should be carried out to determine the degraded soil properties and their distributions. The depth to sound (unaffected) soil

layers needs to be estimated and the bearing capacity evaluated on the basis of shear strength.

Figure 4. Protection from tree line

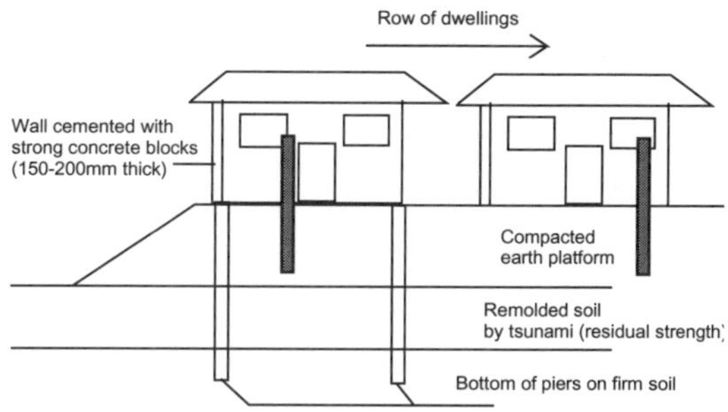

Figure 5. Housing construction on raised earth platform

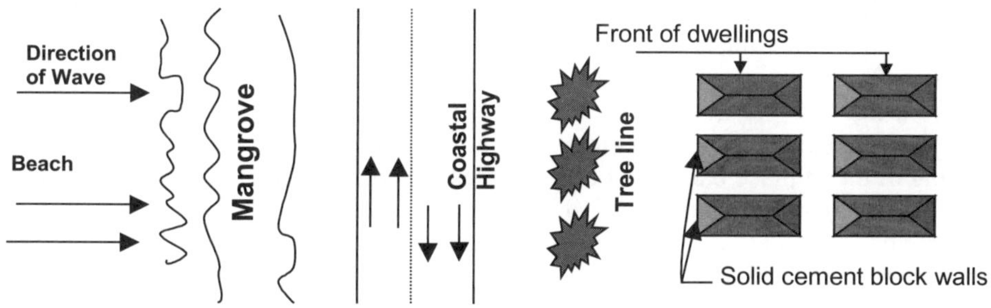

Figure 6. A preferable arrangement for a group of houses

2. Weakened soil conditions in some areas (1-2 m depth) should be stabilized by appropriate ground improvement techniques, as will be discussed later. Removal of transported soft marine sediments up to 200 mm from the soil surface is required, in some areas.
3. In order to provide strong strip footings, at least 250 mm wide trenches may be excavated up to 600 mm deep and the bottom of these trenches compacted to about 95% relative density. Subsequently, a 200-250 mm thick compacted gravel (size 7-10 mm) layer may be placed at the bottom of the trench to provide a good drainage medium as well as to increase the bearing capacity (Figure 7). The gravel layer will rapidly dissipate excess pore pressure and also act as a damping layer in case of earthquake tremors. Concrete piers should be installed about every 3 m and at all corners of the dwellings as shown in Figure 8. These piers may be up to 1.5 m deep, resting on the medium dense sand, affected by tsunami.
4. Finally, the reinforced concrete footings should be placed above the compacted gravel layer to erect the houses (Figures 7 and 8).

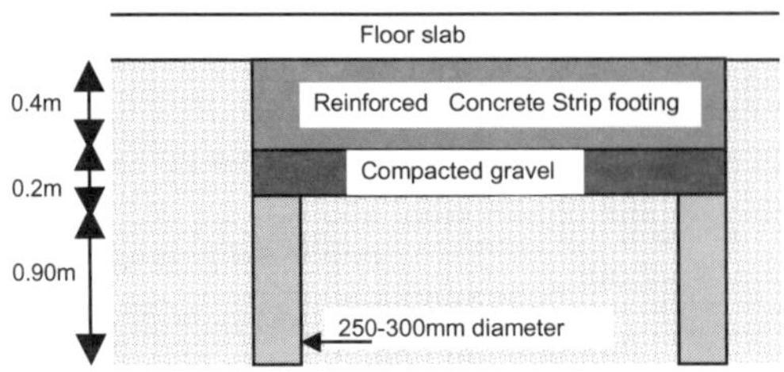

Figure 7. Foundation for soils affected by tsunami

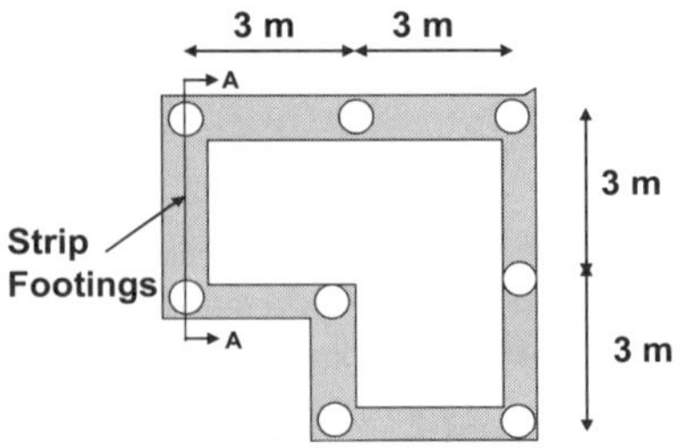

Figure 8. Typical plan view of foundation construction

BALLAST AND TRACK MODIFICATION

Ballast is the largest component of a rail track by weight and volume. While ballast is expected to be low-cost and in adequate supply, it also has to be capable of providing the engineering requirements of the rail track. The main functions of ballast are: distributing of loads received from the sleepers, damping of dynamic loads, producing lateral resistance and providing rapid drainage. Supplying high quality fresh ballast for reconstruction of rail track for entire region affected by tsunami is costly and time consuming. Therefore, the existing ballast of the damaged tracks can be reused, except for some areas that the track substructure has been totally washed out by the waves. Obviously, old ballast should be totally removed, cleaned, sieved and recompacted. Recycling of the in place ballast not only reduces the cost of track rehabilitation, but also contributes to waste reduction and decreases the need for extensive quarrying and environmental degradation. However, recycled ballast has lower internal friction angle due to attrition and reduced angularity, hence, exhibiting higher settlement and lateral deformation compared to those of fresh ballast upon cyclic loading. Research at University of Wollongong indicated that inclusion of geosynthetics improves the geotechnical properties of recycled ballast, as discussed later.

Effect of particle size distribution

The gradation of ballast is a prime consideration for track performance. To evaluate the effects of particle size distribution on deformation and degradation behavior of ballast, large-scale cyclic triaxial tests were conducted on four different distributions of latite basalt at University of Wollongong. Details of the testing apparatus can be found in Indraratna et al. (2003). The gradation and void ratio characteristics of the test specimens are shown in Figure 9. Samples were subjected to an effective confining pressure of approximately 45 kPa, and cyclic loading having a maximum deviator stress of 300 kPa was applied on the ballast specimens at a frequency of 20 Hz. Figure 10 shows the effect of grain size distribution on the axial and volumetric strains of ballast under cyclic loading. The tests results reveal that most uniform to moderately uniform samples give higher axial and volumetric strains. This is attributed to the looser states of the specimens prior to cyclic loading. In contrast, gap-graded and moderately graded distributions provided denser packing with a higher co-ordination number (increased surface contact). Therefore, these gradations provided a higher shear strength as well as reduced settlement.

In terms of deformation and resistance to particle breakage (Figure 11), the test results indicate that moderately graded ballast is far superior to uniform gradations, which is now acknowledged in the current ballast specifications of some countries including Australia. The test results also indicate that moderately graded ballast is still porous enough to maintain sufficient track drainage. Based on these findings, Indraratna et al. (2004) recommended a ballast gradation with a uniformity coefficient exceeding 2.2, but not more than 2.6, in comparison to very uniform (conventional) gradings with C_u= 1.4-1.5. This recommended gradation, which is relatively more well-graded than the current Australian Standards (A S 2758.7 1996) is presented in Figure 12.

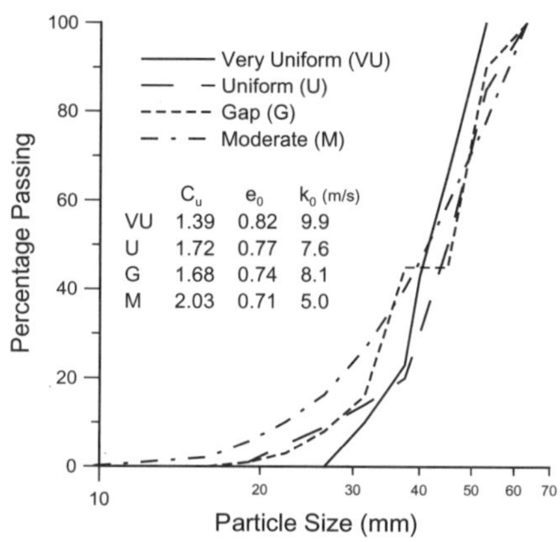

Figure 9. Particle size distributions used in the triaxial tests (Indraratna *et al.* 2004)

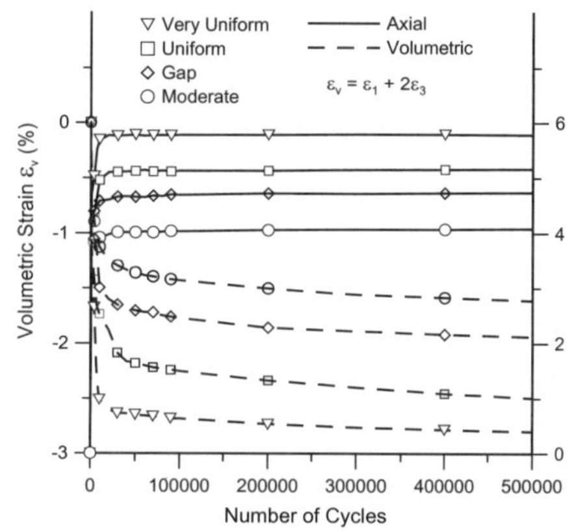

Figure 10. Axial and volumetric strain response of different distributions under cyclic loading (Indraratna *et al.* 2004)

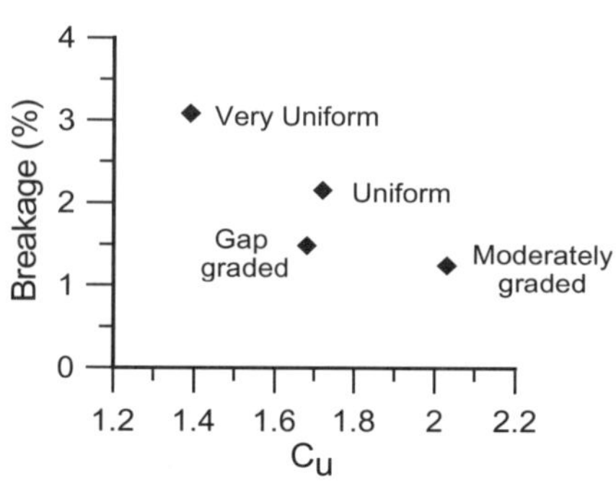

Figure 11. Effect of grading on particle breakage (Indraratna *et al.* 2004)

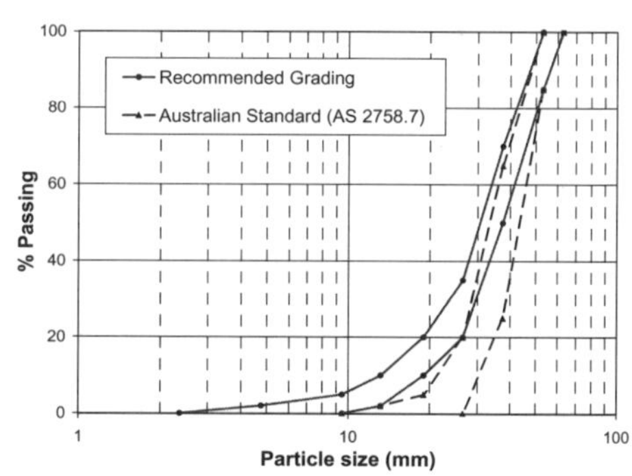

Figure 12. Recommended railway ballast grading in comparison with the current Australian Standard (Indraratna *et al.* 2004)

Effect of confining pressure

The role of confining pressure on ballast performance under cyclic loading has been investigated by Indraratna et al. (2005b). Figure 13 illustrates the effect of confining pressure (σ_3') on the axial and volumetric strains of ballast achieved at the end of 500,000 cycles for a maximum deviator stress of 500 kPa. As expected, the axial strains decreased with the increasing confining pressure. Ballast specimens exhibited dilation at small confining pressure ($\sigma_3' < 30$), but became progressively more compressive as the confining pressure increased from 30 to 240 kPa. The effect of confining pressure on particle degradation is shown in Figure 14. It was found that there is an optimum confining pressure (30-75 kPa) in which the amount of ballast breakage was reduced to its minimum value. Some measures for increasing track confinement include: reducing sleeper spacing, increasing height of shoulder ballast, inclusion of a geosynthetic layer at the ballast-subballast layer interface, widening the sleepers at both ends (Figure 15), and using intermittent lateral restraints at various parts of the track (Figure 16).

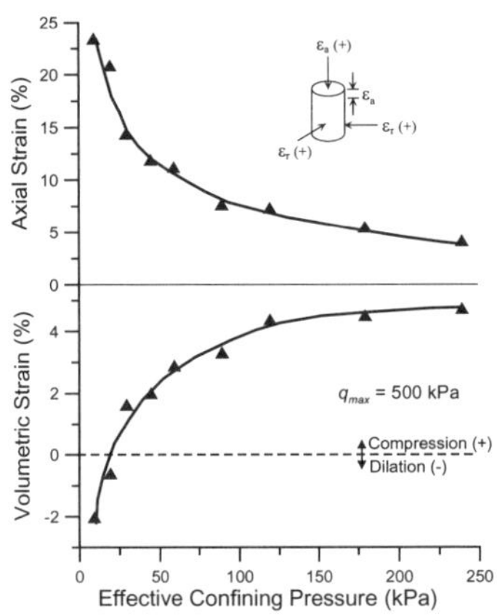

Figure 13. Variation of axial and volumetric strains with confining pressure (Indraratna *et al.* 2005b)

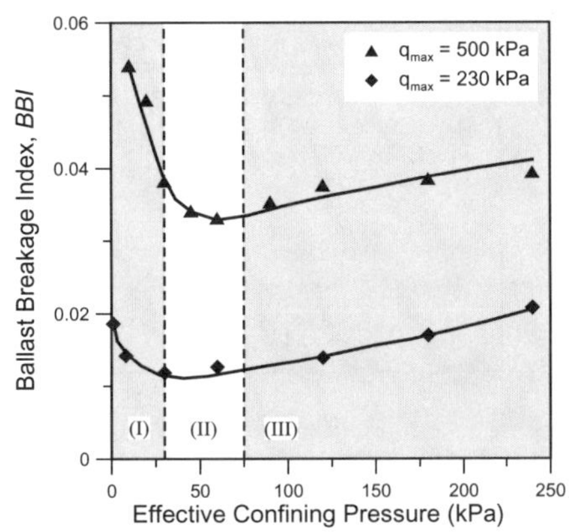

Figure 14. Effect of confining pressure on particle degradation (Indraratna *et al.* 2005b)

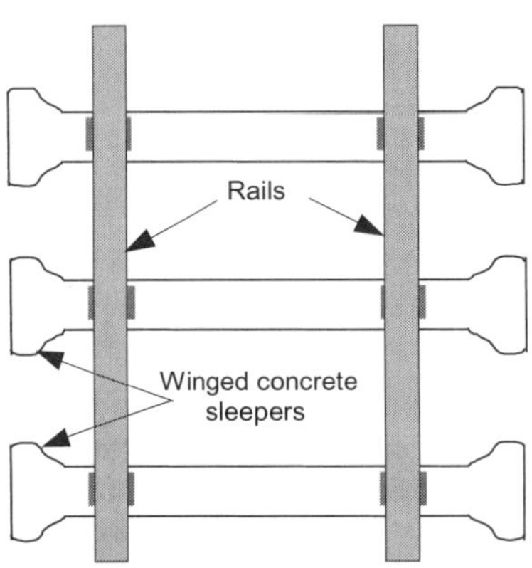

Figure 15. Sleepers with enlarged ends to Increase the confining pressures

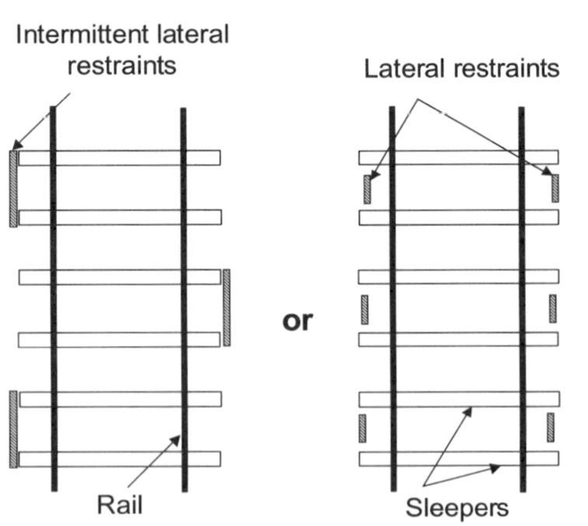

Figure 16. Increasing confining pressure using intermittent lateral restraints

Improvement of recycled ballast using geosynthetics

The deformation and degradation behavior of fresh and recycled ballast was investigated in a large triaxial chamber (Figures 17 and 18) simulating a small track section. This large-scale prismoidal triaxial rig of 800 mm length, 600 mm width and 600 mm height was designed and installed at the University of Wollongong to model the cyclic loading response of ballasted tracks. By allowing the lateral strain of ballast upon loading, the triaxial rig with unrestrained sides provides a reliable facility for physical modeling of ballast. details of the large-scale rig can be found in Indraratna et al. (2003). The stabilization aspects of recycled ballast using various types of geosynthetics were also studied in these model tests. The effectiveness of various geosynthetics in stabilizing recycled ballast was investigated through laboratory model test results. Three types of geosynthetics were used including woven geotextiles, geogrids and geocomposites (bonded geogrids and non-woven geotextiles). The tests were conducted in both dry and wet conditions to study the effects of saturation. The testing procedures together with complete findings and discussions have been reported by Indraratna et al. (2004). Only selected results illustrating the effects of inclusion of geocomposites on ballast settlement and breakage are given in this paper.

Figure 17. Large-scale prismoidal triaxial equipment built at the University of Wollongong

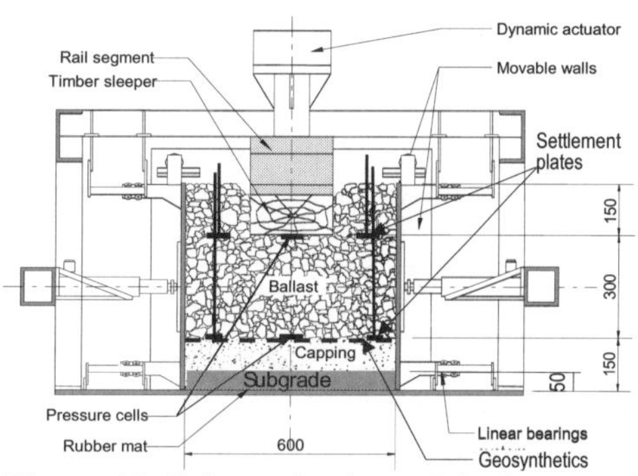

Figure 18. Schematic view of the large-scale prismoidal triaxial apparatus (Indraratna and Salim 2003)

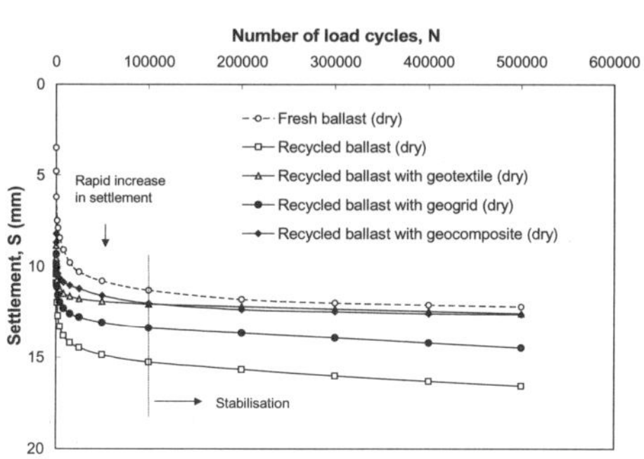

Figure 19. Settlement of dry fresh and recycled ballast with and without geosynthetics (Indraratna *et al.* 2004)

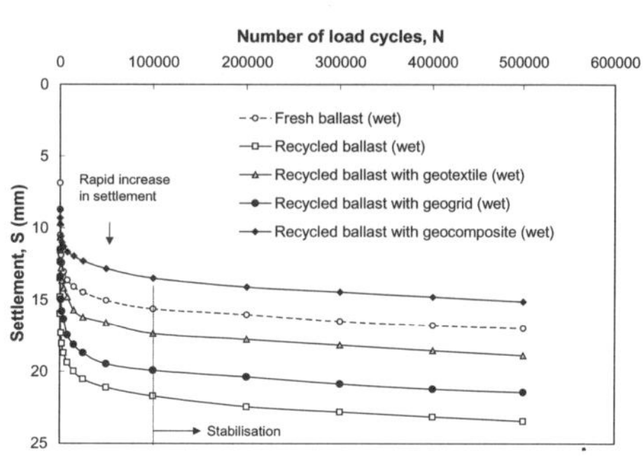

Figure 20. Settlement of wet fresh and recycled ballast with and without geosynthetics (Indraratna *et al.* 2004)

Figures 19 and 20 show the settlement of fresh ballast without any reinforcement and recycled ballast with and without the inclusion of geosynthetics in dry and wet status, respectively. As expected, dry fresh ballast gives the least settlement. It is believed that the higher angularity of fresh ballast contributes to better particle interlock and therefore, causes less settlement. The test results reveal that wet recycled ballast (without any geosynthetic inclusion) generates significant settlement, because, water acts as a lubricant thereby reducing the frictional resistance and promoting particle slippage. Although geogrids and woven geotextiles decrease the settlement of recycled ballast considerably, the geocomposite (geogrid bonded with non-woven geotextiles) stabilizes recycled ballast remarkably well. As described by Rowe and Jones (2000), geocomposites can provide reinforcement to the ballast layer, as well as filtration and separation functions simultaneously. The combination of reinforcement by the geogrid and the filtration and separation functions provided by the non-woven geotextile component (of the geocomposite) reduces the lateral spreading and fouling of ballast, especially in wet conditions. The non-woven geotextile also prevents the fines moving up from the capping and subgrade layers, thus keeps the recycled ballast relatively clean.

To quantify ballast breakage based on Marsal's method (1967), each ballast specimen was sieved before and after testing, and the changes in percentage retained on each sieve size were recorded. The breakage index values of recycled ballast stabilized with geocomposites in dry and wet conditions were almost the same as fresh ballast (without geocomposites), and approximately 50 % lower than those of recycled ballast without geosynthetics. This indicates clearly the benefits of using geosynthetics in the reduction of recycled ballast breakage in both dry and saturated conditions.

SUBBALLAST LAYER

Subballast is the layer of aggregates placed between the ballast layer and the subgrade. This is usually comprised of well-graded crushed rock or a sand/gravel mixture. The subballast layer should be designed to prevent the penetration of coarse ballast grains into the subgrade, and the upward migration of subgrade fines (formation soil) into the ballast layer. Therefore, subballast acts as a filter and a separating layer in the track substructure, which transmits and distributes stress from the ballast layer down to the subgrade over a wider area. It also acts as a drainage medium to dissipate cyclic pore water pressures developed by the passage of trains.

Using geotextiles in conjunction with a sand-capping layer is highly recommended for track subballast. The sand filter in this case prevents the migration of fine soil upwards, and the geotextile acts as a separator between the sand and ballast. Raymond (1986) recommended the non-woven needle punched geotextiles in preference to woven geotextiles, because the woven geotextile is unable to dissipate excess pore pressures quickly enough. This is particularly important in the case of freight trains that are both very heavy and very long.

FORMATION SOIL MODIFICATION

In many tsunami-affected areas in Southern Sri Lanka, the total rail track structure has to be removed before carrying out any reconstruction. Therefore, the subgrade condition should be carefully examined and appropriate measures have to be taken to improve the formation soil performance. One of the most important steps at the preliminary stage was identifying and classifying the subgrade soils. Although limited subsurface exploration has been conducted at the site (Indraratna *et al.* 2005a), no rail rehabilitation mission will be completed without reanalyzing the post-tsunami soils strength and densities along the coastal rail track and comparing this data with the pre-tsunami soil profiles and data, past geological maps and geotechnical reports. In general, two types of subgrade: (a) loose fine sand and (b) soft silty or clayey sediments need special consideration, as they are highly susceptible to failure under cyclic loading, including excessive plastic deformation and piping. Some differential settlement would be inevitable in some parts of the track that required urgent reconstruction for passenger transport without optimum soil improvement, but of course at a reduced train speed. For stabilizing the formation soils, a variety of options of ground improvement are available in Sri Lanka. Some selected key means of subgrade enhancement are discussed in this section.

Loose sandy subgrade

The preliminary site reconnaissance (Indraratna *et al.* 2005a) indicated that the disturbed top layer of the soil includes mixed soils such as loose sand with marine sediments (including dark brown to black organics). Densification of loose granular soils, heterogeneous soils and liquefiable soils can be achieved by surface compaction. The purpose of densification is to increase the relative density and the internal friction angle of the soil (shear strength), thereby reducing the post-construction settlement as well as increasing the bearing capacity. The surface vibratory or pneumatic compaction is regarded as the simplest and the cheapest method for densification of loose, saturated and cohesionless remolded soils in the tsunami affected areas. This method reduces the void volume of remolded soil by forcing the soil particles into a tighter state using dynamic forces. In the field, the soil up to 1 m deep can be compacted using rolling and kneading, of impact vibrating and ramming.

Grouting technology and chemical stabilization can also be used as alternatives to compaction to retard water seepage and to increase the shear strength of loose sand formations. For preventing formation collapse during significant earthquakes, deep mixing method may be employed to increase the liquefaction resistance of loose sediments beneath rail track and improve its stability.

Clay subgrade

In the region of the site of soil exploration, no thick layers of soft clays were found (up to 12-15 m depth), but clayey sands and organic sediments will still show significant compressibility. However, many coastal regions of Southeast Asia contain soft clays (estuarine or marine), which have poor geotechnical properties such as low bearing capacity and high compressibility. In the City of Colombo itself and to the south of Colombo towards the Southern Province that was devastated by the tsunami, compressible clays and peaty soils are often encountered, and some of these areas only just escaped the wrath of the tsunami. If these areas had been flooded, the subsequent settlement of buildings may have been substantial. Transport infrastructure including rail tracks are affected by the settlement and lateral movement of soft formation soils, in the absence of appropriate ground improvement prior to track reconstruction. In Sri Lanka, the improvement of clayey soft soils also requires equal attention as much as the loose sandy deposits in the Southern and Eastern provinces. This is because, towards the City of Colombo not only the coastal population increases, hence the need for taller buildings, but also the extent and the frequency of soft clays and peaty soils increase. Cost effective techniques for ground improvement are now employed for compressible formation soils in Sri Lanka. Some popular methods include: stabilization by chemical admixtures; geosynthetic reinforcement; and preloading and the use of prefabricated vertical drains.

CONCLUSIONS

Sri Lanka, among several other countries, was devastated by the Indian Ocean tsunami, which occurred on 26th December 2004. In this paper, the recommended procedures and suggestions that can be adopted for reconstruction of dwellings (including housing locations and design of foundations), and rail track substructures in the affected costal soils have been explained. The approaches recommended are cost-effective and not time consuming.

Based on the preliminary site investigation conducted by the first author, reconstruction of the affected areas requires reducing the soil void ratio to at least 0.45 to achieve an adequately stiff surface layer prior to the reconstruction process. Reconstructing low-cost dwellings in the affected coastal areas requires the following guidelines. Dwellings should be constructed at a safe distance of at least 100-200 m away from the beach in such away that the strong masonry side faces the direction of waves. Dwellings should also be constructed on a raised earth platform in the order of 0.6-1 m in height. Housing in areas behind densely grown palm trees is encouraging.

To accelerate rail track rehabilitation, selected waste ballast in tsunami-affected areas, when properly cleaned and stabilized with geosynthetics can be reused as load bearing materials. The use of composite geosynthetics at the bottom of the recycled ballast layer is highly desirable due to the heterogeneously mixed soils along the track. Geocomposites also decrease track lateral movement, ballast degradation and subgrade pumping.

It is illustrated that the gradation of ballast plays a significant role in the strength, deformation, degradation, stability and drainage of tracks. Well-graded ballast gives denser packing, better frictional interlock and hence, lower settlement. On the other hand, almost all ballast specifications demand uniform gradation for free draining. However, the uniformly graded ballast gives higher settlement and also more vulnerable to breakage than well-graded ballast. Test results indicate that the use of slightly broader graded ballast than the current Australian Standard gives considerably lower settlement and decreases the extent of breakage while not affecting drainage significantly. It is recommended that the rail tracks in Sri Lanka adopt a less uniform ballast gradation based on these findings. Another factor that affects the performance of ballast is the track confining pressure. Findings based on large scale triaxial testing indicate that a small increase in confining pressure (σ_3' = 30-75 kPa) improves the track stability with less ballast degradation.

ACKNOWLEDGEMENTS

The contents of this paper are mainly taken from the first author's keynote paper at the International Symposium on Tsunami Reconstruction with Geosynthetics held in Bangkok, Thailand. The authors express their sincere gratitude to Cooperative Research Center for Railway Engineering and Technologies (Rail–CRC), RailCorp of NSW and related geosynthetics companies in Australia for their support and cooperation. Support provided by the Minister of Science and Technology of Sri Lanka is greatly acknowledged. The authors also express their special thanks to Dr Wadud Salim (RailCorp of NSW) and Ms Joanne Lackenby (PhD student at UoW) for their assistance and contribution.

REFERENCES

Australian Standards 2758.7 (1996). "*Aggregates and rock for engineering purposes*", Part 7: Railway Ballast.

Indraratna, B., Khabbaz, H., Puswewala, A. and Bandrara, W. (2005a). "Effects of tsunami on coastal ground conditions and appropriate measures for rail track rehabilitation", *International Symposium on Tsunami Reconstruction with Geosynthetics - Protection, Mitigation and Rehabilitation of Coastal and Waterway Erosion Control*, Bangkok, Thailand.

Indraratna, B., Khabbaz, H., Salim, W. and Christie, D. (2003). "Geotechnical characteristics of railway ballast and the role of geosynthetics in minimizing ballast degradation and track deformation", *RAILTECH 2003 - Railway Technology in the New Millennium*, Kuala Lumpur, Malaysia, 3.1-3.22.

Indraratna, B., Khabbaz, H., Salim, W., Lackenby, J. and Christie, D. (2004). "Ballast characteristics and the effects of geosynthetics on rail track deformation", *International Conference on Geosynthetics and Geoenvironmental Engineering, ICGGE*, Bombay, India, 3-12.

Indraratna, B., Lackenby, J. and Christie, D. (2005b). "Effect of confining pressure on the degradation of ballast under cyclic loading", *Geotechnique*, Vol. 55, No. 4, 325-328.

Indraratna, B. and Salim, W. (2003). "Deformation and degradation mechanics of recycled ballast stabilized with geosynthetics", *Soils and Foundations*, Vol. 43, No. 4, 35-46.

ITIC (2000). "Tsunami classification and glossary", *Intergovernmental Oceanographic Commission of UNESCO*, Hawaii, 1-20.

Marsal, R.J. (1967). "Large scale testing of rockfill materials", *Journal of Soil Mechanics and Foundation Engineering*, Vol. ASCE, 93, No. SM2, 27-43.

Raymond, G.P. (1986). "Performance assessment of a railway turnout geotextile", *Canadian Geotechnical Journal*, Vol. 23, No. 4, 472-480.

Rowe, P.K. and Jones, C.J.F.P. (2000). "Geosynthetics: innovative materials and rational design", *Proceedings, GEOENG 2000*, Melbourne, Australia, Vol. 1, 1124-1156.

Wattegama, C. (2005). "The seven tsunamis that hit the isle of Sri Lanka", *Engineering News, Institution of Engineers Sri Lanka*, Vol. 40, No. 3, 6-7.

International Conference on Geotechnical Engineering for Disaster Mitigation & Rehabilitation
Chu, Phoon & Yong (eds) © 2005 World Scientific Publishing Company ISBN 981-256-469-1

NATURAL DISASTER MANAGEMENT IN THE REPUBLIC OF KOREA

SUNGRYUL KIM
Dept. of Civil Engineering, DongA University, #840 Hadan 2-dong, Saha-gu, Busan, Korea

YONGKYUN KIM
Preparedness Division, National Emergency Management Agency, Room 1314, #55 Sejongro, Jongro-gu, Seoul, Korea

MYOUNGMO KIM
Dept. of Civil, Urban and Geosystem Engineering, Seoul National University, Sillim-dong, Kwanak-gu, Seoul, Korea

The Republic of Korea is located east of the mid longitude of the Asian continent. Korea has a temperate climate with four distinct seasons characterized by many arid days in spring, substantial rain in summer, and much snow in winter. Average annual precipitation is 1,274mm with approximately 50 to 60% of the annual precipitation concentrated during the summer. An annual average of seven natural disasters hit Korea, with the annual average property loss and death toll of about KRW634.2 billion and 106 deaths, respectively. Korea spends about KRW 1.0071 trillion for rehabilitation of devastated areas per year. Natural disasters include torrential rain, storms, typhoons, and heavy snow. In terms of frequency, torrential rain, storms, and typhoons make up 78.7% of the natural disasters, accounting for 84.2% of the total damage. The status and characteristics of disasters, the natural disaster management system, the implementation of countermeasures against natural disasters, and the international exchanges and cooperation programs of Korea will be explained in this report.

OCCURRENCE OF NATURAL DISASTERS

Topography and weather

The Republic of Korea is located east of the mid-longitude of the Asian continent (33-43°N, 124-132°E). The Korean peninsula faces the Yellow Sea and the Chinese continent on its west, the East Sea and Japan on its east and the South China Sea and the Pacific Ocean on its south. Its topography is characterized by "a high east and low west" as its western coastal areas consist of gentle sloping terrain while its eastern coastal areas consist of steep gradient. About 50-70% of Korea's annual precipitation flows into seas because rivers have short watercourses and riverbeds have steep gradients.

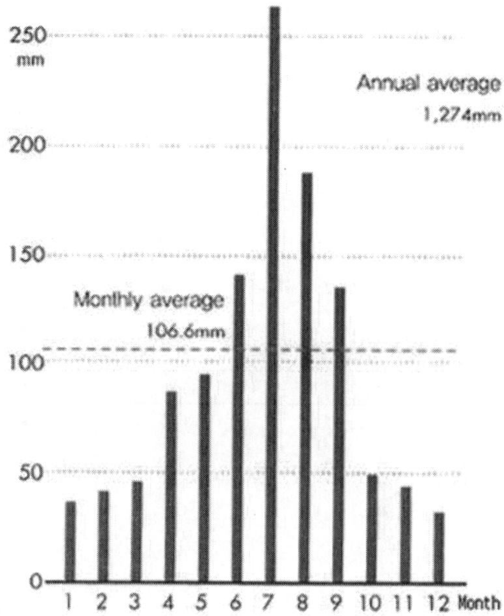

Figure 1. Monthly precipitation

Korea has a temperate climate with the four distinct seasons characterized by many arid days in spring, substantial rain in summer, and much snow in winter. Average annual precipitation is 1,274mm with approximately 50 or 60% of the annual precipitation concentrated during the summer, as shown in Figure 1.

Status and characteristics of disasters

The average number of annual natural disasters hitting Korea is placed at seven, with annual average property loss and death toll standing at KRW634.2 billion and 106 deaths, respectively. Korea spends about KRW1.0071 trillion for rehabilitation of devastated areas per year

Figure 2 shows the damage and rehabilitation costs from natural disasters and Figure 3 shows the annual average damage from each type of natural disaster during the past 10 years. Types of natural disasters include torrential rain, storm, typhoons and heavy snow. In terms of damage, the heavy rain and typhoon make up 83.6% of the total damage.

Heavy rain generally occurs from June to August. About two thirds of the annual average precipitation amounting to 1,274mm is concentrated during this period. Recently, localized heavy downpour, which is highly unpredictable, has been inflicting severe damage to the country. Approximately 26 typhoons occur in the Northwest Pacific Ocean on the average from July to September every year. Among these, two or three hit Korea, as shown in Table 1.

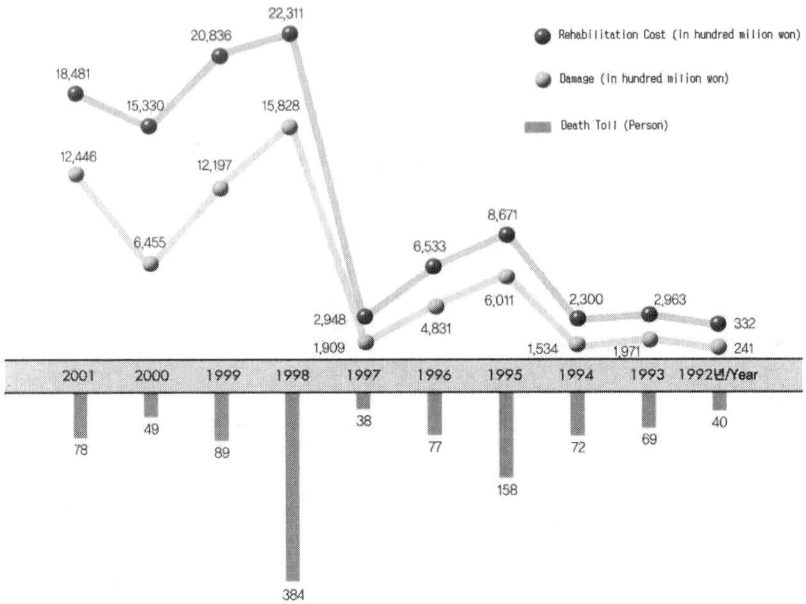

Figure 2. Damage from natural disasters and rehabilitation costs during past 10 years

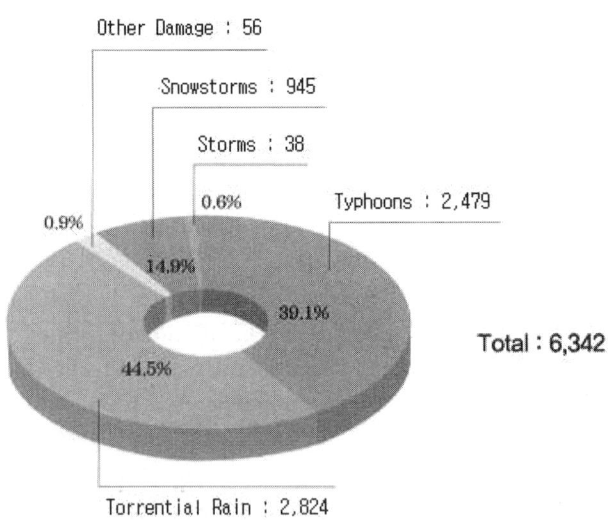

Figure 3. Annual Average Damage from Each Type of Natural Disaster during Past 10 years (unit : hundred million won)

Table 1. Damage from Major Typhoons

Year	Name of Typhoon	Maximum Precipitation per Day (mm)	Death Toll (person)	Damage (hundred million won)
1595.9	SAHRA	168.1	849	662
1987.7	THELMA	216.8	345	3,913
1989.7	JUDY	221.0	20	1,192
1991.8	GLADYS	439.0	103	2,357
1995.8	JAINS	361.5	65	4,562
1998.9	YANNI	516.4	57	2,749
1999.7	OLGA	280.3	67	10,490
2000.8	PRAPIROON	310.0	28	2,521

Storms frequently occur in the ocean from November to February and significantly damage small fishing vessels, coastal fishery and agricultural facilities.

In the winter, heavy snow, which usually occurs from January to February, causes traffic problems in urban areas and damages to vinyl hothouses in rural areas.

Drought often occurs from March to June. It causes a shortage of water needed for agriculture and enormous subsequent damage to crops. It also impedes drinking water supply to island residents.

Korea is hit by an average of 27 earthquakes per annum. Although earthquake frequency is increasing in Korea, the country has not sustained any substantial damage.

NATURAL DISASTER MANAGEMENT SYSTEM

The central government is comprised of 18 "Bu" (ministries), 4 "Cheo" (ministries and agencies) and 16 "Chong" (offices, administrations and services), which individually take relevant measures against possible disasters.

The Ministry of Government Administration and Home Affairs(MOGAHA) manages the overall measures to counter natural disasters in Korea. The Disaster Prevention Officer of the Headquarters of Civil Defense and Disaster Management is responsible for practical affairs.

The National Disaster Prevention and Countermeasures Headquarters (NDPCH) is in charge of prevention and status control of natural disasters, as well as recovery planning, and executes necessary measures related to such disasters. The Minister of Government Administration and Home Affairs heads the NDPCH, while deputy ministers of the Ministry of Construction and Transportation serve as its vice chiefs. Twenty three personnel from 21 government organizations serve on its council.

Laws related with natural disasters include the National Disaster Countermeasures Act, Act on Countermeasures against Agricultural and Fishery Disasters, Disaster Relief Act, etc. The Natural Disaster Countermeasures Act (enacted in 1995) prescribes control of and countermeasures against natural disasters such as torrential rain and typhoons, investigations of relevant damage, rehabilitation costs, etc. The Act on Countermeasures against Agricultural and Fishery Disasters (enacted in 1995) stipulates measures concerning prevention of and countermeasures against disasters affecting agriculture and fishery such as damage from disease, harmful pests and drought. Also, the Disaster Relief Act (enacted in 1962) provides relief to victims of a natural disasters or calamities, etc.

IMPLEMENTATION OF COUNTERMEASURES AGAINST NATIONAL DISASTERS

Basic disaster prevention plans, which are long-term plans against disasters, including disaster prevention systems and relevant countermeasures, are formulated every five years. The sixth Basic Disaster Prevention Plan will go into effect in 2002. In accordance with basic disaster prevention plans, specific action plans are formulated and implemented yearly by the government.

The countermeasures against natural disasters have been executed as follows.

1) With the period from the beginning every March to the end of May designated as the period of preparation against possible disasters, disaster prevention facilities, etc. are inspected and repaired at the level of each administrative unit including province, city, county and district as a precautionary measure against disasters. From February to April each year, disaster prevention training and education programs are provided to working level government employees in charge of disaster prevention in each province, city, county and district. During the period from March to May every year, disaster prevention training, including mapping and computerized exercises along with training based on regional characteristics, is carried out concurrently with civil defense drills against simulated massive natural disasters comparable to those that actually occurred in the past.

2) Areas prone to disasters are managed with special cares. A total of 537 "disaster-risk" areas, which are prone to disaster, are under special management. These areas are classified and managed by type, grade, management entity, size, etc.

3) Regarding risky small rivers that can cause disasters, precautionary measures are taken, along with steps to convert such rivers into nature-friendly ones.

4) Disaster Prevention Facilities, such as multi-purpose dam, river and banks, and others, are managed. Multi–purpose dams built in major basins contribute to flood control as well as secure and supply water for agricultural, industrial and daily use. With regard to rivers, project to permanently protect houses and farming land from flooding are actively pursued. A total of 8,054 facilities vulnerable to disasters including water gates, reservoirs, breakwater and seawalls are under special management.

5) There is the disaster impact assessment system. This system is aimed at fundamentally eliminating potential causes of disasters inherent in various development projects in advance and ultimately protecting the lives and properties of the people. The disaster impact assessment system is implemented when the area of targeted development

is not less than 300,000 m^2. With respect to small-and medium-sized development projects (150,000 m^2 to less than 300,000 m^2 in size), each city and province has introduced a regional disaster impact assessment system.

6) Automated high-quality alert facilities, automatic voice notice system regarding the status of disasters and IT system for disaster prevention are in operation.

Municipal, provincial, county and district authorities maintain close cooperation with relevant government agencies with NDPCH at the center in order to control and respond to disasters in organized and timely manner. In principle, the officer in charge of natural disaster prevention at a damaged facility takes responsibility for taking measures for rehabilitation. When substantial damage is inflicted by a disaster, NDPCH investigates such damage and formulates recovery plans.

INTERNATIONAL EXCHANGES AND COOPERATION

The Republic of Korea actively participates in a variety of international organizations and conventions to support international cooperation in countering natural disasters, as follows.

1) Typhoon Committee
As a member of the Typhoon Committee, the Republic of Korea dispatches delegations to annual meetings of the committee which deliberate such matters as weather, hydrology, and disaster prevention. Providing data on disaster prevention, countermeasures and relevant statistics, the Republic of Korea endeavors to elevate the stature of the committee and promote cooperation among its members.

2) Asia Disaster Reduction Center
As a member of the Asia Disaster Reduction Center organized for the purpose of accumulating and sharing information on disaster prevention, emergency aid, etc., the Republic of Korea participates in its annual meetings to enhance the status of the center. In addition, Korea has been annually paying dues for its operation since 1999.

3) Korea – Japan Disaster Prevention Council
The Korea-Japan Disaster Prevention Council was formed in the wake of the summit talk between the two countries in October 1998 as a part of the joint declaration agreed to during the summit. The two countries with similar weather conditions share information on disaster prevention, hosting annual meetings of the council in turn.

4) ASEAN Regional Forum
As a member of the ASEAN Regional Forum, a security forum of the members of ASEAN (Association of Southeast Asian Nations) and major countries in the Pacific region, the Republic of Korea is actively engaged in activities to promote cooperation among its members by exchanging information and experience in connection with relief activities and disaster prevention.

International Conference on Geotechnical Engineering for Disaster Mitigation & Rehabilitation
Chu, Phoon & Yong (eds) © 2005 World Scientific Publishing Company ISBN 981-256-469-1

EVALUATION OF SEISMIC DISPLACEMENTS OF QUAY WALLS

SUNGRYUL KIM
Dept. of Civil Engineering, DongA University, #840 Hadan 2-dong, Saha-gu, Busan, Korea

INSUNG JANG
Korea Ocean Research and Development Institute, Sa2-dong, Sanglok-gu, Ansan-si, Kyoungi-do, Korea

CHOONGKI CHUNG
School of Civil, Urban and Geosystem Engineering, Seoul National University, San 56-1, Shinlim-dong, Gwanak-gu, Seoul, Korea

MYOUNGMO KIM
School of Civil, Urban and Geosystem Engineering, Seoul National University, San 56-1, Shinlim-dong, Gwanak-gu, Seoul, Korea

This paper is the summary of "Evaluation of Seismic Displacements of Quay Walls", which was printed in "Soil dynamics and earthquake engineering" Vol. 25, pp.451-459.

A new simplified dynamic analysis method is proposed to predict the seismic sliding displacement of quay walls by considering the variation of wall thrust, which is influenced by the excess pore pressure developed in the backfill during earthquakes. The method uses the Newmark sliding block concept and the variable yield acceleration, which varies according to the wall thrust, to calculate the quay wall displacement. A series of 1g shaking table tests was executed to verify the applicability of the proposed method, and a parametric study was performed. The shaking table tests verified that the proposed method properly predicts the wall displacement, and the parametric study showed that the evaluation of a realistic wall displacement is as important as the analysis of liquefaction potential for judging the stability of quay walls.

INTRODUCTION

The simplified dynamic analyses based on the Newmark sliding block concept are widely used for preliminary designs of quay walls because they can easily evaluate the wall displacement by using basic design parameters such as the weight of the wall, the internal friction angle of the backfill, and the frictional coefficient at the bottom of the wall. Richard and Elms (1979) and Whitman and Liao (1985) proposed simplified dynamic analyses based on the Newmark sliding block concept to evaluate the seismic displacement of a quay wall. In their methods, yield acceleration is defined as the wall acceleration, when the factor of safety of the wall for sliding becomes 1.0, and the wall displacement is presumed to occur if the ground acceleration exceeds the yield acceleration. However, these analyses does not consider the variation of wall thrust due to the development of excess pore pressure in the backfill when they determine the yield acceleration; therefore, previous analyses are inappropriate for the design of quay walls with saturated backfill soils, where high excess pore pressure can develop during earthquakes.

Several researchers suggested degrading yield acceleration models, in which yield acceleration decreased as a function of shear deformation for geosynthetic cover analyses (Matasovic et al., 1998) or as a function of the magnitude of the excess pore pressure for saturated slope analyses (Giovanni et al., 2001). To account for this excess pore pressure in the calculation of wall sliding displacement, we propose a new simplified dynamic analysis method, which still utilizes the Newmark concept but varies the yield acceleration according to the varying wall thrust.

A parametric study is performed to analyze the effect of the input parameters of this new method on the seismic wall displacement, and the proposed method is verified by comparing the predicted displacements with the results of a series of shaking table tests.

DEVELOPMENT OF NEW DISPLACEMENT CALCULATION METHOD

Assumptions

The following assumptions were used in the proposed method.

① Quay walls always fail in a sliding mode. This assumption is valid only for dense foundation soils.

② Wall displacement occurs in a forward direction only. This should be a reasonable assumption for most cases since the wall can hardly move toward the backfill soils during shaking.

Newmark sliding block method

The Newmark sliding block method defines the yield acceleration as the amplitude of the block acceleration when the factor of safety for sliding becomes 1.0, and evaluates the block displacement by double integration of the ground acceleration, which exceeds the yield acceleration. The integration method by Wilson and Keeper (1983) was used to calculate the block displacement by integration of the ground acceleration.

Determination of yield acceleration

The method proposed in this study evaluates the yield acceleration according to the varying wall thrust and therefore, this method is distinct from the previous ones. The wall thrust on a quay wall (F_{TH}) is the resultant of diverse force components, as shown in Figure 1 (Kim et al., 2004) : static water forces acting on the back and front of the wall, inertia force of the wall (F_I), dynamic water force on the front of the wall (F_{FWD}), static thrust on the back of the wall before shaking (F_{ST}), and dynamic thrust on the back of the wall, which develops during shaking (F_{DY}). In this research, we assumed that the water levels on both sides of the wall were the same, and thus, the static water forces acting on the both sides of the wall were not considered. The wall thrust F_{TH} can be obtained by summing the various force components, as shown in Equation (1). The resisting force (F_R) of the wall, which comes from the frictional force between the bottom of the wall and the foundation soil, can be calculated by Equation (2).

$$FTH = FI + FFWD + FST + FDY$$
$$FR = cB \cdot L + W \cdot \tan \phi B$$

where, L = length of contact surface between bottom of wall and foundation, W = weight of wall per running unit length, c_B = cohesive stress between bottom of wall and foundation, and ϕ_B = interface friction angle between bottom of wall and foundation

The wall displacement begins to occur when the wall thrust F_{TH} exceeds the resisting force F_R (Equation (3a)). The inertia force of the wall (F_I) at this point can be obtained by multiplying the mass of the wall (M) by the yield acceleration of the wall (a_y) (Equation (3b)). Finally, a_y is obtained by Equation (3c). However, a_y is also needed to calculate F_{FWD} and F_{DY} on the right side of Equation (3c). Therefore, the yield acceleration can only be determined by an iterative calculation.

$$F_{TH} (= F_I + F_{FWD} + F_{ST} + F_{DY}) \geq F_R \tag{3a}$$
$$F_I = F_R - (F_{FWD} + F_{ST} + F_{DY}) = M \times a_y \tag{3b}$$
$$a_y = \frac{F_R - (F_{FWD} + F_{ST} + F_{DY})}{M} \tag{3c}$$

where, M = mass of wall

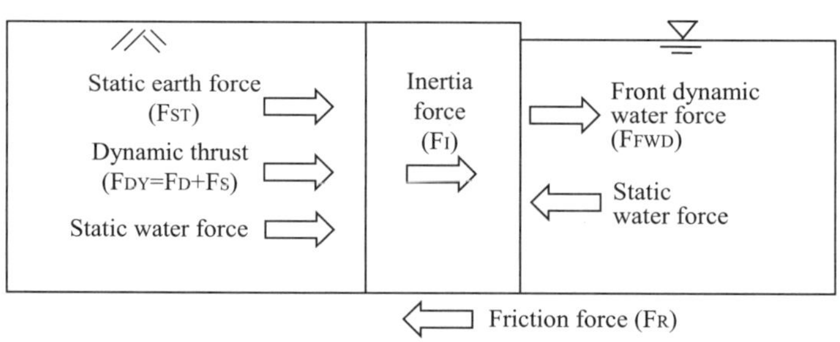

Figure 1. Orientation and location of force components (Kim et al., 2004)

Determination of force components acting on wall

The time histories of force components acting on walls have to be evaluated to determine the time history of the yield acceleration, as was shown in Equation (3). The evaluation methods of force component are summarized in Table 1, as suggested by Kim et al.

(2004). Kim's method requires the time histories of the wall acceleration and the excess pore pressure ratio, r_u in the backfill as input for the evaluation of the force components. The latter can be evaluated by either a simple empirical formula like Equation (4) or a dynamic analyses or a laboratory test such as a shaking table test.

Table 1. Evaluation of force components acting on quay walls (Kim et al., 2004)

<table>
<tr><th colspan="2">Force component</th><th>Evaluation method</th></tr>
<tr><td colspan="2">Inertia force of wall (F_I)</td><td>mass of wall × time history of input acceleration</td></tr>
<tr><td colspan="2">Dynamic water force acting on front side of wall (F_{FWD})</td><td>Westergaard equation</td></tr>
<tr><td colspan="2">Static thrust (F_{ST})</td><td>Coulomb method</td></tr>
<tr><td rowspan="2">Dynamic thrust (F_{DY} $=F_D+F_S$)</td><td>Fluctuating component (F_D)</td><td>$F_D = F_{DI} \times (1-r_u) + F_{DF} \times r_u$
$F_{DI} = F_{WD} + F_{ED} - (F_I + F_{FWD})$
$F_{DF} = \frac{7}{12} k_h \gamma_{sat} H^2$

where, F_{DI} : fluctuating component of dynamic thrust without excess pore pressure
F_{DF} : fluctuating component of dynamic thrust after liquefaction
F_{ED} : fluctuating component of dynamic earth force
acting on back side of wall
F_{WD} : fluctuating component of dynamic water force acting on back side of wall
r_u : excess pore pressure ratio in the backfill
H : wall height
k_h : horizontal seismic coefficient
γ_{sat} : the saturated unit weight of soil</td></tr>
<tr><td>Non-fluctuating component (F_S)</td><td>$F_S = -\frac{1}{2}\gamma_{sub} K_{AS} H^2 r_u + \frac{1}{2}\gamma_{sub} H^2 r_u$

where, γ_{sub} : submerged unit weight of backfill soil
K_{AS} : static active earth pressure coefficient calculated by Coulomb method</td></tr>
</table>

$$\mathbf{r_u = \frac{1}{2} + \frac{1}{\pi}\sin^{-1}\left[2\left(\frac{N}{N_L}\right)^{\frac{1}{\theta}} - 1\right]} \tag{4}$$

where, N_L = number of loading cycles required to produce initial liquefaction corresponding to the cyclic stress ratio, CSR (=cyclic stress required to initiate liquefaction / vertical effective stress in soil) and θ = constant representing soil properties and test condition (commonly, θ=0.7)

The CSR can be calculated by Equation (5) (Seed and Idriss, 1971).

$$\mathbf{CSR} = \frac{\tau_{cyc}}{\sigma_v'} = \frac{a_{max}}{g}\frac{\sigma_v}{\sigma_v'} r_d \qquad (5)$$

where, τ_{cyc} = cyclic shear stress, a_{max} = maximum amplitude of ground acceleration, σ_v , σ_v ' = total stress and effective stress in backfill, respectively, and r_d = stress reduction factor

PARAMETRIC STUDY

Input parameters

A parametric study was performed to analyze the wall behavior under various combinations of input parameters, using the proposed method. Figure 2 shows an example quay wall, which was used for the parametric study. The backfill was made with sand, whose effective grain size was 0.10 mm. The coefficient of permeability was 1.0×10^{-4} m/sec. Relative densities of the backfill (D_r) were varied to be 40 %, 60 %, and 70 %. The interface friction angles between the bottom of the wall and the foundation soil (ϕ_B) were varied to 25, 30, 35 and 40 degrees. Input acceleration was a 1 Hz sine wave. The number of loading cycles of the input acceleration was set to 10, which corresponds to the design earthquake magnitude of 6.5 in Korea. The amplitude of the input acceleration (a_{max}) was varied to 0.072 g, 0.10 g, 0.12 g, 0.14 g, 0.15 g and 0.20 g. The 0.072 g was obtained by converting the amplitude of irregular earthquake wave of 0.11 g into the amplitude of regular sine wave. To obtain the excess pore pressure ratio, r_u which is one of the input parameters, Equation (4) was used. Figure 3 shows the cyclic strength curves, which was obtained by the cyclic triaxial tests for the backfill sands of various relative densities. The CSR (Cyclic Stress Ratio) at mid-depth of the backfill (depth=7.5 m) was calculated by Equation (5) by inputting 1.0 for r_d. Figure 3 also shows how N_L is obtained after CSR is calculated by Equation (5).

Figure 4 shows the time histories of the calculated excess pore pressure ratios, using Equation (5) and Figure 3, for various relative densities of backfill at a_{max}=0.10 g. Liquefaction occurred for D_r=40 % and D_r=60 %, whereas, the liquefaction did not occur for D_r=70%.

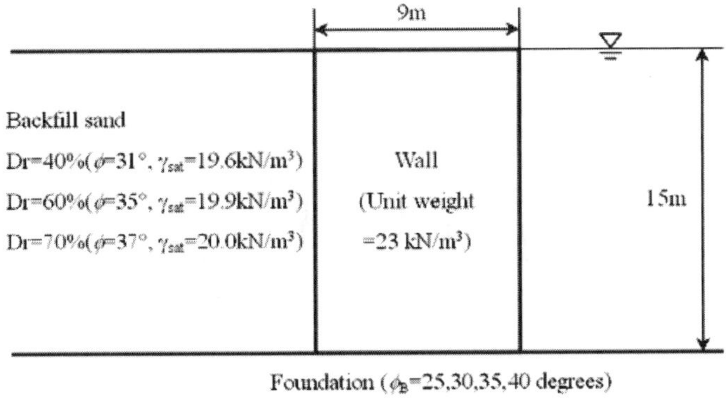

Figure 2. Example quay wall for parametric study

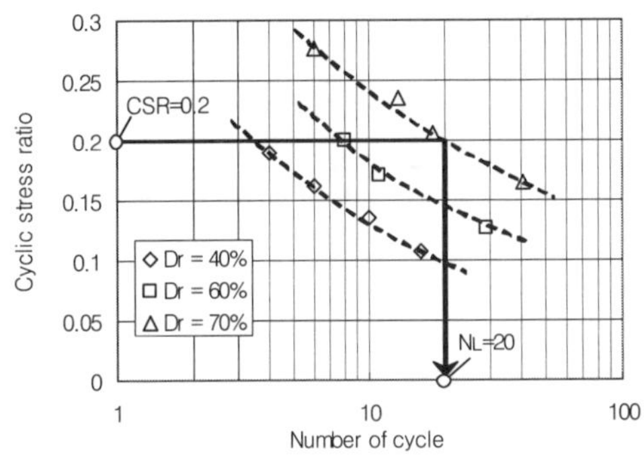

Figure 3. Cyclic strength curve

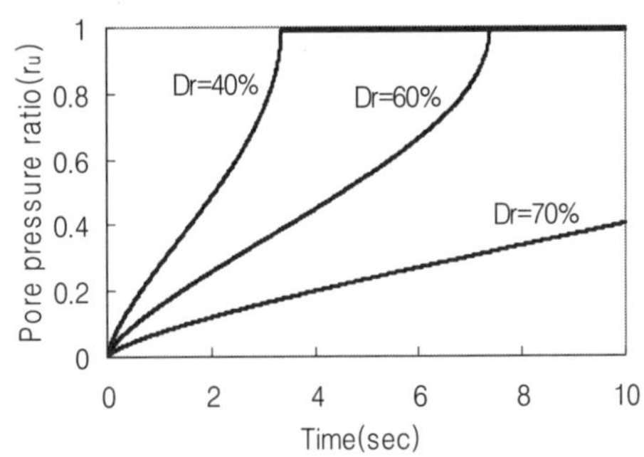

Figure 4. Excess pore pressure ratios (a_{max}=0.10 g)

Results of parametric study

Figure 5 shows the time histories of the input acceleration and the yield acceleration at a_{max}= 0.10 g. As cyclic loading continues, the yield acceleration decreases and finally becomes smaller than the input acceleration at 3.3 sec for D_r=40 % and at 7.3 sec for D_r=60 % after the backfill soil liquefies. If the effect of excess pore water pressure is not considered in the determination of the yield acceleration as is in the existing methods, the yield acceleration does not change from the initial value as is shown in the same figure for D_r=40 % and r_u=0. Thus, previous methods will predict zero wall displacement for the example quay wall, which is obviously an erroneous result.

Figure 6 shows the final displacement of the wall for various combinations of input parameters. Even if the backfill liquefied at a_{max}= 0.072 g with D_r=40 %, the final displacement of the wall was only 9.6 cm at ϕ_B=35 degrees and 0.7 cm at ϕ_B=40 degrees. On the other hand, the wall displacement of 59 cm occurred at a_{max}= 0.12 g with D_r=70 % and ϕ_B=25 degrees, where liquefaction did not occur in the backfill. In addition, the displacements calculated by the proposed model were very sensitive to the interface friction angle. Therefore, the frictional resistance between a wall and foundation must be properly evaluated.

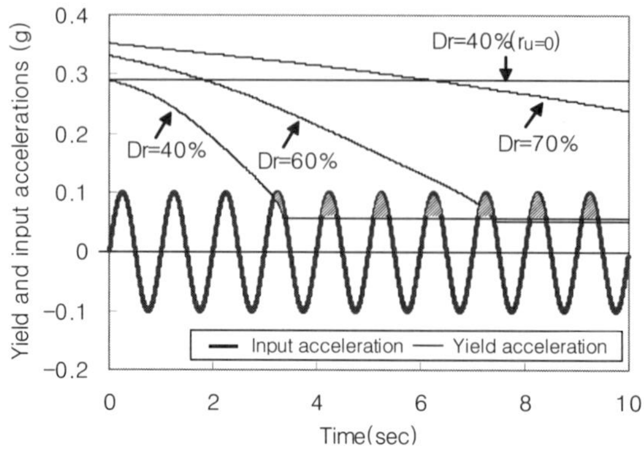

Figure 5. Time histories of input and yield accelerations (a_{max}=0.10 g, ϕ_B=40 degrees)

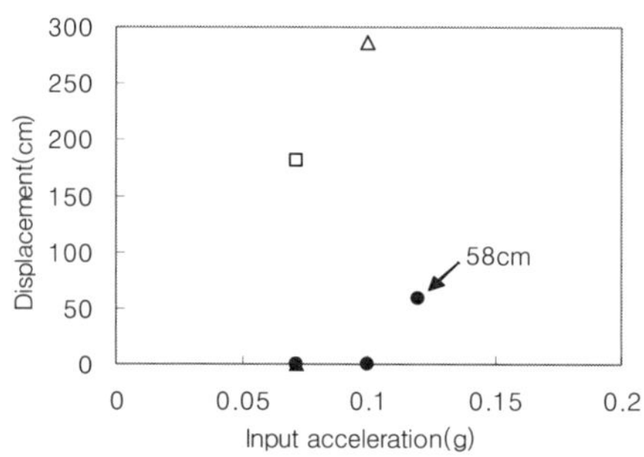

(a) ϕ_B=25 degrees

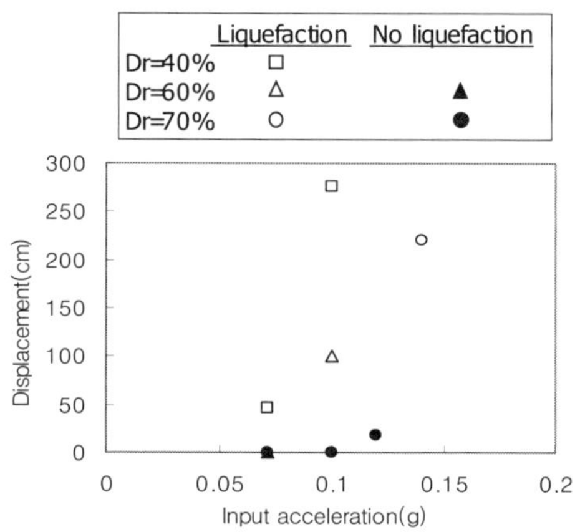

(b) ϕ_B=30 degrees

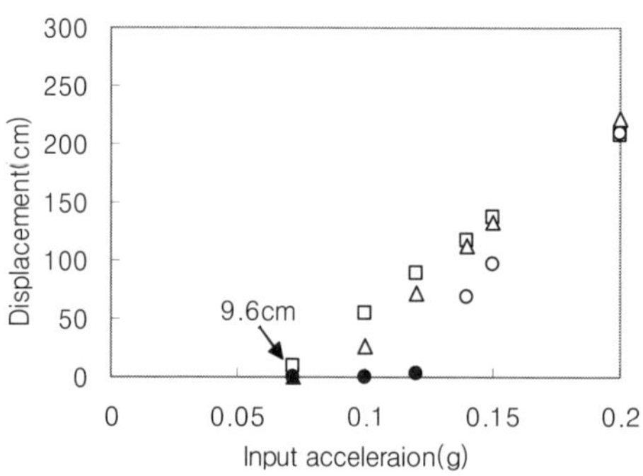

(c) ϕ_B=35 degrees

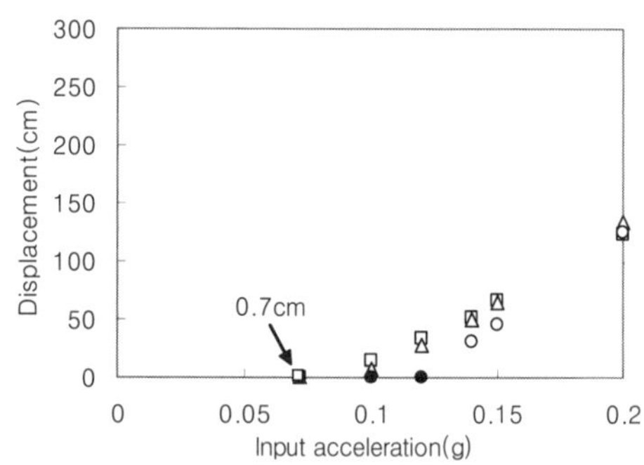

(d) ϕ_B=40 degrees

Figure 6. Final displacements of wall with respect to amplitude of input acceleration, relative density of backfill and interface friction angle at bottom of wall

VERIFICATION OF PROPOSED METHOD

Test set-up and procedure

The proposed method was verified by comparing the calculated wall displacements with the results of 1g shaking table tests. The dimension of the soil box was 194 cm long, 44 cm wide, and 60 cm high, and the model wall was 17.5 cm long, 42.0 cm wide, and 26.4 cm high. Figure 7 shows the test section and the instrumentation. The amplitude of the sinusoidal input motion at 5 Hz was increased linearly up to 0.2 g during the initial 5 sec, and the final amplitude was maintained for the next 5 sec.

The model soil was Joomoonjin sand, whose average particle size was 0.55 mm and uniformity coefficient was 1.37. The maximum and minimum dry unit weights of the sand were 16.7 kN/m^3 and 13.9 kN/m^3, respectively. The loose backfill of 20 % relative density was prepared by the water sedimentation method. The internal friction angle and the saturated unit weight of the backfill soil were about 30 degrees and 18.9 kN/m^3, respectively. The permeability coefficient of the backfill soil was measured to be 4.1×10^{-4} m/sec by the constant head permeability test. A dense foundation layer was made by preshaking the foundation soil. The relative density and the internal friction angle of the foundation layer were about 90 % and 40 degrees, respectively. The interface friction angle between the foundation soil and the bottom of the wall was estimated by pulling tests. The interface friction angle increased with the wall movement velocity. The average value of the interface friction angle for the velocity range of the wall movement in the shaking table tests was about 28 degrees. The shaking table tests were performed for two walls of identical geometry but of different unit weights, 23.0 kN/m^3 and 25.7 kN/m^3. In the latter, water was situated in the space where load cells were installed (Figure 7).

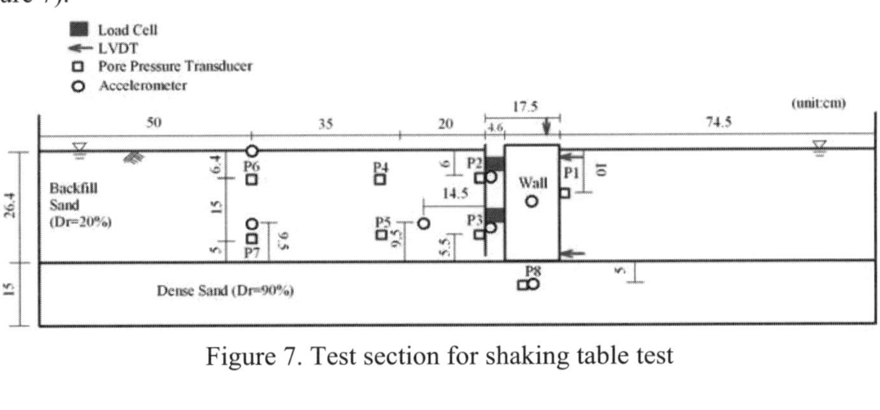

Figure 7. Test section for shaking table test

Comparison with results of shaking table tests

The time history of the horizontal seismic coefficient k_h was obtained by non-dimensionalizing the time histories of the input acceleration. The average excess pore pressure, which is the average value of the excess pore pressures measured from the two water pressure transducers (P2 and P3) installed on the back side of the wall, was used to obtain the time history of the excess pore pressure ratio in the backfill r_u. The time histories of k_h and r_u measured during the tests were used to predict the displacements of the wall.

Figure 8 shows the time histories of the excess pore pressure ratio. Figure 9 shows the comparisons between the measured and the predicted displacements of the walls of two different unit weights. The measured wall displacement started to occur at around 4 sec which is the time when the excess pore pressure ratio increased rapidly and reached almost its maximum value (Figure 8). The final differences of the horizontal displacements at the top and the bottom of the wall were about 0.4 cm (0.9 degrees to vertical) for the wall of unit weight of 25.7 kN/m^3 and 1.0 cm (2 degrees to vertical) for the wall of unit weight of 23.0 kN/m^3, which are small compared with the final horizontal displacements of 5.5 cm and 8.3 cm, respectively. Therefore, this observation satisfies the assumption of the proposed model in that only the sliding failure of walls occurs. The calculated final displacements of the walls compared very well with the measured values : 5.2 cm for the wall of unit weight of 25.7 kN/m^3 and 7.6 cm for the wall of unit weight of 23.0 kN/m^3.

Thus, it is believed that the proposed method predicted the quay wall behavior properly in terms of its cumulative displacement with time.

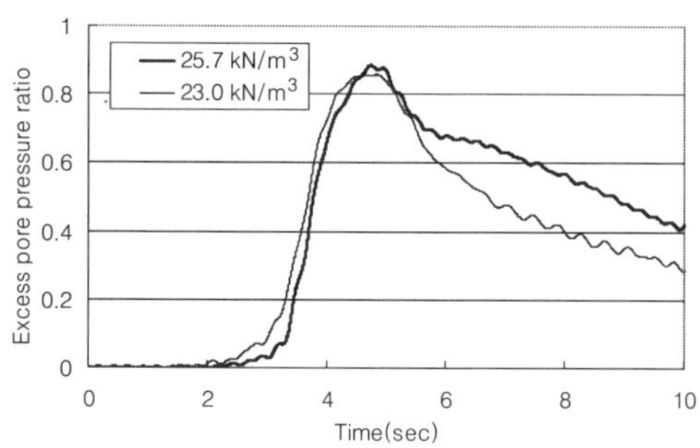

Figure 8. Measured excess pore pressure ratios in backfill soil

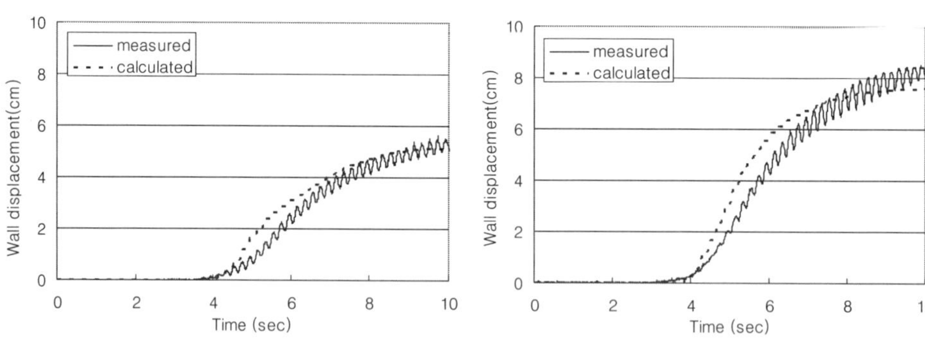

(a) wall with unit weight of 25.7 kN/m^3 (b) wall with unit weight of 23.0 kN/m^3

Figure 9. Comparisons between measured and predicted wall displacements

CONCLUSIONS

The conclusions of this study are as follows.

1. A new displacement calculation method was proposed which considers the effect of the excess pore pressure developed in backfill. This method basically uses the Newmark sliding block concept but varies the yield acceleration according to the varying wall thrust.
2. The parametric study showed that the evaluation of realistic wall displacements under earthquakes is as important as the analysis of liquefaction potential for judging the stability of quay walls.
3. It was verified from a series of 1g shaking table tests that the proposed method properly predicts the wall displacement.

ACKNOWLEDGEMENT

The financial support from the Ministry of Maritime Affairs and Fisheries (MOMAF) in support of this work is gratefully acknowledged.

REFERENCES

Giovanni B, Ernesto C, Michele M. Seismic response of submerged cohesionless slopes. In: Proceedings of the 4th International Conference on Recent Advances in Geotechnical Earthquake Engineering, San Diego, California, 2001, Paper No. 7.07.

Matasovic N, Kavazanjian E Jr, Giroud JP. Newmark seismic deformation analysis for geosynthetic interfaces. Geosynthetics International 1998, Special Issue on Geosynthetics in Earthquake Engineering; 5(1-2):237-264.

Kim SR, Kwon OS, and Kim MM. Evaluation of force components acting on gravity type quay walls during earthquakes. Soil Dynamics and Earthquake Engineering 2004; 24(11):853-866.

Richards R, Elms D. Seismic behavior of gravity retaining walls. Journal of the Geotechnical Engineering Division 1979;105(4):449-464.

Whitman RV, Liao S. Seismic design of retaining walls. Miscellaneous Paper GL-85-1, U.S.Army Engineer Waterways Experiment Station, Vicksburg, Mississippi. 1985.

Wilson RC, Keefer DK. Dynamic analysis of a slope failure from the 6 August 1979 Coyote Lake, California, earthquake. Bulletin of the Seismological Society of America 1983;73(3):863-877.

International Conference on Geotechnical Engineering for Disaster Mitigation & Rehabilitation
Chu, Phoon & Yong (eds) © 2005 World Scientific Publishing Company ISBN 981-256-469-1

OBSERVATIONS ON SOME GEOTECHNICAL ISSUES RELATING TO HAZARD AND DISASTER MITIGATION

FRED H. KULHAWY, Honorary Member, ASCE
School of Civil and Environmental Engineering
Cornell University, Hollister Hall
Ithaca, New York 14853-3501, USA

A review is made of many natural hazards, subsequent disasters, and issues of mitigation, which go well beyond the technological. Geotechnical matters are prominent throughout. As discussed, for the future, we must focus more on working with natural processes.

INTRODUCTION

At least one type of natural hazard exists nearly everywhere on our planet. No one is immune to these hazards, and entire historical civilizations have been destroyed by various natural disasters. The great many life forms on this planet adapt, adjust, or conform to their natural surroundings, or exploit them in some basically passive manner. Only humans approach their environment consistently in an active, aggressive manner, adapting and modifying the environment to suit them. Nearly always, these modifications are for immediate or short-term purposes, without considering the intermediate to long-term implications. In so doing, fundamental natural processes often are thwarted, and solutions are proposed that modify or attempt to contradict these processes. Humans still do not realize that the natural processes will prevail eventually. In this paper, some natural hazards and disasters are discussed, along with the geotechnical issues involved and how they relate to the broader issue of disaster mitigation. Some basic suggestions are made for future implementations.

SOME BASIC DEFINITIONS

Hazards to humans are of many types, and they can be classified as: (a) natural, from atmospheric and/or geologic processes, (b) medical/social, such as epidemics and famine, and (c) technological, such as biological and chemical agents, hazardous materials, transportation accidents, utility disruption, etc. The focus herein is on natural hazards.

A natural hazard can be defined as "a naturally occurring event that poses danger to humans and/or their assets". More quantitatively, a natural hazard can be defined as "the probability of occurrence, within a period of time and locale, of a potentially dangerous natural phenomenon". In 2001 alone, natural hazards killed over 25,000 people and caused some USD 36 billion in damage worldwide (Natural Hazards 2005).

Table 1 shows a generalized summary of natural hazards. For convenience, they are grouped first into the relatively common geologic and atmospheric hazards, which need to be addressed in normal hazard planning. Then the extreme natural hazards are listed, more for general interest than any other reason, because these approach "doomsday" scenarios that can not be addressed in normal hazard planning.

All but one or two of the geologic hazards are known well within the geotechnical engineering community. They are studied at our universities, and they are known well in practice. We may not have good solutions for all of the related design problems, but the fundamentals are understood, at least conceptually. Drought is a specialized case that can lead to extreme changes in effective stress state and subsequently all geotechnical properties, and to desertification in the extreme. Geomagnetic storms do not affect the geotechnical conditions directly, but they disrupt the geomagnetic field which, in turn, disturbs communications, power, computers, instrumentation, etc. In other words, most support systems are influenced negatively.

The atmospheric hazards also are studied and are known well. They often have a symbiotic relationship with the geologic hazards, in many cases triggering the geologic hazard. Climate change and the el niño / la niña phenomena are longer-term issues that influence the frequency and intensity of the other atmospheric hazards.

The extreme natural hazards are exactly that - extreme! They are of interest and are important, but their frequency and intensity are so far into the extreme tails of any design distributions that they can not be considered in plausible design scenarios.

For the International Decade for Natural Disaster Reduction, which was a United Nations program for the 1990s, the focus was on earthquakes, windstorms, tsunamis, floods, landslides, volcanic eruptions, wildfires, grasshopper and locust infestations, and drought

Table 1. Geologic, atmospheric, and extreme hazards.

Geologic Hazards	
drought	sinkhole
earthquake	slide (natural slope, constructed slope, snow/ice avalanche, volcanic lahar)
erosion (coastal, river/stream bank)	subsidence (local, regional)
expansive soil	tsunami
flood	volcano
geomagnetic storm	
Atmospheric Hazards	
climate change	tornado
el niño and la niña	tropical cyclone (hurricane/typhoon, storm surge)
fog	wildland fire
heat wave	winter storm/blizzard (wind, snow, ice)
summer storm/thunderstorm (lightning, wind, rain, hail)	
Extreme Natural Hazards	
ice age	supervolcano
extraterrestrial impact	megatsunami
solar flare	

and desertification.

Within the United States, all of the hazards listed in Table 1 are plausible and do occur, and the great majority are common, unfortunately.

Figure 1 is an interesting 1978 graphic. Unfortunately, I have not seen any update. However, the relative values should still be approximately valid. As shown, the mean annual economic impact of "expansive soil" is the largest, by far, followed by floods and then slides. Mean annual loss of life is dominated by floods, with "earthquakes" a distant second. This figure shows an unfortunate reality in perception and in research financing. Although "expansive soil" is by far the largest mean annual loss in the aggregate, it is something that occurs in small increments and therefore is not perceived as more than a local nuisance. Significant earthquakes, on the other hand, are few and far between, but their effects are large and visible, and they receive a great deal of high visibility press coverage. So where does the majority of the research funding go? To earthquakes, of course! Perhaps our priorities need some re-direction.

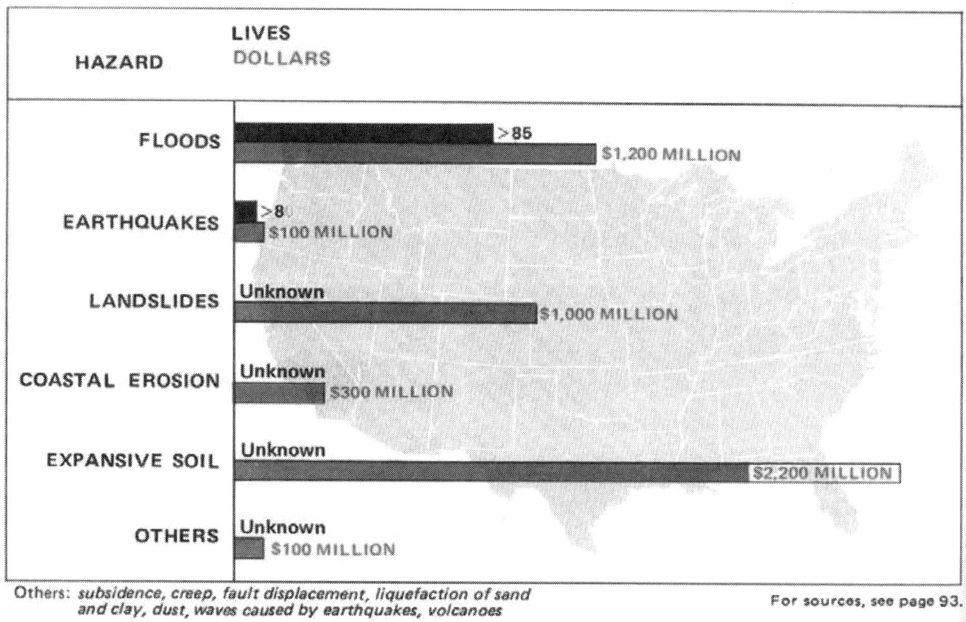

Figure 1. Mean annual losses in U.S. from geologic hazards (Robinson & Spieker 1978).

NATURAL DISASTERS

A natural disaster can be described as a natural event with catastrophic consequences. Or "a disaster is a social disruption that can occur at the level of the individual, the community, or the state" (Kreps 1986). Or "disasters occur when hazards meet vulnerability" (Blaikie et al. 1994). Within the last year, three incredible natural disasters occurred - the December 2004 tsunami in Southeast Asia, the August and September 2005 hurricanes in the south-central U.S., and the October 2005 earthquake in northeast Pakistan. In these cases, either extreme loss of life occurred, numbering in the tens of thousands, and/or extreme loss of assets occurred, in excess of USD 100 billion. Can these disasters be prevented? At this time, the answer is no. Can the impact of these disasters be minimized? The answer is yes. Traditionally, two approaches have been used: (a) mitigation, which includes actions taken before, during, and after to minimize the impact, and (b) response, which includes short-term emergency efforts by first-responders and longer-term efforts to provide human needs (food, shelter, etc.) and to re-establish the infrastructure. In the past, response efforts have predominated. However, considering these recent disasters, it is clear that much is yet to be done in this area. For the future, there will be increasing use of sound hazard mitigation strategies that include risk management and engineering.

HAZARD MITIGATION

Hazard mitigation is the process of minimizing, or possibly eliminating, the risk to people and/or their assets from natural hazards, usually through careful planning. Solutions can range from engineering design, such as earthquake-resistant design, to limiting or eliminating human activities in potentially hazardous areas, such as hurricane-prone coastal areas and floodplains. Hazard mitigation ideally is dedicated to breaking the cycle of repeated damage and reconstruction.

As described by the Board on Natural Disasters (1999), "comprehensive mitigation planning includes: (a) determining the location and nature of the potential hazards, (b) characterizing the population and structures (present and future) that are vulnerable to specific hazards, (c) establishing standards for acceptable levels of risk, and (d) adopting mitigation strategies based on an analysis of realistic costs and benefits. In practice, mitigation may be difficult to implement, both politically and economically."

The Board on Natural Disasters (1999) also noted that there are five high-priority tasks to be pursued. First, improve risk assessments. Much has been done in modeling risk within the engineering community. Realistic models exist, but calibrations are somewhat limited. Second, implement mitigation strategies. Implementation can be tricky because it involves comprehensive land-use planning. These measures are rarely adopted locally without mandates from higher levels of government, and the local people must be convinced that the hazard is a real threat. However, relatively straightforward matters, such as implementing sound building codes and ensuring that they are followed, can do much in minimizing losses. Third, improve technologies that support warnings and the dissemination of, and response to, warnings. Forecasts and warnings are vital for mitigation, and they must be as good as possible. Implementation, maintenance, and monitoring are potentially weak links in this important issue. Fourth, improve the basis for natural disaster insurance. This enormous financial issue needs to be addressed better by linking disaster policy, risk-reduction strategies, and incentives to minimize risk. And fifth, assist disaster-prone developing nations. This humanitarian and technological effort is necessary to ensure that these nations actually can develop into productive members of the worldwide community. After all, the interests of all countries extend worldwide at this time.

Figure 2 presents many of these issues graphically. As shown, there are many issues, from many sectors, that all play key roles in reducing hazard loss. Moving in clockwise fashion, the first two issues relate to society, individuals, and education. People, collectively and individually, need to make informed decisions about risks and hazards. The third and fourth issues relate to insurance and economics and how they play a role in setting reduction and reconstruction strategies. The fifth and sixth issues are largely technological, such as quantifying uncertainty and risk and developing mitigation measures.

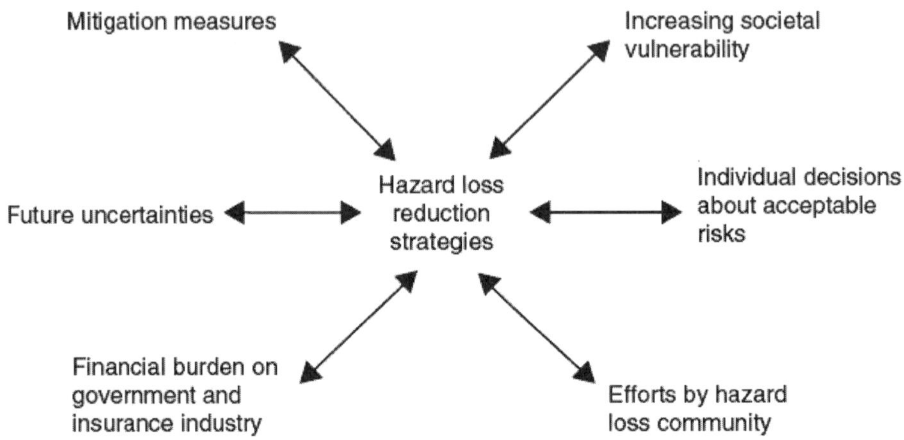

Figure 2. Factors influencing hazard loss reduction (Meade & Abbott 2003).

SOME GEOTECHNICAL ISSUES

Geotechnical issues pervade many aspects of hazards, disasters, and mitigation. As noted previously, most of the geologic and atmospheric hazards are known and studied. We can calculate the ideal behavior of these hazards, but we can not assess the risks well. Considering the three recent disasters noted previously, it is clear that our concepts of "acceptable risk" need re-evaluation. It is also clear that infrastructure maintenance is sorely lacking on a global basis, which means that the risk of a disaster clearly is greater than it was at the time of design and construction. Is this increased risk acceptable?

Let us think about some natural geotechnical systems. Most natural systems evolve gradually, and in nature there is a "geo-redundancy" that provides some protection for these systems. For example, natural slopes normally are broken or benched, vegetation grows most everywhere, and the types of vegetation vary in height, lateral expanse, and root structure. These all represent natural "energy dissipators" that will minimize the atmospheric loading functions and provide additional geotechnical stability. Consider now the effects of urbanization in this terrain. Slopes are re-shaped, and their grades are made uniform. Vegetation is removed, and paving is done for streets and parking. Well-trimmed lawns and localized decorative vegetation are installed. Suddenly the loading functions become more severe because the runoff increases dramatically and the infiltration into the natural vegetation decreases. In addition, the new vegetation systems (i.e., lawns) provide very little assistance in slope stability. The result is that loading is increased, resistance is decreased, and therefore the risk increases.

Many other examples can be cited, such as: (a) housing within littoral drift zones, (b) modification of barrier islands, (c) construction within actively-creeping fault zones, (d) construction on creeping slopes, (e) river channelization and/or levee construction that virtually eliminates beneficial flooding that replenishes soil nutrients and maintains vegetation, (f) channelization that results in increased storm surge, etc. The list continues almost indefinitely. The point to be made is that we need to work with natural processes, adapt them to our needs, and maintain or enhance the geo-redundancy. Most policies to date are the opposite and are directly confrontational to the natural processes. Sooner or later, nature will win the battle. Instead, we need to follow the adage quoted below.

"Nature to be commanded must be obeyed"
Francis Bacon, Novum Organum, 1620

CONCLUDING COMMENTS

Hazards exist everywhere. This paper reviewed the types of natural hazards, subsequent disasters, and issues of mitigation, which go well beyond the technological. Geotechnical matters are prominent within all of these. As discussed, for the future, we really must work more with and within natural processes.

REFERENCES

Blaikie, P, Cannon, T, Davis, I & Wisner, B (1994). *"At Risk: Natural Hazards, People's Vulnerability, & Disasters"*, New York: Routledge.

Board on Natural Disasters (1999). "Mitigation Emerges as Major Strategy for Reducing Losses Caused by Natural Disasters", *Science*, 284, 18 Jun 99, 1943-47.

Kreps, GA, Ed. (1986). *"Social Structure & Disaster"*, Newark: Univ. Delaware Press.

Meade, C & Abbott, M (2003). *"Assessing Federal Research & Development for Hazard Loss Reduction"*, Santa Monica: RAND.

NaturalHazards.org website (2005).

Robinson, GD & Spieker, AM, Eds. (1978). "Nature to be Commanded", *Prof. Paper 950*, U.S. Geological Survey, Washington.

International Conference on Geotechnical Engineering for Disaster Mitigation & Rehabilitation
Chu, Phoon & Yong (eds) © 2005 World Scientific Publishing Company ISBN 981-256-469-1

DYNAMIC PROPERTIES OF FILL MATERIALS AND ASEISMATIC ANALYSIS FOR HIGH EARTH AND ROCK FILL DAMS

H. L. LIU

Geotechnical Research Institute (GeoHohai), Hohai University, 1 Xikang Road Nanjing, 210098, China

A number of high earth and rock fill dams have been under construction as a result of the recent Western Development Project in China. Most of these dams are located in the high seismic activity area of the southwestern China. The dynamic properties of rock fill materials and dynamic response analysis of the high earth and rock fill dam are the essential research topics for situation under strong seismic motion. This paper presents firstly the current status of high earth and rock fill dams in China, including concrete faced rock fill dams and clay core rock fill dams. Then, the cyclic loading dynamic triaxial testing apparatus and its testing techniques, the research on the dynamic properties of coarse granular soils, the dynamic response analysis, the seismic permanent deformation and the aseismatic measures of high earth and rock fill dams are reviewed. Finally, the development of a seismic safety assessment for the earth and rock fill dam and an expert system for disaster prevention are proposed.

INTRODUCTION

Earth and rock fill dam is a essential dam type in hydraulic and hydropower engineering, since it has following advantages such as, easy to be constructed, low geology requirements, comparable low cost, and abundance materials in site. This is a long construction history of earth and rock fill dam engineering. Statistically, there are about 86000 earth and rock fill dams above 15m have been constructed in China since 1950s. With the rapid economic development, the construction of hydraulic and hydropower engineering advanced rapidly too. In recent years, a number of clay core seepage prevention earth and rock fill dams are at the height of 100m or even 200m. The Xiaolangdi dam is inclined core earth and rock fill dam with the height of 167m and constructed on a cover layer with the thickness of 70m. It is the most typical project in clay core earth and rock fill dam construction. The Nuzhadu dam in Lancangjiang River will be clay core rock fill dam with the height of 265m, which will be the highest dam in China.

Since1980s, concrete faced rock fill dams have been under construction and developed very fast in China. Most of the dams, such as Xibeikou, Gouhou, Chengping (phase I) and Zhushuqiao Dam are all in the height of 70m and some of them above 100m. Recently, there are a lot of concrete faced rock fill dams above 100m and parts of them are close or above 200m. For examples, Jiangshanxi Dam in Zhejiang province is constructed with excavated rock fill materials at the height of 132.5m, Qiezishan dam in Yunnan province is constructed with the granite at the height of 107m. Gongboxi aconcrete faced rock fill dam on the Yellow River is at the height 133m in seismic intensity VIII area. Tianshengqiao (phase I) dam is at the height of 178m. Shuibuya dam on Qingjiang River in Hubei province is under construction and its height will be 233m.

A number of high earth and rock fill dams have been under construction as a result of the recent Western Development Plan in China. Most of these dams are located in the high seismic activity area of the southwestern China. The dynamic properties of rock fill materials and dynamic response analysis of the high earth and rock fill dam are the essential research topics for situation under strong seismic motion. This paper presents state of arts on the dynamic properties of fill materials and aseismatic analysis for high earth and rock fill dams

LABORATORY DYNAMIC TESTS AND DYNAMIC PROPERTIES OF FILL MATERIALS

Laboratory dynamic tests

The development of dynamic properties of coarse granular soil mostly relies on the improvement of testing technique. There are three kinds of approaches on studying dynamic properties of coarse granular soils in laboratory: (1) test wave velocity and attenuation properties based on wave theory (ultrasonic and pulse technique); (2) test soil coefficients of stiffness and damping based on vibrate theory (resonant column triaxial test); (3) obtain the relationship of stress-strain directly based on cyclic loading tests such as cyclic triaxial test, hollow torsional test, simple shear test, torsinal simple shear test etc. The advantages of cyclic loading test are prone to controlling initial stress condition, dynamic stress condition, drainage condition. Therefore, it is widely used and developed in the past few years and there are a few new progresses were achieved in this field in recent years.

The field stress condition can be reasonably reflected through stress path laboratory test on dynamic deformation properties of coarse granular soil. The direction of main stress in conventional cyclic triaxial test is fixed and the deformation condition is axial symmetrical which is greatly different from the field stress condition. But the cyclic torsional simple shear laboratory test can simulates many conditions such as the torsion of main stress axis, cyclic simple shear deformation, the uniformity distribution of vertical shear strain etc. It can also reflect the field loading condition. So it is used recently. Attempt experiment study on the shear strength through true triaxial apparatus was performed by Guo Qingguo which will be a base and trend in coarse granular soil mechanics. It is obvious that dynamic properties experiment study under complex loading conditions is still under development.

Each test method has different strain level and test index, so it is difficult to generally compare the test results. The dynamic properties can be measured in relatively wide strain range, especially under microcosmic strain condition by introducing high precise microcosmic displacement transducer. Then the continuous measure method of dynamic properties can be put into a substantiality stage for the rock fill materials specimen. It not only predigests the testing method of characteristic parameters but also degreases the measure cost and enhances the testing precision.

Dynamic deformation properties of rock fill materials

Under all kinds of load effect, the dynamic deformation properties is mostly affected by the factors of strain level, stress condition, testing measures, grain properties and drainage condition. The studies of coarse granular soil at present is mainly based on testing the rule of shear modulus ,which is calculated in equivalent and linear expression ,and damping ration non-linear changing with strain value. In the following it will be concluded from the experience description and influence factors of dynamic shear modulus and dynamic damping ratio.

Dynamic shear modulus of rock fill materials

The void ratio and relative density are also important influence factors on dynamic shear modulus of rock fill materials besides the shear strain value, secondly the grain size (especially maximum grain size), lithology, sediment history and cycle of cyclic stress etc. According to strain magnitude it can be divided into dynamic shear modulus under microcosmic strain condition and dynamic shear modulus under middling strain condition.

(1) Dynamic shear modulus under microcosmic strain condition and its influence factors

The dynamic shear modulus under microcosmic strain condition which is also called initial shear modulus (G_{max}) is an important parameter to description dynamic properties of rock fill materials. It keeps constant under microcosmic strain condition (≤10-4%). Normally it is obtained by resonant column triaxial test in laboratory or field wave velocity test. In recently using cyclic loading test to test initial shear modulus becomes true because of the use of microcosmic strain transducer test apparatus.

Confined pressure and the density of specimen are the main influence factors on shear modulus of rock fill materials under microcosmic strain condition. Many laboratory test results indicate that the modulus present the increasing trend with the increase of confined pressure, while the modulus gradually degrease with the increase of void ratio. Considering the main factors, all kinds of experimental relationships of initial shear modulus are proposed. The experience expressions combined the field testing results by Seed et al. (1970) and combined the lab test results by Hardin et al. were two typical examples.

The effects of grain properties (lithology, grain size etc), sampling method and test measure were also studied besides two main factors reflected by experience expressions. It was indicated from the test on rock fill materials with the same density but different coarse content that the initial modulus increases obviously with the increases of coarse content. The effect of the grain diameter of rock fill materials on initial modulus was studied by Kong Xianjing et al. (1988) from another view of point through large model dam in laboratory vibrating table. It was pointed out that when shear strain held the line the modulus increased with the increase of average grain diameter but the rise rate tend to a smooth way when the grain diameter increased to a certain value. Liang Yongxia (1990) indicated that initial shear modulus decreased with the decrease of grain diameter of testing material with the same density and also suggested that it should times a correct coefficient between 1.0-1.5 when the test results of rock fill materials was applied into practice. The test results from high precise and large hydraulic pressure serve triaxial apparatus by Jia Gexu (1998) indicated that elastic modulus of rock fill materials was very close to each other by all kinds of test measures under microcosmic strain condition (less than 10-3%).

(2) The rule of shear modulus should be considered among a relatively wide strain value for the soil dynamic problem of dynamic shear modulus and its effects under moderate strain conditions. In order to describe the attenuation of dynamic shear modulus with the increase of strain value the shear modulus G in a certain strain value was commonly normalized the maximum shear modulus G_{max}. The shear modulus ration is denoted as $\delta = G/G_{max}$. It is difficult to collect and manage the parameter in the model because the test data obtained from dynamic deformation soil test are scattered. However based on the before-mentioned initial experience expression of shear modulus it can be expressed in a wide strain range by the variety of modulus parameter K and index m. The experience expression was proposed by Wu Xingzheng et al. considering the comparability between the curve changing trend of modulus ration and shear modulus in semi-logarithm coordinate and properties curve of non-saturated soil water proposed by Fredlund et al.

It is indicated that there aren't obvious difference of the degressive curve of modulus between saturated sample and dry sample of rock fill materials. The reason is probably that the change of pore water pressure had little effect on sample made up of large grain diameter. So the field shear modulus G could be obtained by the laboratory anisotropic consolidation test only the magnitude of average main stress to be corrected.

As for the shear modulus normalized curves from cyclic triaxial test and torsional simple shear test, it indicates that the two curves are basically the same. While Iida et al. (1984) concluded that the shear modulus normalized curve from trosional simple test was lower than the curves from dynamic triaxial test, which should be verified by more test studies.

Furthermore, specimen lithology and sediment history also affected the shear modulus of rock fill materials.

The accumulated residual deformation properties of rock fill material are also experimental studied.

Dynamic damping properties of rock fill materials

The soil damping is mainly hysteresis damping. It is described by the corresponding equivalent viscosity damping ratio ξ that ratio of the soil losing energy in a cycle and the maximum elastic shear strain energy. The test data is much scatter as the friction of test apparatus is sometimes great and difficult to obtain precisely. It becomes more and more deeper for the properties of dynamic damping of rock fill materials with the development of some high precise and microcosmic displacement test apparatus

There are commonly two experience relationships to describe the dynamic damping properties of coarse granular soil. The first one is to directly set up the relationship between the dynamic modulus and damping ratio. The second one is to directly set up the experience relationship considering all kinds of effect factors of dynamic damping ratio.

The main effects on rock material damping ratio are shear strain, void ratio or relative density and confined pressure.

Dynamic strength properties of rock fill materials

Hatanaka et al. (1988) firstly conducted the cyclic undrained test and studied the dynamic strength. Yasuda et al. (1997) also studied the dynamic strength properties of rock fill materials systematically, and used the large scale torsional single shear apparatus first time.

Characteristics of dynamic strength properties of rock fill materials

The results of cyclic undrained loading tests indicate that compare with sand, the dynamic strength of coarse granular soil are different in the following two aspects, (1) the increasing pattern of pore water pressure is different from dense sand remarkably. In early stage of cyclic test, with the times of cyclic loading increasing, the pore water pressure of the compression direction increases up to almost 100% of confining pressure, the effective stress reduces to zero, the maximum ratio of pore water pressure reaches up to 0.9 or more. (2) As rock fill materials considered, with the times of cyclic loading increasing, the strain increases gradually. Even the cyclic time reaches 100, sudden increment of strain will not happen in the liquefaction process of loose sand. It has been studied that the rock fill materials have higher cyclic deformation properties, but the accumulative deformation increases with the cyclic times increasing, which can not be ignored as for the aseismatic stability of rock fill materials structures which is dominated by the deformation control.

The effect and reasonable description of rapid increasing pore water pressure of saturated rock fill materials on the dynamic deformation and strength properties should be taken into full consideration.

Description methods of dynamic strength properties of rock fill materials

For strength properties of coarse granular soil, there is no uniform description method so far. Usually one of ratio of cyclic shear stress, peak value of dynamic strain and vibration times is considered as criterion, different description patterns will be achieved with the other two considered. For example, (1) The relationship of cyclic shear stress ratio and corresponding vibration times at a certain strain peak value. (2) The cyclic shear stress ratio at certain vibration time at certain strain peak value. (3) The vibration time at certain cyclic shear stress ratio to the appointed strain peak value etc.

Influence factors of dynamic strength of rock fill materials

The influence factors of dynamic strength mainly include relative density, confining pressure, initial shear stress, test method, sampling pattern and lithology etc.

The results of undrained cyclic test of different relative density of rock fill materials indicate that the relative density increases with the cyclic shear stress ratio. The cyclic shear stress depends on confining pressure at relative higher density. With the average stress decreasing, the cyclic shear stress ratio increases, which is consistent with the conclusion that cyclic shear stress depends on confining pressure. While the cyclic shear stress ratio of rock fill materials doesn't depend on initial confining pressure at relative lower density. The cyclic shear stress ratio increases with the initial shear stress increasing at a relative higher density. The result of cyclic shear stress of torsional single shear test is lower than that of triaxial compression test, while the mean principal stress is dependent of initial shear stress. With the same relative density, the cyclic shear stress ratio of disturbed sample is 50% of undisturbed sample.

The results of vibrating table tests with different contents of gravel indicates that the sandy gravel have aseismatic properties when the content of gravel reaches up to a certain limit. It is also indicates that grained material can not form framework when the gravel content is lower than limit content. The permeability of sandy gravel depends on sand and liquefaction performance is also close to sand.

In addition, the dynamic strength of coarse granular soil is also remarkably influenced by the sediment history and lithology of samples.

The key technique study on high earth and rock fill dam of the 8th 5-year plan special topic is combined with 5 earth and rock fill dams at the height of 200m such as Pobogou, Xiaolangdi, Jilintai, Zipingpu, Daliushu. The relationship of dynamic stress and strain is conducted with 3 different kinds of fill materials, 1 kind of cushion material, 2 kinds of transition materials and dam foundation sandy gravel. The main results on cyclic loading dynamic triaxial apparatus and testing technique and the dynamic properties of coarse grained soil are as follows:

(1) China Institute of Water Resources and Hydropower Research (IWHR) developed the wave velocity testing apparatus and testing demarcates technique of laboratory large-scale triaxial apparatus, which improved the testing precision up to 10^{-4}s to 10^{-6}s, and also improved and perfected the testing method for dynamic deformation properties of coarse granular soil from a range of small strain to large strain under max confining pressure, 1Mpa with the wave velocity testing apparatus of large-scale dynamic triaxial test combined with testing large-scale samples, proposed in the 7th 5-year plan period,

(2) In Dalian University of Technology, Dynamic deformation properties of sand from small strains to large strains are tested on a medium and a small dynamic triaxial apparatus with testing set up for local axial strain at the stress level of max confining

pressure from 0.2MPa to 0.4MPa, which is proved feasible.

(3) The equivalent visco-elastic model parameters of former several kinds of dam material are achieved from tests, which can be used for the dynamic response analysis in the mentioned projects. When testing the relationship of stress and small strain, the wave velocity and local axial microcosmic deformation testing technique are applied.

(4) With the mentioned test apparatus, consolidate drained cyclic loading tests on air-dry materials and saturated materials are conducted in China Institute of Water Resources and Hydropower Research and the relationships of dynamic stress and residual shear strain of air-dry saturated material are presented. It is also indicated the static strength is approximately equal to the dynamic strength for the kind of saturated sand. The relationships of dynamic shear stress ratios and residual shear strains of rock fill materials from Guanmenshan and Pubugou are obtained in the consolidated drained or air exhausted cyclic loading tests in Dalian university of technology, which can be used for calculating the permanent deformation of earth and rock fill dam under seismic action.

(5) In order to reduce the seismic deformation of dam, it is better to study out reliable engineering measures in the aspects of high quality fill materials, enhancing fill density, increasing the confining pressure, and seepage control and reducing saturated water zoon.

ASEISMATIC ANALYSIS OF EARTH AND ROCK FILL DAM

According to the constitutive model, the methods of seismic response analysis can be divide into two kinds, one kind is equivalent-linear method based on the equivalent visco-elastic model, and another was real nonlinear analysis based on elastic-plastic model. The first model is convenient for application and abundant test data and project experience, which can be accepted for engineering field, had been accumulated in the aspect of parameter determination and model application. While the second model, which had more reasonable theory, could approach well actual response of soil, and could directly calculate the residual deformation of dam.

According to the whether considering pore water pressure in the earthquake process, the methods of seismic response analysis may be divide into total stress method and effective stress method. The effective stress method also has two kinds; one does not consider the dissipation and diffusion of pore water pressure, and another considered it. The closed relevant research, such as dynamic pore water characteristics, dynamic strength and liquefaction properties, had achieved important development. The liquefaction mechanism theory has been established, and the relationship and difference of soil liquefaction, failure and limiting equilibrium state also had obtained thoroughly knowledge. All these established the base for the earth and rock fill dam earthquake disaster mechanism research.

From present national and international research situation, the dynamic response analysis method of earth and rock fill dam gradually developed from, equivalent linear, total stress method to three-dimensional, real non-linear, the dissipation and diffusion of pore water pressure considering, effective stress method. And thorough research work should be performed in the aspects, such as reservoir water-dam-foundation couple nonlinear analysis, nonlinear constitutive model under complicated stress condition, pore pressure calculating mode, earthquake residual deformation calculating method, interface modeling and boundary condition processing, seismic input, high-precision numerical simulation, nonlinear calculating method and so on. The equivalent static method is used to carry on anti-slide stability analysis and safety evaluation of earth and rock fill dam in project. But this traditional method could not well consider the internal stress-strain relationship and real work state of soil, which is related closely with seismic characteristics, and the obtained safety factor was only so-called factor along the supposed potential slip surface. It is unable to obtain actual internal force and determinate the deformation, and forecast the occurrence and the developing process of failure. Moreover, it could not consider the effect of partial deformation on the stability of the dam. Therefore, dynamic response analysis method of earth and rock fill dam and ground are gradually developed.

Based on the nonlinear earthquake responds analysis, combining some research methods, such as laboratory test, in-situ test and so on, studies of hazard mechanism, seismic security and preventing measures, maybe the essential questions of the high earth and rock fill dam during the earthquake. For the research of failure mechanism under the earthquake, the nonlinear problems should be thoroughly studied, including nonlinear material properties, nonlinear failure parameters, nonlinear seismic respond characteristics, nonlinear solution theory and laboratory test methods. And the research of these aspects at present is very imperfectly, the innovative research work is required. According to the structure failure characteristic and the project destruction mechanism, the reasonable seismic design methods and preventing measures of high earth and rock fill dam should be proposed, and the function mechanism of preventing measures should be studied. The decrease of worry about the earthquake is the key point and the goal of seismic research work. It's important to value the pertinence, the reliability, the usability and the efficiency of the measures in the research.

Aseismatic key technique research on high earth and rock fill dam of special topic of the 8th 5-year plan achieved main results in the aspect of the seismic respond analysis method and the computational technique are as follows,

(1) The structural nonlinear visco-elastic-plastic analysis method or named incremental linear visco-elastic dynamic respond analysis method was proposed by Beijing Hydraulic Research Institute. The main characteristic is using incremental method and superposition method in turn, which can control the error accumulation of incremental method and make the analysis result to be more reasonable.

(2) Adopting the cyclic shear testing results of the interface between the cushion material and the concrete, which was conducted by Hohai University using the vibrated drag plate installation and refitting dynamic simple shear apparatus, the dynamic model of the interface between the face slab and the cushion is used to compare the effect of element shape on the calculating results by Dalian University of Technology. The results

of comparison show that it is feasible to use slab interface element in the face slab dam analysis. Comparing Goodman element with thin element, it indicates that the maximum relative error between the two calculating results does not exceed 1%, when the thickness of thin element is about 10cm. Therefore, it is suggested to use anisotropic thin element with certain thickness to substitute for the Goodman element for simplified calculation.

(3) The development of permanent deformation of earth and rock fill dam.

The structure nonlinear visco-elastic-plastic analysis method, which proposed by Beijing Hydraulic Research Institute, regards the soil as visco-elastic-plastic material, and considers the mathematical pattern of soil elastic-plastic deformation rule under dynamic loading as the combination of initial load curve, skeleton curve and hysteretic loop. The tangent shear modulus is calculated by established nonlinear elastic-plastic shear strain model, and the secant shear modulus and (secant or tangent) damping ratio is calculated by nonlinear elastic model. The secant Young's Modulus is calculated by supposing the damping ratio equals const. Then, the permanent deformation of dam can be computed by the structure nonlinear visco-elastic-plastic analysis method. During the research, the semi-empirical methods of permanent deformation, such as the whole deformation analytic method (mainly is nodal force method) and rigid body slide method have been improved significantly.

(4) Development of the computational method of dynamic hydraulic pressure.

The interaction of dynamic hydraulic pressure and the concrete faced rock fill dam is studied by Hohai University. When calculating dynamic hydraulic pressure 3 conditions (water is not compressible, water is compressible, and calculating with Westergaard formula) are considered. The analyze results show that it is reasonable to use the numerical model with uncompressible water to perform dynamic couple analysis. For the low dam, it is wise to use Westergaard formula which calculating dynamic hydraulic pressure is considered as additional mass to the couple dynamic analysis. The workload is too much for the model with compressible water, which is not recommended.

(5) Influence of traveling wave to the earth and rock fill dam.

The studies by Sichuan United University showed that the influence mainly displayed in the earthquake direction and the occurrence of the strengthen area. The traveling wave velocity and direction had the significant influence to seismic respond: when the traveling wave input along the axes of the dam, a greater vertical acceleration, displacement difference, and horizontal stress occurred. The respond of traveling wave had enhancement under the most situations for ditch earth and rock fill dam, and the enhancement would be about 30%.

(6) Development of computer programme.

Nanjing Hydraulic Research Institute developed the total stress 3D static and dynamic analyze general programme (TOSSD3), and effective stress 3D static, dynamic analyze programme (EFFESD3). Hohai University developed the dynamic finite element programme (WWCC2D) and (WWCC3D). Sichuan United University developed the 3D nonlinear dynamic analyze programme with traveling wave input function.

EARTHQUAKE INDUCED PERMANENT DEFORMATION OF EARTH AND ROCK FILL DAMS

At present, it is regulated in national specifications of seismic actions on hydraulic structures that surface sliding method is applied in aseismatic stability analysis of rock fill dams. However, harzards and investigations indicate that safety factor against sliding can not correctly reflect whether rock fill dams are failure after earthquake. Rock fill dams may be instantaneous failure or instantaneous limit equilibrium state under strong seismic actions. However, the instantaneous instability of dams may not induce engineering failure. Deformation of dams induced by instantaneous instability is permitted if it has no great influence on engineering function. Thus, it is obvious that the safety factor of dams under seismic actions is not the direct parameter. Therefore, researchers started to look for more reasonable method to evaluate dam stability. For concrete faced rock fill dam, safety standard and design basis mostly depend on the deformation of the dam body, so it is widely accepted that permanent deformation is applied to assess the dam safety. For many years, the researchers have studied in this field and several simplified analysis methods are proposed through the results of dynamic response analysis based on numerical analysis. The method for earthquake induced permanent deformation includes two respects of determinacy and randomness. The so-called determinacy analysis of earthquake induced permanent deformation is a method that permanent deformation is calculated through dynamic stress and dynamic displacement which is obtained by numerical integrating one or several seismic wave spectrums. It can also be classified to sliding block deformation and whole deformation analysis.

Sliding block deformation analysis

It is firstly proposed by Newmark (1965) based on limit equilibrium theory and assuming that permanent deformation is the sliding displacement of sliding body along the most possible sliding surface during instantaneous failure under seismic actions. Sliding surface will start to move when the acceleration of some point exceeds the critical value. The permanent deformation can be calculated by integrating the value of the ground acceleration minus the critical value twice. Based on Newmark method, lots of other researchers (Markdisi and Seed, Chen Shengshui and Shen Zhujiang et al.) also developed the method.

Whole deformation analysis

The basic assumption of whole deformation analysis is regarding soil deformation as continuous medium. The method is developed with applying visco-elastic model is in constitutive law and using FEM for calculation combined with experimental study. According to the mechanism of permanent deformation, it includes two kinds of methods, (1) Softening modulus method. It is based on the assumption that permanent deformation is due to a depressing of static shear modulus by seismic stress. The permanent deformation is calculated by twice finite element based on the difference of static strain induced by depressing shear modulus and before earthquake. It is firstly proposed by Lee and several methods, such as initial approximate estimations, linear modified modulus and nonlinear modified modulus, were proposed by Serff et al. (2) Equivalent node force method. It is suggested that the influence of seismic induced deformation is replaced by a batch of static nodal force (equivalent node force) acting on element nodes. The permanent deformation is obtained by using FEM calculating superimposed deformation under equivalent node force according to the relationship between residual deformations and dynamic stress received from experiment. Taniguchi and Whitman proposed some typical method and equivalent nodal method were proposed by Zhang Kexu and Liu Hanlong et al.

The stochastic analysis of permanent deformation also includes sliding block deformation analysis and whole deformation analysis. (1) Stochastic analysis of Sliding block deformation. Lin (1984) firstly proposed the method of permanent deformation of earth and rock fill dams. Later on Lin (1986) introduced particularly the process through which the permanent deformation of perfect sliding block was calculated. Wu Zaiguang et al. (1991) carried out further studies based on Lin's research. (2) Stochastic analysis of Continuous deformation. At present, less investigation is carried on in this aspect and only Wu Zaiguang et al. deals with some preliminary study. Wu introduces the accumulative damage model of liquefaction analysis into the softening modulus approach under stochastic seismic excitation. Liu Hanlong established a stochastic analytical method of permanent deformation based on the equivalent nodal force scheme and seismic randomicity.

RESEARCH ON ASEISMIATIC MEASURES OF THE HIGH EARTH AND ROCK FILL DAM

In order to meet the aseismiatic requirements of earth and rock fill dam some measures have to be taken. When there are fine sand crystals in the shallow layer of the foundation, vibro replacement stone column is commonly used. In general, the slope of the dam is slow. For example, the grade ratio of a concrete panel sandy gravel dam in Jilingtai, Xinjiang province is 1:1.7~1:1.9, the design earthquake intensity of which is 8 degree. For the same example, when the rock fill of which the antiskid ability is better is used in the outer dam slope, the grade ratio could be 1:1.4 or little slower. The aseismiatic performances of most rock fill dams with face slab are good, which are related to the attributes of the rock fill and earth. Furthermore, in order to reduce the reclamation materials and meet the aseismiatic requirements, concrete sashes or bar meshes are installed in the dam.

(1) The cracking at the top of the impermeable barrier in the core wall rock fill dam is one of the common phenomena of earthquake damage. For the sake of preventing the cracking caused by earthquake, it's important to select better reclamation materials, take notice of the reclamation quality, and optimize the fracture surface design. In

addition, the designs of the inverted filter and transition layer of the impermeable barrier at the upriver and downriver surfaces are much more important, especially at the downriver surface [10].

(2) The earthquake damages for rock fill dam with face slab are mostly at the top of dam, such as the cracking of the face slab, the deformation of the dam body, the land slip of the downriver slope, and so on. In addition, the fractional damages of the peripheral joint and the watertight seal parts always appear. Based on the height of dam and the earthquake response, different measures can be taken, such as: slowing the top slope, the addition of the packways at the proper altitude, the setting of the face slabs at the top of the dam and at the upriver and downriver surfaces, the setting of tension braces and geotextile at the top of the dam, the adoption of reclamation with roller compacted concrete, and so on. When it's necessary, the horizontal seams can be added in upriver face slab at the proper altitude, and the antiseep ability of the cushion should be increased at easily-cracked parts.

RESEARCH ON EXPERT SYSTEM OF ASEISMIATIC SAFETY EVALUATION AND DISASTER PREVENTION MEASURES OF HIGH EARTH AND ROCK FILL DAM

Regarding the earthquake damage data and the data of reclamation materials dynamic behavior as knowledge base, considering the constitutive models applied in the earthquake response analysis of earth rock and fill dam and earthquake response analytic methods as model base, taken the aseismiatic safety evaluation criteria and method for evaluation standard base, regarding the research system of aseismiatic reinforcement measures as countermeasure base, and based on the artificial intelligence theories and expert system tools, the aseismiatic safety evaluation and disaster prevention countermeasure expert system of high earth and rock fill dam can be founded, as shown in Fig1. If the proper initial conditions have been given, the aseismiatic safety evaluation conclusion of dam could be achieved with the expert system, no matter the dam is being constructed, have been built or will be built. Furthermore, the corresponding aseismiatic measures and the further work could also be suggested. At the same time, the use of the artificial intelligence will make the system learn the new knowledge, so that the system will self-perfect and develop continuously during the application of the system in the process of decision, scientific research, design and construction of the earth and rock fill dam projects.

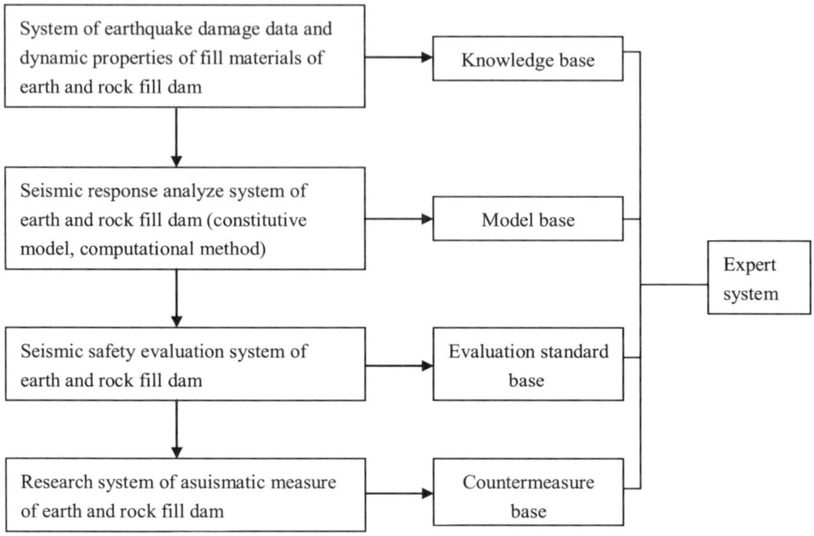

Fig 1 expert system of aseismiatic safety evaluation and disaster prevention countermeasure of high earth and rock fill dam

CONCLUSIONS

In order to meet the requirements of the economic and water resources development, it is necessary to construct high dams in strong seismic area. As for the construction of high earth and rock fill dam in strong seismic area, the aseismiatic problems are always the controlling factors. Once high earth and rock fill dams were failure, the result will be an awful disaster. On the other hand, so far, there aren't any cases and field data of earthquake failure of high earth and rock fill dam. Therefore, seismic actions on high earth and rock fill dams should be studied. Further researches are suggested as follows,

(1) Research on the mechanism of the earthquake failure of high earth and rock fill dam. In order to prevent and mitigate disaster effectively, the mechanism of the earthquake failure should be clearly understood.

(2) Research on the dynamic properties of fill materials and the acquisition and management of earthquake failure data of high earth and rock fill dam. The reliable dynamic properties of fill materials and earthquake failure data are the bases of the research on mechanism of the earthquake failure.

(3) The dynamic hydraulic pressure and dynamic residual displacement can be calculated directly by elasto-plastic method, so 3-D elasto-plastic seismic response analyze method should be developed.

(4) Research on nonlinear problems of high earth and rock fill dam, such as nonlinear material propertiesr (constitutive laws), nonlinear failure parameters, nonlinear seismic response characteristic, nonlinear solve theory and experiment method, etc.

(5) The evaluation theory and method of the aseismiatic safety of high earth and rock fill dam should be established, according to the structure failure characteristic and failure mechanism.
(6) Research on the engineering measures of resisting earthquake damage of high earth and rock fill dam and the mechanism of the measures.

REFERENCES

Chang, Y.P. (1998). "Key technical study on earthquake-resistance of high earth-rock dams", *Journal of Hydroelectric Engineering*, No. 3, 36-44.

Chen, S.S. and Shen, Z.J. (1990). "Earthquake induced deformation analysis method of rockfill dams in strong motion areas", *Journal of Hohai University*, 20(2),117-120.

Finn, W.D. L., Lee, K. W. and Martin, G.R. (1977). "An effective stress model for liquefaction", *Proc. ASCE*,103 (GT6).

Guo, Q.G. (1998). "*Project propertiesistic and application of coarse granular soil*", Zhenzhou: Yellowriver Water Conservancy Press.

Hardin, B.O. and Richart, F.E. (1972). "Elastic wave velocities in granular soils", *Journal of the Soil Mechanics and Foundations Division*, ASCE , 98(7),667-692.

Jia, G.X. (1998). "*New technical study on geotechnic triaxial test*", Master's degree thesis of Dalian University of Technology.

Kong, X.J., Han, G.C. and Lin, G. (1988). "Size influences on dynamic shear modulus", *Journal of Southwest Water Conservancy and Power Generation*, 4(4), 9-15.

Liang, Y.X. (1990). "Simulating dynamic deformation testing and application of rockfill materials", *The 3rd national academic conference symposium of soil dynamics*, Shanghai, 121-124.

Lee, W.H. and Wang, W.S. (1993). "Dynamic shear strain model of non-cohesive soil", *Journal of Water Conservancy*, (9).

Liu, Y. and Xie, J.F. (1984). "*Shock liquation of sand clay*", Beijing: Earthquake Industry Press.

Liu, H.L., Lu, Z.Z. and Qian, J.H. (1996). "Earthquake Induced permanent deformation of earth and rock fill dams", *Journal of Hohai University*, 24(1),91-96.

Liu, H.L. (1996). "Permanent deformation of foundation and embankment dam due to stochastic seismic excitation", *China Journal of Geotechnical Engineering*, 18(3), 19-27.

Makdisis, F.I. and Seed, H.B. (1978). "Simplified procedure for estimating dam and embankments earthquake induced deformation", *Proc. ASCE, JGED*, 104(GT7),849-867.

Prevost, J. H. ,Abdel - Ghaffar, A. M. and Lacy, S. J. (1985). "*Nonlinear dynamic analyses of an earth dam*", Proc. ASCE , JGED, 111 (7).

Seed, H. B. and Idriss, I.M. (1970). "*Soil moduli and damping factors for dynamic response analyses*", In: Report No.EERC70-10, Earthquake Engineering Research

Center, University of California, Berkeley, California.

Serff, N., Seed, H.B. and Markdisi F.I. (1976). "Earthquake induced deformation of earth dams", *Report No. EERC76-4*, University of California, Berkeley.

Shen, Z.J. (2000). "*Theoretical soil mechanics*", Beijing:China Water Conservancy and Hydroelectricity Press.

SL203-97. (1997). "*Earthquake-resisitance design norm for hydraulic structures*", China Water Conservancy and Hydroelectricity Press.

SL274. (2002). "*Design norm for rolled earth and stone dams*", China Water Conservancy and Hydroelectricity Press.

Wang, B.L., Liu, Y.Z. and Wu, H.H(2005). "New development of China earth and stone dam project construction", *Water Power*, Vol. 31, No. 1, 63-65.

Wang, W.S. (1999). "*Selecting from geotechnic problems theses of academician Wang Wen-shao*", China Architecture and Building Press.

Wang, W.S. (1997). "*Soil dynamic strength and liquefied propertiesistics*", Beijing:China Power Press.

Wang, W.S. ,Jin, C.P. and Wang, K .C.(1987). "Earthquake-resistance calculation of rock and stone dams and model test as well as prototype measure", *Journal of Water Conservancy*, (12).

Wu, X.Z. (2001). "*Static and dynamic constitutive model of rockfill materials with its application in concrete panel rock-fill dams*", Doctoral degree thesis of Dalian University of Technology.

Wu, Z.G., Han, G.C. and Lin, G. (1991). "Risk analysis on Earthquake-Induced permanent deformation of earth and rock fill dams", *China Journal of Geotechnical Engineering*, 13(2).

Xie, D.Y. (1988). "*Soil dynamics*", Xi'an: Xi'an Jiaotong University Press.

Xu, Z.Y. and Shen, Z.J. (1981). "FEM dynamic analysis for the creation and diffusion as well as dissipation of pore water pressure in earth-dams earthquake", *Journal of Hohai University* , (4).

Zhang, J.M. and Xie, D.Y. (1993). "Advances in theoretical and applied research on seismic pore water pressure of saturated sand", *Advances in Mechanics*, 23(2).

Zhang, K.X. and Xie, J.F. (1989). "*Soil Dynamics*", Beijing: Earthquake Press.

Zhao, J.M. (2000). "*Study and analysis on 3-D authentic nonlinear dynamic CFRD*", "95" national key scientific and technological report(96 -221 -02 -03 -02), China Institute of Water Resources and Hydropower Research.

Zhao, J.M., Wang, W.S., Chang, Y.P. and Chen,N.(2003). "3-D authentic nonlinear method for dynamic analysis of high CFRD", *Journal of Water Conservancy*, (9).

Zhao, J.M., Wang, W.S. and Zhang, C.W.(2000). "Study on influence factors of seismic pore water pressure in earth-rock dams", *Journal of Water Conservancy*, (5).

Zhao, J.M., Chang, Y.P. and Chen, N.(2004). "Significance and prospects about earthquake resistant studies of high earth-rockfill dams", *World Earthquake Engineering*, Vol. 20, No. 1.

International Conference on Geotechnical Engineering for Disaster Mitigation & Rehabilitation
Chu, Phoon & Yong (eds) © 2005 World Scientific Publishing Company ISBN 981-256-469-1

DESIGN WITH CONSIDERATIONS OF RELIABILITY AND RISK

S.R. LO
School of Aerospace, Civil and Mechanical Engineering,
University of New South Wales at ADFA,
Canberra, ACT 2600, Australia

Consideration of reliability provides the necessary theoretical basis for incorporating safety in geotechnical design in a rational and consistent manner. Some of the problems of reliability-based design are examined. These problems are sometimes due to avoidance of probabilistic framework, or to a restrictive interpretation of probabilistic method. A reliability-based risk approach is then discussed. In such an approach, the reliability and consequence need to be quantitatively considered. This is particularly relevant in the design for extreme events related to natural hazards. However, the determination of failure consequences is not necessarily a non-engineering matter.

INTRODUCTION

From a probabilistic point of view, three levels of probabilistic analysis can be defined in a design process (Li and Lee 1991, Lo et al 1992). Level 1 analysis does not involve any explicit reliability analysis. Level 2 analysis, loosely speaking, is based the calculation of a reliability index (Hasofer and Lind 1974, Li and Lee 1991). Level 3 analysis is a complete evaluation of the probability of failure. Clearly design codes based on either allowable stresses of unclear origins, or overall safety factors (which may then infer allowable stresses), or a series of partial factors are all level 1 methodology. The geotechnical design community has been pre-occupied with the transition to design codes based on partial factors within the so-called Limit State framework. The structural design community has also been attending to performance-based design aimed at meeting specific goals, some of which may be related to extreme events such as natural hazards.

As illustrated in Lo et al (1992), a geotechnical design with a higher safety factor does not necessarily imply a safer design. This points to the need for reliability analysis as the theoretical basis of our design methodology. This paper does not argue for any specific means of introducing reliability analysis into the geotechnical design process. It can take a back-stage role for establishing partial factors that target a consistent reliability index. Or it may be an explicit, full-blown reliability analysis. The focal point is reliability considerations need to be embraced rather than avoided. This paper deliberately use relatively simplistic, which may be seen as crude and dated, probability theory to substantiate the argument. It is the principle rather than the specific refinements that are central to this paper. For unusual design scenarios or designing for extreme events, consequences of a failure also need to be considered. This leads to the need for consideration of risk, which can be loosely defined as the product of probability of failure and consequence.

CASE FOR RELIABILITY-BASED DESIGN

The use of partial factors no doubt will contribute to better control on reliability. This can be easily illustrated via an example of a foundation design with loading dominated by the self-weight of the structure. The use of partial factor will automatically lead to a lower overall safety factor. This is a more logical outcome compared to design based on a constant overall safety factor (or a prescribed allowable stress) as the uncertainty attributed to loading is less. However, Phoon et al (2003) argued that the benefits from a re-distribution of an existing overall safety factor to a series of partial factor may be limited and call for a shift in mindset to reliability-based design (RBD).

The most common objection against RBD, from a practical perspective, is the need for most input data. Geotechnical engineers are familiar with difficulties in obtaining geo-parameters even for relatively simple design. The challenge presented by the scarcity of statistical data for RBD has to be confronted and not avoided. One also needs to recognize that the failure probabilities of most infrastructures are low that it is rarely possible to verify a computed probability of failure (P_f) or reliability index (β). This leads to the concept of relative reliability (or relative failure probability). The paradigm is to compare the computed β or P_f of different designs based on the same reliability model and input parameters. The computed values of β are thus not necessarily correct, but can be used as a quantitative and objective basis of assessing the relative "safety" of different design. This approach also enables us to develop "deterministically looking" design codes that deliver design outcomes with approximately consistent β (Lo et al 1992, 1994), ie having RBD as the theoretical basis for deriving partial factors of limit state design codes.

Having confronted the issue of scarcity of statistical data, the case for RBD still needs to be argued. As discussed by Whitman (1984) and Christian et al (1992), the probability of failure (or reliability index) of our design is not clear. Human intuition is generally not effective for reasoning with uncertainties and some quantitative measures are needed for effective risk management and communicating risk to our colleagues outside our specialization. As a profession, we also need to communicate with the society about risk level of our design (Ellingwood 1999). Although such arguments appear to be philosophical, it has immediate relevance for code harmonization. Geotechnical engineers are well aware of the challenges involved in harmonizing geotechnical and structural code. An important factor contributing to the difficulty is the unclear reliability implied by our conventional geotechnical design process. The advent of WTO has initiate interest to code harmonisation across different countries and this is even more challenging. To facilitate such a harmonization process, it is essential to communicate reliability (or risk) within a transparent and rational framework (Frank 2002, Honjo and Kusakabe 2002). RBD also facilitate risk-sharing among various parties in a major project, especially based on a DC (Design and Construct) or BOT (Built Operate and Transfer) or partnering contracts. When we are have to design for extreme events such as earthquake, extreme rainfall in terms of both duration and intensity, the need for a reliability-based framework becomes obvious as we have little precedence to rely upon.

CHARACTERISTIC VALUE

The definition of characteristic value of a geotechnical parameter has been a controversial topic (Green and Becker 2001, Simpson and Driscoll 1998). The author believes much of the controversy is due to either misunderstanding or confusion as a result of avoiding inevitable statistical concepts. Recognising the variability of ϕ', the meaning of $\phi' = 35$ is not clear. Does it represent the mean, the mode, or the value at certain fractile? Using terms such as "representative value" only conveys some ill-defined meaning. In order to control the reliability of a design to a consistent value, any input material parameter must have a statistical meaning. For structural strength parameter X, characteristic material strength, X_k, is defined by the 5% lower-tail fractile, and is commonly calculated with the following equation.

$$X_k = X_\mu - 1.645 X_\sigma \tag{1}$$

where subscripts μ and σ denote the population mean and population standard deviation respectively. EC-7, Part 1, (ENV 1997-1 : 1995) defines characteristic values as "a cautious estimate of the value affecting the occurrence of the limit state". But it also states that "if statistical methods are used, ... should be derived such that the probability of a worse value governing the occurrence of a limit state is not greater than 5%". Although these two statements in EC-7 are often considered as confusing, the only consistent explanation of these two statements is that cautious estimate corresponds to 5% fractile; but it is not mandatory for the engineer to use statistical methods to establish this value. The engineer is allowed to exercise judgement on what value corresponds to 5% fractile. This is indeed the definition used in R57 (2002), which is the design specification for the design of reinforced soil structures by the Roads and Traffic Authority, NSW, Australia.

Sometimes, test data are available for calculation a characteristic geotechnical strength parameter using equation (1). However, such an exercise often gives unreasonable characteristic values for geotechnical parameters. This leads to concerns on embedding statistical concept in the definition of characteristic value, and more generally the sensibility of using RBD. Its important to note that equation (1) above is based on three assumptions: X following a normal distribution, X_μ and X_σ are population values estimated reliably from an adequate number of data points, and strength measured in a test specimen is the value affecting the occurrence of collapse. For geotechnical engineering problems, it is quite rare that all three assumptions are satisfied, and hence equation (1) in most cases is not valid One therefore cannot expect it can give a reasonable characteristic value.

Another important point in the EC-7 definition is the phrase "governing the occurrence of the limit state". In structural engineering, collapse can be triggered by a small quantity of structural material being "under-strength". Hence the strength given by a test specimen governs the occurrence of ultimate limit state (ie collapse). For a ductile form of geotechnical ultimate limit state, failure occurs when the "average strength" of a large domain of materials is under-strength. Statistically, this means the strength measured with a test specimen represent the strength of a point in space, and the variability of perfect test data represents the spatial variability (from point to point). This can be illustrated by an idealized slope stability problem illustrated in Figure 1. The material idealization is a constant cohesion material and **v** is the curvilinear distance along the potential slip surface. The cohesion value has a population mean c_μ and a large point-to-point scatter as indicated by the zigzag line. However, what governs the slope stability, the occurrence of the limit state in EC-7 language, is the cohesion averaged along the potential slip surface. This consideration is also stated in EC-7 except that its not expressed in statistical language.

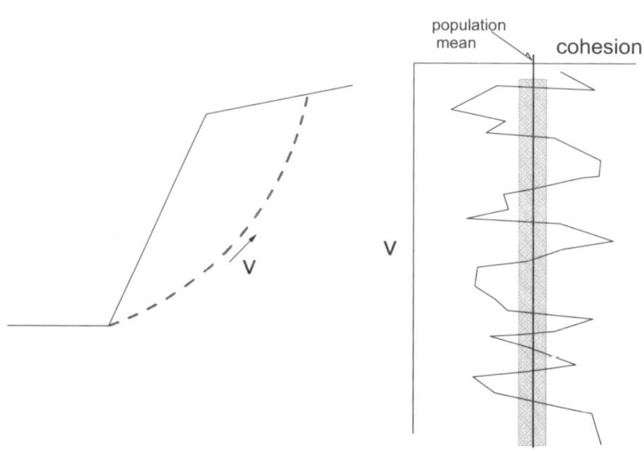

Figure 1. Variability and spatial average

Due to statistical averaging, this spatially averaged value will manifest a considerably lesser variability as indicated by the shaded zone of Figure 1. Such a variability reduction can be mathematically derived using the Random Field Model (Li and White 1987, Vanmarcke 1977), and can be expressed as

$$Var(\tilde{X}) = \Gamma^2(D).Var(X) \qquad (2)$$

where Var(X) is the variance of X, $\tilde{X}$ is the spatial average of X over the domain "D", and Γ^2(D) is the variance reduction factor. Note that $\tilde{X}$ is also a random variable, and the expectation of $\tilde{X}$, $E(\tilde{X})$, equals to X_μ. Therefore Equation (2) can be expressed as

$$COV(\tilde{X}) = \Gamma(D).COV(X) \tag{3}$$

where COV(X) is the coefficient of variation of random variable X, and $\Gamma(D) \leq 1$. The value of Γ(D) is dependent on the size of the domain "D" relative to the auto-correlation distance, a parameter in a Random Field Model (Vanmarcke 1977, Li and Lee 1991). The domain "D" loosely speaking is the failure zone or the failure surface. It is not unusual, say for a slope stability problem, for Γ to be as low as 0.1 (Li and White 1987). This means the variability of the spatially averaged strength parameter, which is the value governing the occurrence of a limit state, is reduced to one tenth of he point value. This begs the question why we need to model variability of soil parameters at all.

However, one has to realize the population mean needs to be estimated. In general, the number of data point available is small and there is significant uncertainty in the sample mean, $\bar{X}$, calculated as the arithmetic average of the data points. If one assume the distribution of the point data is normal, the equation for estimating mean value with less than 5% chance of an inferior worse population mean occurring is given by the following equation.

$$\tilde{X}_5 = \bar{X} - K\left(\frac{X_S}{\sqrt{N}}\right) \tag{4}$$

where X is the parameter (point value), N is number of data points, X_S is the sample standard deviation, and $\tilde{X}_5$ is the estimated population mean at 5% fractile, K is a constant dependent on N but in general $K \leq 1.645$. A high variability of X will lead to a high X_s, and hence higher uncertainties in $\tilde{X}$. Noting the assumption of normal distribution is usually worse than reality, equation (4) may be approximated by equation (5) below, which use the lower bound value of K.

$$\tilde{X}_5 \approx \bar{X} - 1.645\left(\frac{X_S}{\sqrt{N}}\right) \tag{5}$$

The above rationale was tested for the determination of characteristic peak friction angle for the select fill (in the reinforced zone) of a reinforced soil wall. The soil is a silty sand with fines content ≤ 15%. Five good uncorrelated test results were selected from direct shear box testing of the fill source. These five good data points gave a sample mean of 40.8° and a sample standard deviation of 3.7°. Therefore, equation (5) yields $\tilde{\phi}_5 \approx 38°$. A careful examination of the failure mechanism of a reinforced soil wall also revealed that $\Gamma < 0.1$ and hence the variability of spatial average, $\tilde{X}$, relative to uncertainties in the estimation of population mean is negligible. Therefore, $\tilde{\phi}_5$ can be considered as the characteristic value. The above procedure was also used in a European trial on estimating characteristic value of a stiff clay from ten reliable test results as reported in Ovesen (1995).

It also need to emphasized that sometimes, the test results need to be converted to represent the behaviour of the material in the ground. In such a scenario, X should be post-converted value, a point clearly stated in EC-7.

The above discussion does not argue that the EC-7 definition of characteristic value is a necessity. Rather, it highlighted that most of the alleged problems to the use of a characteristic value and assigning a statistical meaning to characteristic value is in fact due to avoidance (or partial avoidance) of a statistical framework. The corollary is that geo-parameters need to be considered from a statistical (probabilistic) point of view.

GEOTECHNICAL UNCERTAINTIES

Many attempts have been made in classifying, or identifying the sources of, geotechnical uncertainties. A detailed discussion is contained in Li and Lee (1991). Different researchers tend to classify the uncertainties differently, and therefore a relatively detailed classification is listed below.

- Innate variability
- Sampling uncertainties
- Testing uncertainties
- Model Uncertainties
- Human uncertainties

The innate variability of soils is well discussed in Li and Lee (1991). One important consideration is the concept of a Random Field Model (Vanmarcke 1997). However, the uncertainties of ground water condition also need to be considered. In a lot of geotechnical problem, this may have a major impact on the reliability of the design.

Sampling uncertainty is due to the fact that only finite, and often small number of, uncorrelated data points are available. This implies that population parameters (such as population mean) required for probabilistic modeling cannot be precisely known.

Soil testing inherently is an error prone process and following testing standard does not necessarily give a non-biased outcome. This aspect of uncertainties, in the opinion of the author, is often overlooked.

The first three types of uncertainties may be considered under parameter uncertainty in the classification of Morgentern (1995). Coefficients of variations of some soil parameters have been reported in a number of literatures. The fact that parameter uncertainties can be attributed to three different sources means one need to exercise care is adopting these reported statistical values.

The uncertainty of our prediction model is well documented (Kay 1993). However, model uncertainty is not restricted to prediction model. There can be significant uncertainty in our material model. For example, a c-ϕ material model is simply a linearization of a generally curved failure surface (Au 1998, Lo 2005). Uncertainties in material models and design models are often coupled. Furthermore, uncertainties in material modeling and parameters uncertainties are also coupled, and this aspect is often overlooked in reliability modeling.

SEMI-EMPRICAL DESIGN MODEL

A number of apparently theoretical design models in geotechnical engineering are in fact of a semi-empirical nature. One such example is pile design. Consider the calculation of side resistance of a pile in clay with the so-called alpha design equation. The side resistance R_s is given by

$$R_s = \xi L_p(\alpha c_u) \tag{6}$$

where ξ and L_p are the perimeter and penetration respectively of the pile, c_u is the undrained cohesion and α is an empirical parameter correlated to c_u or c_u/σ'_{vo} and σ'_{vo} is the insitu overburden stress. It appears cohesion c_u is the only soil parameter involved; but in fact the empirical parameter α is dependent on a number of soil parameters in a very complicated and unknown manner. Thus any attempt to perform a reliability analysis relying on modelling the uncertainties of c_u, or considering parameter and model uncertainties separately, may be relatively ineffective. This becomes more obvious when one considers other design models such as the beta model, which some regarded as theoretically better than alpha model, or the evidently empirical Hiley driving formula.

This dilemma can be handled by defining a Bias Ratio, B, by

$$B = R/R_{pu} \tag{7}$$

where R is the actual ultimate pile resistance and is a random variable. R_{pu} is the pile resistance predicted by the calculation model and is thus deterministic. Therefore, B is also a random variable, and we have:

$$R_\mu = B_\mu R_{pu} \tag{8a}$$

$$COV(B) = COV(R) \tag{8b}$$

This means B_μ and COV(B) can be used to model the statistical attributes of a particular method of calculating R_{pu} under a given set of project/design scenario. A value of B_μ differing from unity implies that the design model gives a biased prediction, ie, on the average fails to predict the actual pile capacity. For a given design scenario such as the alpha design method, load test data published in the public domain enables us to establish data points for such a design method and hence establish its B_μ and COV(B) values.

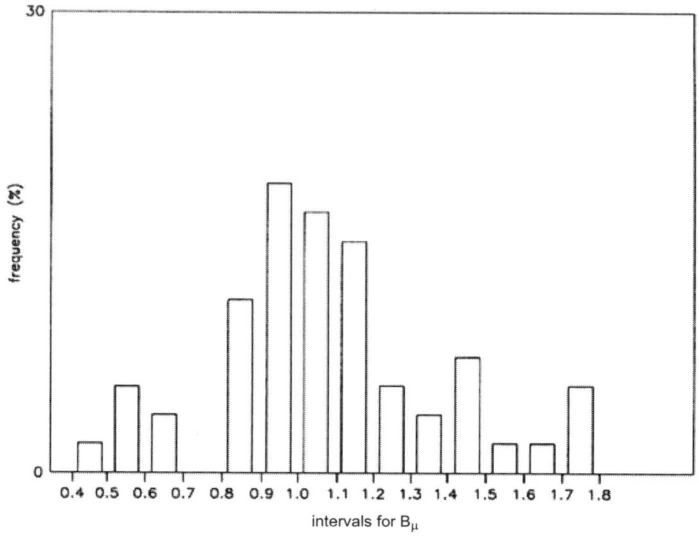

Figure 2. Histogram of B_μ

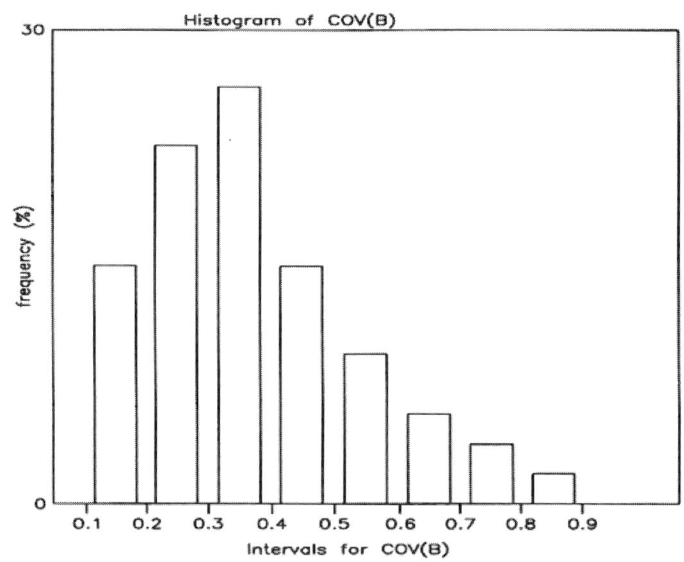

Figure 3. Histogram of COV(B)

Cameron (1992) conducted an extensive search of published load test data and the findings were analysed by Lo et al (1995) to form a database for a range of design scenarios. In essence, a measured failure load, in conjunction with R_{pu} obtained from the geo-parameter(s) given in the same publication, give a sample point for B. An adequate number of sample points of B for the same design scenario give B_μ and

COV(B) which are specific to this design scenario. Data for a range of design scenarios give the database. We thus arrived at the histograms of B_μ and COV(B) presented as Figures 2 and 3.

Figures 2 and 3 showed some unexpected results. B_μ has a central tendency of about unity. This means we cannot assume that our design predictions in most cases are conservative. COV(B) is in general of high value. This is due to both the limitations of our design model and the input parameter may not have a consistent statistical meaning.

Although such an approach may be seen as crude, it can be used to deliver design outcome that achieve a consistent reliability index as explained in Lo et al (1994). Such an approach can also be used for downstream calibration of partial factors in a Load Resistance Factored Design format as explained in Lo and Lee (2002). A brief summary is also given in the Appendix.

RISK

Accepting reliability analysis as the theoretical basis of incorporating safety in design inherently implies that failure, although unlikely, can happen. As discussed in Lumb (1991), when the failure probability is less than a threshold value, we may take no action. It was suggested in Lumb (1991) that the threshold value is about 10^{-6} on individual scale and 10^{-15} for world community scale. These numbers were extremely low and may often be unachievable. Furthermore, Lumb (1991) also quoted examples on how the threshold probability changes as a society prospers. Therefore, the consequence of a failure needs to be examined. The importance of considering consequence is well illustrated by slope stability issues in Hong Kong. Before Hong Kong developed into an international city, minor slope failures had little consequences and were often unreported. As Hong Kong is now a highly urbanized international city, a very minor slope failure (ie the failed soil mass is of a small volume and manifests a small travel distance) leads to considerable economic consequences with possibility of fatalities. Therefore significant resources have to be channeled to address slope stability issues.

Before having a mathematical definition of risk in the context of reliability-based risk analysis, it is interesting to note that our colleagues in project management have been considering consequence as an integral part of risk management. This is often done with a risk ranking matrix as illustrated in Figure 4a below. The likelihood of occurrence equates to our probability of failure. If the issue under consideration is of high likelihood of occurrence and high consequence, corrective action(s) must be undertaken. One the other hand, monitoring or further review is often considered as acceptable even for high likelihood if the consequence is classified as low. Although the exercise may involve the assignment of a ranking number and referred to as quantitative, it is in fact qualitative. A similar approach is also use to categories slopes for the purpose of priority ranking in stabilization work (Baynes et al 2002). In such an exercise, however, the likelihood ranking is based on a site inspection, desk study, and expert opinion. One may argue then that such a procedure is quantitative.

The objectivity of such a process can be improved further by conducting level 2 reliability analysis based on a common reliability model and input parameters determined from a common set of criteria. With the computing power available these days, the reliability indices, β, of the slopes to be categorized can be easily computed. We can compare the risk of different designs with the plot presented as Figure 4b, where the designs being examined are indicated by the filled-symbols. It is important to emphasise that the approach discussed so far gives relative, not absolute, risk ranking. However, this is both meaningful and effective as an engineer has to compare design options or a manager has to prioritise upgrading works with limited funds.

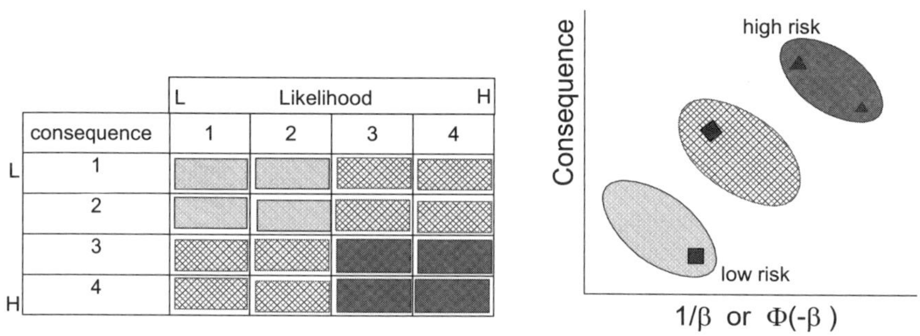

(a) ranking matrix

(b) consequence Vs reliability

Figure 4. Consideration of Consequence

RELIABILITY-BASED RISK ANALYSIS

A clear definition of risk is needed in a reliability-based risk analysis, sometime also referred to as probability based risk analysis or quantitative risk analysis. However, this is not an easy task because a range of definitions exists in the literatures. The author came across situations where the same researcher used two very different definitions in two different publications.

For damages to properties, Risk is often defined as

$$\text{Risk} = \text{Probability x Vulnerability x Property Value} \tag{9}$$

where vulnerability is defined as the extent of damage if failure occurs. By definition, $1 \geq \text{vulnerability} \geq 0$. Although vulnerability appears to be a deterministic quantity, it may depend on the response of the building to extreme loading, and hence it is may be an expected value in a probabilistic analysis.

If loss of life is the prime consequence under consideration, then a slightly different definition as expressed below may be used.

$$\text{Risk(L)} = P(f) \times P(S|f) \times P(T|S) \times L|T \qquad (10)$$

where R(L) = risk in term of life loss, f = failure, S = spatial impact (ie the building under consideration being affected), T= temporal impact (ie the building being occupied), and L|T = life loss given temporal impact. In this definition, the consequence is failure is clearly probabilistic as it is given by product of three terms, two of which are conditional probabilities.

Equations (9) and (10) can be further generalized by including a summation function if multiple properties or buildings are involved. Despite the many forms that Risk can be defined, for the purpose of reliability-based risk analysis, it can all be expressed in a very simple form as equation (11) below (Stewart 2005).

$$\text{Risk} = P(f) \times \text{Consequences} \qquad (11)$$

Depending on the scenario, the consequence is may be cost or equivalent life loss. Furthermore, Equation (10) showed that the consequence of a failure may be probabilistic, and in fact can be altered by suitable management measures, for example reducing P(T|S) by an effective evacuation system.

One may also compare the risk values of a range of options if a given hazard does happen. In such a case, its is valid to replace P(f) by the conditional probability P(f|H) where H denotes hazard occurrence.

CONSEQUENCE ANALYSIS

The simplest consequence analysis may be the assessment of the vulnerability of a building subject to extreme loading as a consequent of certain adjoining geo-failure. This can be handled by examining the probability of various limit states corresponding to partial collapse or damages that require expensive repairs. Fragility analysis, which is well-developed in the structural reliability area (Rosowsky and Ellingwood 2002, Stewart and Netherton 2005), may provide an objective basis for such an assessment.

The importance of a careful consequence analysis is illustrated in two similar-looking cofferdam designs presented in Figure 5. In Figure 5a, the cofferdam is for basement excavation in reclaimed land with high water table. The main activating force is the differential water pressure. If the old water main bursts, the water table may rise to an extent that the cofferdam may collapse. If there is no building next to the excavation, the consequence of collapse may appear to be just the economic cost associated with road closure. However, if the excavation is in a CBD area and there are gas and electricity lines below the road, excavation collapse may lead to gas leaks and power loss. This will trigger off a chain event that leads to closure of a number of block in a business district and the associated economic cost will be many times that of road closure. The author did come across such a disastrous scenario.

In Figure 5b, the cofferdam is part of an embankment levee that protects a heavily populated low-lying area. Extreme storm surge may leads to rising of the water level. Even large deformation of the cofferdam may trigger off collapse. This is because large quantity of water may leak into the "dry-side" of the cofferdam and undermine its foundation. The consequence of such a cofferdam failure is evidently unacceptably huge in a developed society. If one inherited the as-constructed infrastructure as illustrated in Figure 5b and it is economically not feasible to upgrade the design, then the reliability of the design has to be assessed when a storm approaches. This is because the probability of water level rising above a certain value knowing a storm is coming has to be higher than the unconditional probability of the water level exceeding the same elevation. Evidently, evacuation may be needed based on the latest value of $P_f|$storm.

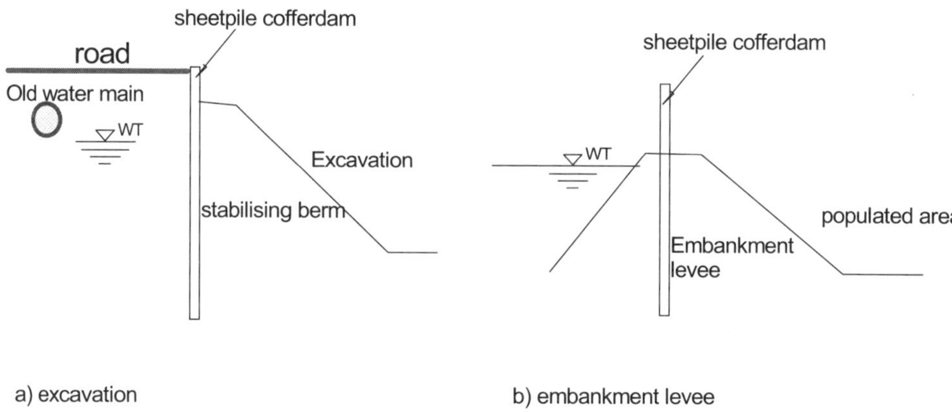

Figure 5. Cofferdam designs

The geotechnical behaviour of the failing soil mass also affects the consequence as illustrated in the slope-road-building problem as illustrated in Figure 6 below. The slope above the road is a cutting into an old fill slope. A conventional slip failure will only affect the road but not the housing estate below. Therefore, $P(S|f) \rightarrow 0$ in the context of equation (10). However, if the failing soil mass manifest static liquefaction, then the housing estate will be affected. Obviously the difference in consequences is too significant to be left to chance and a study of the old fill material may be a more rational approach. Alternatively, the ground water table need to be monitored in some way during heavy or extend period of rainfall so that the conditional probability of liquefaction failure can be assessed. This may then be used as a basis of triggering off an evacuation, if needed.

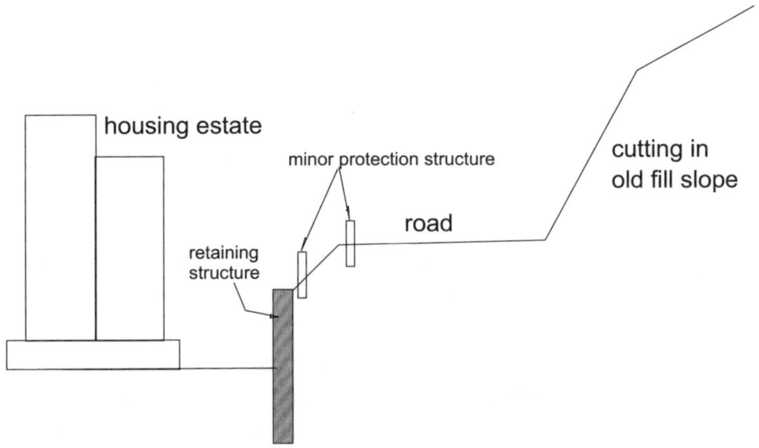

Figure 6. Slope-road-building

SUMMARY AND CONCLUSIONS

In order to incorporate safety in the geotechnical design process in a rational and consistent manner, consideration of reliability is a theoretical necessity. Some of the problems of reliability-based design are sometimes due to avoidance of probabilistic framework, or to a restrictive interpretation of probabilistic method.

A reliability-based risk analysis factors into both reliability and consequence(s) of failure. This methodology is particularly relevant in the design for extreme events. However, the consequences of a failure can be complicated and can be altered by well-engineered human intervention.

Although natural hazards causing failures may be considered as acts of nature, disaster are at least partly preventable as human intervention based on good engineering can reduce the disastrous consequence.

ACKOWLEDGEMENT

The issues on reliability analysis addressed in this paper originated with the many stimulating discussions made with my former colleague Dr K.S. Li. His contributions to this paper, and indeed all my work on reliability analysis, are gratefully acknowledged.

REFERENCES

Au S.W.C. (1998). "Rain induced slope instability in HK". *Engr Geology*, 51, 1-36.

Baynes F.J., Lee I.K., and Stewart I.E. (2002). "A study of accuracy and precision of some landslide risk analysis". *Australian Geomechanics*, Australian Geomechanics Society, 37(2), 149-156.

Cameron, T. (1992). "The Limit State Design of Pile Foundations". *MEngSc thesis, University College, University of New South Wales, Canberra.*

Christian J.T., Ladd C.C., and Baecher G.B. (1992). "Reliability and probability in stability". *In stability and Performance of Slope Embankments-II, Geotechnical Special Publication No. 3, Proc ASCE Specialiaty Conference*, Berkeley, 1071-1111.

ENV-1997-1(1995). "Geotechnical design - Part 1, General rules". *Eurocode 7, European Committee for Standardization (CEN).*

Ellingwood, B. R. (1999). "Probability-based structural design: prospects for acceptable risk bases". *Proc. 8th Intl Conference on Application of Statistics and Probability,* Vol. 1, Sydney, 11-18.

Frank, R. (2002). "Development and future of Eurocode 7 – Geotechnical design". *Proc. Intl Workshop on Foundation Design Codes and Soil Investigation in view of Intl Harmonization and Performance Based Design,* Tokyo, Japan, 51 – 58.

Green, R. & Becker, D. (2001). "National report on limit state design in geotechnical engineering: Canada". *Geotechnical News*, 19(3), 47-55.

Hasofer, A.M. and Lind, N.C. (1974) Exact and invariant second moment code format. *J. of Engr. Mechanics, ASCE*, 100[1], 111-121.

Honjo, Y. & Kusakabe, O. (2002). "Proposal of a comprehensive foundation design code: Geo-code 21 Ver. 2". *Proc. International Workshop on Foundation Design Codes and Soil Investigation in view of International Harmonization and Performance Based Design,* Tokyo, Japan, 95-103.

Kay J.N. (1993). "Probabilistic design of foundations and earth structures". *Proc of Conference on Probabilistic Methods In Geotechnical Engr.* Li & Lo (Eds), Balkema, 49-62.

Lo S.R., Li, K.S., and Cameron T. (1995). "Modelling of bias and non-normal distribution for Limit State Design of piles", *Australian Civil Engr Transactions,* CE37[4], 315-323.

Lo S.R., Li K.S. & Lee (1992). "Limit state design of pile foundation: a probabilistic appraisal". *6th ANZ Conference on Geomechanics*, Christchurch, New Zealand, Feb, 1992, 452-457.

Lo S.R. (2005). "Characterisation of Hong Kong Soils: Laboratory Testing and Geotechnical Properties". *2nd Hong Kong Soils Symposium.* Hong Kong Centre of Professional Development and Reseach. In Print.

Li K.S. and White W. (1987). "Probabilistic characterization of soil profiles". *Research Report No. 19,* Dept of Civil Engr, University College, University of New South Wales, 30p.

Li K.S. and Lee (1991). "The assessment of geotechnical safety". *Selected topics in Geotechnical Engineering: Lumb Volume.* Editor K.S. Li, University College University of NSW, 195-229.

Lumb P. (1991). "Statistical soil mechanics". *Selected topics in Geotechnical Engineering: Lumb Volume.* Editor K.S. Li, University College University of NSW, 155-194.

Morgentern N. R. (1995). "Managing risk in geotechnical engineering". *In Proc 10th Pan Americal Regional Conference in Soil Mech and Fdn Engr.*, Guadalajara, Mexico.

Oveson, N.K. (1995). "Eurocode 7 for geotechnical design". *Proc Broms Symposium on Geotechnical Engr.*, 333-360.

Paloheimo, E. & Hannus, M. (1974) Structural design based on weighted fractiles. *J. of Structural Div., ASCE*, 100[7], 1367-1378.

Phoon K.K., Becker D.E., Kulhawy F.H., Honjo Y., Oversoen N.K., and Lo S.R. (2003). "Why consider reliability analysis for geotechnical limit state design?". *In Proc of Intl workshop on Limit State Design in Geotechnical Engr Practice: LSD2003.* Jun 2003, MIT, Massachusetts, published by Word Scientific, ISBN 981-238-784-6(pbk),in CD

R57 (2002). "Design of reinforced soil walls". Quality Assurance Specification, *Road and Traffic Authority, New South Wales*, Australia.

Rosowsky D.V. and Ellingwood B.R. (2002) "Performance based engineering of wood frame housing: a fragility analysis methodology". *J of Structural Engr, ASCE*, 128[1], 32-38.

Semple, R.M., and Rigden, W.J. (1984). "Shaft capacity of driven piles in clay". *Analysis and Design of Pile Foundations, ASCE symposium,* San Francisco, 1984, 59-79.

Simpson, B. & Driscoll, R. (1998). "Eurocode 7: a commentary". *Construction Research Communications Ltd.*, Watford, Herts.

Stewarts M.G and Netherton M.D. (2005). "Risk assessment and mitigation of blast damage to built infrastructure". *Recent advances in counter-terrorism technology and infrastructure protection. Proc of 2005 Science, Engr, and technology Summit.* Ed. Mendis, Lai and Dawson, Australia Homeland Security Research Centre. 67-82.

Vanmarcke E.H. (1997). "Probability modeling of soil profiles". *J of Geot Engr, ASCE,* 103[11], 1227-1246.

Whitman, R. V. (1984). "Evaluating calculated risk in geotechnical engineering". *Journal of Geotechnical Engineering, ASCE*, 110(2), 145 - 188

APPENDIX

The action, S and resistance, R, of a pile is treated as random variables. The limit state equation is expressed as:

$$\gamma_S \mu_S \leq \gamma_R \mu_R \tag{12}$$

where μ_X and V_X denote mean and COV of random variable X. To achieve a target reliability index, β, the pertinent equations based on normal distribution are:

$$\gamma_R = 1 + \alpha_R \beta . V_R \tag{13}$$

$$\gamma_S = 1 + \alpha_S \beta . V_S \tag{13a}$$

$$\alpha_R = \frac{-F_0 . V_R}{\sqrt{F_0^2 V_R^2 + V_S^2}} \tag{14}$$

$$\alpha_S = \frac{V_S}{\sqrt{F_0^2 V_R^2 + V_S^2}} \tag{14a}$$

$$F_0 = \frac{1 + \beta\sqrt{V_S^2 + V_R^2 - \beta V_R^2 V_S^2}}{1 - \beta^2 V_R^2} \tag{15}$$

Combining equations (8a) and (12) leads to the following equations for calculating partial safety factor, f_R, for pile resistance.

$$\gamma_R \mu_R = \gamma_R B_R R_{pu} = \frac{R_{pu}}{f_R} \tag{16}$$

$$f_R = \frac{1}{\gamma_R B_R} \tag{17}$$

For non-normal distribution, the PH method of normal tail approximation (Paloheimo and Hannus 1974) will be used to transform a non-normal distribution into an equivalent normal distribution of the same probability context in terms of β. In lieu of V_R, an equivalent value, V'_R as defined below, is to be used in equations (14) to (17).

$$V'_R = \frac{1 - \exp(1 - [\beta . \varepsilon + \varepsilon^2/2])}{\beta} \tag{18}$$

For a given design scenario, we have COV(B) and B_μ. One can calculate V'_R with equation (18) as V_R = COV (R) = COV(B). For a given target β, we can calculate the partial factor, f_R, from the above equations.

International Conference on Geotechnical Engineering for Disaster Mitigation & Rehabilitation
Chu, Phoon & Yong (eds) © 2005 World Scientific Publishing Company ISBN 981-256-469-1

DENSIFICATION EFFECT OF GRANULAR PILES IN LIQUEFACTION MITIGATION

M. R. MADHAV
Professor Emeritus, Jawaharlal Nehru Technological University
Hyderabad 500 072, India

A. MURALI KRISHNA
Research Scholar, Indian Institute of Science
Bangalore 560 012, India

In this paper liquefaction mitigation of the ground treated by granular piles (GP) is assessed considering the pore pressure generation and dissipation accounting for both the densification and drainage effects of GPs. Seed and Booker (1977) approach for pore-water pressure generation and dissipation is applied, with some modifications for evaluating the densification effect of GPs, incorporating modifications in coefficients of volume change and permeability, to the analysis of columnar gravel drains under a variety of earthquake conditions.

INTRODUCTION

Deposits of loose granular soils get affected by ground vibrations induced by earthquakes or cyclic loading, resulting in large total and differential settlements of the ground. Saturated granular material, when subjected to cyclic loading involving the reversal of shear stress, tends to compact or densify. In cases where the soil consists of loose granular materials and high water table, the tendency to get densified may result in the development of excess hydrostatic pore water pressures of sufficient magnitude to cause liquefaction of the soil, resulting in settlements and tilting of structures. A qualitative understanding of the mechanism underlying the liquefaction of saturated sands, subjected to cyclic loading, such as that induced by earthquakes, had been recognized widely since first being examined by Casagrande in 1936 (Martin et al. 1975). For a saturated sand under undrained conditions, the tendency for volume reduction during each cycle of loading results in a corresponding progressive increase in pore-water pressure. If the magnitude of pore-water pressure generated equals the confining pressure, the effective stress becomes zero, the soil is said to have liquefied. Pore pressure buildup leading to liquefaction may be due to static or cyclic stress applications and the possibility of its occurrence depends on the void ratio or relative density of sand and the confining pressure (Seed 1979). A possible method of stabilizing a soil deposit susceptible to liquefaction is to install a system of gravel or rock drains (Seed and Booker 1977).

It is observed that granular piles are effective in mitigating liquefaction damage due to the reinforcement effect and drainage facility. Adalier and Elgamal (2004) reviewed the current state of the stone column technologies as a liquefaction countermeasure. Due to the installation of gravel drains, the generated pore water pressure due to repeated loading may get dissipated almost as fast as they are generated. In addition to these reinforcement and drainage effects, densification of the ambient soil around the rammed granular pile (RGP) should also be considered for total and better evaluation of improvement. Granular piles are of the displacement type and hence densify in situ ground during the process of installation (Madhav, 2001). The effect of densification is manifested through an increase in the coefficient of earth pressure at rest and in the values of modulus of deformation of the soil (Ohbayashi et al. 1999). Densification by RGP causes increase in deformation moduli and decreases in the coefficients of permeability and volume change. Further, the densification effect decreases with distance from the center of the compaction point and may become negligible at the periphery of unit cell. Thus, the coefficient of horizontal permeability, $k_h(r)$, and volume change, $m_v(r)$, of the soil around the granular can be considered to vary with distance, r, from the center of the granular pile.

In this paper the pore pressure generation and dissipation of the treated ground under earthquake conditions is analyzed to quantify the densification effect of ground treatment with RGP. Theory of pore-water pressure generation and dissipation developed by Seed and Booker (1977) is applied, with some modifications for evaluating the densification effect of RGPs, to the analysis of columnar gravel drains under a variety of earthquake conditions. Linear variations of the coefficients of permeability and volume change with distance are adopted.

RAMMED GRANULAR PILES – DENSIFICATION EFFECT

Ground improvement by means of granular piles/stone columns/geopiers, which is associated with partial substitution of the *in–situ* soil, originated in sixties. Granular pile ground provides increased bearing capacity, significant reduction in settlement, free drainage, increase of liquefaction resistance, *etc*. Granular piles are installed by vibro-compaction, vibro-replacement, cased bore hole (rammed stone columns/RGP) or by simple auger boring methods. RGP are installed into the ground by partial or full displacement methods and by ramming in stages, using a heavy falling weight, within a 'pre-bored casing' or 'driven closed end casing', retracting the casing pipe stepwise. In the latter case driving of closed end tube itself densifies the surrounding soil. Ramming action of the falling weight tends to ram the stone into the sides of the hole, which further densifies and reinforces the ground. Densification and reinforcement effects cause modifications in the properties of the in situ soil. This densification and modification in the soil parameters of the ground are not uniform over the entire zone of surrounding ground and is function of the distance from the point of densification. During the process of installation of RGP, the soil adjacent to and in the vicinity of the point of treatment gets densified most. This densification effect decreases with the distance form the point of densification. Thus, the densification effect of the ground, due to the installation of granular pile, is maximum near the periphery of the granular pile and decreases with the distance from the granular pile. Densification by RGP causes increase in deformation moduli and decrease in the coefficient of permeability and coefficient of volume change.

Sujatha (1998) analyzed several field test data and postulated linear and exponential variations of the stiffness of soil with distance as limits for the densification effect. Similar variations are considered for the reductions of flow and deformation parameters of the in situ soil (coefficients of permeability and volume change). This reduction will be most at the point of densification and decays with distance towards the periphery of the unit cell reaching the original in-situ values at the farthest end of the unit cell.

Linear variation of the flow parameters

A linear variation of the coefficient of horizontal permeability, k_h, with distance from center of RGP is considered with k_h minimum at the edge of RGP (r = a) and maximum at the periphery of the unit cell (r = b) (Fig. 1). The general expression for the variation of the coefficient of permeability at a distance 'r' from the centre of the granular pile considering linear variation can be expressed as

$$k_h(r)=\left(\frac{k_{sb}-k_{sa}}{b-a}\right).(r-a)+k_{sa} \tag{1}$$

where k_{sa} and k_{sb} are the coefficients of permeability at the edge of RGB (r=a) and at the edge of the unit cell (r=b) respectively. Non-dimensionalising with k_{hi}, initial in-situ value of coefficient of permeability

$$R_k(r)=\left(\frac{R_{kb}-R_{ka}}{b-a}\right).(r-a)+R_{ka} \tag{2}$$

where $R_k=k_h/k_{hi}$, $R_{kb}=k_{sb}/k_{hi}$ and $R_{ka}=k_{sa}/k_{hi}$

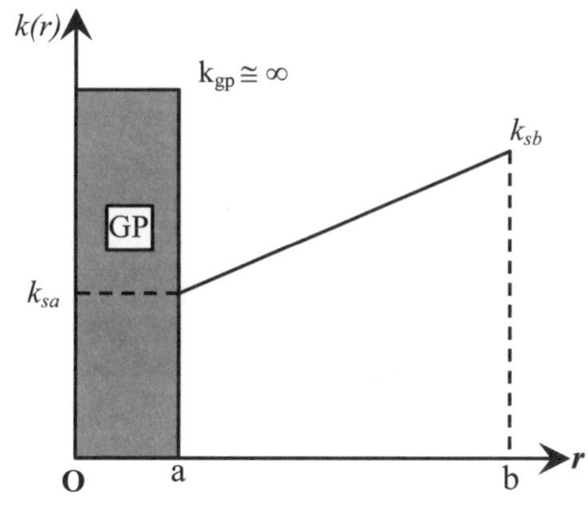

Figure 1. Variation of coefficient of permeability of soil with distance.

Similarly, the expression for linear variation of coefficient of volume change, m_v, is

$$m_v(r)=\left(\frac{m_{vb}-m_{va}}{b-a}\right).(r-a)+m_{va} \tag{3}$$

where m_{va} and m_{vb} are respectively the coefficients at distances 'a' and 'b' from the centre. Non-dimensionalizing with 'm_{vi}', initial in situ value of coefficient of volume change, one gets

$$R_{mv}(r)=\left(\frac{R_{mb}-R_{ma}}{b-a}\right).(r-a)+R_{ma} \tag{4}$$

SEED & BOOKER (1977) MODEL

In most practical cases, the horizontal permeability of sand/gravel drains will be several times greater than its vertical permeability and spacing between vertical drains is closer than the distance required for pore water to drain vertically towards the free surface. Furthermore, many natural deposits of sand are interspersed with narrow horizontal layers of relatively impermeable silt, which may severely inhibit vertical drainage. For these reasons, the dominant mechanism in the operation of a gravel drain system often is one of pure horizontal drainage (Seed and Booker 1977).

For flow into gravel drain, assuming pure radial flow, and constant coefficients of permeability (k_h) and volume compressibility (m_v), the governing equation for the phenomenon can be written as (Seed and Booker 1977):

$$\frac{k_h}{\gamma_w.m_v}\left(\frac{1}{r}\frac{\partial u}{\partial r}+\frac{\partial^2 u}{\partial r^2}\right)=\frac{\partial u}{\partial t}-\frac{\partial u_g}{\partial N}.\frac{\partial N}{\partial t} \tag{5}$$

where u is the excess pore pressure at a radial distance r, from the centre, t is time, γ_w the unit weight of water, u_g = peak excess hydrostatic pore-water pressure generated by the earthquake; and

$$\frac{\partial u_g}{\partial N}=\frac{\sigma'_o}{\alpha\pi N_l}\frac{1}{\sin^{2\alpha\text{-}1}\left(\frac{\pi}{2}r_u\right)\cos\left(\frac{\pi}{2}r_u\right)} \tag{6}$$

where $r_u = u/\sigma'_o$ = the pore pressure ratio, $/\sigma'_o$ = the initial mean bulk effective stress for axi-symmetric conditions or the initial vertical effective stress for simple shear conditions; N_l is the number of cycles required to cause liquefaction; and α = an empirical constant which is a function of the soil properties with typical average value of 0.7. The irregular cyclic loading induced by an earthquake is converted to an equivalent number, N_{eq}, of uniform cycles at an amplitude of 65% of the peak cyclic shear stress, i.e. $\tau_{cyc} = 0.65.\ \tau_{max}$, occurring over a duration of time, t_d, and

$$\frac{\partial N}{\partial t}=\frac{N_{eq}}{t_d} \tag{7}$$

NEW MODEL CONSIDERING DENSIFICATION OF GROUND

The governing equation of Seed and Booker (1977) (Eq. 5) is modified to include the densification effect of RGP in dissipating the excess pore-water pressures. In the ground treated with RGP, coefficient of permeability (k_h) and coefficient of volume compressibility (m_v) are considered to depend on the distance, r, instead being constant values. Considering an element of soil, in polar coordinates as shown in Figure 2 through which laminar flow occurs. Using Darcy's law, from the expression for the flow into radial direction the basic flow equation is obtained as,

$$\frac{k(r)}{r}\frac{\partial h}{\partial r}+k(r)\frac{\partial^2 h}{\partial r^2}+\frac{\partial k(r)}{\partial r}\frac{\partial h}{\partial r}=\frac{S}{1+e}\cdot\frac{\partial e}{\partial t} \tag{8}$$

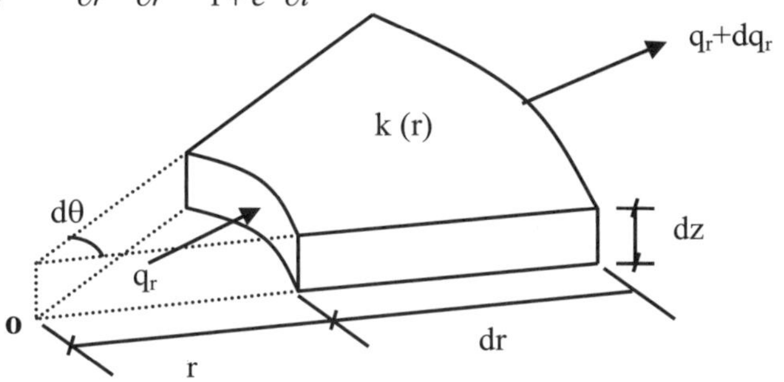

Figure 2. Flow through soil element.

From the stress-strain relations and definitions of coefficient of compressibility (a_v) and coefficient of volume change (m_v) and for S =1(fully saturated),

$$\frac{1}{1+e}\cdot\frac{\partial e}{\partial t}=\frac{m_v}{1+e}\cdot\left(\frac{\partial u}{\partial t}-\frac{\partial \sigma}{\partial t}\right) \tag{9}$$

From Eq. 9 and with $h = u/\gamma_w$; (u = excess pore pressure; γ_w = unit weight of water) Eq. 9 is rewritten as:

$$\frac{k_h(r)}{\gamma_w.m_v(r)}\left(\frac{1}{r}\frac{\partial u}{\partial r}+\frac{\partial^2 u}{\partial r^2}\right)+\frac{1}{\gamma_w.m_v(r)}\cdot\frac{\partial(k_h(r))}{\partial r}\cdot\frac{\partial u}{\partial r}=\frac{\partial u}{\partial t}-\frac{\partial \sigma}{\partial t} \tag{10}$$

where $k_h(r)$ and $m_v(r)$ are defined in Eqs. (1) and (3) respectively. Considering rate of change of total stress $\frac{\partial \sigma}{\partial t}$ as the rate of pore pressure generation

$$\frac{\partial u_g}{\partial t}=\frac{\partial u_g}{\partial N}\cdot\frac{\partial N}{\partial t} \tag{11}$$

The final form of Eq. 10 is,

$$\frac{k_h(r)}{\gamma_w.m_v(r)}\left(\frac{1}{r}\frac{\partial u}{\partial r}+\frac{\partial^2 u}{\partial r^2}\right)+\frac{1}{\gamma_w.m_v(r)}\cdot\frac{\partial(k_h(r))}{\partial r}\cdot\frac{\partial u}{\partial r}=\frac{\partial u}{\partial t}-\frac{\partial u_g}{\partial N}\cdot\frac{\partial N}{\partial t} \tag{12}$$

Eq. 12 is a more generalized equation considering all forms of non-homogeneity. In non-dimensional form, with normalized pore pressure W = u/σ'_o, which is the same as the pore pressure ratio r_u (Seed and Booker 1977) and $W_g = u_g/\sigma'_o$, Eq. 12 becomes

$$\frac{k_h(r)}{\gamma_w . m_v(r)}\left(\frac{1}{r}\frac{\partial W}{\partial r}+\frac{\partial^2 W}{\partial r^2}\right)+\frac{1}{\gamma_w . m_v(r)}.\frac{\partial(k_h(r))}{\partial r}.\frac{\partial W}{\partial r}=\frac{\partial W}{\partial t}-\frac{\partial W_g}{\partial N}.\frac{\partial N}{\partial t} \tag{13}$$

Non-dimensionalising the terms 'r' with 'R' (=r/b) and 't' with 'T' (=t/t_d), Eq. 13 becomes

$$T_{bd}.\frac{R_k(R)}{R_{mv}(R)}\left(\frac{\partial^2 W}{\partial R^2}+\frac{1}{R}\frac{\partial W}{\partial R}\right)+\frac{T_{bd}.}{R_{mv}(R)}\frac{\partial(R_k(R))}{\partial R}.\frac{\partial W}{\partial R}=\frac{\partial W}{\partial T}-\frac{\partial W_g}{\partial N}N_{eq} \tag{14}$$

where

$$\frac{\partial W_g}{\partial N}=\frac{1}{\alpha\pi N_l}.\frac{1}{\sin^{2\alpha-1}\left(\frac{\pi}{2}r_u\right)\cos\left(\frac{\pi}{2}r_u\right)} \tag{15}$$

and

$$T_{bd}=\left(\frac{k_{hi}}{\gamma_w}\frac{t_d}{m_{vi}}\right).\frac{1}{b^2} \tag{16}$$

Eq. 14 is solved numerically using a finite difference approach, discretizing the unit cell radially into a number of elements, for the appropriate boundary and initial conditions (Murali Krishna 2003).

Boundary Conditions

The material in the drains is far more permeable than the surrounding sand layer. So the excess pore-water pressure in the drain is effectively zero i.e., at r = a or R = a/b, u = 0 or W= 0. (Gravel drain is infinitely permeable). At the outer boundary of the unit cell, due to symmetry, rate of pore-water pressure in the radial direction is zero, i.e.,

$$\text{at } r = b \text{ or } R = 1, \quad \frac{\partial u}{\partial r}=0 \text{ or } \frac{\partial W}{\partial R}=0 \tag{17}$$

Initial Condition

At t = 0 or T = 0, pore pressures at all the nodes in the soil are equal to the average of pore water pressure generated over the initial time period of 'dt' (or 'dT'). i.e average of pore-water pressure generated over an initial cycle interval, dN.

$$W_{g\ at\,T=0}=\left(\frac{u_g}{\sigma'_o}\right)_{at\,t=0}=\frac{1}{2}\left(\frac{2}{\pi}\text{arc sin}\left(\frac{dN}{N_l}\right)^{1/2\alpha}\right) \tag{18}$$

RESULTS AND DISCUSSION

Under the assumption of purely radial flow the pore pressure ratio, w= u/σ'_o, throughout the sand depend on the dimensionless parameters: a/b = a ratio representing the geometric configuration of the RGPs; N_{eq}/N_l = cyclic ratio characterizing the severity of the earthquake shaking in relation to the liquefaction characteristics of the sand; T_{bd}, relating the duration of the earthquake to the consolidation properties of the sand; α, a parameter characterizing the shape of the pore pressure generation curve. It is found that for many materials $\alpha = 0.7$ fits the experimental data well, and this value is adopted (Seed and Booker, 1977); $R_k(r)$ & $R_{mv}(r)$, functions of radial distance, r, defining the variations of coefficient of horizontal permeability and volume compressibility of soil respectively.

If coefficients of permeability and volume change of the surrounding soil are unchanged due to installation of RGP, the condition represents the 'No Densification' case. In this case the ambient soil will have the initial k_h and m_v (i.e., $R_{ka}= R_{kb}= R_{ma}= R_{mb}= 1$). The effect of a/b ratio on maximum pore water pressure is shown in Figure 3 for cyclic ratio (N_{eq}/N_l) of 2 and a range of a/b (0.1 to 0.4) with T_{bd} value of 1. Maximum pore pressure ratio $W_{max}(T)$, i.e. maximum value of u/σ'_o throughout the layer is plotted against T (t/t_d). These results are for no densification effect and agree closely with results of Seed and Booker (1977). For cyclic ratio, (N_{eq}/N_l) equals to 2, if no drains are present (a/b = 0) the soil liquefies at $T_l = 1/(N_{eq}/N_l)$ i.e., at T = 0.5. For a/b = 0.1, initial liquefaction is deferred for a period of 'T' but eventually occurs at about T=0.6, the liquefied state, $W_{max} = 1$, continues until the end of the period of strong shaking. Thereafter it decreases as pore-water pressure dissipates. It can be observed that at higher a/b values, initial liquefaction may be prevented and the maximum pore pressure ratio decreases with increase in a/b value.

Effect of cyclic ratio on maximum pore pressure ratio, without densification effect, is shown in Figure 4. A range of cyclic ratios, N_{eq}/N_l, 1 to 5 for a/b of 0.3 and T_{bd} of 1 is considered. As cyclic ratio increases, pore water pressure increases and leads to liquefaction depending on the generation and dissipation of pore pressures. The time at which the ground liquefies decreases with increase of cyclic ratio as per T_l ($=1/(N_{eq}/N_l)$). If attained, the liquefied state continues until the end of the period of strong shaking. i.e., T=1. Thereafter pore pressure ratio decreases gradually. It can be observed from the Figure 4 that the peak value of W_{max} increased from about 0.27 to 0.52 for an increase in N_{eq}/N_l from 1 to 2. For N_{eq}/N_l values 3, 4 and 5, W_{max} reaches a value of one, at T of about 0.62, 035 and 0.24 respectively.

Densification effect with respect to coefficient of volume change only (i.e., no change in coefficient of permeability ($R_{ka}=R_{kb}= 1$)) on maximum pore pressure ratio, W_{max}, is shown in Figures 5 and 6. Figure 5 shows the effect of densification at the near end, R_{ma}, on W_{max}, for no modification in the coefficient of volume change at the farthest end ($R_{mb}=1$). A range of R_{ma}(1-0.1) for $T_{bd} = 1$, a/b = 0.3 and $N_{eq}/N_l = 2$ is considered. It can be observed that the W_{max} decreases with decrease in R_{ma}. Maximum pore pressure ratio, W_{max}, decreased from 0.525 to 0.375 for a decrease in R_{ma} from 1 to 0.1. The effect of densification at the farthest end (R_{mb}) is shown in Figure 6, considering the densification at the near end as R_{ma}=0.3. A range of R_{mb} 1 to 0.5 for $T_{bd} = 1$, a/b = 0.3 and $N_{eq}/N_l = 2$ is considered. It can be observed that W_{max} decreases from 0.41 to 0.25 for a decrease in R_{mb} from 1 to 0.5. It can be concluded from Figures 5 and 6 that densification effects with respect to coefficient of volume change reduce the pore pressures and mitigates the liquefaction potential.

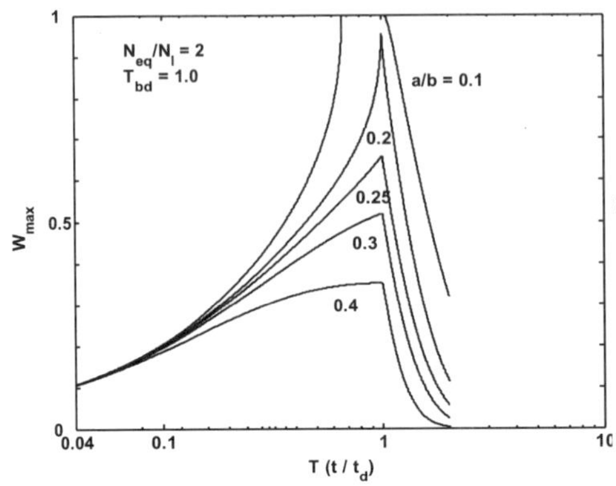

Figure 3. Effect of area ratio on W_{max} for 'No Densification' Case

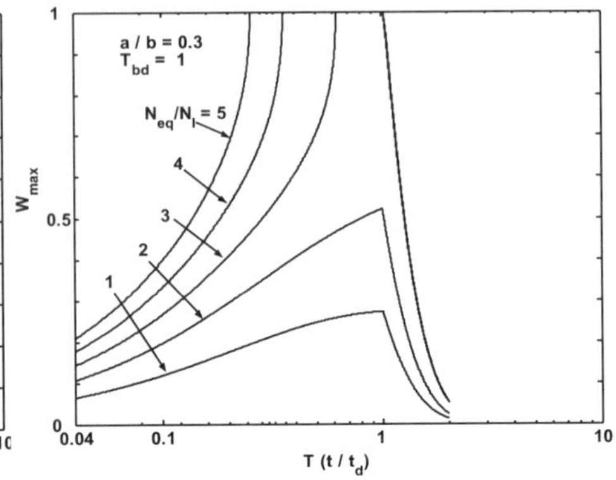

Figure 4. Effect of N_{eq}/N_l ratio on W_{max} for 'No Densification' Case

Densification effect in terms of coefficient of permeability only, i.e., no change in coefficient of volume change (R_{ma}=R_{mb}= 1), on maximum pore pressure ratio is shown in Figures 7 and 8. The effect of densification at the near end, R_{ka}, on W_{max} for no modification in the coefficient of permeability at the farthest end (R_{kb}=1) is shown In Figure 7. A range of R_{ka} (1-0.5) for T_{bd} = 1, a/b = 0.3 and N_{eq}/N_l = 2 is considered. The W_{max} increases from 0.525 to 0.675 for a decrease in R_{ka} from 1 to 0.8. For R_{ka}=0.6 the maximum pore pressure ratio reaches 1, i.e. liquefied state. Further decrease in R_{ka} will reduce the time to attain the liquefaction state. Densification effect at the farthest end (R_{kb}) on maximum pore pressure ratio is shown in Figure 8 considering the densification at the near end as R_{ka}=0.6. A range of R_{kb} 1 to 0.6 for T_{bd} = 1, a/b = 0.3 and N_{eq}/N_l = 2 are considered. It can be observed that the maximum pore pressure ratio is slightly affected by the densification at the farthest end, R_{kb}. It can be concluded from Figs. 8 and 9 that densification effects with respect to coefficient of permeability increase the pore pressures. This observation is revealing in that the densification effect with respect to coefficient of permeability alone results in a negative effect.

Figure 9 shows the densification effect on both the coefficients of permeability and volume change at near end on maximum pore pressure ratio, W_{max}. In this case the coefficients of permeability and volume change at the farthest end remain unaffected (R_{kb}= R_{mb}=1). A range of R_{ka} and R_{ma} (1-0.3) for T_{bd} = 1, a/b = 0.3 and (N_{eq}/N_l) = 2 is considered. Equal densification effect is considered for both R_{ka} and R_{ma}. Maximum pore pressure ratio increases from 0.525 to 0.62 for a decrease in R_{ka} and R_{ma} from 1 to 0.8. W_{max} almost reaches 1, i.e. attains liquefaction state, at T =1 for R_{ka}=R_{ma}=0.6. Further decease in R_{ka} and R_{ma} reduces the T corresponding to W_{max}=1.

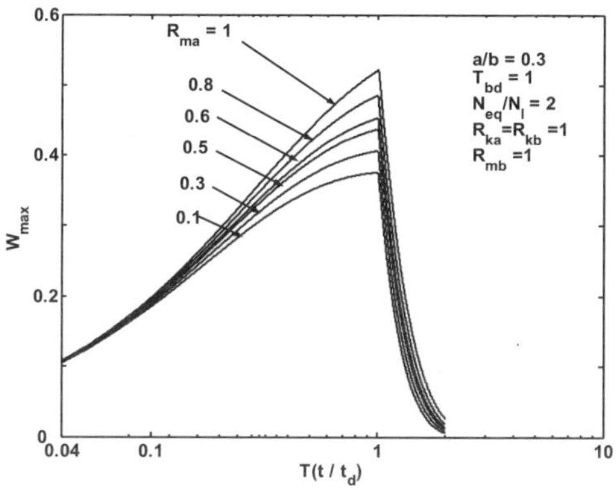

Figure 5. Effect of R_{ma} on W_{max}

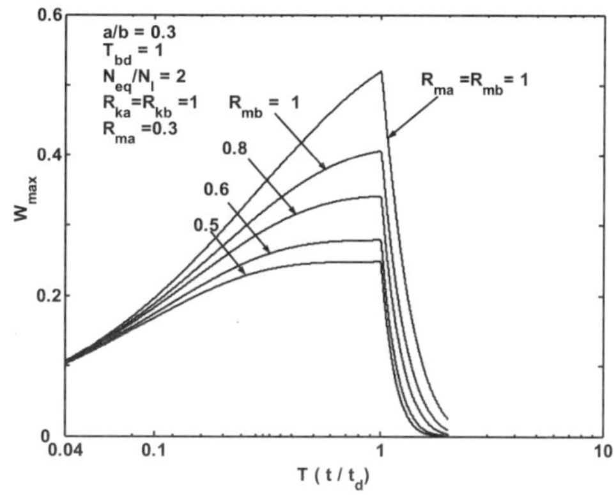

Figure 6. Effect of R_{mb} on W_{max}

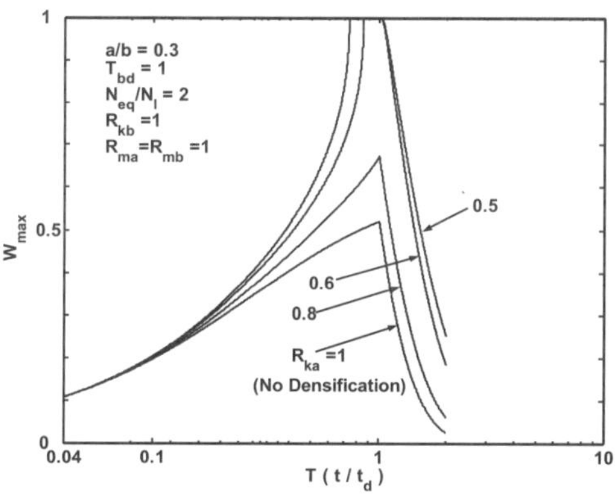

Figure 7. Effect of R_{ka} on W_{max}

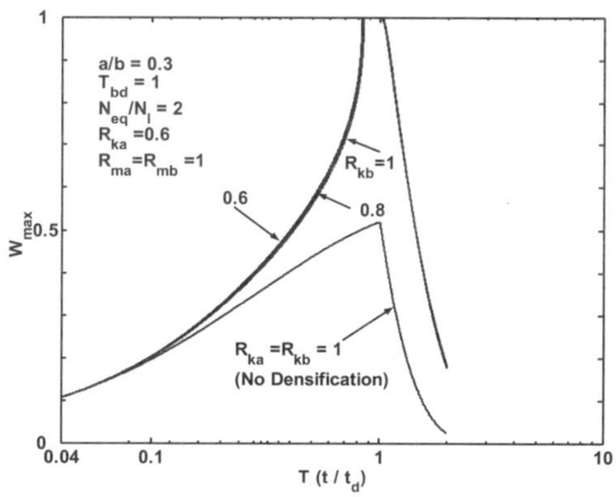

Figure 8. Effect of R_{kb} on W_{max}

The effect of densification at the farthest end with respect to both coefficients of permeability (R_{kb}) and volume change (R_{mb}) on maximum pore pressure ratio, W_{max}, is shown in Figure 10, considering densification at the near end as R_{ka}= R_{ma}=0.6. Same densification effect is considered for both R_{kb} and R_{mb} (1 – 0.7). W_{max} reaches 1 indicating liquefaction state, at T =1 for R_{kb}= R_{mb}=1 i.e. for no densification at the farthest end. Densification at the farthest end, i.e. reduction in R_{kb} and R_{mb} (from 1 to 0.7) reduce W_{max} (from about 1 to about 0.57), and thus establishes that liquefaction of ground is prevented by overall densification of the ground.

CONCLUSIONS

An analysis of ground treated with gravel drains (stone columns) considering the effect of installation in densifying the ground is proposed. Both the coefficients of volume change and permeability are considered to be affected. In fact they decrease because of

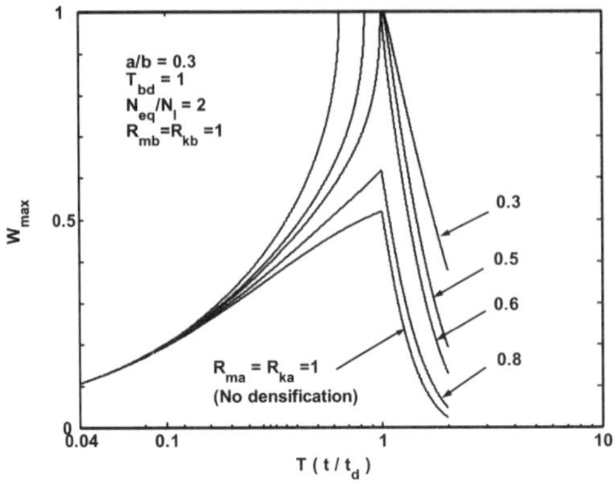

Figure 9. Effect of R_{ma} and R_{ka} on W_{max}

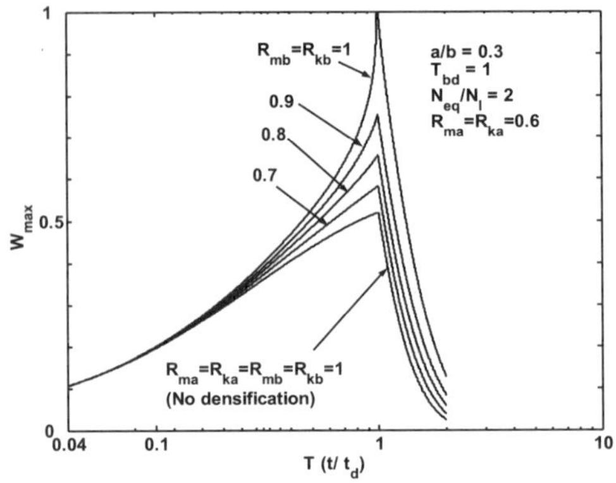

Figure 10. Effect of R_{mb} and R_{kb} on W_{max}

densification. Maximum pore pressure ratio decreases with closer spacing of gravel drains and increases with increase in cyclic ratio (N_{eq}/N_l). Densification effect on the coefficient of volume change alone is positive in that the maximum induced pore water pressure ratios get reduced. Densification effect on the coefficient of permeability alone increases maximum pore water pressure ratios giving a negative effect. Densification effect on both coefficients of permeability and volume change gives a slightly negative or positive effect depending on the extent of densification.

REFERENCES

Adalier, K. and Elgamal, A. (2004). "Mitigation of liquefaction and associated ground deformations by stone columns", *Engineering Geology*, Vol. 72, No. 3-4, 275-291.

Madhav, M.R. (2001). "Engineering of ground for earthquake disaster mitigation", *Proc. Indian Geotechnical Conference 2001*, Indore, India, 29-34.

Martin, W.D.L. Finn and H.B. Seed. (1975). "Fundamentals of liquefaction under cyclic loading", *J. of the GE Division*, ASCE, 101(GT5), 425-438.

Murali Krishna, A. (2003). "*Liquefaction mitigation of loose sand deposits by rammed granular piles*", M.Tech Thesis, Indian Institute of Technology, Kanpur, India.

Ohbayashi, J., Harda, K. and Yamamoto, M. (1999). "Resistance against liquefaction of ground improved by sand compaction pile method", *Earthquake Geotechnical Engg.,* ed. by Seco e Pinto, Balkama, Rotterdam, 312p.

Seed,. H.B. (1979). "Soil liquefaction and cyclic mobility evaluation for level ground during earthquakes", *J. of the GE Division*, ASCE, Vol. 105, No. 2, 201-255.

Seed. H.B., and John R. Booker. (1977). "Stabilization of potentially liquefiable sand deposits using gravel drains", *J. of GE Division*, ASCE, Vol. 103, No. 7, 757-768.

Sujatha, K.P. (1998). "*Analysis of reinforcement and densification effects in improved ground*", M.Tech Thesis, Indian Institute of Technology, Kanpur, India.

International Conference on Geotechnical Engineering for Disaster Mitigation & Rehabilitation
Chu, Phoon & Yong (eds) © 2005 World Scientific Publishing Company ISBN 981-256-469-1

PUBLIC EDUCATION OF TSUNAMI DISASTER MITIGATION AND REHABILITATION PERFORMED IN JAPANESE PRIMARY SCHOOLS

HIDEKI OHTA, THIRAPONG PIPATPONGSA AND TORU OMORI
Department of International Development Engineering, Tokyo Institute of Technology, Tokyo 152-8552, JAPAN

Tsunami disasters are not necessarily very few events in the Asian coastal areas and yet they are not well recognized by the people living in the coastal areas. The tsunami disaster in 2004 triggered by Sumatra Earthquake proved that the people in the coastal areas should have reasonable degree of basic knowledge what tsunami is all about. The public education performed in Japanese primary schools during a period of 1937-1947, to introduce the children what the traditional disaster mitigation and rehabilitation measure are, helped the Japanese people have an idea how to save themselves when tsunami could sweep the shores. This paper describes an event related to a tsunami disaster in one of the coastal areas in Japan and how it was introduced to the primary school children in Japan.

INTRODUCTION

In order to minimize the sad events possibly triggered by the tsunami disasters, people living in the coastal areas should have some basic knowledge of enormous tidal waves, tsunami, which could strike them even from the other side of an ocean, for examples, the tsunami caused by the Sumatra Earthquake in 2004 which struck Sri Lanka, India and Africa and the tsunami caused by the Chile Earthquake which occurred 4:10 am (Japanese local time), 23rd May, 1960 and reached the eastern coast of Japan 22 hours and 30 minutes after earthquake occurrence. The tsunami caused by the Chile Earthquake traveled 17,000 km in one day and destroyed 142 human lives in the coastal area of Sanriku which periodically suffered from tsunami in 1611, 1677, 1896, 1933, 1960 and 1978.

It is not only Sanriku but also many other Japanese coastal areas that sequentially suffered from tsunami disaster. In one of those coastal areas, there was a village called Hiro which eventually became very famous among Japanese people because of the mitigation and rehabilitation measures taken by the people in Hiro village after severely suffering from a tsunami in 1854, a picture of which painted by Eisho Furuta is shown in Fig. 1. Their mitigation measures were effective enough to protect the main part of the village from another tsunami in 1946 that is about 100 years after the previous tsunami. The authors believe that their mitigation and rehabilitation measures are still worthy to be known by Asian people living in costal areas.

Figure 1. Picture of tsunami painted by Eisho Furuta in 1854
(By courtesy of Eshu Kikumoto, Yogen-ji)

SETTING FIRE TO RICE-STACKS READY FOR TRANSPORTATION

What happened in Hiro village at the time of tsunami in 1854 is partly described by a naturalized Japanese Yakumo Koizumi (English name before naturalization: Patrick Lafcadio Hearn 1850-1904) in an article "A living God" (1897).

Summary of the article is as follows: A man called Gihei (Gohei) Hamaguchi, the most influential resident of Hiro village, was at home on a small plateau overlooking a bay about which most of the village houses were. He became aware of something unusual in the offing when he saw the sea running away from the beach after he felt an earthquake. He knew all the traditions of the coast and realized that tsunami would be likely to come soon. But it was too late to send a message to the village before tsunami strikes the village. Then he made up his mind to set fire to a number of rice-stacks which were the crystal of farmers' labour of the year. Because of the fire, village people came up to the plateau in a hurry and found themselves saved from the violence of tsunami.

Fig. 2 shows a picture painted on a concrete wall constructed aiming at protecting Hiro village from possible tsunami disasters in the future. The picture explains what was done by Gihei Hamaguchi whose bronze statue standing in the village school garden is shown in Fig. 3.

Figure 2. Burning rice-stacks picture on a protective wall in Hiro

Figure 3. Gihei Hamaguchi overlooking the school garden in his village

Tsunezo Mitsuhashi Nakai (1907-1994), a primary school teacher, introduced this story in a schoolbook for 5th grade pupils of primary schools all over Japan. The schoolbook compiled by the state was read by all of the Japanese children during a period of 1937-1947. This helped Japanese people have an idea how to save themselves when tsunami could sweep the shores. On the board standing at a school garden in Hiro village, a replica of the schoolbook is displayed as shown in Fig. 4.

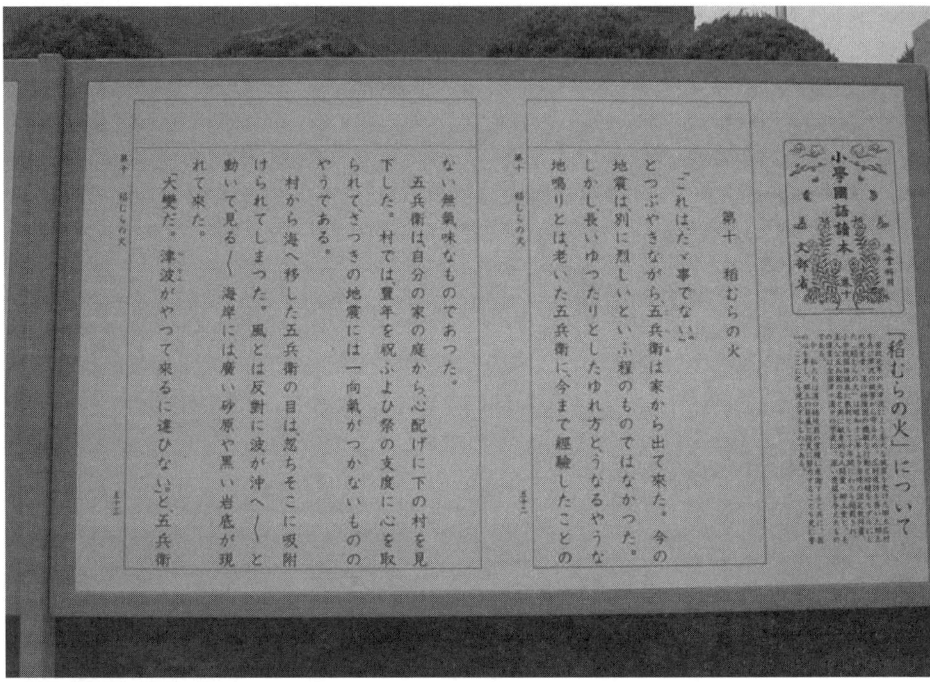

Figure 4. Replica of the schoolbook displaying the 1st and 2nd pages of the story written by Tsuneo Mitsuhashi Nakai (1907-1994)

Masanori Hamada, who is currently the president of the Japan Society of Civil Engineers, and his colleagues in JSCE tried to introduce the above mentioned story of Hiro village to the Indonesian school children immediately after the tsunami disaster as reported by Hamada, Kiyono, Kunizaki and Suzuki (2005). The authors believe that trials such as made by Hamada et al. are needed. The authors also believe that it is worthy to produce some booklets written in major languages spoken in the Asian region aiming at introducing what tsunami is all about in a way impressive and understandable to the school children in Asian coastal areas. Visual teaching aids in local languages should also be of importance.

PROTECTIVE EMBANKMENT AND TREE PLANTING

After the tsunami that swept main part of Hiro village in 1854, Gihei Hamaguchi sponsored village people for construction of a protective embankment 5m high and 20m wide at the base. The embankment was placed along the shoreline for a length of 600m protecting main part of the village as shown in Fig. 5. The construction work took a period from February, 1855 through December, 1858 continuously giving part-time jobs to the village people.

Figure 5. Protective embankment along the shoreline

In front of the embankment (sea side), they had an old stone wall (3.5m high and 2m wide at the top, not shown in Fig. 5) which was re-built by the village people. In a narrow space (9m wide) between the old stone wall and the newly constructed embankment, they planted black pine trees which could bear the sea breeze. Spindle trees were planted on the sea side slope of the embankment and wax trees on the other slope. The village people were wise enough to get some cash out of these trees afterwards. The cross section of the protective embankment in Hiro village is shown in Fig. 6.

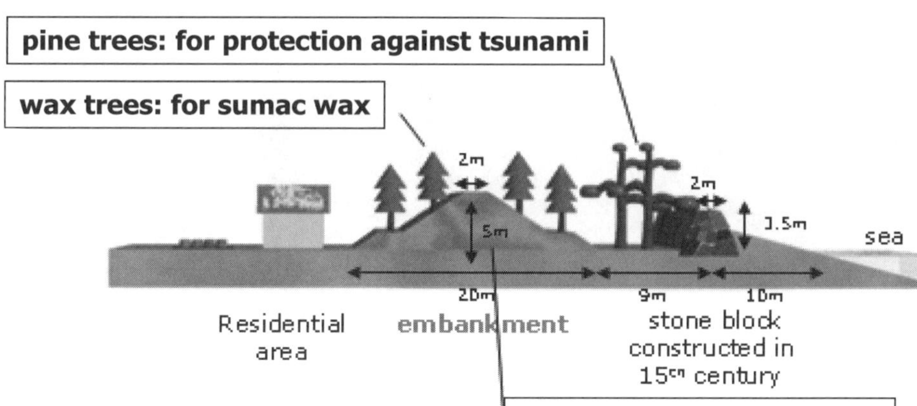

Figure 6. Cross section of the protective embankment in Hiro village

PROTECTIVE EMBANKMENT SAVING THE VILLAGE AFTER 100 YEARS

The protective embankment completed in 1958 was found to be of service even after 100 years. Another tsunami struck Hiro village in 1946, but the 100 years old embankment and trees were effective enough to protect the built-up main area of the village. Even now, the village people annually have a special gathering (i) to thank Gihei Hamaguchi's contribution to the village, (ii) to let the school children know what happened in the past and what were done by the people to protect themselves and (iii) to clean and maintain the embankment and trees by offering their hands working with the school children. Figs. 7 (a) and (b) show how the embankment looks like at present.

Figure 7. (a) and (b) Protective embankment in Hiro covered with pine trees

The above mentioned example of Hiro village is introduced in a form of colourful pamphlet with schematic diagrams and pictures edited by Kenshiro Tsumura (2003), see Fig.8.

Figure 8.Cover page of "Fire of rice-stacks and protective embankment in Hiro village" (Kenshiro Tsumura, 2003)

PROPOSALS

The example introduced in this paper is one of many case histories in Japan. Tree planting and embankment along the shoreline are Japanese traditional measures of tsunami disaster mitigation and rehabilitation. Figs. 9 (a) and (b) show a small-scaled embankment possibly placed along the shoreline to protect a small village, while Figs. 10 (a) and (b) show a large-scaled one to protect a seaside resort.

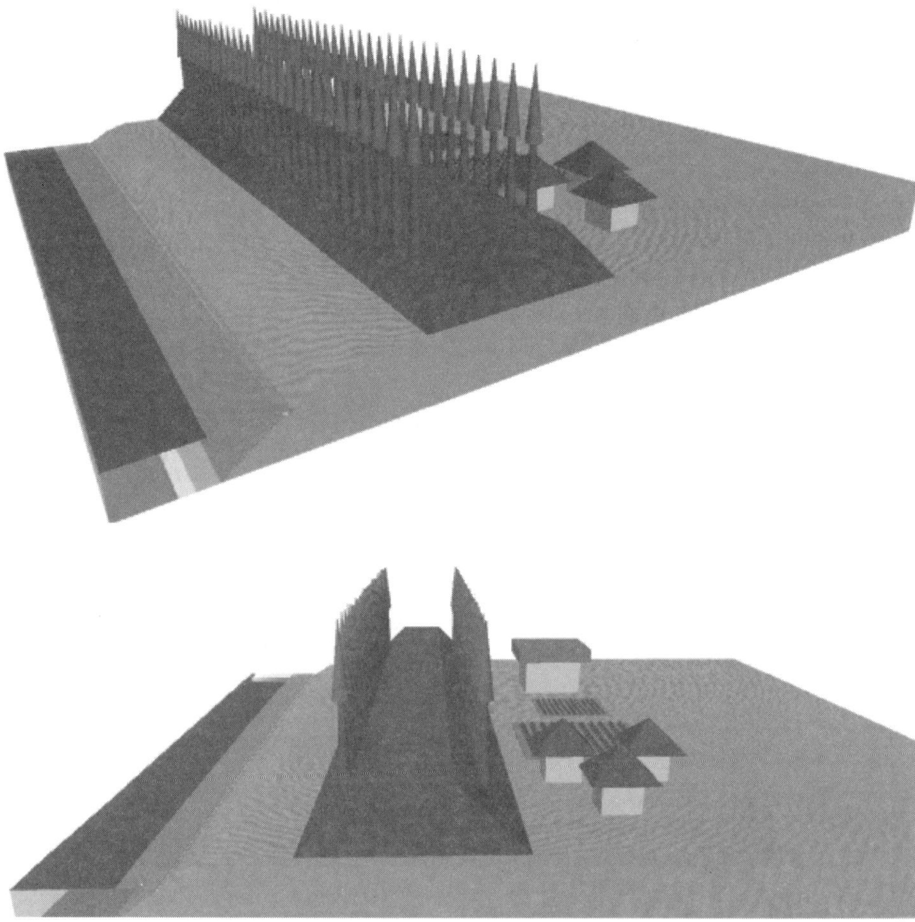

Figure 9. (a) and (b) A small-scaled embankment to protect a small village

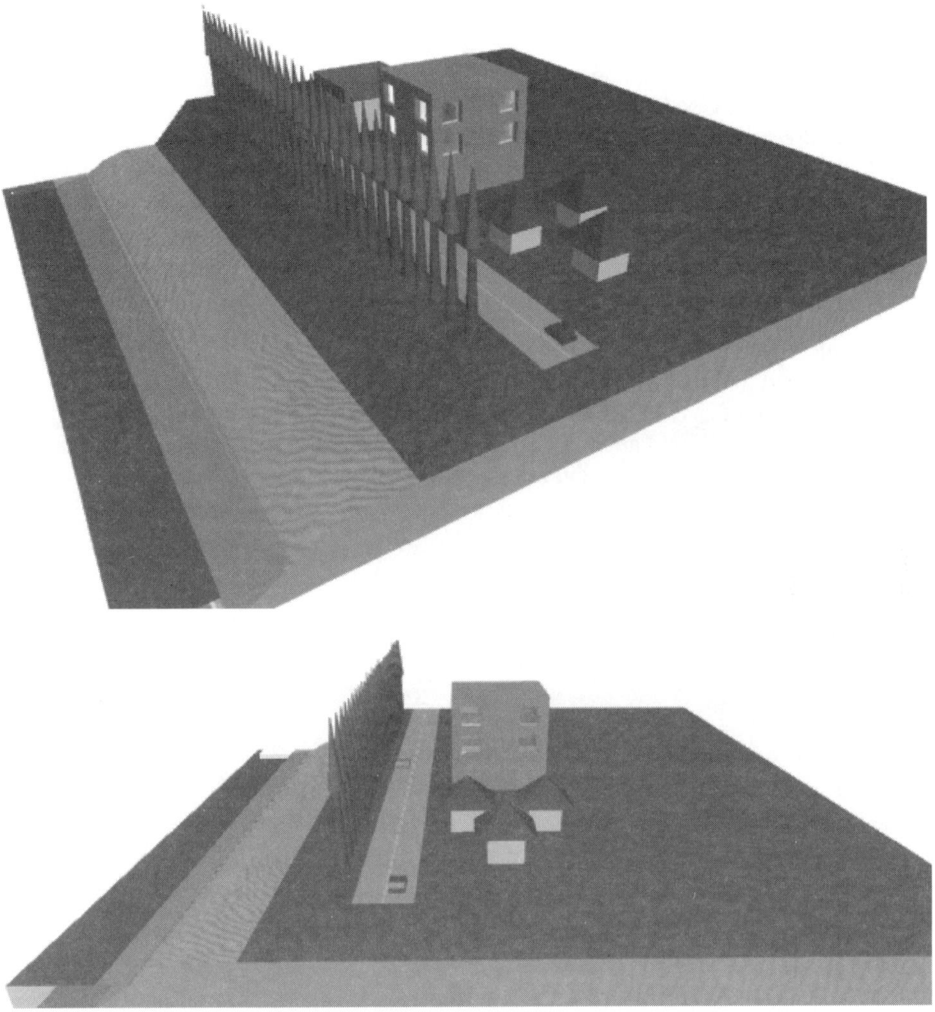

Figure 10. (a) and (b) A large-scaled embankment to protect a seaside resort

As the concluding remarks, the authors propose to take the case of Hiro village as a typical teaching material for the school children in the coastal areas in Asian region. The authors also propose to conduct a continuous series of campaign targeted on the school children as part of the joint project of geotechnical community in Asian region.

REFERENCES

Furuta, E.:*Ansei Monroku*, Yogen-ji, 1854 (in Japanese)

Hamada, M., Kiyono, J., Kunizaki, N. and Suzuki, T.: Why didn't you tell us earlier? – "Fire of rice-stacks", Support activities for the education of disaster mitigation in Banda Aceh-, Journal of the Japan Society of Civil Engineers, Vol. 90, No. 6, 43-46, 2005

Hearn, P. L.: A living God, included in *Gleanings in Buddha-Fields*, Houghton Mifflin Company, Boston & New York, 1897
introduced in http://www.inamuranohi.jp/english.html

Nakai, T. M.: Fire set to rice-stacks, in Japanese schoolbook for 5th grade pupils compiled by the state, 1937-1947
(introduced in http://www.inamuranohi.jp/inamura/index.html)

Tsumura, K.: Fire of rice-stacks and protective embankment in Hiro village (in Japanese), Symposium on West Pacific Ocean Earthquake and Tsunami Disaster Mitigation, Japan Meteorological Agency, 2003

International Conference on Geotechnical Engineering for Disaster Mitigation & Rehabilitation
Chu, Phoon & Yong (eds) © 2005 World Scientific Publishing Company ISBN 981-256-469-1

REPORT ON SOME MAJOR GEOTECHICAL DISASTERS IN MALAYSIA

T.A. OOI

TAO Consult Sdn Bhd, 17A, Jln Awan Hijau,Taman Overseas Union Batu5, Off Jln Klang Lama, 58200 Kuala Lumpur, Malaysia

W. H. TING

Zaidun Leeng Sdn Bhd , Tingkat 5, Jalan Bukit Nanas, 50250 Kuala Lumpur, Malaysia

Flooding, Landslides, Debris Flow and Tsunami are some of the disasters experienced in Malaysia. The flooding of Kuala Lumpur in the 1970s caused serious damages to lives and properties and called for the flood mitigation scheme in Kuala Lumpur. Over the years despite the repeated dredging and canalization of floodwater in Kuala Lumpur, there were repeated incidences of severe flooding of the city centre. As part of the overall solution to the frequent flooding problem, the diversion tunnel known as the SMART tunnel was constructed and will be completed in 2006. The SMART tunnel is a dual-purpose tunnel designed to cater to flow of water and ease traffic congestion in the Kuala Lumpur city. The Landslide that caused the collapse of Block 1 of the Highland Tower in December 1993 claimed 48 lives. The landslide occurred during 10 days of incessant rainfall. In November 2002, another landslide occurred and buried the bungalow at the foothill within the vicinity of the Highland Tower site. The incidence also occurred during the period of incessant rainfall and 8 people were killed. Debris flow occurred at the Genting Highlands area emerging from the mountain side flanking the access road and causing debris to flow onto the highway on 30th June 1995 and caused temporary closure of Kuala Lumpur- Karak highway. In the incidence 20 people were killed and 23 people were injured. Debris flow also occurred in the Gunong Tempurong area along the North-South highway, causing debris comprising boulders, timber logs and mud to impact on the beams of a bridge, necessitating closure of a stretch of the highway. Though not as common as conventional landslides, the unexpected nature of debris flow has cause problems, when the flow traversed facilities such as roadways and settlements. The Tsunami that struck Pulau Langkawi, Kuala Muda and Penang on 26 December 2004 took Malaysia by surprise and caused great concern to the public as to the safety of the buildings to withstand the tremors and the need to design for such forces generated, particularly for the high-rise buildings in Malaysia. This paper reports the incidences and look into the need of mitigation measures such as the design aspect of buildings for seismicity in Malaysia and other rehabilitation works.

INTRODUCTION

The Institution of Engineers, Malaysia (IEM), in its September 2005 bulletin devoted the entire issue to report on Disasters (IEM, 2005). Registered Engineers for Disaster Relief (REDR) was founded by Peter Guthrie, a Civil Engineer in 1979 whilst serving his stint at the Vietnamese refugee camps in Malaysia. It is a voluntary register of engineers who can be seconded to the front line relief agencies on notice (Wong, 2005). Marlinda (2005) gave an outlined description of the Disasters Management System in Malaysia. These are essentially a governmental effort in dealing with the occurrence of disasters in general. What is of concern to the geotechnical engineer is the prevention and mitigation of disasters as well as rehabilitations in geotechnical engineering. In Malaysia, major disasters arising out of geotechnical failures are: the flooding of Kuala Lumpur due to uncontrolled Earthworks, the collapse of Highland Towers Condominium in 1993, the Genting Highlands access road debris flow in 1995, landslide buried the bungalow at the foothill within the vicinity of the Highland Tower site in 2002, debris flow in the Genting Highlands area along the Kuala Lumpur highway and the tsunami that struck Pulau Langkawi, Kuala Muda and Penang on 26 December 2004. Malaysia has been considered as a non-seismic risk country until this tsunami incidence. In 1982, the Public Works Department, Malaysia stated in the Keynote address to the Conference on Tall Buildings held in Kuala Lumpur that "Malaysia is actually located in the 'ring of fire' that marked the areas affected or likely to be affected by earthquakes and cautioned that Malaysia will experience in the future severe earthquake arising from the Sumatra and Andaman fault that runs the length of Sumatra and the Andaman Sea to the north of the Straits of Malacca" (Yunus, 1982). The interest in earthquake design amongst Malaysian engineers in fact already started at that time. The Komtar Towers in Penang and the Penang Bridge built in the 1980s in fact were designed for seismic forces. This paper describes some of the incidence of the disasters and gives examples of some of the mitigation and rehabilitation works.

THE FLOODED CONFLUENCE OF KUALA LUMPUR

Kuala Lumpur (KL) was founded in the late 19th century as a tin mining settlement at the river confluence of the Klang and Gombak Rivers. KL is situated in the mid upper reaches of 120km long Klang river that drains a catchment of some 1288 sq. km. Situated in a flood plain, KL has experienced a number of major flood events with the worst recorded being the 1926 flood. The 1971 disastrous flood in the Business District at the centre of Kuala Lumpur then initiated the action by the government to look into the long term solution to the flooding problem in Kuala Lumpur City. By mid 1970s, a flood mitigation Master Plan had been developed that incorporated a number of engineering options including upstream storage; ponding and pumped drainage; and improving the drainage capacity of the Klang River and its major tributaries. A decade later, this plan had been enlarged to cover the whole of Klang Valley and modified to take into consideration to cancel the proposed Gombak Dam. During the 1990s, the pace of development increased with more green areas being urbanized and this resulted in a three-fold increase in the annual discharge of the Klang River at the city centre. Entering the millennium, the flooding of the city has become an annual event. The ongoing works under the Kuala Lumpur Flood Mitigation Project is designed to reduce the occurrence of floods in the city. The Storm Water Management and Road Transport System (SMART) tunnel is a dual-purpose tunnel designed to cater for flow of water and ease traffic congestion in the Kuala Lumpur city. Fig. 1 shows the alignment of the SMART Tunnel. The alignment is notorious as it passes through the treacherous Kuala Lumpur Limestone Formation. The complex subsurface formation is as shown in Fig. 2.

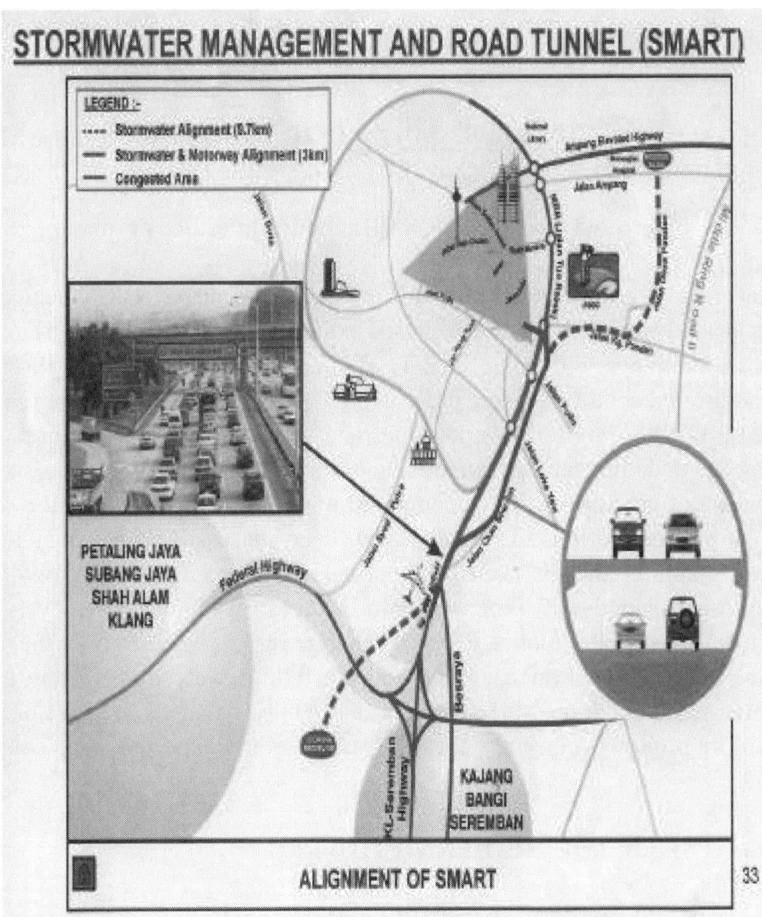

Figure 1. Stormwater Management And Road Tunnel (SMART)

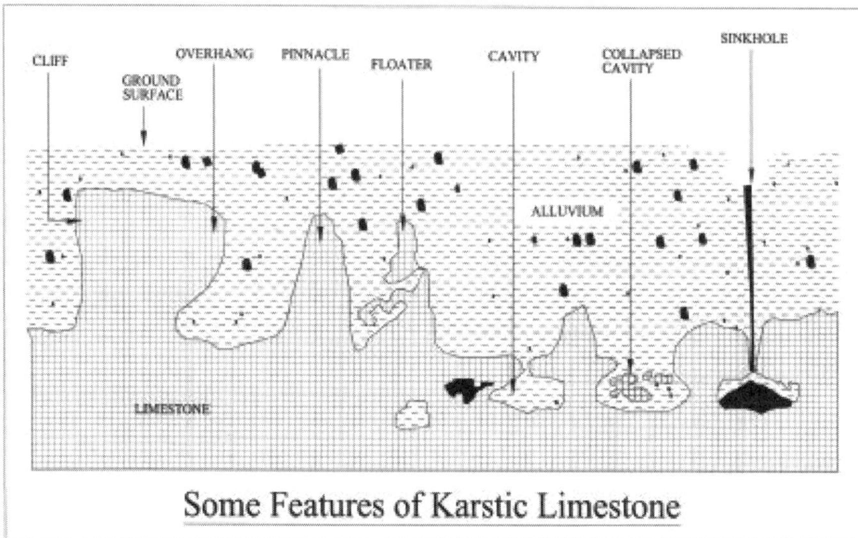

Figure 2. Treacherous Kuala Lumpur Limestone Formation.

The lowering of ground water table for the assembly of the TBM machine at about 18m below ground was known to have caused extensive sinkhole formations to the roads and cracks on buildings within the vicinity of the site. Some aspects of the tunnelling methods and ground conditions were presented by Ting, et al. (2005a). The use of slurry shield type of TBM boring through the limestone bedrock and the overlain problematic alluvial deposits and mine tailings comprising of loose sands/soft slimes was successfully done by the use of pressurized slurry composed of bentonite and air pressure. This is an illustration of how a notorious site condition has been successful managed and overcome with forward planning and the use of the correct and innovative technology. The total storm water tunnel length is 9.7km with 3km of motorway having two levels of traffic deck within the storm water tunnel. Pre-cast concrete linings were used for the tunnel and the voids between the excavated ground and the linings were immediately filled with cement mortar grout under pressurized conditions to ensure that all voids are properly and fully filled. The project is currently ahead of schedule and expected to be completed by end of 2005.

THE COLLAPSE OF THE HIGHLAND TOWERS

In December 1993, during a period of 10 days of incessant rain, Block 1 of Highland Towers Condominium collapsed resulting in loss of 48 lives and the loss of use of the remaining 2 Blocks that were still standing. Fig. 3 shows the collapsed Block 1 of the

Condominium Tower. The Ampang Municipal Council (MPAJ, 1994) set up a Technical Committee of Enquiries and the findings were as follows:

1. The Highland Towers Condominium was sited mainly on fill ground over granitic formation. The maximum depth from the ground surface to bedrock is about 19m. Granitic rocks found in and around the areas were not highly soluble minerals to adversely affect the stability of the foundations.

Figure 3. Collapse of Block 1 Highland Towers Condominium

2. Soils overlying the granitic bedrock were very loose to loose silty sand and highly permeable.

3. All 3 Blocks were on rail piles designed to take only vertical loads.

4. Surface drainage system provided was not in accordance to approved plan. Situations worsen when earthwork activities changed the drainage pattern on

hill-slope behind the Condominium Blocks and available drainage system not maintained.

5. Clearing of trees on upper catchments resulted in increased runoff that flowed down the terraced hill-slope immediately behind the towers

6. Retrogressive slides progressively moved uphill starting from loss of toe mass at the back of the Condominium Block 1 (see Fig. 4).

7. The fallen debris accumulated behind the back terrace of Condominium Block 1 caused the slip to occur beneath the entire rail pile foundation that brought down the Condominium Towers within minutes of the landslide occurrence.

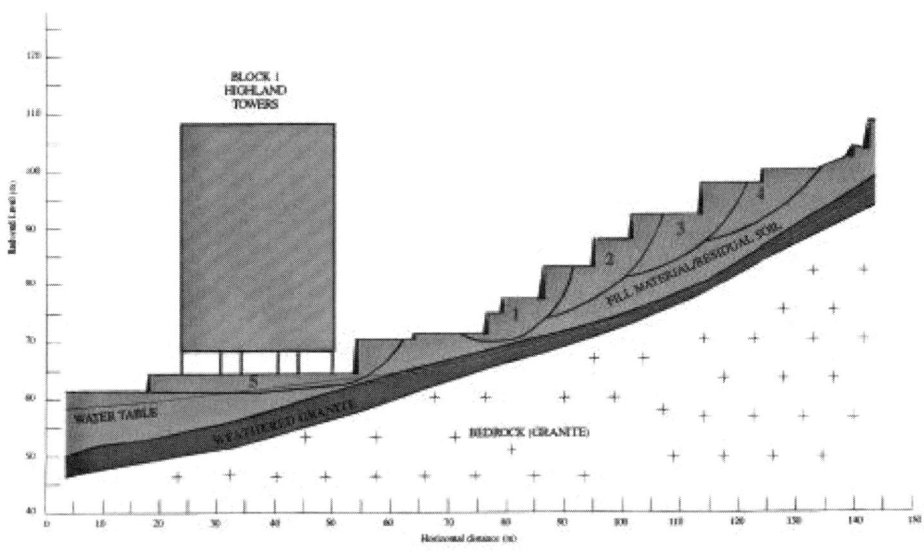

Figure 4. Retrogressive Slope Failure and Slip Circle Failure below building foundation

The case has several important implications for Developers, Building Professionals, Local Authorities, absentee landlords and developers of neighbouring properties in Malaysia. The Resident Association took the case to Court (Steve Phoa Chen Loon & Ors v Highland Properties Sdn Bhd & Ors, 2000) and the Court found that the landslide that brought down Block 1 of the Highland Towers Condominium was due to a rotational retrogressive slide emanating from a high retaining wall behind the 2nd tier of the 3 tiered car parks. Water was found to be the principal factors that caused this high wall to fail. Lessons learned from the decision of the High Court for an engineer are as follows:

The Engineer was liable in negligence for (i) not having taken into account the hill or slope behind the Towers, (ii) not having designed and constructed a foundation to accommodate the lateral loads of a landslide or alternatively to have ensured that the adjacent hill-slope was stable, (iii) for not having implemented that approved drainage scheme, (iv) for colluding with the First (Developer) and Second (Architectural Draughtsman) Defendants to obtain CF without fulfilling the conditions imposed by the Fourth Defendant (Local Authorities) and also in nuisance as he was an unreasonable user of land.

Ooi (2004) in his Special Lecture on Earthworks Practice in Malaysia at the MGC2004 in Kuala Lumpur presented many case histories of earthwork failures and their rehabilitations. Table 1 shows some of the landslide incidences. In general, water has been the principal cause of many geotechnical failures. The design should take in account of suitable surface and subsurface drainage of slopes and embankments. The use of tipped fill on slope and embankment should not be allowed under any circumstances.

Table 1. Case histories of earthworks failures due to water

Date	Location	Landslide Details
A. January 1971	Bukit Gasing, Petaling Jaya	• Gasing Height Development • Flow slide of tipped fill damaged two Government Quarters completely during one week of incessant rain • The Quarters were subsequently rebuilt • Case settlement out of court • PWD internal report (Ooi, 1971)
B. September 1988	Ulu Kelang, Ampang Jaya	• Slope failure due to toe erosion at river bank and excavation on neighbouring land during raining season • Damaged to Bungalow and Swimming Pool • Court decided Engineer has a duty of care that he owed to the house owner • House owner awarded economic loss amounting to about RM360,000.00
C. December 1993	Ulu Kelang, Ampang Jaya	• Collapse of Block I of Highland Towers on 11 December 1993 during the raining season

Date	Location	Landslide Details
		• 48 people were killed • Technical committee report of investigation by MPAJ (MPAJ, 1994) • Engineer failed in his duty of care he owed to the plaintiff
D. June 1995	Genting, Selangor	• Debris flow, Genting Highlands on 30 June 1995 caused closure of Kuala Lumpur – Karak Highway • 20 people were killed • 23 people injured • Economic losses, deconstruction of several vehicles, deconstruction and disruption of roadway • No inquiries
E. May 1999	Ulu Kelang, Ampang Jaya	• Bukit Antarabangsa fill slope failures during raining season • Access road to Bukit Antarabangsa cutoff • Residents of the area were evacuated • Economic loss and anxieties
F. November 2002	Ulu Kelang, Ampang Jaya	• Landslide buried the bungalow at the foothill during raining season. 8 people were killed.
G. November 2003	Bukit Lanjan, NKVE	• Rockslide caused closure of NKVE Highway at Bukit Lanjan • Continuous heavy rainfall in the month of October and November before the rockslide

DEBRIS FLOW

Debris flow is a natural Hydraulic-Geotechnical phenomenon that has been reported in various parts of the world. In Malaysia, debris flows do not occur as frequently as landslides; as a result of which there is no clear recognition of the problem. The impact of the flow becomes apparent when it traverses or is obstructed by a roadway or some human settlements. The immediate mechanism that distinguishes one from the other is that debris flow emerges from a streambed of a hill/mountain site, whilst a landslide is a slip of the flank of a hill or bank.

It can be said that debris flow results from the dam break of a barrier across the stream that has been "naturally" built up; releasing significant amount of water at high velocity and carrying along with it the debris upstream and downstream of the natural barrier. The "debris" besides comprising materials ranging between boulders and mud also contains significant amount of timber logs and other vegetable matters. Although of natural occurrence, certain hydraulic and geotechnical factors that may have contributed to the debris flow phenomenon could be identified. From the factors, certain mitigating measures may be recommended. Hydraulic factors dominate the impact of the debris flow, whilst geotechnical factors determine the formation of the natural barrier and the materials of the debris.

TSUNAMI ON 26 DECEMBER 2004—A MALAYSIAN EXPERIENCE

The tsunami that struck the Indian Ocean on 26 December was a global catastrophe that caught many in the world by surprise. There were no tsunami early warning systems established in the countries in and around the Indian Ocean. Malaysia was not spared the destruction caused by tsunami in places like Langkawi, Penang and Kuala Muda in Kedah. The actual death toll caused by the tsunami must necessarily exceed that reported figure of about 280,000 people. In Kuala Muda, the last time that tsunami struck was in 1880s according to record passed down by words of mouth. The December 2004 tsunami damaged about 1000 houses with about 4000 people directly affected. Fig.5a-5c shows the scenes of damage to properties.

Figure 5(a).

Figure 5(b).

Figure 5(c).

Figure 5(a) – 5(c). Damages to properties due to 26 Dec 2004 tsunami

The victims were housed in longhouse type temporary accommodations some few kilometers from the site. According to the Chief Resident Engineer of the Muda Dam Project, Ir. Ooi Choon Aun, who felt the tremor in Sungei Petani at about 9.00am local time but did not expect anything unusual had happened until he was in Balik Pulau in Penang later that day at about 1.30pm to witnessed the horror that had happened in the streets of Balik Pulau. People were dashing about in panic and there police with dead bodies laid by the roadside. Most of the tragedy occurred at Pasir Pangjang beach in Balik Pulau where people were there for picnic. The site faces the Straits of Malacca and when the seawater receded very far from the shorelines a lot of picnickers were excited and went after the fish that were left behind by the receding seawater. When the seawater rose again many were swept away with serious injuries resulting in numerous casualties. Tibballs (2005) in his book on the chapter on ' What is a Tsunami' stated that ' Often the only warning of an impending tsunami is the sight of the tide suddenly going out as exposed on the dry seabed but the eager fishermen or curious tourists who seek to far as the horizon........Experts say that a receding ocean may give people as much as five minutes' warning to escape to high ground. As the water is suck out to sea, fish are investigate this freak nature invariably pay with their lives when the water suddenly comes rushing back in as giant waves.'. This seemed to be the case that had happened at the Pasir Pangjang Beach.

On the other hand, there were very few casualties at the Batu Ferringhi site since they were able to clear most of the people on the beach in time. This according to informed sources was due to the early warning that they received from the hotel operators who has the experience of Tsunami in Langkawi earlier on 26 December 2004. The Batu Ferringhi site faces the Andaman Sea and therefore has the most direct effect from the Tsunami that destroyed Bandar Acheh. Interestingly, the floating Mosque at the adjacent Tanjung Bungah Beach was virtually untouched while the buildings on the beach suffered severe damages from the effect of Tsunami (see Figs. 6 and 7). Professor Shuto in his Lecture Tour of Southeast Asia on Tsunami, told his audiences in Penang in April 2005 that 'it is important to educate the people and provide timely information so that they know how to deal with the arrival of Tsunami'.

SEISMICITY DESIGN IN MALAYSIA

Traditionally, Malaysia has been considered as a non-seismic risk country. In 1982, the Public Works Department, Malaysia stated in the Keynote address to the Conference on Tall Buildings held in Kuala Lumpur that Malaysia is actually located in the 'ring of fire' that marked the areas affected or likely to be affected by earthquakes and cautioned that Malaysia will experience in the future severe earthquake arising from the Sumatra and Andaman fault that runs the length of Sumatra and the Andaman Sea to the north of the Straits of Malacca (Yunus, 1982). The tsunami that occurred on 26 December 2004 could be said to be the prediction that came true.

Figure 6. Mosque spared by the wrath of Dec2004 tsunami

Figure 7. Building damages on beach Tanjung Bungah, Penang, Dec 2004 tsunami

The interest in earthquake design amongst the Malaysian engineers in fact already started at that time. The Komtar Towers in Penang and the Penang Bridge built in 1980s were already designed for seismic forces. This is despite the fact that there is no requirements in the Uniform Building Bye-law for seismic forces for buildings. Ting et al (2005b) discussed some aspects of seismic design for slope stability of dams in Malaysia. Table 2 shows the different g values used for various dams. The designed g value applied in each dam site is dependent on the iso-seismal maps from Malaysian Meteorological Service (MMS) that are used as an indicator for zones that received different level of recorded intensity on Modified Mercalli (MM) scale (Mohd Rosaidi Bin Che Abas, 2001). Figure 8 shows the iso-seismal map for Peninsular Malaysia. The trend to use higher design g value in dam depended on the better information received and the increase in the intensity of seismic activities in the region. Of foremost importance, dam is a live threatening structure and needs to be fail proof. Judging from the increased frequency and intensity of tremors being felt in Malaysia in recent years, it may be prudent to allow for design of seismic forces for structures of higher level of public risks such as that of long span bridges, public and private buildings of higher than five storeys.

Table 2. Different g valuues used in dam slope stability designs in Malaysia

	Dam Location	Design g value
1.	Gemencheh, Negeri Sembilan (1980s)	0.10g
2.	Sungai Selangor, Selangor 1990s)	0.10g
3.	Sg. Kinta, Perak (2000s)	0.15g
4.	Bakun, Sarawak (2000s)	0.20g
5.	Kelalong, Sarawak (2000s)	0.30g

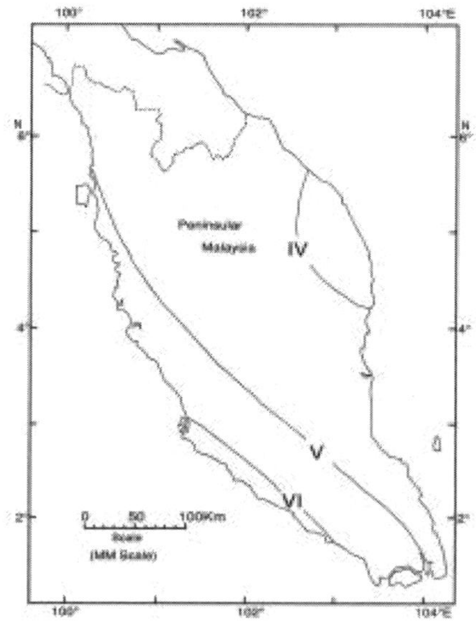

Figure 8. Maximum Observed Earthquake Intensity in MM scale for Peninsular Malaysia.

CONCLUSIONS

An attempt is made to highlight some of the major geotechnical disasters in Malaysia. Mitigation, Prevention and Rehabilitation of disasters are important considerations for any geotechnical disasters. The SMART tunnel project was born out of the disastrous flooding of Kuala Lumpur and is believed to be the first of its kind in the world where the tunnel will be used as a dual purpose tunnel for both flood control and to ease the traffic congestion of the Kuala Lumpur City Centre. The Highland Tower Condominium collapse shows that it is important for the designer to consider all aspect of foreseeable possible danger to the buildings in relation to the environment including future maintenance. The need to get rid of surface and subsurface water from site is clear from this tragedy. The stability of slope and the structural foundation of the building are integral in the design analysis process. Debris flow failures will increase with more development in the Highlands and mitigating measures recommended must consider hydraulic factors that dominate the impact of the debris flow, whilst geotechinical factors determine the formation of the natural barrier and the materials of the debris. The tsunami catastrophe that struck the Indian Ocean on 26 December 2004 brought about the urgent need from the geotechnical community in the region and Malaysia in particular to look at the mitigation by way of design provision and consideration. The programme of public education in awareness and training in the handling of such disaster must also be implemented.

REFERENCES

IEM (2005). *"Bulletin, Institution of Engineers, Malaysia"*, No. 9.

Marlinda, B. A.M. (2005). "Disasters Management System in Malaysia", *Bulletin IEM, No*. 9, 10-12.

Mohd Rosaidi bin Che Abas. (2001). "Earthquake Monitoring in Malaysia", *Seismic Risk Seminar 2001, Malaysia.*

MPAJ (1994). "Report of the inquiry committee into the collapse of Block 1 and the stability of Block 2 and 3 Highland Towers Condominium, Hulu Klang, Selangor Darul Ehsan." *Majlis Perbandaran Ampang Jaya.*

Ooi, T.A. (1971). "Report on landslides at government quarters nos 1276 and 1280 at Section 5, Petaling Jaya, Selangor" *PWD internal report.*

Ooi, T. A.(2004) "Earthworks Practice in Malaysia", *Proc. Conf. MGC2004, Kuala Lumpur, 45-58*

Steven Phoa Cheng Loon & Ors v Highland Properties Sdn Bhd & Ors (2000) *Current Law Journal*, Vol 4, p.508–602.

Tibballs, G. (2005). "Tsunami The World's Most Terrifying Natural Disaster", *Carlton Books Ltd., London*

Ting, W.H., Ooi, T. A., Ahmad, A. and Tan, B. K. (2005a) " Comparison of Tunnelling Methods in Granitic and Karstic Limestone Bedrock in Kuala Lumpur.", *Journal Engng Geology,Geomechanics and Tunnelling, Felsbau 23*,No.2, 32-37.

Ting, W.H., Ooi, T. A., Ahmad, A. and Tan, B. K. (2005b) " Seismicity in Slope Stability Practice in Malaysia." *Proc. Int. Conf. Geotech. Engng. For Disaster Mitigation & Rehabilitation,Chu, Phoon & Yong (eds) © 2005,* World Scientific Publishing Company.

Wong, G. (2005). "Disaster Relief Engineers", *Bulletin IEM*, No. 9, 6.

Yunus, M. M. Y. (1982). "Keynote Address- Building High", *Proceedings Asian Regional Conf. on Tall Buildings and Urban Habitat, Kuala Lumpur*, Institution of Engineers, Malaysia., K-1 to K-8.

International Conference on Geotechnical Engineering for Disaster Mitigation & Rehabilitation
Chu, Phoon & Yong (eds) © 2005 World Scientific Publishing Company ISBN 981-256-469-1

GEOTECHNICAL ENGINEERING ASPECTS RELATED TO ACEH'S TSUNAMI AND EARTHQUAKE DISASTER AND THE NEED FOR ITS MITIGATION STRATEGY

I WAYAN SENGARA
Geotechnical Engineering Division - Civil Engineering Department and Center for Disaster Mitigation, Institute of Technology Bandung, Jl. Ganesha 10, Bandung-40132, INDONESIA

HAMZAH LATIEF
Oceanography Division – Geophysics and Meteorology Department and Center for Disaster Mitigation, Institute of Technology Bandung, Jl. Ganesha 10, Bandung-40132, INDONESIA

MUNIRWANSJAH
Civil Engineering Department, University of Syiah Kuala-Banda Aceh
Jl. T Syech Abdul Rauf 7, Banda Aceh - Indonesia

SPR WARDANI
Disaster Prevention and Rehabilitation Research Centre, Soil Mechanics Laboratory, Civil Engineering Department, Engineering Faculty, Diponegoro University, Jl. Prof. Soedarto SH , Semarang 50275 - Indonesia

Mega Sumatra earthquake and tsunami disaster has caused catastrophic damages to city of Banda Aceh and Meulaboh, Indonesia. Results of reconnaissance survey and preliminary study of the building and infrastructure damages related to geotechnical engineering aspects are presented. Many building and infrastructure damages due to earthquake ground motions at both cities are caused by strong seismic motion causing soil liquefactions, lateral spreads, slope failures, and settlements. Whereas, soil erosion and foundation failures for most single to two-story buildings close to shorelines were due to high tsunami wave force. Preliminary microzonation and liquefaction occurrence maps based on the reconnaissance survey have been developed. Seismic back analysis of the Aceh's subduction seismic source as well as preliminary probabilistic seismic hazard analysis post-Aceh earthquake for city of Banda Aceh is presented. Liquefaction potential analysis based on estimated peak ground acceleration was conducted. Geotechnical and foundation engineering aspects related to infrastructure and residential housing damages due tsunami are presented. Some descriptive countermeasures related to geotechnical engineering aspects and mitigation strategies for rehabilitation and reconstruction are provided.

INTRODUCTION

Extremely strong earthquake (Mw=9.0) followed by tsunami has occurred on December 26th, 2004. The earthquake magnitude was M=9.0 at hypocenter depth of 30 km beneath the seabed. The epicenter of the earthquake is 3.316^0 N, 95.854^0E, about 149 km south of

Meulaboh city and 250 km south of Banda Aceh city, Nanggroe Aceh Darussalam (NAD) (Source: USGS). The earthquake was of type subduction mechanism, where Indian-Australian plate subduct to the Eurasian plate. This main Earthquake in Aceh is followed by a powerful aftershock in Nicobar Islands (M=7.5) and in the Andaman Islands (M=6.2). These earthquakes have stricken most areas in NAD province and felt strongly in Sumatra Utara (SUMUT) province. The earthquakes were followed by tsunamis that cause extremely high casualties and destruction of buildings, houses, and lifelines of cities and small islands along North-West Sumatra shorelines as well as many areas in Asia and Indian shorelines.

The Aceh earthquake and tsunami has caused tremendous casualties and catastrophic damages in the affected areas, especially in city of Banda Aceh and Meulaboh. In Indonesia, from the total pupulation of 468,823 exposed to the disaster, the earthquake and tsunamis have caused at least 228,948 people deaths and missing in Indonesia. The tsunami has killed 49,000 people in Sri Lanka, 16,423 in India, 8,476 in Thailand, 109 in Maldives, 74 in Malaysia, 64 in Myanmar, 2 in Bangladesh , 150 in Somalia, 10 in Seychelles, 136 in east Africa, 2 in South Africa, and 1 in Kenya. Tsunamis also occurred on the coasts of Cocos Island, Kenya, Mauritius, and Reunion.

Damage Estimates

The earthquake has also caused damages of buildings, infrastructures, and lifelines. Thousands of building damages have been reported. Many school buildings, government buildings, hospitals, and public housings have been damaged. Total number of more than 175,000 unit houses need to be rehabilitated and reconstructed. Approximate length of road damage is estimated 800 km and 1,664 bridges need to be fixed. The highest damages occurred in Meulaboh city, since it is closer to the earthquake epicenter. Infrastructures such as roads, bridges, and ports are damaged. The casualties in both cities were mainly due to the tsunami. The tsunami has caused catastrophic damages. Most of the buildings along the shorelines in both cities swept and wiped out by the tsunami. The tsunami penetrates as far as 3-4 km to the land in the city of Banda Aceh. Most of powerlines were not functioning after the earthquake and tsunami. Diesel power plant swept out by tsunami, some other diesel power were damaged, hundreds of km of transmission lines was not functioning. Estimated telecommunication system damage: automatic central telephone 22, damaged unit telephone connection 27,095. The total losses of Aceh earthquake is roughly estimated more than 4 billions USD.

TECTONIC SETTING OF THE SUMATRA REGION AND EARTHQUAKE HISTORY

West coast of Sumatra is area of convergence of the Indian-Australian plate moving relative to Eurasian plate. The subduction zone that extends from western part of Andaman Island in the northwest and Banda Island in the east is the most active seismic source in the Indonesian region. The high seismicity in Sumatra is affected by Sunda Arc source zone. City of Banda Aceh is approximately located 100-200km from the subduction zone. In addition, the city is also in the line of Sumatra Fault Zone (SFZ). Figure 1 shows a map of the principal tectonic elements of the Sumatran plate boundary and fault system.

Earhquake History in Sumatra Island

Many large earthquakes have been reported and recorded that were originated from subduction source zone along south-west of Sumatra island. Among large earthquakes were occurred in 1833 (M=9.2), 1861 (M=8.2), 1907 (M=7.6), Bengkulu earthquake in 2000 (M=7.8), and lately in Aceh (M_w=9.0). Many earthquakes occurred along interface of the subduction zone and the rate of this subduction is estimated about 60 mm/year. Figure 2 shows seismological summary of Sumatra source regions of large earthquakes from historical records (after Shieh and Natawidjaya, 2000). According to Indonesian seismic zonation (SNI-1726, 2002), Aceh province along the west coast lines is classified as zone 5 to 6 (highest) with peak baserock acceleration (PBA) in the range of 0.3 to 0.4g and about 0.25-0.3g for city of Banda Aceh.

Attenuation of the Aceh 2004 Earthquake

Aceh earthquake is one of the largest earthquake that occurred in "*megathrust*" part of seismic subduction zone where Indian-Australian subducts Eurasian plate. To estimate peak baserock acceleration (PBA) at some damaged cities in Aceh province, preliminary seismic wave attenuation of the earthquake has been conducted herein. Appropriate attenuation functions need to be adopted, since there is no attenuation relationship available that is derived from earthquake recording in this specific subduction region. Therefore, for the Aceh earthquake, attenuation functions that represent similar mechanism are used. In this case, attenuation function proposed by Crouse (1991) and Youngs et al. (1997) to represent subduction source zone mechanism are adopted. Result of earthquake attenuations from hypocenter to some distance is presented in Figure 3.

It is indicated that PBA is approximately 0.23-0.27g at city of Banda Aceh which is located about 250km from the epicenter and about 107km rupture distance. The preliminary attenuation analysis shown in the plot also indicates that PBA of approximaterly 0.30-0.35g is estimated for Meulaboh city which is located about 149km from the epicenter and 54 km rupture distance. Due to its large magnitude, the earthquake is felt to a large distance to as far as 1000km. The earthquake to large distance is estimated to be rich in low frequency content that could cause high amplification in soft soil deposits. The Aceh earthquake has damaged many buildings, infrastructures, and lifelines in city of Banda Aceh and Meulaboh.

HISTORICAL TSUNAMIS AROUND SUMATERA AND TSUNAMI SIMULATION

Historically, since 1700 at least 18 tsunamis have been occurred around the Sumatera region before the 2004 Aceh Tsunami as listed in Table 1. Tsunamis in this area mostly due to the submarine earthquakes which the epicenters (tsunami sources) were located in the western part of Sumatera, another tsunami potential is the submarine Krakatau eruption at the Sunda Strait generated huge tsunami in 1883.

Latief et al., (2005) have made preliminary simulation of the Aceh tsunami. An initial profile as tsunami source due to a crustal deformation is calculated by using Mansinha and Smylie's method (1971). The fault parameter of 400 km length x 80 km width and seismic moment of 3.5×10^{27} dyne-cm have been estimated from seismological analysis based upon the focal mechanism of the Harvard CMT Solution and Fault Parameter of the ERI-Tokyo University. The preliminary numerical simulation results are shown in Figure 4.

It can be seen from the preliminary tsunami simulation that city of Banda Aceh and Meulaboh were hit by tsunami 20 to 30 minutes after the earthquake. The wave travel and has caused tsunamis in Indian coasts at many region in Indonesia, Malaysia, Myanmar Thailand, Sri Lanka, India, Bangladesh, Maldives, Somalia, Seychelis, South Africa, Kenya, Cocos Island, Mauritius, and Reunion. From the field tsunami survey is found that the tsunami height at the northern part of Banda Aceh city (Uleulhe) is around 11 up to 17 m and the western part of Banda Aceh City (Lhok Nga) is around 30-35 above MSL.

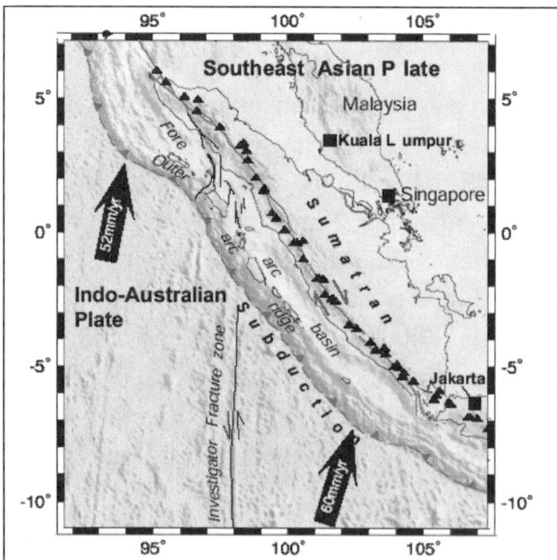

Figure 1. Map of the principal tectonic elements of the Sumatran plate boundary (after Natawidjaya, 2002)

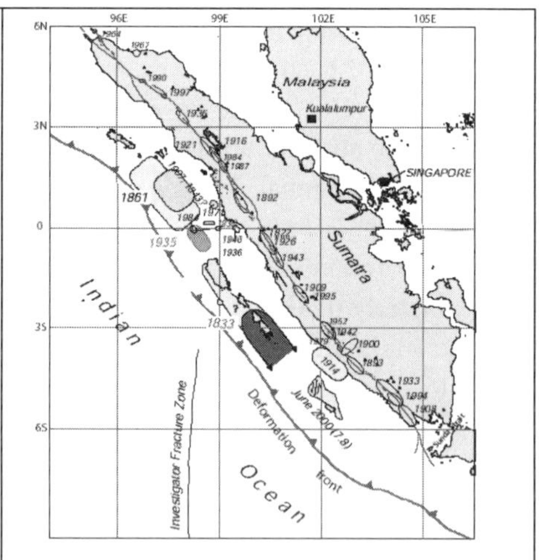

Figure 2. Seismological summary of Sumatra. Source regions of large earthquakes from historical records (after Shieh and Natawidjaya, 2002)

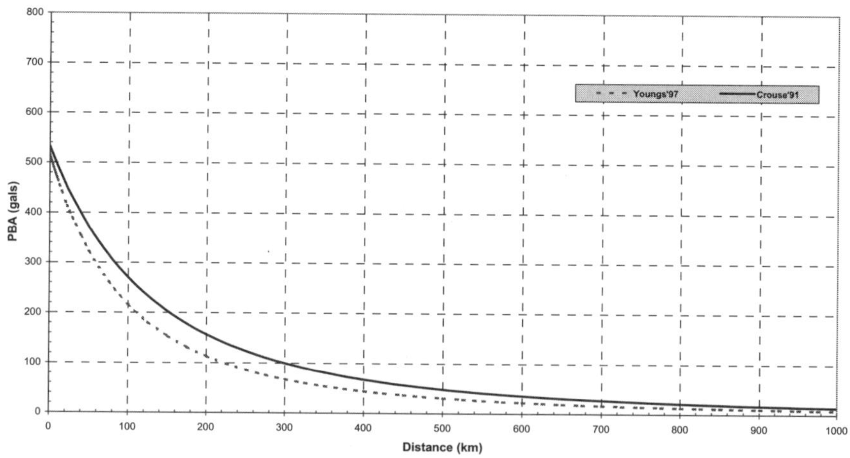

Figure 3. Plot showing 26 December 2004 Aceh earthquake PBA attenuation (Sengara et al., 2005)

RECONNAISSANCE SURVEY

A reconnaissance survey on earthquake and tsunami disaster has been conducted by ITB task force not long after the earthquake under the coordination by the first author to identify the affected area and the degree of damages on buildings, infrastructures, and residential houses and other elements of the city. In the reconnaissance survey two types of damages was identified, that is damages due to tsunami and that due to the earthquake.

During the survey, measurement of tsunami height covering different locations within the city from the very edge of the affected area to the shoreline has been conducted. There is a great variation in the tsunami height across the city, and therefore degree of damages of buildings and infrastructures are closely associated with this tsunami height variation. From the survey, the tsunami inundation area, tsunami direction, wave height, wave run-up distribution measured from ground surface is shown in Figure 5 to Figure 7. It is indicated that the tsunami wave height at the very north of city of Banda Aceh is approximately 15 meter. The highest tsunami wave height is observed at Lhok Nga at west direction of the shoreline where the tsunami hit the hardest. The tsunami wave height at this location is about 30m. Based on the tsunami wave run-up survey within the tsunami affected area of Banda Aceh city and the surrounding area, a contour showing distribution of inundation wave run-up was developed as shown in Figure 8.

Tabel 1. List of tsunami in the Sumatera region for period from 1700 to 2005 (Modified from Latief et al., 2002)

Year M D h m	Lat. Lon.	M/Dp	Tm	H(m)	Dead/ Injure	Observed Area Province: location
1797 02 10	0.58 100,2	8.4		Strong	300	W. Sumatera, Padang, Sipora-Pagai
1799				16		Sumatera
1818 13 18	-3.5 100.5					Bengkulu
1833 01 29		9				W. Sumatra: Padang, Priaman, Sipora-Pagai
1833 11 24						Bengkulu
1843 01 5-6	2.08 98.23			Large	Many fatalities	N. Sumatera: Barus Is.. G. Sitoli, Nias
1847						Great Nicobar island
1861 02 16		8.4			50	Sumatra: Batu Is., Nias Is.
1861 09 25	-2.04 100.6		1		Several thousand	Sumatra: Padang, Indrapura, Nias
1864						W. Sumatra: Padang, Batu
1881 12 31						Andaman, Nicobar island
1883 08 26	-5.8 106.3	Volc		35	36000	Sumatera: Sunda St., Java,
1904 07 04						W. Sumatra: Siri-siri
1907 01 -					Several thousand	Sumatra : Western Coast, Simelue_nias
1908 02 06	-2.0 100.0					W. Sumatra
1909 06 03 18 41U	- 2.5 101.5	7.3			200	Sumatra: Kerinci-Jambi
1928 03 26	-5.8 106.3	Volc				S. Sumatera: Sunda St., Lampung,
1935 12 28	-0.3 97.9					N Sumatera: Batu Island
1936 03 01	-	7.0			110/479	Sumatra:
1941						Andaman
1967 04 12	5.3 96.5	6.5		big		N Sumatera: Sigli
2004 12 26		9			300.000	Aceh, Andaman, Nicobar
2005 3 28						Nias

Figure 9 to Figure 11 show how catastrophic and destructive the tsunami was that destroy almost entire single-story residential houses and infrastructures due in the area close to the shorelines. These-single story buildings were generally light with mostly shallow foundations such that almost entirely swept by the strong tsunami wave. Many foundations of the buildings were also eroded by the tsunami wave. Only a few 2-story buildings and some mosques survive under the tsunami wave. Among these surviving buildings most of them experience structural damages. Observation shows that among

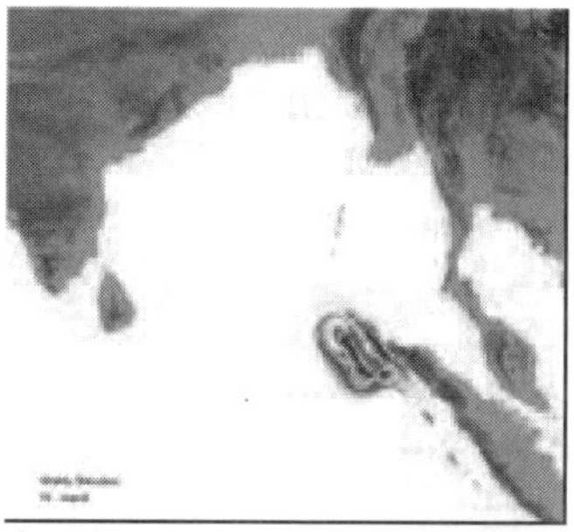

10 minutes

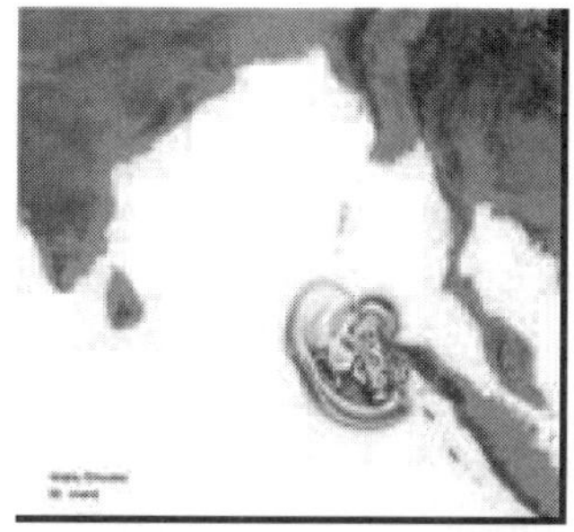

30 minutes

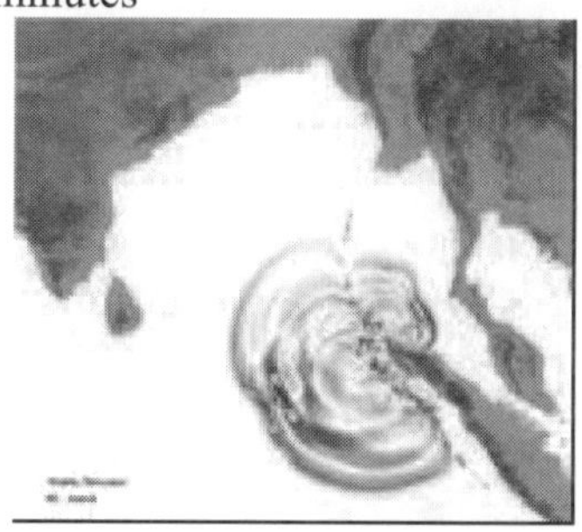

60 minutes

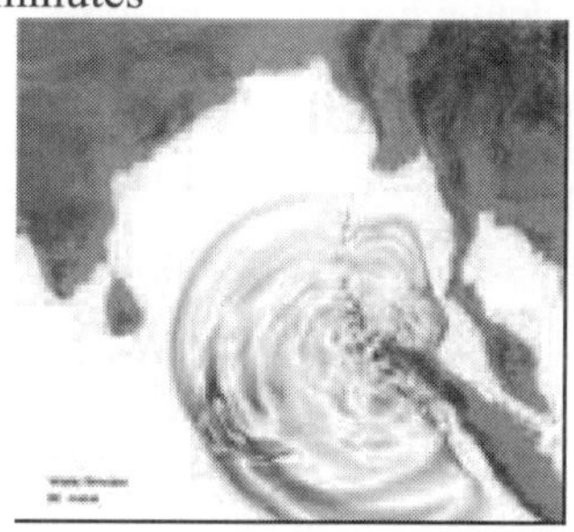

90 minutes

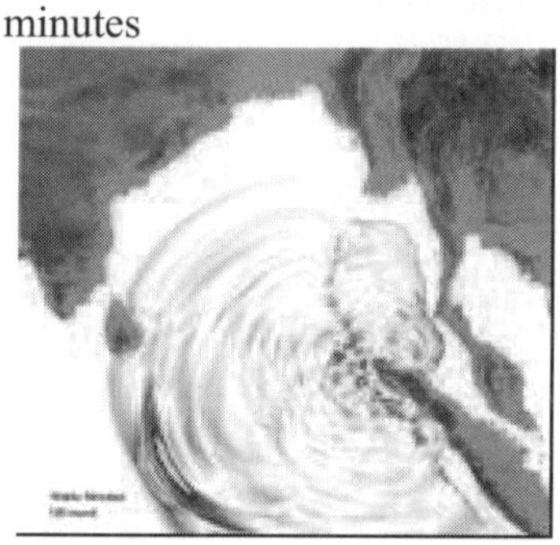

120 minutes

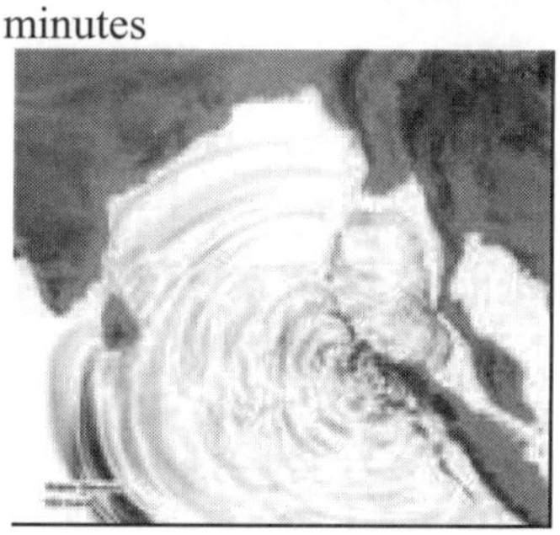

150 minutes

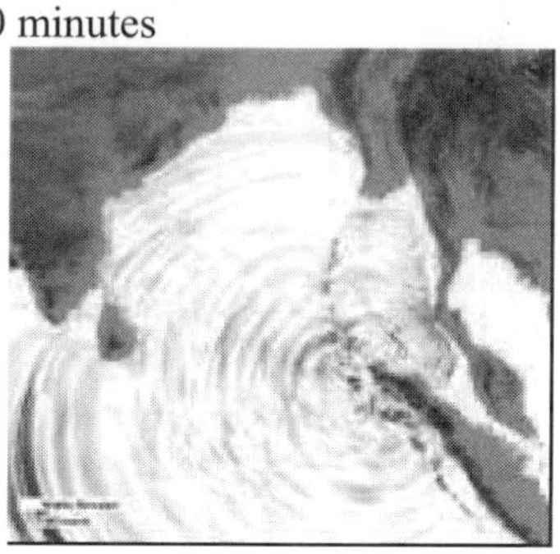

180 minutes

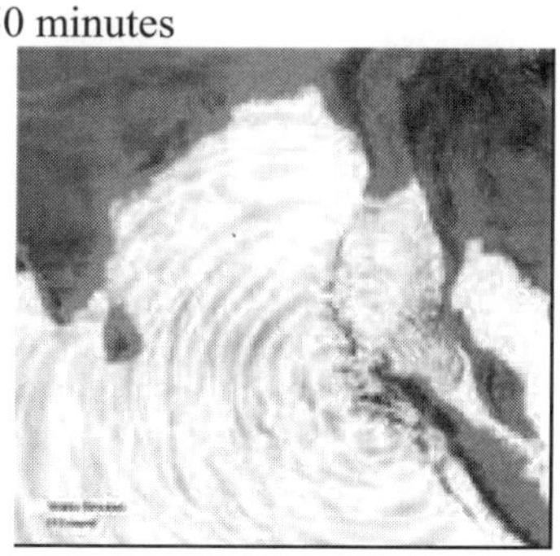

210 minutes

Figure 4. Preliminary simulation results of Aceh tsunami (Latief ,et.al, 2005)

buildings in the corridor approximately 300-500m away from close to the shoreline, two-story or higher buildings with some openings with strong foundation survive under the high and strong tsunami wave. Some two-story reinforced concrete buildings about 200-300m from the shoreline at Meulaboh city completely collapsed under high (>25m) and very strong tsunami wave.

Building Damages and Geotechnical Failures due to Earthquake

Many buildings and infrastructures damages due to Mega Aceh earthquake have been identified during the site reconnaissance survey. Mega Aceh earthquakes had destroyed many vulnerable buildings due to relatively high ground motion and in some areas due to failure of the building foundations caused by liquefaction. There were great variations in building and infrastructures collapse and damages observed. The observed damages vary from slightly non-structural damage to totally collapsed. Many 3 to 4-story buildings were totally collapsed, while other buildings only experience only slight to non-structural damages. Also many residential houses located on relatively strong ground condition do not experience even non-structural damage.

Figure 12 shows two extreme of totally collapsed buildings and building with only experience slight to non-structural damages. Under relatively high PBA with possibly low seismic amplification, some buildings only experience non-structural damage, whereas other similar buildings, with the same PBA at other locations with possibly softer local ground conditions or rest in a slope were completely collapsed. There could be many factors associated with this variation in level of damages. One contributing factor would be the effect of local ground condition on seismic amplification from base-rock to the ground surface. Figure 13(a) shows collapse of some buildings located at edge of a river with soft ground condition. The variation in the level of damages is considered to be due to variation in vulnerability (in terms of the design and construction) of the buildings and their foundations, variation in the peak ground acceleration of possibly different local ground condition, and buildings or structures those might experience liquefaction induced foundation failures or ground settlement. Figure 13(b) shows settlement of a building caused by building rocking during earthquake ground shaking. Buildings failures seem to be contributed by more due to poor structural integrity than problems on their foundations.

In general, infrastructure damages such as embankments of bridges and roads also vary. The variations seem to be most controlled by location or possible variation in seismic amplification due to the different local ground conditions. In addition, liquefaction induced some of infrastructures failures were also observed. Figure 14 shows total collapse of a bridge estimated due to liquefaction induced foundation failures. This figure also shows liquefaction induced retaining wall failures.

For bridges and embankments, failures seem to be more contributed by more due to poor embankment and foundation design and construction rather than their structural integrity. Many embankment and foundation failures of the bridges were estimated due to liquefactions induced failures. Shoreline retaining walls and bridge abutment failures

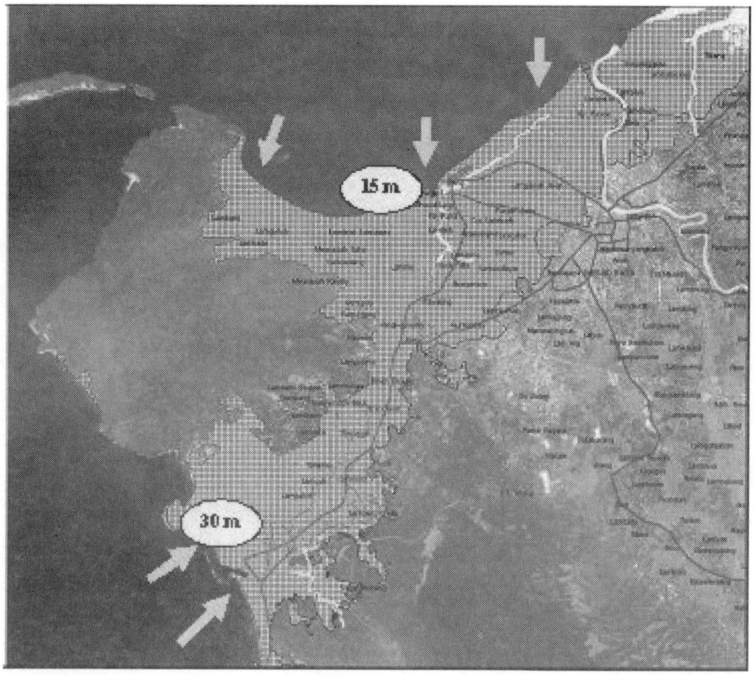

Figure 5. Tsunami direction, wave height, and the area boundary of tsunami inundation.

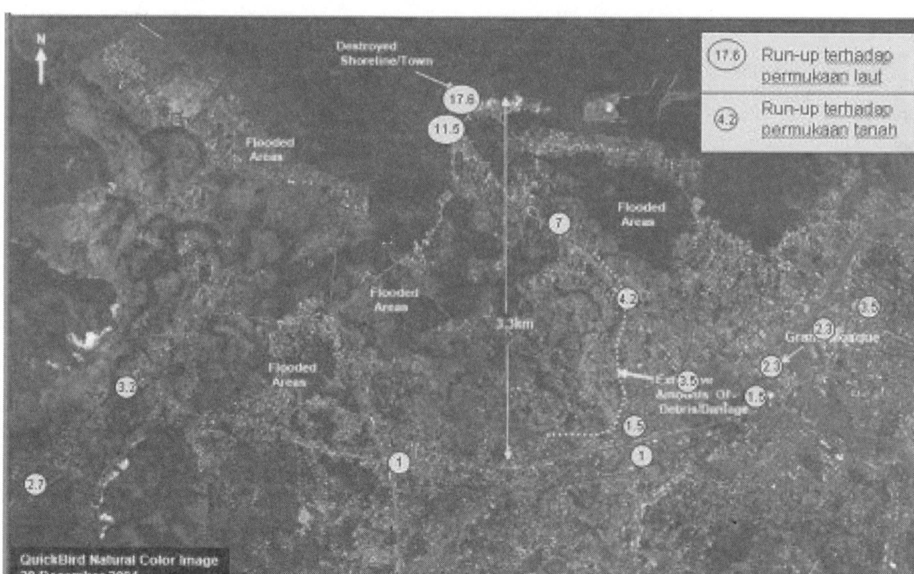

Figure 6. Post-tsunami survey showing tsunami wave run-up from existing ground level

Figure 7. Tsunami direction observation and tsunami height measurement survey at city of Banda Aceh

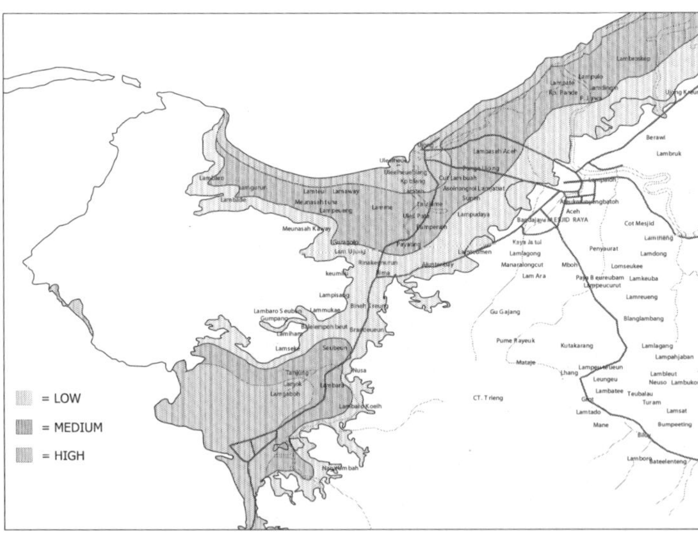

Figure 8. Tsunami wave height distribution due to Dec 26, 2004 tsunami at city of Banda Aceh and surrounding area

Figure 9. Transported residential house that was separated from its foundation by the very strong tsunami wave close to Lhok Nga beach (*From: ITB task force*)

Figure 10. Some 2-3-story buildings that survive among entirely destroyed residential houses by the strong tsunami wave at Lampaseh, Aceh (*From: ITB task force*)

Figure 11. Bridge embankment failures induced by tsunami wave force and ground shaking (*From: ITB task force*)

(a) (b)

Figure 12. (a) Total collapse and (b) four story buildings that only experience non-structural damage due to Aceh earthquake, December 2004.

(a) (b)

Figure 13. (a) Failures of vulnerable residential houses due to high amplification of seismic ground motion inducing foundation and slope failures; (b) Settlement caused by building rocking during earthquake ground shaking.

(a) (b)

Figure 14 (a) Total collapse of a bridge estimated due to liquefaction induced foundation failures (b) Liquefaction induced retaining wall failures seem to be due to strong active seismic force causing tilting/instability of the wall and also due to global slope stability under strong ground shaking.

Observation-based Seismic Microzonation Map

There is an indication of distribution in ground shaking or seismic amplification of ground motion within the city. Geological map and visual observation reconnaissance survey-based microzonation map of city of Banda Aceh was developed as shown in Figure 15. This microzonation map presents approximate distribution of seismic ground shaking from low to high, as interpreted from observation of ground condition, distribution and level of damages of similar building, information geological map as well as limited available soil investigation data.

General geotechnical conditions for city of Banda Aceh is divided by Krueng Aceh river. To the shore consist of gravel alluvium formation, sand, mud, etc. Subsurface condition to the hill (high land) consists of clay and massive limestone. At some area, volcanic sediment and limestone is encountered. To the direction away from the city, older terrace deposits, partly volcanic gravels and sands is encountered. Baserock is estimated to consist of calcareous sandstone, conglomerate, approximately 1000m from the ground surface. General geotechnical subsurface conditions (at grand mosque) is q_c = 0-10kg/cm^2 at depth 0-1m, q_c =10-40 kg/cm^2 at depth 1-3m, and 40-150 kg/cm^2 at depth 3-9m. Below this layer from 9-21m, q_c < 30 kg/cm^2 and again to reach q_c > 120 kg/cm^2 below 20m. The ground water level at this location is approximately 1.5m from the existing ground surface. At nearshore close to Ulee Lheu harbor area, 0-3m consists of silty sand, 3-6.5m consists of silty fine sand (N=27); 6.5 – 10m consists of silty fine sand (N=43); 10-12m consists of silty sand fine to course, very dense (N=56); 12-20m consists of silty fine sand and sandy clayey silt (N=39). Groundwater level is about 1 m from the ground surface.

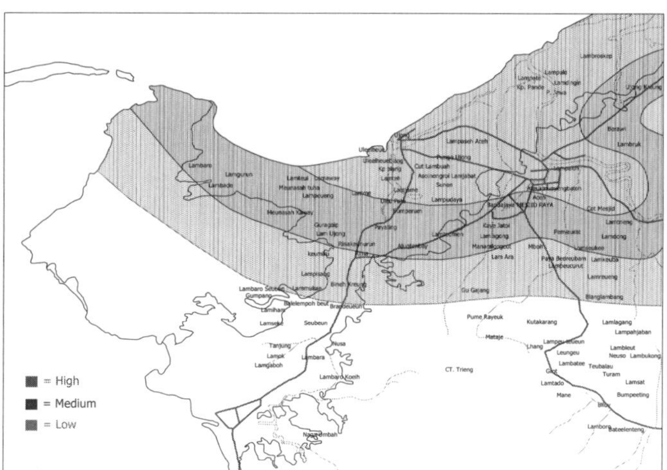

Figure 15. Observation-based microzonation map of city of Banda Aceh

Liquefaction Occurrence

Based on the reconnaissance survey, some occurrences of liquefactions and ground fractures in the area close to the shore were observed as indicated in Figure 16. Considering the general geotechnical conditions where medium to dense silty sand dominate the profile, it is considered that liquefaction did occur in the area with strong shaking and loose to medium dense silty sand condition. Since there is very limited geotechnical data in the city, existence of more sandy like material in the lowland area to the shore direction is anticipated. Within approximately 500m corridor of west coast of Sumatra from city of Banda Aceh to Meulaboh, some trace of sand boils caused by liquefactions identified after the Mega Aceh earthquake. This trace of sand boils is shown in Figure 17. Countermeasures need to be provided for infrastructures developed within this area.

GEOTECHNIAL ENGINEERING MITIGATION STRATEGIES AND MEASURES

Aceh earthquake and tsunami disaster demands provision of geotechnical earthquake engineering mitigation strategies and measures to minimize risk of future disasters. Many failures and damages associated with lack of geotechnical earthquake engineering input to the development of the city. Some mitigation measures to be presented herein are seismic hazard analysis, site-specific response analysis, development of microzonation map, and liquefaction potential analysis. Some aspects related to geotechnical earthquake engineering is an input in the effort to formulate risk mitigation strategies for sustainable development process of the city.

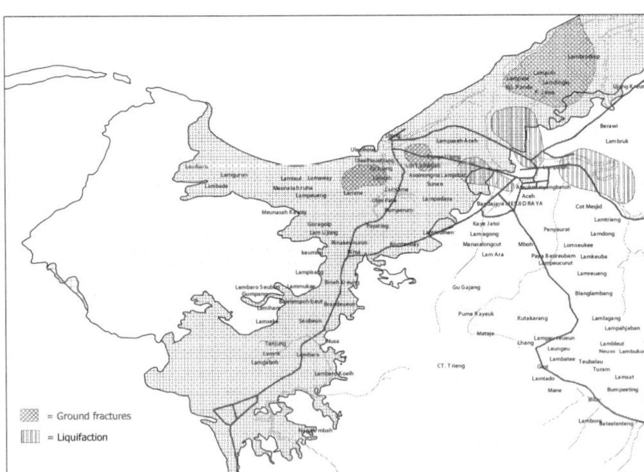

Figure 16. Liquefaction and ground fractures occurrence map of city of Banda Aceh estimated from visual observation reconnaissance survey

Figure 17. Trace of sand boils caused by liquefactions identified at some locations along the west coast of Aceh to Meulaboh after the Mega Aceh earthquake, December 2004.

Preliminary Seismic Hazard Analysis for City of Banda Aceh

Indonesian seismic code prior to the year 2002 for building was based on seismic hazard criteria of 200 years return period. Current Indonesian seismic codes is based on seismic hazard criteria of 500 years return period as stipulated in most international seismic codes (10% probability of not being exceeded in 50 years). This level of probability is developed from the basic concept that earthquake resistant design of buildings or infrastructures is to provide protection particularly to prevent loss of human life and personal injury in addition to minimize damage to property and ensuring that vital services are maintained in the event of earthquake. The phylosophy is to prevent the earthquake hazard to become a disaster. Based on the current codes, the PBA for city of Banda Aceh is about 0.25-0.3g. Seismic potential post-Aceh earthquake may need to be re-evaluated for input to the current seismic load. Existence of Sumatra Fault Zone (SFZ) that passes to the city increases the concern of imminent future big earthquake originated from this SFZ rather than from the subduction source.

Preliminary probabilistic seismic hazard analysis (PSHA) for city of Banda Aceh post Aceh mega earthquake using EZ-FRISK (Risk Engineering, Inc, 2004) computer program has been conducted (Sengara and Hendarto, 2005). We consider 3-dimensional seismic source zones and differentiate attenuation functions for subduction and strike-slips source zones. Young's (1997) attenuation function was adopted for subduction zones, and Boore-Joyner-Fumal (1997) and Sadigh (1997) were adopted for the strike-slips seismic sources. Result of the preliminary PSHA indicates that the PBA is about 0.25g. Although existence of SFZ has been considered in the PSHA, more attention needs to be made on the potential of this near source fault to generate big earthquake in the future and to be considered in the seismic building code for Aceh. Figure 18 shows seismic source zones used in preliminary PSHA.

Development of Seismic Design Spectra and Seismic Microzonation Map

Research related to geotechnical earthquake engineering such site specific response analysis need to be conducted for developing more representative seismic design criteria for buildings and infrastructures following large earthquake. Understanding the characteristics of ground motions of potential future earthquake needs to be studied. Installation of strong-motion records to capture earthquake ground motion characteristics is needed. Effect of local ground conditions represented by dynamic soil properties need to be considered so that seismic load to the buildings appropriately represented. Seismic design spectra representing the effect of local ground condition and characteristic of seismic ground motions from anticipated far field (subduction) and near field (SFZ) seismic sources need to be represented for improvement of in the seismic building codes for the city.

Seismic microzonation map developed from collection of geotechnical subsurface data would identify site with different seismic amplifications. Microzonation map for city of Banda Aceh would need to be developed using available methodologies such as site-classification-based map or site-specific response analysis based-map such as described in Sengara (2005). Seismic microzonation map would be useful as an input as guidance for the local government or building authority that seismic load has been represented. The map would also an essential input in the improvement of city spatial planning. Furthermore, the microzonation map combined with vulnerability maps of elements at risk within the city, is also needed in seismic risk assessment to formulate risk mitigation strategies for development of the city.

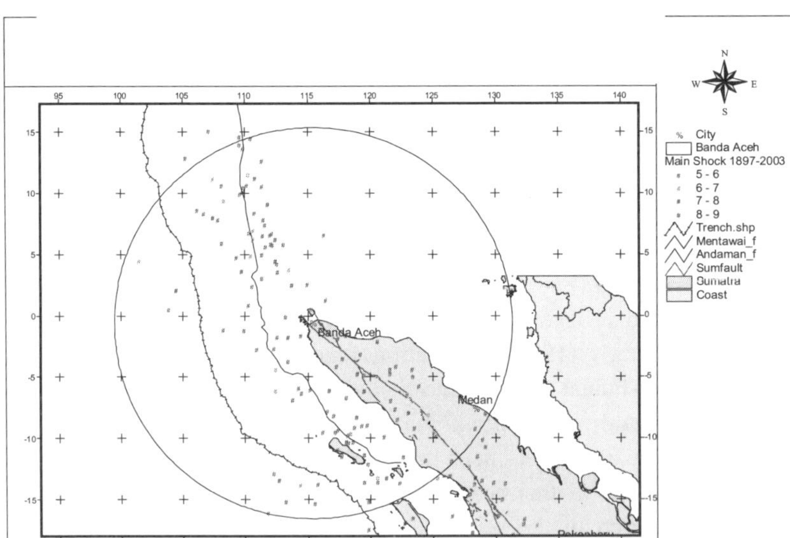

Figure18. Seimicity and source zones within radius of 500km from city of Banda Aceh

Liquefaction Potential Analysis

Recognizing general subsurface condition at site location near grand mosque at city of Banda Aceh and at neashore area close to Ulhe Lheu harbor, a simple liquefaction potential analysis was conducted. Based on the general soil condition using the CPT and SPT data it could be interpreted that the site classification is approximately fall into medium in Indonesian seismic code or S_D according to UBC97). The PGA at baserock from the preliminary seismic hazard analysis is 0.25g. There would be some seismic amplification from base-rock to the ground surface of approximately 1.3, so that peak ground surface acceleration is approximately 0.325g. Calculating Cyclic Stress Ratio (CSR) for potential incoming earthquake of M_w=9.0 from subduction or M_w=7.5 from Sumatra Fault Zone, it is possible to estimate the minimum relative density of saturated sand layer not to liquefy by adopting Stark and Olson (1995) methodology. Table summarizes the minimum relative density (D_r), minimum N_{SPT}, and minimum q_c to prevent soil liquefaction. Some range of PGA of 0.25 – 0.325g and 5 m depth of sand layer is assumed. Based on this simple analysis, it is recommended to have soil improvement and compaction to achieve minimum relative density of approximately > (70-80)% (depend upon fine content of the sand) to avoid liquefaction under future potential earthquakes at Banda Aceh city. Further research on the peak acceleration and effect of local ground condition as well as characteristics of seismic ground motion is suggested for reconstruction of infrastructures of the city.

Table 2. Summary of a simple liquefaction potential analysis for city of Banda Aceh

a_{max}	M_w	r_d	CSR	Minimum $(N_1)_{60}$ to prevent liquefaction			Minimum N_{field} to prevent liquefaction			Minimum relative density to prevent liquefaction - D_R (%)		
g				FC<5%	FC=15%	FC=35%	FC<5%	FC=15%	FC=35%	FC<5%	FC=15%	FC=35%
0.250	7.5	0.92	0.363	27	23	19	17	15	12	83	76	69
0.275	7.5	0.92	0.399	28	24	20	18	16	13	84	78	71
0.300	7.5	0.92	0.436	29	25	20	19	16	13	86	79	71
0.325	7.5	0.92	0.472	30	26	20	19	17	13	87	81	71
0.250	9.0	0.95	0.375	28	24	19	18	16	12	84	78	69
0.275	9.0	0.95	0.412	28	24	20	18	16	13	84	78	71
0.300	9.0	0.95	0.450	29	25	20	19	16	13	86	79	71
0.325	9.0	0.95	0.487	30	26	20	19	17	13	87	81	71

a_{max}	M_w	r_d	CSR	Minimum q_{c1} to prevent liquefaction - **kPa**			Minimum $q_{c\text{-}field}$ to prevent liquefaction - **kPa**			Minimum $q_{c\text{-}field}$ to prevent liquefaction - **kg/cm²**		
g				FC<5%	FC=15%	FC=35%	FC<5%	FC=15%	FC=35%	FC<5%	FC=15%	FC=35%
0.250	7.5	0.92	0.363	14000	11500	6200	8944	7347	3961	89	73	40
0.275	7.5	0.92	0.399	14200	11800	6400	9072	7539	4089	91	75	41
0.300	7.5	0.92	0.436	14500	12000	6500	9264	7667	4153	93	77	42
0.325	7.5	0.92	0.472	15100	12200	6600	9647	7794	4217	96	78	42
0.250	9.0	0.95	0.375	14100	11700	6300	9008	7475	4025	90	75	40
0.275	9.0	0.95	0.412	14400	12000	6500	9200	7667	4153	92	77	42
0.300	9.0	0.95	0.450	15000	12200	6700	9583	7794	4281	96	78	43
0.325	9.0	0.95	0.487	15200	12500	6800	9711	7986	4344	97	80	43

Some other Mitigation Measures and Strategies:

Based on preliminary study and reconnaissance survey, some general descriptive mitigation measures and strategies could be formulated herein as a general for rehabilitation/reconstruction and future development of city of Banda Aceh and affected cities after Aceh earthquake and tsunami disaster. Some of the mitigation strategies and measures would need to be implemented. Learning from the disaster, general guidelines would be to limit buildings/infrastructures reconstruction along the shorelines from tsunami hazard. Whenever or wherever required that buildings/infrastructures need to be constructed along the high risk tsunami hazard, then those buildings shall be relatively heavy (> 3 story) with piled foundation or could be of type piled-raft. The first and second floor of the buildings should have as many openings as possible. The pile and the columns need to be capable of resisting moment generated by the anticipated tsunami wave force. Tsunami resistance housings could be designed and the foundations need to be of types deep buried footing, piled , or piled-raft foundations to avoid erosion and shall have sufficient resistance to relatively high lateral tsunami wave force. Research is needed to correlate tsunami wave height and force on stability of foundations and other geotechnical related structures to tsunami wave force.

Some other general descriptive mitigation strategies related to geotechnical engineering aspects would be among others are:

- develop geotechnical seismic design guidelines
- proper practice in design, construction, and quality assurance/quality control in foundations/geotechnical constructions to minimize risk of foundation failures, liquefactions, lateral spreads, excessive settlements, landslides, and so forth.
- capacity building through dissemination of seismic building codes and geotechnical seismic design guidelines and trainings of engineers, consultants, contractors, and government officials on geotechnical engineering guidelines related to seismic loading and design of foundations, embankment and excavation slope stability, retaining systems, soil improvements, and so forth.
- Education and trainings of building authority/government officials on the potential geotechnical hazards and geotechnical problems such as liquefaction potential, landslides, and other potential instability problems such that to increase their capacity in enforcement process of building designs, permits, and constructions.

CONCLUSIONS AND RECOMMENDATIONS

There was great variation in buildings damage from totally collapsed to those only having slight to non-structural damage during the Aceh earthquake. This damage variation could be contributed by combination of building vulnerability and distribution of seismic amplifications associated with variation in local ground condition within the city. There were also some liquefactions induced failures of bridges embankment, and retaining walls. Liquefaction or strong ground shaking induced settlements was identified. Liquefaction occurrences were identified in some areas and trace of sand boils liquefaction at west coast along Banda Aceh to Meulaboh were also identified. Most single-story buildings or light residential houses within 500-1000m distance from shorelines were completely destroyed by tsunami and erosion and failures of the building foundations occurred. For city development therefore, limit new buildings construction in this corridor, or tsunami and seismic resistance buildings that with soil improvement to prevent liquefaction potential need to be designed.

Preliminary probabilistic seismic hazard analysis has been conducted in this study, however more detailed research needs to be conducted on Sumatra Fault Zone potential to generate future big earthquake. Simplified procedure adopted for liquefaction potential analysis based on peak ground acceleration ranging from 0.25-0.325g demands compaction of saturated sandy or silty sand layer to achieve above (70-80%) relative density.

Some geotechnical and earthquake engineering mitigation measures and strategies such as installation of strong motion records to understand characteristics of seismic ground motions from subduction and strike slips sources potential to hit the city in the future and development of design response spectra need to be implemented to provide to input for improvement of building codes. Geotechnical seismic design guidelines need to be prepared and dissemination of the guidelines is essential for proper practice in the reconstruction of the affected area. Microzonation map and other hazard maps need to be developed for city planning and further formulation of risk mitigation strategies.

Acknowledgements

Materials presented in this paper are partly results of preliminary study and reconnaissance survey of ITB task force team members for earthquake and tsunami disaster. Contribution by the team members is highly appreciated. Preliminary study for seismic hazard analysis of Banda Aceh city was conducted at Geotechnical Engineering Laboratory (GEL) and Center for Disaster Mitigation (CDM) – ITB This study is partially sponsored by Indonesian Ministry of Research and Technology. The authors thankful to Mr. Engkon Kertapati who has involved in providing geological data and join reconnaissance survey and mapping. Also contributions to this paper by Hendarto, P. Sumiartha, and A. Purba as research assistants at GEL and CDM-ITB are appreciated.

REFERENCES

Day, R.W., (2002). "*Geotechnical Earthquake Engineering Handbook*", McGraw-Hill, New York.

Ishihara, K., (1993), "*Manual For Zonation on Seismic Geotechnical Hazards*," Technical Committee for Earthquake Geotechnical Engineering, TC4, The Japanese Society of Soil Mechanics and Foundation Engineering.

Ishihara, K., (1992), "Mitigative Measures Against Ground Damage During Earthquakes", Proceeding IDNDR International Symposium on Earthquake Disaster

Reduction Technology, Building Research Institute, Ministry of Construction, Tsukuba, Japan.

Indonesian Seismic Building Codes, SNI-1726, (2002), Indonesian Department of Public Work.

Kertapati, E, (1999), "Probabilistic Estimates of the Seismic Ground Motion Hazard in Indonesia", Proceeding of National Conference on Earthquake Engineering, Bandung.

Kramer, (1996), "*Geotechnical Earthquake Engineering*", Prentice-Hall Inc. Upper Suddle River, New Jersey.

Latief, H., N.T. Puspito, F. Imamura (2000), "Tsunami Catalog and Zoning in Indonesia", Journal of Natural Disaster, Japan.

Latief, H, Aditya A.G, Haris, S. (2005), "Preliminary Report of Numerical Simulation of the 2004 Aceh Tsunami", to be published.

Risk Engineering, Inc, (2004), EZFRISK, Software for In-depth Seismic Hazard Analysis, Boulder, Colorado, USA.

Sengara, IW., Munaf, Y., Aswandi, and Susila, IG. M., (2001), "Seismic Hazard and Site Response Analysis for city of Bandung-Indonesia", Proceeding of Geotechnical Earthquake Engineering Conference, San Diego.

Sengara, IW. et al. (2005). "Report on Preliminary Study and Reconnaissance Survey of Aceh Earthquake and Tsunami disaster", December 2004 Aceh's Earthquake and Tsunami Disaster Task Force, Institute of Technology Bandung, Indonesia, February.

Sengara, IW., (2005), "Need of City Microzonation as a Mitigation Strategy against Seismic Disaster", Proceeding of the 3rd International Conference on Geotechnical Engineering Combined with 9th Yearly Meering of the Indonesian Society for Geotechnical Engineering, Semarang- Indonesia, August.

Sengara, IW. and Hendarto, (2005), "Probabilistic Seismic Hazard Analysis for Sumatra", Preliminary Research Report, Center for Disaster Mitigation, Institute Technology Bandung.

Shieh, K. and Natawidjaja, D., (2002), "Neotechtonic of the Sumatra Fault Indonesia, Journal of Goephysical Research, Vol. 105, No. B12, 28295-28326.

Stark, T.D. and Olson, S.M. (1995), "Liquefaction Resistance Using CPT and Field Case Histories", Journal of Geotechnical Engineering, ASCE, Nol 121. No. 12, pp. 856-869.

Watabe, M. (1992), "Design Seismic Force", Proceeding IDNDR International Symposium on Earthquake Disaster Reduction Technology, Building Research Institute, Ministry of Construction, Tsukuba, Japan.

Youngs, R. R., Chiou, S. J., Silva, W. J., Humphrey, J. R., (1997), "Strong Ground Motion Attenuation Relationship for Subduction Zone Earthquake", Bulletin of Seismological Society of America Vol. 68, No. 1.

International Conference on Geotechnical Engineering for Disaster Mitigation & Rehabilitation
Chu, Phoon & Yong (eds) © 2005 World Scientific Publishing Company ISBN 981-256-469-1

GEOTECHNICAL ASPECTS OF NATURAL HAZARDS : SRI LANKAN EXPERIENCE

H.N. SENEVIRATNE
Department of Civil Engineering, University of Peradeniya, Peradeniya, Sri Lanka

H.G.P.A. RATNAWEERA
Department of Civil Engineering, The Open University of Sri Lanka Nawala, Sri Lanka

R.M.S. BANDARA
Landslide Division (LSSD), National Buildings Research Organisation 99/1, Jawatta Road, Colombo 5, Sri Lanka

There had been little concern within Sri Lanka of natural disasters until the Tsunami of December 2004. Cyclones, floods, landslides, and droughts have been the more frequent natural disasters experienced by Sri Lanka, however their impact has been considered lightly by people and government. Relief and some rehabilitation had been the solution adopted to tackle these disasters. Recent research and happenings indicate that tsunamis and earthquakes (direct impact) are the low frequency events which may have a catastrophic influence on the country. This paper describes geotechnical aspects of landslides and seismic hazards, and briefly on tsunamis, all disasters having a major geotechnical influence within the country. This paper discusses the nature of the landslide problem, research studies carried out to identify its characteristics, landslide hazard mapping, and other mitigation strategies. It also presents the seismic hazard and research studies carried out to determine the seismic status of Sri Lanka. Long term and short term geotechnical impacts that arise from tsunami hazards are also described.

INTRODUCTION

Sri Lanka, is located in the Indian Ocean approximately 24 km. to the southeast of India between northern latitudes of 5°51' to 9°51' and longitudes of 79°40' to 81°55'. It occupies an area of nearly 65,000 sq.km. spanning 435 km from North to South and 224 km from West to East.

The landform and geology of Sri Lanka are well documented (e.g. Cooray (1967), Vitanage(1994)). The island is made of three well-recognized platforms referred to as peneplains. The lowest peneplain consists of coastal lowlands with elevations from sea level to 300 m above mean sea level (MSL) except for a few isolated inselbergs. The middle peneplain with elevations from 100 m to 1000 m above MSL consists of ridge and valley topography and highly dissected plateaus, occupying nearly three-tenth of the island. The highest peneplain with a series of well defined high plains and plateaus comprises central highlands with elevations above 1000 m increasing to 2500 m above MSL.

The climate of Sri Lanka varies quite considerably over its area. This variation due to central highland region being surrounded by an extensive low land area is influenced by two monsoonal and two inter-monsoonal periods. The highest annual rainfall of 3000 mm to over 5000 mm is received in the wet zone to the south-west of the island. In the lowland dry zones to the north-west and northeast-southeast, rainfall averages less than 2000 mm annually. The wet zone is separated from dry zone by an intermediate zone having an annual rainfall of 2000 to 3000 mm. The mean annual temperature is 27° C in the lowlands and is 15° C in the central highlands.

Sri Lanka forms the southern continuation of Indian Precambrian Shield and crystalline metamorphic rocks underlie over 90 percent of the country (Cooray, 1967). The rest consists of sedimentary rocks of predominately Miocene age and some of the Jurassic age, mainly confined to the northwestern and eastern coastal belts and to the Jaffna archipelago. The Precambrian rocks of Sri Lanka are made of three major groups, namely the Highland Series, the Vijayan Complex and the Southwestern Group. The Highland Series occupies mainly the central part of the island, consisting of pyroxene granulites, facies rocks made of garnet sillimentite gneisses, graphite schist, charnockites, quartzites, and marbles. The Vijayan Complex rocks consists of mainly granitic gneisses, hornblende biotlite gneisses, and migmatites formed under amphibolites facies, and are mainly confined to the eastern sector. The Southwestern Group consists of cordierite granulite facies rocks made of cordierite gneisses, calc granulites, and metasediments.

Foliation and jointing are the two most important structural properties of Sri Lankan rocks, particularly in highlands. Flatish minerals such as mica are arranged parallel to each other giving a planar structure to the rock mass often leading to schist or gneiss formations. Banding of light and dark colour minerals is common. Jointing plays an important role, making the rocks more permeable, a cause of deep weathering. When weathering products remain insitu throughout, the overburden thickness may be as high as 30-50 m; overburden is separated from underlying bedrock by a partly weathered horizon. However, weathering products are often transported by gravity, wind, or water giving rise to formations such as colluvium. In addition to these, in the lowland coastal areas, estuarine deposits, such as peat and organic soil, aeoline and deltic deposits, and coastal sediments are found.

NATURAL DISASTERS IN SRI LANKA

In comparison to other countries in the Asian region, Sri Lanka has been considered for a longtime to be relatively free from major natural disasters. However, the cyclone induced floods of May 2003, and more importantly the tsunami of December 2004 demonstrated the vulnerability of the country to natural disasters of low frequency and high impact. Natural disasters in Sri Lanka are mainly hydro-meteorological phenomenal events such as floods, landslides, cyclones, tidal waves, and droughts causing damage to the people and property disrupting social and economic development; some of these disasters are aggravated by human intervention.

Table 1 shows the extent of the loss of life and property due to natural disasters during the period 1993-2001 in Sri Lanka. Cyclones and landslides have caused the highest number of deaths during this period; floods and cyclones damaged the largest number of houses. Relief expenditure and the number of families affected were substantial for droughts, cyclones, and floods. This picture with a moderate disaster record was drastically amplified with cyclone induced floods and landslides that took place around 17th May 2003, and the tsunami disaster of 26 December 2005. May 2003 disaster caused over 240 deaths and over 200,000 people displaced. It was estimated that total loss to the economy due to May 2003 disaster is around US$ 76.7 million. Earthquakes, though with a low frequency of occurrence had caused little damage with regard to human life and property (one death and a house destroyed) during the last decade. Sri Lanka experienced its worst natural disaster on the 26th of December 2004. The massive earthquake registering 9.0 on the Richter scale that struck off the coast of Sumatra, set off a series of tsunami waves that reached coastal areas of many countries including Sri Lanka. Sri Lanka was one of the hardest hit countries estimated losses at more than 31000 lives lost, over 4000 missing and 1 million affected.; almost half of the affected lost their livelihoods. The estimated economic loss is estimated at 8.0% of the GDP.

Table 1. Number of Deaths and Families Affected by Natural Disasters and Amount of Relief Distributed by the Government 1993-2001

Type of Disaster	No. of Deaths	No. of houses damaged/100	No. of families affected/100	Expenditure for relief /million Sri Lankan Rs
Floods	46	1308	9411	265
Landslides	75	9	29	11
Sea Erosion	1	91	122	18
Cyclone	38	893	4366	580
Drought	0	0	8976	859
TOTAL	86	2301	22904	1733

Floods contribute to about 50% of total disaster occurrence in the country; this is the commonest of natural disasters causing damage to life and property. Between 1984 and 2003, the records show that about 310 people lost their lives due to floods, and about 2.5 million being displaced in addition. Sri Lanka is frequently hit by droughts. Short duration droughts of local significance are experienced every year. Regionally significant droughts occur once in every 3 or 4 years and severe droughts of national significance occur occasionally. Costal erosion is still a severe problem in Sri Lanka, affecting fishing communities. The loss of coastal land each year has been estimated to be around 300000 - 500000 square meters. The most vulnerable areas are the southern and south-western costal belts.

The eastern, northern, and north-central regions are vulnerable to cyclones which occur as a result of climatic changes in Bay of Bengal and in the Arabian Sea. Cyclones are not frequent in Sri Lanka but high gales and wind storms accompanied by monsoonal rains have often caused considerable damage to people's livelihood and the environment. According to information available, the 1978 cyclone was the second most tragic disaster in Sri Lanka ever recorded. It claimed more than 1000 lives and over 250,000 houses were damaged in eastern, northern and north central regions of the country. Out of the natural disasters affecting Sri Lanka, geotechnical problems are associated mainly with landslides, earthquakes, and to a lesser extent with tsunamis. Therefore, the remaining part of this paper is confined to geotechnical aspects of these disasters.

LANDSLIDE HAZARD

Occurrence

An area of nearly 20,000 sq.km. of Sri Lanka is prone to landsliding. The incidence of landslides and the frequency of its occurrence are growing by the year. The present landslide density is estimated to be in the order of 1-2 landslides per sq.km, within the hilly landslide prone districts of the country. The landslide density is reported to increase in regions where slope degradation processes are at an advanced stage, resulting in adding one landslide per every six sq.km. every year. The changes resulting from undue human interventions such as bad land use and cropping practices, settlements in unstable areas, changes to natural drainage paths resulting in blockages, and non-engineered development and expansion have caused a phenomenal increase in the incidence of landslides in hilly areas.

A cyclonic depression in the Bay of Bengal during the period 19 – 21 May 2003, led to torrential rains in Southern coastal and highland regions of Sri Lanka. The heavy rains led to flash floods and landslides causing over 190 deaths and displacing nearly 200,000 families. About 3000 families were severely affected by landslides and their dwellings destroyed. A comprehensive study of 316 landslides during this disaster period was carried out by OPC (2003), a voluntary team of geotechnical experts drawn from different government organizations. The team inspected 316 landslides during a two weeks period and identified 135 high risk areas and 69 medium risk areas, and 831 householders to be evacuated.

Figure 1. Landslide at PothipitiKanda

The severity of the rainfall was unusually high in this disaster; unlike the normal monsoonal rain which triggers the failure of slopes with marginal factors of safety; unusually intense rain caused the failure of slopes hitherto considered to be safe from landslide hazard. Degradation of slopes due to improper land use practices, gem mining, and unplanned development also contributed to the large number of failures. Early warnings were not heeded. Most of the failures occurred on scarp slopes, particularly along river valleys. The biggest slide shown in Figure 1 at Pothipitikanda resulted in a debris flow which formed a dam across the stream below; the subsequent breach of the dam created a flood wave which devastated the downstream settlements. This landslide is unusual in the sense that debris flow split and rejoined downstream giving a garland like appearance to the slide. The largest number of deaths numbering 62 occurred on a scarp slope failure at Abeypura village.

Hazard zonation

Landslide hazard zonation mapping was carried out by National Building Research Organisation of Sri Lanka under assistance from United Nations Organizations. By the end of year 2004, 4360 sq.km. of area within landslide prone districts of Sri lanka has been covered (Bandara, 2004); this is about 20% of the total landslide prone area of the country. The zonation maps were prepared at 1:10,000 scale; they were based on topographic maps of the Survey Department of Sri Lanka. Landslide Hazard Zonation Maps have been compiled using basic data gathered from field surveys and desk studies. The factors taken into consideration in preparation of the maps were slope category, bedrock geology and structure,, past landslides and overburden deposits, landform, land use and management, human settlements and infrastructure, and hydrology. A set of "derived maps" is then deduced from these maps using an appropriate scoring (point) system, which are later analyzed by the use of a computer program. A Landslide Hazard Zonation Map is produced by integrating the effect of all the categories of maps mentioned above with the use of GIS Software (eg. PC ARC/INFO). Cumulative attribute values associated with the hazard map are categorized into four different ranges shown in Table 1 below, depending on the degree of the hazard.

Table 1. Criterion for Landslide Hazard Zonation

Overall Hazard Rating (R)	Hazard Zone	Description
R< = 40	I	Safe Slopes
40<R<=55	II	Landslides not likely to occur
55<R<=70	III	Modest level of landslide hazard exists.
70<R	IV	Landslides are expected.

Monitoring and Early Warning

In Sri Lanka, the major triggering factor of landslides has been intensive heavy rains. The research conducted by National Building Research Organization, the government organization responsible for landslide hazard studies, has indicated that if the rainfall exceeds a threshold value of 200 mm within three consecutive days and the rain continues, the probability of landsliding should be considered high. A slow moving landslide at Nawalapitiya, Sri Lanka was monitored by Seneviratne and Ilmudeen (1990). The settelements of the landslide area were regularly monitored by levelling of markers buried on the site area; the fluctuation of the watertable was carried out by monitoring the response of standpipe piezometers installed on the site. The results obtained over a period of six months indicated a good correlation between the fluctuation of watertable and observed settlements. The average rate of land subsidence in the sliding area increased from 13.5 mm/year in the dry season to 100 mm/year in the wet season. According to Varnes (1958), this corresponds to a slow moving landslide. This landslide subsequently failed totally during a heavy rainy period, illustrating the of use of 100mm/year settlement rate as the figure for early warning.

Monitoring of another landslide at Beragala, Sri Lanka was carried out by a joint research programme between, Hiroshima University, Japan, and University of Peradeniya, Sri Lanka. Here the settlement of the landslide area was monitored using extensometers and through levelling of surface markers. The fluctuation of the watertable within the sliding area was monitored using standpipe piezometers installed insitu. The measurements carried out over a period of 75 weeks showed that subsequent to a heavy rainfall, the landslide has reactivated along a different central axis (Dissanayake, et.al., 1999). The reason for reactivation along a different axis was due to the effect of recently installed horizontal subsurface drains reducing the piezometric pressures within the original slide area. The subsidence data showed better correlation with weekly rainfall than with two weekly rainfall or three weekly rainfall; a threshold weekly rainfall of 210 mm was proposed by the authors as an early warning figure for this landslide. The extensometer measurements clearly indicated the extent of the landslide region, which was helpful in identifying the potential slip surface.

Monitoring of Watawala landslide was carried out by National Buildings Research Organization of Sri Lanka. Watawala landslide has been one of the most important landslides as it affects the main railway line to central highlands. A major failure occurred in 1993 June and continued for a period of about six months during which a staggering movement of about 20 m took place. During heavy rainy periods, recorded rates of displacement at certain locations were around 8m per month. The movements monitored using surface markers and inclinometers helped to identify the potential slip surface and showed that the movement of the sliding area was directly related to piezometric observations and rainfall (Jayaratne, et. al., 1994). Stability analyses carried out assuming the potential slip surface and shear strength of the colluvium mass overlying the sliding surface gave high factors of safety. However, the stability analyses carried out using the residual shear strength on the slip surface as determined from field shear box tests showed that with the rise of piezometric pressures (e.g. due to heavy rain) the slope approaches failure (Rajaratnam and Bhandari, 1994).

Mitigatory measures

In Sri Lanka, disaster mitigation is integrated in the environmental impact assessment process. This is a requirement in landslide prone areas and it encompasses mini hydro-power projects, quarries, and large scale housing schemes. This is also carried out by certain local government authorities using the landslide hazard zonation maps, by enforcing some control through construction permits for houses etc. A systematic incentive scheme to discourage dwellers or cultivators on unstable slopes was introduced by Sri Lankan Government in 2003 by compensating the hillside dwellers to abandon the affected slopes and also to resettle them on suitable alternative land. Although Insurance scheme do exist to cover certain disasters, natural disasters are not covered by insurance; some initiatives taken to persuade insurance companies to play a role in disaster mitigation has met with little success. The difficulty to encourage dwellers on unstable slopes to abandon them, especially those dogmatic and the conventional hillside cultivators, still exists and unless the necessary investment can be made little can be expected in that direction.

In the area of structural mitigation, National Housing Development Authority now design their housing schemes on hill slopes to suit the terrain conditions by including retaining structures and appropriate drainage networks. Among some community-based mitigation works carried in landslide risk areas are sites in Nawalapitiya and Kandy Districts, which include drainage network development, forestation, and excavation of trenches. Among tangible examples of structural intervention in risk mitigation works in the past, a major milestone is the detailed study of the Watawala landslide described earlier in this paper. Following the detailed monitoring and investigation work carried out by the National Buildings Research Organization, lowering of ground water table in the sliding area was carried out by an Australian company in 1992/1993 using directional drilled drains. This project is considered a success in stabilizing the movements and the railway line has been functioning without any problems since this intervention.. This has been a landmark case in which directional drilling has been used to install subsurface drains along the potential slip surface, to lower the piezometric pressures on the slip surface (BDP YEARBOOK, 1998).

Other interventions include landslides on trunk roads leading to Beragala, Pussellawa, and Koslanda, where Sri Lankan Road Development Authority executed road repairs creating the necessary surface and subsurface drainage network thus lowering the ground water table to ensure road stability. Some landslides were stabilized by using low cost techniques such as used tyre retaining walls in which sliding area was anchored using nylon ropes; tyres were also used as a facing element in the wall. Model and analytical studies in this regard were carried out by (Sumanarathna et al, 1998).

SEISMICITY

Sri Lanka is considered to be in an aseismic zone away from major plate boundaries or any active faults. However, in the past, there had been many devastating intra-plate earthquakes reported in so called Stable Continental Regions of the world causing enormous damage. It is observed that the damage caused by such earthquakes tend to be very high due to lack of preparedness against them. A classic example of this is 1993 Maharashtra earthquake in India which caused death of more than ten thousand people

(Mandal et al. 2000). Therefore, the risk of being struck by an earthquake cannot be overlooked merely due to hard evidence to the contrary.

From 1600 A.D. to date there are about 100 earth tremors reported in and around Sri Lanka (Abayakoon, 1998). Some of these events are described in historical records and more recent events have been identified by institutions such as United States Geological Survey (USGS). Seismic activity close to Sri Lanka during the period 1800 to September 2001 is shown in Figure 2. Many of these earthquakes have been categorized as shallow earthquakes. There are records of ten earthquakes to have occurred within Sri Lanka during the last 400 years. Seven of these earthquakes had epicenters close to Colombo, the heavily populated business capitol of Sri Lanka, than any other major city in the country. The reported damage and human casualties due to these earthquakes have been moderate. However, the first documented earthquake in Sri Lanka occurred in 1615 and had an estimated magnitude of 6.5 on the Richter scale; it severely damaged Colombo killing thousands of people and destroying hundreds of homes.

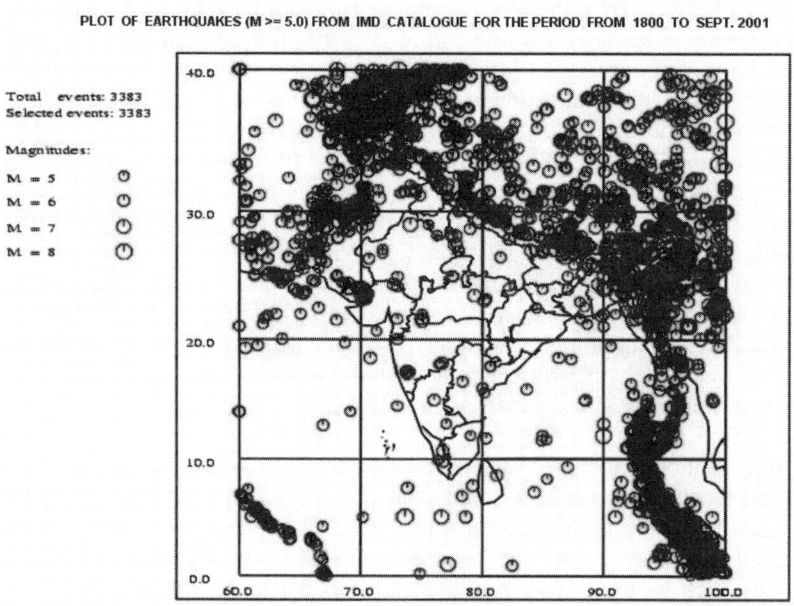

Figure 2. Seismicity around Sri Lanka (Source:USGS)

Given the lack of preparedness and the absence of predisaster mitigatory measures, the occurrence of even a moderate earthquake close to urban centres such as Colombo may have disastrous consequences. Therefore, the study of seismic effects on Sri Lanka has become an important concern, particularly after the experience of 2004 tsunami, the result of another low frequency high impact hazard.

Abayakoon (1996) carried out a seismic risk analysis of the region around Sri Lanka based on reported seismic events in the region. This analysis focused on developing seismic risk contours based on the peak ground acceleration (PGA) values estimated around Sri Lanka. A contour map of PGA having return periods ranging from 100 years to 300 years was proposed. Abayakoon (1998) analyzed the seismic responses of soil profiles in low-lying areas in Colombo. These investigations had been focused on predicting the attenuation characteristics of the base rock motion through the soil profiles using acceleration history records; in the absence of local data, two scaled-down acceleration history records of Chiba EW (Japan) and El Centro NS (U.S.A.) with peak accelerations of 0.1g and 0.2g respectively were used in the analyses. The computer program SHAKE (Schnabel et al. 1972) was used to demonstrate the site responses for five broadly identified soil sub-series in Colombo. Out of the five sub-series, the input motions tend to amplify in one series and damp to about 50% in two other series. As the categorization into sub-series was fairly broad based, a more detailed analysis was recommended for further study.

An important study of seismic status of Colombo is reported in Mohadevan (2003). The data from 300 boreholes at 200 locations within Greater Colombo area have been used to determine the stratigraphy of the study area. The depth to bedrock in the study area varied from 20 m to 25 m. Based on the dominant soil type close to ground surface, the study area was classified into three main groups viz. locations dominated by sand, clay, and peat respectively. Based on the SPT values, sand deposits were classified as loose and dense. The dynamic properties for modeling the soil behaviour have been found from SPT data using empirical relationships available in the literature.

Fifteen different scenarios representing earthquakes of three different Richter magnitudes of 5.0, 6.0, and 7.0 and five epicentral distances ranging from 10-160 km have been used to analyse responses at 123 locations within the study area. In the absence of any data from earthquakes close to region, the North-South component of motion recorded in El Centro during 1940 Imperial Valley Earthquake was used in the analysis. The analyses were conducted using ProShake (EduPro Civil Systems, Inc, 1998). The analyses revealed that all the fifteen levels of input motions have been modified by a similar trend in all the locations. The amplification ratio, i.e. the ratio of peak acceleration at the ground surface divided by peak acceleration of input at base rock level, ranged from 0.2 to 10. Further, for a particular level of ground shaking, the ratio of maximum to minimum amplification ranged from 10 to 20; the degree of amplification was higher in weak motions.

The soil liquefaction potential in the study area was evaluated using the procedure given by Seed et al. (1970). According to the analysis, 20% of the locations are highly susceptible to liquefaction even at a low level of shaking; 50% of the locations required moderate to high levels of shaking for liquefaction to occur, and 30% of the locations were not susceptible to liquefaction.

The weakness of a building against seismic motion depends on the strength of the seismic action and the ability of the building to resist such actions. In general the buildings which are vulnerable to a moderate earthquake are considered as weak; most of the buildings in Colombo constructed prior to 1960 fall into this category. A field assessment focusing on weak buildings was carried out to investigate the structural setting and seismic vulnerability at some selected locations within the study area. The information reported by Lewcock (1980) has been used to select the buildings for evaluation.

Forty old buildings (pre 1960 era) identified as weak, and ten buildings constructed after 1960s' were considered for the seismic evaluation. These fifty buildings include twenty seven un-reinforced masonry buildings, twenty two concrete frame buildings and one light metal building. Most of the buildings in Colombo fall into the second category. The buildings in the first category, i.e. those constructed prior to 1960, lack reinforcement. and are in a poor condition having no capacity to resist any seismic loading.. According to Lewcock (1980) a few hundreds of such buildings are found in the Greater Colombo area.

The seismic vulnerability of the structures was determined based on the evaluation procedures recommended by the FEMA (Federal Emergency Management Agency of United States of America). In this procedure, each building is given a set of basic and modifying scores to account for various structural conditions affecting the seismic stability. The analysis was carried out on all the structures against three levels of shaking, namely low, moderate, and high level as reported by FEMA[1] (1988). The performance modification factor to account for the effect of local soil profiles is carried out as given by FEMA[2] (1988). The sum of the hazardous scores of each of the buildings indicates the vulnerability of the structure against the level of shaking considered.

For the case of high level of shaking all the buildings considered except the light metal structure registered poor seismic performance with more than 10% of the buildings exposed to major damage. About 75% of the buildings considered recorded less than 1.0 structural hazard score for moderate level of shaking. All the 27 un-reinforced masonry buildings registered critical hazard scores for all the three levels of shaking. Out of the 22 cases corresponding to concrete framed buildings, 15 satisfied the cut-off score (2.0) for the low level of shaking. The light metal building exhibited best performance with high structural hazard scores.

Results of the research carried out so far illustrate the vulnerability of Colombo against an earthquake. It is important to concentrate research into areas such as, investigation of seismic activity associated with the emerging plate boundary below Sri Lanka, to determine the characteristics of earthquakes which may happen in the future, and to develop an effective management strategy to reduce earthquake risk in the future.

GEOTECHNICAL ASPECTS OF TSUNAMI

Introduction

Sri Lanka experienced its worst natural disaster on the 26th of December 2004. The massive earthquake registering 9.0 on the Richter scale that struck off the coast of Sumatra, Indonesia at 0059 GMT set off a series of tsunami waves that reached coastal areas of many countries including Sri Lanka. Sri Lanka was one of the hardest hit countries in terms of loss of life, infrastructure, and assets. Between two to three hours after the earthquake, waves reached more than two thirds of the coastal area of Sri Lanka. Final estimated losses due to this disaster stand at more than 31000 lives lost, over 4000 missing and 1 million affected.; almost half of the affected lost their livelihoods. Sri Lanka with 1809 persons killed per 1 million inhabitants had the highest number of per capita casualties from all natural disasters in the world in 2004. These are staggering numbers for any country, but especially for Sri Lanka when they are compared in relative terms and the capacity to recover (Herath, 2005).

The tsunami affected many coastal cities and villages along the eastern, southern and western regions of Sri Lanka. The eastern and southern coastal areas were subject to an onslaught caused by direct waves generated from the earthquake, whilst the western region was damaged by reflected waves. The boundary between the continental crust and the oceanic crust is located at close proximity to the present shoreline and hence wave implication and energy dissipation took place within the landmass. In certain locations where this boundary had caved in towards inland had experienced massive destructions. Many natural and artificial harbours and confluences were subject to damage and resulted in inland flooding. Sri Lanka was not prepared to brace itself against a tsunami hazard since such occurrences had taken place a few generations before and such knowledge and experience is not passed over or documented. This event illustrated the necessity for studying the country's vulnerability to hazards of low frequency and high impact.

In Sri Lankan coastal zones, erosion takes place in the south and south-western regions, long-term, while sediment deposition takes place in the eastern and south-eastern regions. Sri Lanka's coastal zone has been gazetted as a no-built zone for a 100m distance inland from the shoreline. Successive governments have not been able to implement this regulation mainly due to socio-political reasons. Many dwellings, schools, hospitals, hotels, and other government and commercial buildings are located within this region and have experienced damage or total destruction during the recent tsunami. Urban Development Authority has proposed a 100m – 200m no built zone from the shoreline for new construction. However, despite the tsunami impact this has not been found wide acceptance required for effective implementation.

Geotechnical impacts on buildings

During the recent tsunami many buildings were subjected to the onslaught of the wave, the direct impact causing damage to the super-structure and the foundation. Many buildings in the eastern and south-eastern districts were washed away or damaged by wave action.

Figure 3. Scouring of building foundations

These foundations were subject to the direct wave with a significant force on the super-structure. Many school buildings in the southern district of Matara had the foundation washed away completely. These structures perhaps more than 100 years old had weaker foundations made of rubble masonry. The remnants show that some of these structures had weathered granite or beach rock that may have decayed over its lifetime. It is also observed that addition of lime to cement mortar may have also weakened the bond. In contrast to these observations newly built two-story school buildings had sustained the impact even though the furniture housed within the building was swept away to the sea. These framed structures were found to be adequate to resist the impact and the usual 1m deep strip foundations were able to transfer the horizontal force safely. Many single-story housing units that weren't adequately braced had experienced significant damage. There were instances where scouring below foundations and floor had made the super-structure to collapse. Scouring effect due to receding waves and occasional foundation failure due to differential settlements caused by rising watertable are observed particularly along the eastern coast where surface soil deposits mainly consists of sand (Figure 3).

Damages at river mouths and other locations

River mouths and those of local waterways were subject to erosion damaging some bridge abutments (Figure 4). The southern artery that passes next to the shoreline in south-western districts experienced damage and was repaired immediately. These districts lie within the low-lying coastal peneplain. Drawdown failures occurred along river ways along which water advanced with the incoming wave and receded afterwards; this is visible clearly in the southern railway soil embankments. Certain inland locations where coral mining had resulted in open pits had caused those segments to breach.

Figure 4. Damages to a bridge abutment in the Eastern Province

Geo-environmental Hazards

The recent tsunami had resulted in long-term geo-environmental hazards. High salinity levels in groundwater and within the soil root zone are observed in most parts of the coastal regions. Rural and semi-urban households obtain their potable water, through open dug wells. These are used for domestic and home-gardening needs. The increased soil salinity within the root zone also has damaged home-garden agriculture. High salinity levels are still observed in many wells that were rehabilitated during post-tsunami efforts. Many wells were over pumped during rehabilitation, causing high salinity levels. Overpumping and alteration to beach profile has changed the fine balance between fresh ground water overlying brackish water regime connected direct to sea. Considering the scarcity of potable water within south and south-eastern coastal zones, the effect of tsunami on ground water resources and long-term remediation may be necessary. Tsunami has also led to accumulation of seabed sediments in some areas; however, these accumulations are usually not more than 150 mm in thickness and therefore, may be treated as part of the top soil as far as the foundations are concerned. However, these deposits may have some impact on ecology and environment along with tsunami debris and biological waste.

The scale of the disaster does not warrant technological interventions and hence remediation should take place through natural processes. In this regard, monitoring is an important aspect to facilitate the necessary transformation. The success of restoring pre-tsunami conditions or new post tsunami equilibrium of natural forces, depends largely on community participation. Post-tsunami efforts to date have not addressed such long-term concerns, and hence no organization has embarked on such a study. However, one exception may be ground water contamination and salinity intrusion for which research studies have already been initiated.

REFERENCES

Abayakoon, S.B.S. (1996). "*Seismic risk analysis of Sri Lanka*", Journal of the Geological Society of Sri Lanka, Vol. 6, 65-72.

Abayakoon, S.B.S. (1998). *"Seismic response of low lying areas in Colombo Sri Lanka*", Engineer, Journal of Institution of Engineers, Sri Lanka, Vol. 28, No-02, 29-36.

Bandara, R.M.S. (2004). *Sri Lanka Country Report,* Asian Program for Regional Capacity Enhancement for Landslide Impact Mitigation (RECLAIM), September 2004, Bangkok, Thailand.

BDP Yearbook (1998), *"Stabilisation of Watawala Landslide Sri lanka",* Association of Consulting Engineers, Australia, pp. 29-39.

Celebi, M., Dietel. C., Prince, J., Onate, M.,and Chaverz, G. (1987). *"Site amplification in Mexico City"* (Determined from 19 September 1985 Strong –Motion Records and from Recordings of Weak Motions), 3rd International Conference on Soil Dynamics and Earthquake Engineering. Princeton University, U.S.A, 141-151.

Cooray, P.G. (1967). *"An introduction to geology of Ceylon", A Publication by National Museum of Ceylon.*

Dissanayake, A.K., Seneviratne, H.N., Edirisinghe, H.J. and Abayakoon, S.B.S., (1999), *"Monitoring of surface movements of Beragala landslide in Sri Lanka",* Eleventh Asian Regional Conference on Soil Mechanics and Geotechnical Engineering, Hong et al. (eds), Balekama, Rotterdam.

EduPro Civil Systems, Inc. (1998). *"Proshake, Ground response analysis program, Ger 1.1, Users Manual"*. Redmond, Wasjngton, 53

Federal Emergency Management Agency, (1988[1]). *"Rapid visual screening of buildings for potential seismic hazards; A handbook"*. FEMA 154, July, Earthquake Hazard Reduction Series 41, 185.

Federal Emergency Management Agency, (1988[2]). *"Rapid visual screening of buildings for potential seismic hazards; Supporting documentation"*. FEMA 155, September, Earthquake Hazard Reduction Series, 42, 137.

Herath, Srikantha. (2005) *"Post Tsunami Survey forHazard Map Preparation in Sri Lanka"*, Paper prepared under Environment and Sustainable Development Programme, United Nationas University, Tokyo.

Jayaratne, V.S., Jayarani, H.M., Dissanayake, D.M.A.S., Fernando, T.R.J. (1994), *"The meaning of surface movements monitored at the Watawala Landslide",* National Symposium on Landslides in Sri Lanka, organized by National Buildings Research Organisation, Colombo, March .

Mandal. P., Rastogi, B.K. and Gupta, H.K (2000). *"Recent Indian earthquakes"*, Current Science, Vol 79(9), 1334-1345.

OPC (2003), Operational Professional Combine Report on "Landslide Disaster May 2003", Central Engineering Consultancy Bureau, Colombo.

Rajaratnam, K. and Bhandari, R.K. (1994) *"Back Analysis of the Watawala eartslide in terms of effective Stress",* National Symposium on Landslides in Sri Lanka, organized by National Buildings Research Organisation, Colombo, March .

Schnabel, B., Lysmer, J. and Seed, H.B. (1972). *"SHAKE A computer program for earthquake response analysis of horizontally layered sites"*, Earthquake Engineering Research Center, College of Engineering, University of California, Berkeley, 100

Seed, H.B. and Idriss, I.M., (1972). *"A Simplified procedure for evaluating soil liquefaction potential"*, College of Engineering, University of California, Berkeley, 25

Seneviratne, H.N. and Ilmudeen, M.S.M. (1990). *"Subsidence and Stability Analysis of a Landslide in Sri Lanka",* Tenth Southeast Asian Geotechnical Conference, April, Taipei.

Sumanarathna, I.H.D., Mallawarachchi, D.P., and Kulathilake, S.A.S. (1998), *"Stabilization of Slopes by Anchored Tyre Retaining Structures"*, Proceedings of the Sri Lankan geotechnical Society Seminar on Earth Retaining Structures, Colombo.

Varnes, D.J. (1958), *"Landslide Types and Processes"*, Landslides and Engineering Practice, Ed. by Eckel, E.B., Highway Research Board, Special Report 29, Washington.

Vitanage, P.W. (1994). *"Morphology and lineaments as factors in landslide studies in Sri lanka"*, National Symposium on Landslides in Sri Lanka, March, Colombo, Sri Lanka, 23-30.

International Conference on Geotechnical Engineering for Disaster Mitigation & Rehabilitation
Chu, Phoon & Yong (eds) © 2005 World Scientific Publishing Company ISBN 981-256-469-1

APPLICATIONS OF TWO COSTAL PRETECTION STRUCTURE CONSTRUCTION METHODS

SHUWANG YAN
Geotechnical Research Institute, Tianjin University, Tianjin, 300072, China

JIAN CHU
School of Civil and Environmental Engineering, Nanyang Technological University, Singapore 639798

Two methods that can be used for the construction of costal protection structures are introduced in this paper. The first is the geotextile bag method which uses either sand or clay to fill geotextile tubes or bags to form dikes or breakwaters. The second is to a method for using prefabricated semicircular reinforced concrete caissons to construct offshore dikes or structures on soft soil. Two case studies are presented to illustrate the applications of the two methods.

INTRODUCTION

The importance of costal protection has been highlighted during the devastating tsunami in December 2004. However, a large amount of resources are required to construct costal defence or protection facilities. This is particularly true when the coast to be protected is long. Therefore, the selection of the most cost-effective costal protection structures and the construction techniques becomes important in reducing the cost of the project. When the dikes or other types of costal protection structures are long, a small improvement in the design could result in a significant amount of saving. Therefore, it is beneficial to review as many methods as possible so the most cost-effective methods can be identified or to develop new methods that suit the local conditions the best. In this paper, two methods that can be used for the construction of costal protection structures or costal disaster rehabilitation works are introduced. The first is the geotextile bag method which uses either sand or clay to fill geotextile tubes or bags to form dikes or breakwaters. The second is the use of prefabricated semicircular reinforced concrete caissons to construct offshore structures or dikes on soft soil. Both methods have been used successfully in costal protection projects in China. Two case studies are presented to illustrate the applications of the two methods.

GEOTEXTILE BAG METHOD

The traditional method of constructing shoreline structures is to use rock or prefabricated concrete units. In recent years, several methods have been developed to use geotextile materials for the construction of coastal structures such as breakwaters and dikes.

One of the methods is to use geotextiles acting as formwork for cement mortar units cast in situ (Silvester and Hsu, 1993). The mortar mix need be only of sufficient compressive strength to support the weight above, plus the moment from the side force of the waves. Since the flexible membrane is required to hold the mixture in place until it sets, any subsequent deterioration due to UV rays or other conditions is of little concern. Thus, the method tends to be cheaper than the conventional methods. Applications of the mortar filled geotextile tubes are illustrated in Figs. 1 and 2. Details are referred to Silvester and Hsu (1993).

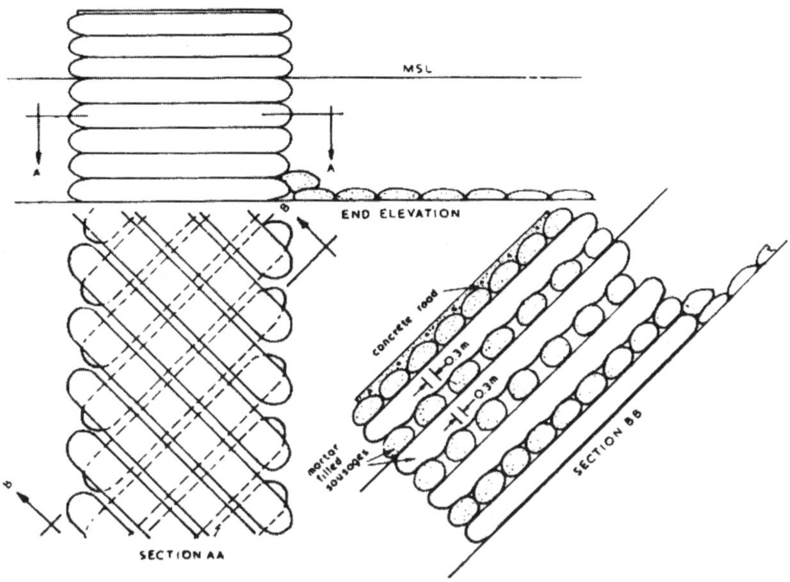

Figure 1. Plan and section of breakwater trunk using the mortar filled geotextile units (After Silvester and Hsu, 1993)

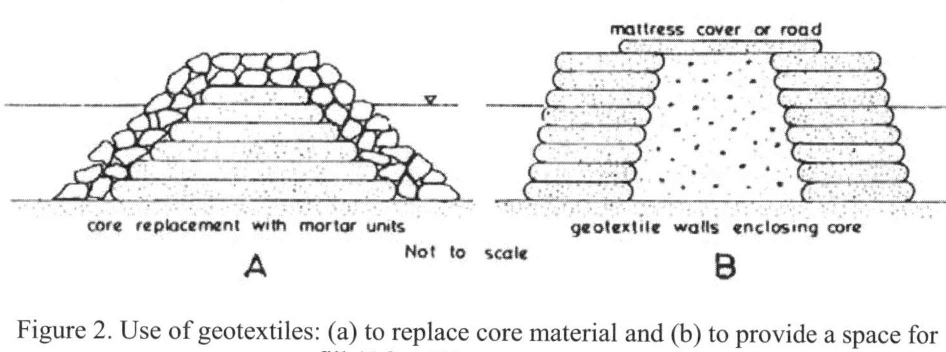

Figure 2. Use of geotextiles: (a) to replace core material and (b) to provide a space for core fill (After Silvester and Hsu, 1993)

Similar methods, but using sand or dehydrated soil as the fill material have also been used for dike construction (Kazimierowicz, 1994; Miki et al., 1996). Sand or sandy soil is the most ideal fill material for this purpose. For near shore or offshore project, a suction dredger can used to pump sand from the seabed or a sand pit directly into the geotextile tubes. In case sand is not readily available, silty clay or soft clay may also be used. In this case, the clayey fill would have to be in a slurry state in order to be pumped and flow in the tube. The slurry would have to be dewatered in the geotextile bags or tubes under an ambient pressure. Then the selection of the geotextile used for the bags or tubes becomes important. The geotextile has to be chosen to meet both the strength and filter design criteria. Some analytical methods have been developed to estimate the required tensile strength for the geotextile (Kazimierowicz, 1994; Miki et al., 1996). The apparent opening size (AOS) of the geotextile needs to be selected to allow the pore pressure to dissipate freely and yet retain the soil particles in the bags.

One technique of using clay slurry fill geotextile bags for dike construction was developed in Tianjin, China, and used for one land reclamation project along the coast of Tianjin. The cross-section of the dike is illustrated in Fig. 3 and a picture showing the alignment of the bags is given in Fig. 4. It can be seen that flat geotextile bags, instead of tubes, are adopted in this method.

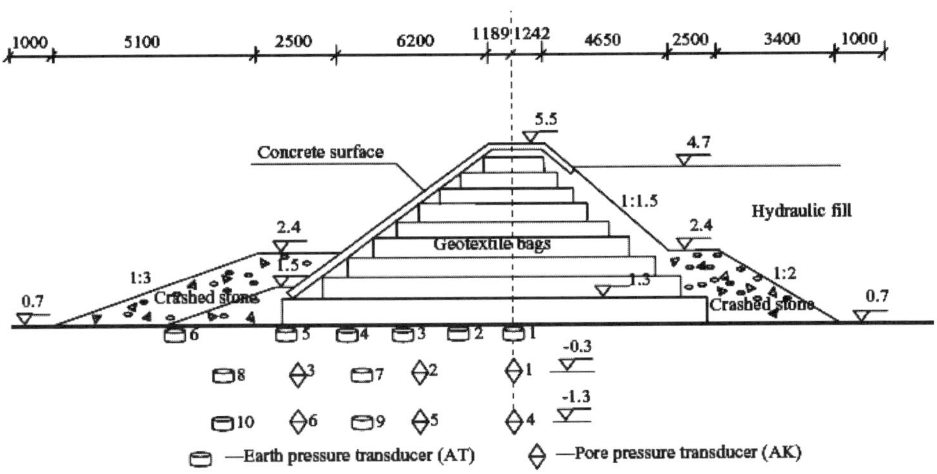

Figure 3 Schematic illustration of the dike made of clay slurry filled geotextile bags

As shown in Fig. 3, the designed height of the dike was 4.8 m with base and top elevations at 0.7 m and 5.5 m respectively. The top width of the dike was 2.43 m. The water levels were at 4.7 m elevation during high tide and at nearly 0.7 m elevation during low tide. The outer and inner slopes of the dike were chosen to be 2L:1H and 1.5L:1H, respectively. For the bottom bag, the dimension used was 30 m in circumference. Clay slurry was dredged from the seabed of a selected area and pumped directly into the bags through an injection hole. The height of the bag after consolidation was around 0.5 m. Nine layers of geotextile bags were used. The surfaces of the slopes formed by the geotextile bags were to be covered with a cast-in-situ concrete layer of 25 cm. The concrete was cast-in-situ using moulds formed by geotextile bags, a technique which is commonly used in China (Chi, 1991). As shown in Fig. 3, berms were used to enhance the stability of the dike and to protect the toes of the slopes. The berms were made of crushed stones of 50 to 80 kg. The slopes of the berms were 2L:1H for the inner side and 3L:1H for the outer side. A 4 m thick of hydraulically filled slurry was to be placed behind the dike after the dike was constructed. An instrumentation scheme was also suggested as shown in Fig. 3. A woven polypropylene geotextile with a mass density of

120 g/m^3 was chosen for the bags. It had a tensile strength of 28 kN/m in the longitudinal and 26 kN/m in the transverse direction respectively. The AOS (O_{95}) of the geotextile was 0.145 mm. The bags are sewn together using sewing machines on site. The soil used to fill the bags was classified as SC-CL according to the Unified Soil Classification System (USCS), that is, a borderline case between sandy clay and low plasticity clay. The liquid limit and plastic limit of the soil were 20.4% and 8.9% respectively and the plasticity index was 8.9%.

Figure 4 A picture showing the formation of a dike using clay slurry filled geotextile bags

PREFABRICATED CAISSON METHOD

The geotextile bag method may only be feasible when dikes are to be constructed in relative shallow or quiet water. When water is too deep or the wave is rough, gravity retaining structures using prefabricated reinforced concrete segments or caissons may be a better option. The use of caisson or concrete segments is not new. However, the most cost-effective design methods are still yet to be established. These concrete segments or caissons have to be tall enough to match the water depth and heavy enough to provide stability against the waves. However, when the concrete segments or caissons are too heavy, they cause settlement or bearing capacity problems. This is particularly the case when the foundation soil is soft. The weak foundation soil can be improved. However, it is difficult and costly to treat soil offshore and over a large area or distance. Therefore, it has become a challenge on how to construct large size gravity structures on soft soil.

In one of costal protection projects along Yangtze Estuary, some dikes for navigation purposes needed to be constructed. The method of using prefabricated reinforced concrete caissons was adopted. The dike was to be constructed at 40 km away from the coast. The water depth ranged from 5.0 to 8.5 m. The design wave height was 3.32 to 5.90 m with a return period of 25 years. The total length of the dike was about 17 km.

The design of the dike is schematically shown in Fig. 5. The caisson used was prefabricated reinforced concrete hollow segment. It was semicircle shaped, as shown in Fig. 5. The radius of the semicircle was 5.7 m. The advantage of using a semicircular cross-section is that the direction of the resultant wave force on the semicircular shaped structure will always pass through the center of the circle, which will greatly improve the loading condition of the structure. The hollow caisson would be filled with sand after installation through a 600 mm diameter hole on top of the caisson. In order to prevent the foundation soil from scouring, a geotextile sheet was used to cover the seabed. A cushion which acted as the foundation bed was placed on top of the geotextile. The cushion was 1 to 2 m high. It was made of crushed stones of 1 ~ 100 kg for the centre and 200 ~ 400 kg for the edge. After the caisson was placed, berms were placed on two sides of the caisson. The berms were made of 400 ~ 600 kg crushed stones.

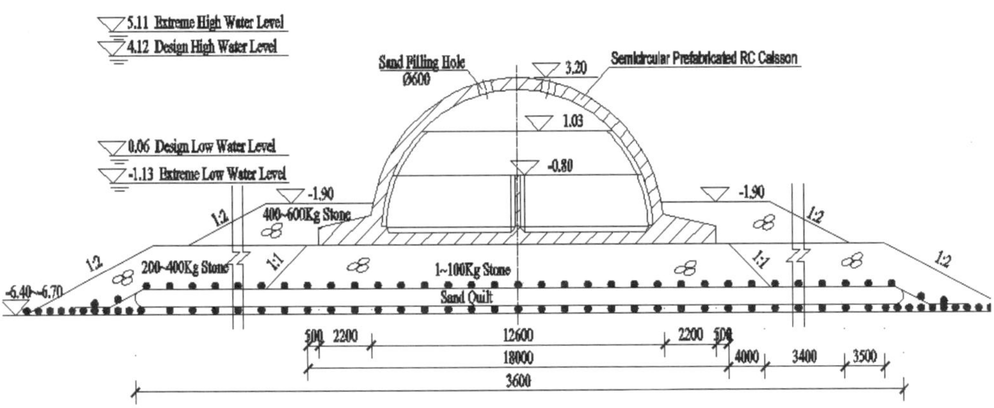

Figure 5 Cross-section of the dike

The soil profile below the dike consisted of a 1.5 to 3.5 m thick silt sand followed by 2 to 4 m thick clay mud and a roughly 30 m thick soft clay underlying the mud. The basic properties of the soil are given in Table 1. Although the soil below the dike was weak, with the use of geotextile and the cushion, the bearing capacity was estimated to be sufficient. However, the shear strength of the soil could be weakened by the wave action, which in turns would affect the stability of the dike. Nevertheless, due to time and other constraints, one section of the dike was constructed without improving the soil first. As the dike was a strip load, the load would only be distributed to a certain depth. It was hoped that under the surcharge of the crushed stone and the caissons, the geotechnical properties of both the upper silty sand layer and the top few meters of lower soft clay layer would be improved with time, and thus the stability of the dike would have been enhanced with time.

However, this proved to be a mistake. Only 2 months after the construction of the dike, an unusual strong storm took place in December 2002. The dike failed during the storm. Large settlements incurred suddenly and some caissons large lateral movements. A picture of the dike after the storm is shown in Fig. 6. It can be seen that the caissons had undergone either large lateral or vertical movements. The settlement versus time curve is shown in Fig. 7. The wave height versus time curve is also plotted in Fig. 7. It can be seen that a sudden settlement took place at the time when there was a surge in the wave height. The dike had become unstable under the wave action. Apparently, the soil needed to be improved. How to improve the seabed soil which was 5 to 8.5 m below the sea level with a wave height of 3 to 6 m in a cost-effective way became the key issue.

Table 1. Soil properties

Soil Stratum	Water content (%)	Unit weight (kN/m^3)	Void ratio	Liquid limit	Plastic limit	Undrained shear strength[1] (kPa)	Compre-ssion Index[2] MPa^{-1}
Silt sand	29.3	19.0	0.827	-	-	-	-
Mud	57.5	16.4	1.672	45.2	23.6	13.4	1.59
Soft clay	51.5	16.8	1.470	45.8	23.8	22.0	1.31

1. The undrained shear strength was measured by unconsolidated undrained tests.
2. The compression index was measured by oedometer tests within the stress range of 100 to 200 kPa.

Figure 6 Displaced caissons after a storm

Several soil improvement methods were considered. The soil replacement method was not feasible as the amount of fill required and the amount of excavated soil to be disposed were too excessive. The method to accelerate the consolidation process of the soft clay using prefabricated vertical drain (PVD) was considered the most economical. This method was also relatively easy to implement. The influence zone of the strip load was estimated to be 7.0 m below the seabed. Therefore, it was only necessary to install the PVD to a depth of 10 m deep. The weight of the crushed stone cushion was considered sufficient in providing surcharge. A special vertical drain installation barge was used to install the PVD offshore. The procedure for improving the soft soil was as follows: (1) PVDs was installed from the PVD installation barge at a spacing of 1.0 m to a depth of 10 m below the seabed; (2) The crushed stone cushion was laid on the seabed as a surcharge to consolidate the soil below; (3) The caisson segments were only placed after an average degree of consolidation of 80% was achieved, which took about 90 days after the placement of the crushed stone.

In addition to soil improvement, anti-slippery rubber pads were also used for the caissons. The rubber pads were embedded into the base of the caisson during the casting stage. Pins were also precast into the base of the caisson to enhance the anchoring effect. Large scale laboratory tests have shown that the use of rubber pads and pins can enhance significantly the resistance of the caisson against sliding.

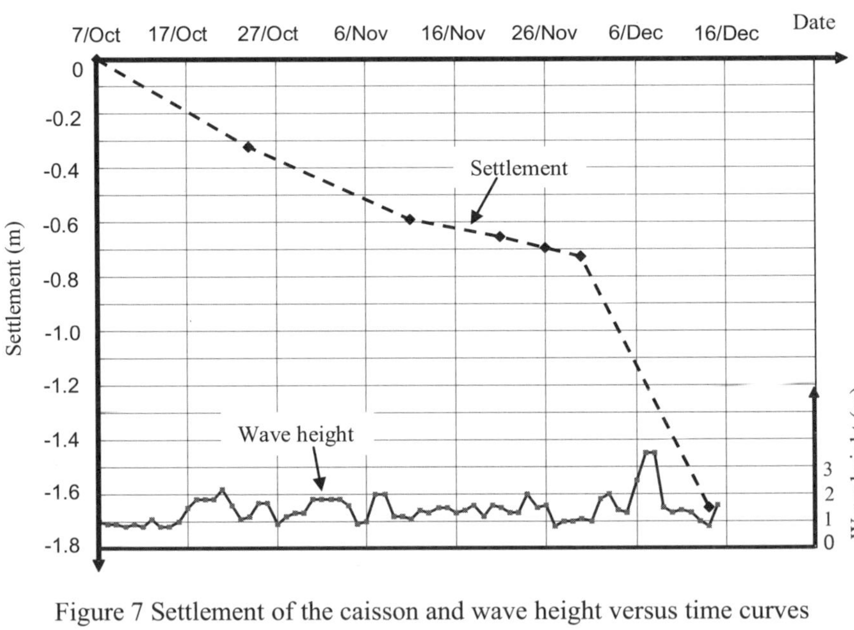

Figure 7 Settlement of the caisson and wave height versus time curves

To assess the effectiveness of soil improvement, the undrained shear strength profiles obtained from vane shear tests conducted before and after the preloading are compared in Fig. 8. The duration of consolidation was 90 days, It can be seen that the vane shear strength increased by almost 2 folds as a result of consolidation using PVDs.

It can also be seen from Fig. 8 that there were little improvement in the soil where PVDs were not reached. However, soils at these depths were not affected by the cyclic wave loads.

The construction of the dike was completed in Dec 2003. The settlements of the dike were monitored after the construction. The settlements measured at six different sections with time are shown in Fig. 9. The wave height versus time curve is also plotted in Fig. 9. It can be seen from Fig. 9 that the dike had experienced several strong storms caused by typhoons. Some of them were even stronger than the one that caused the failure as shown in Figs. 6 and 7. Even through, the dike was stable and there was no additional settlement caused by the storms. The settlement also stabilized in 5 to 6 months. The total settlements were within the expected range. Therefore, the use of PVDs has proven to be a successful method for this project. The dike after construction is shown in Fig. 10.

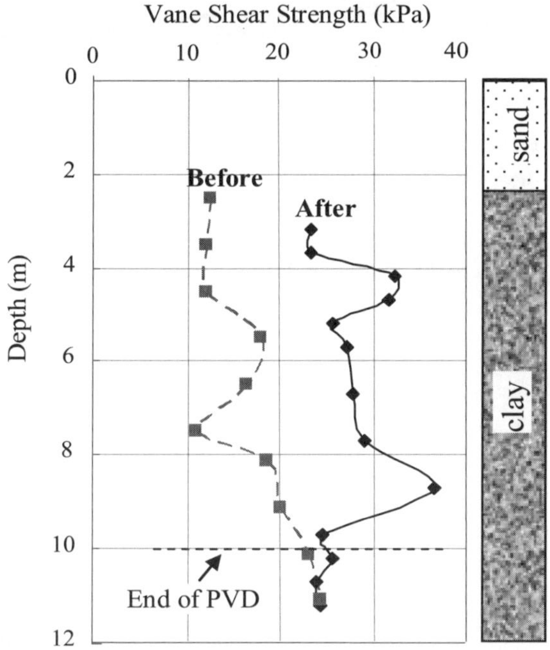

Figure 8 Field vane shear strength profiles before and after preloading using PVDs

CONCLUSIONS

Two methods that can be used for the construction of coastal protection structures or for costal disaster rehabilitation works are introduced in this paper. The first is the geotextile bag method which uses either sand or clay to fill geotextile tubes or bags to form dikes or breakwaters. The second is to use prefabricated reinforced concrete caissons to construct offshore dikes or structures on soft soil. The applications of these methods are illustrated using case studies. The following conclusions can be made on the two methods:

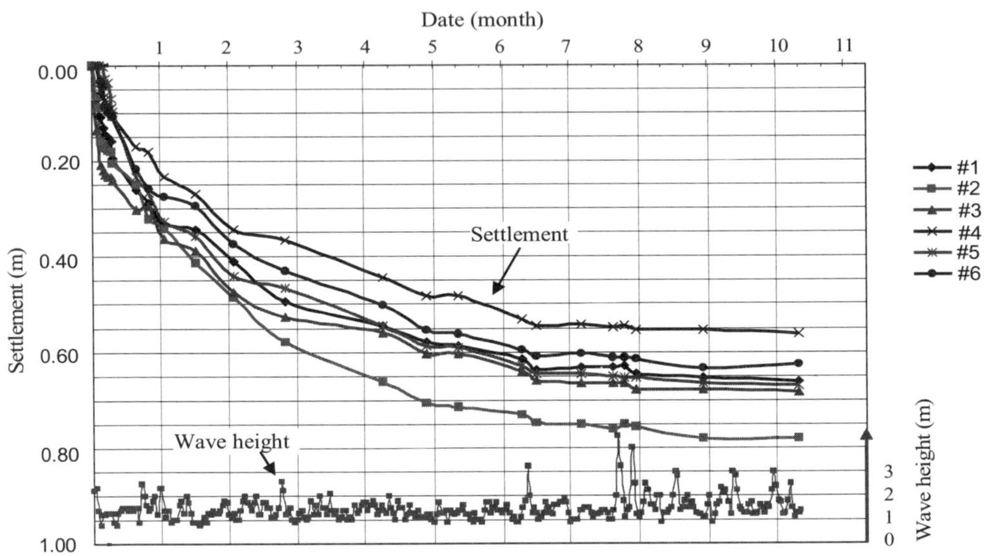

Figure 9 Settlement and wave height versus time curves

Figure 10 the constructed dike

(1) The geotextile bag or tube method is suitable for the construction of breakwaters or dikes in shallow water. The bags can be fabricated using geotextile on site into any desirable dimensions. Although sand or lean concrete mortar are normally used as the fill materials, the presented case study has shown that clay slurry dredged from the seabed can also be use as a fill material to fill the bags. This method is cost-effective, particularly when sand or rocks are not readily available as fill material.

(2) A method to use prefabricated reinforced concrete caissons to construct dikes of offshore structures on soft clay is presented using a case study. The study shows that it is necessary to improve the soft seabed soil before constructing the dike even though a thick cushion made of crushed rocks is used. The dike built on soil without improvement failed during a heavy storm. For the case presented, the soil can be improved by simply installing PVDs to accelerate the consolidation process of the seabed clay. With the use of PVDs, the soft soil would consolidate much faster and gain sufficient strength quickly to maintain the stability of the caissons.

REFERENCES

Chi, J. K. (1991). *Technical Report on the Technique of Pumping Concrete into Geotextile Mould.* Shanghai Geotechnical Research Institute (in Chinese).

Kazimierowicz, K. (1994). "Simple analysis of deformation of sand-sausages." *Proc. 5th Int. Conf. Geotextiles, Geomembranes and Related Products*, Singapore, 5-9 Sept., Vol. 2, 775-778.

Miki, H., Yamada, T., Takahashi, I., Shinsha, H., and Kushima, M. (1996). "Application of geotextile tube dehydrated soil to form embankments." *Proc. 2nd Int. Conf. on Environmental Geotechnics*, Osaka, 5-8 Nov, 385-390.

Silvester, R. and Hsu, J. R. C. (1993). "*Costal Stabilization – Innovative Concepts*", Prentice-Hall Inc.

SPECIAL SESSIONS

International Conference on Geotechnical Engineering for Disaster Mitigation & Rehabilitation
Chu, Phoon & Yong (eds) © 2005 World Scientific Publishing Company ISBN 981-256-469-1

REDUCTION AND REHABILITATION OF SEISMIC DAMAGE TO RIVER DIKES IN JAPAN

YASUSHI SASAKI
Executive y Adviser, Japan Institute of Construction Engineering
3-12-1 Toranomon, Tokyo 105-0001, Japan

Seismic failure of river dikes and current trend of its rehabilitation in Japan are summarized. Serious deformation of dikes was mostly brought by soil liquefaction. Stretching type of failure observed where liquefaction took place in shallow depth of ground was recently noticed. Treatment against liquefaction is essential to mitigate earthquake inducing failure of dikes.

INTRODUCTION

Global trends of urbanization to lowland may cause serious damage to society once a river dike fails due to natural hazards such as typhoon, earthquakes, tsunami, and storm surge, as was seen in the Hurricane Katrina case in 2005. As intense rainfall tends to drop in locally limited area due to global warming adding to the recent trend of urbanization in Japan, the importance of dike role is increasing particularly in the urbanized lowland area where population and infrastructure are concentrating.

Not only the rainfall inducing disaster but earthquake inducing disaster can not be forgotten because the seismicity around Japan is recognized to have become an active period (Oike 1995). As the hinterland was highly urbanized as seen in Figure 1, the Yodo

Figure 1. Restoration work of damaged section of the Yodo river dike

river dike damage during the Kobe earthquake in 1995 gave a big threat to the society.

The river dikes have long history in their construction stage, so it is general that the mechanical properties of embankment material and ground conditions are not fully known. This makes it difficult to find out the failure mechanism and governing factors of seismic damage to dikes when it happens. However, the accumulation of damaged experiences is revealing the complex mechanism of seismic failure of river dikes. This paper summarizes the river dike damage recently seen in Japan (Sasaki et. al 2004).

TYPICAL DAMAGE TO DIKES DURING PAST EARTHQUAKES

It was known that earthquake inducing failure to river dikes often caused extensive decrease of its function, and in the worst case, the height of the embankment became to 1/4 of original height. Failed sections of dikes were mostly accompanied by longitudinal cracks where step-like discrepancies were often caused.

Such an extensive damage was frequently seen on liquefied ground. It was also known that sand boiling was often observed along toes of dikes resting on old river courses. This experience implies that liquefaction susceptibility of subsoil could be evaluated from geomorphologic information such as the old river courses even when enough boring data are not available.

Table 1 shows recent earthquakes which caused damage to river dikes in Japan. Those cases can be divided into two groups from the rehabilitation point of view. No soil improvement technique had been conducted in the restoration works before 1993. But from the case of Kushiro-oki earthquake, remedial treatment of foundation ground (Oshiki and Sasaki, 2001) was accepted as a part of restoration works by the Ministry of Budgeting. The causative reason and mechanism of seismic failure of dike became to be revealed from the cases after 1993. Two typical damage are as follows.

Although subsoil liquefaction had been known as a main cause of dike damage, but it was not recognized until the case in 1993 that the large deformation of dike is induced by the liquefaction of bottom part of dike in a particular situation.

Table 1. Recent earthquakes which caused damage to river dikes

Name of earthquakes	M	Date	Damaged dikes	Restoration
'68 Tokachioki	7.9	1968/5/16	Tokachi River, Mabechi River	-
Nihonkai-chubu	7.7	1983/5/26	Hachirogata	CW, Drain
Kushiro-oki	7.8	1993/1/15	Tokachi River, Kushiro River	SCP, SCP
Hokkaido-nansei-oki	7.8	1993/7/12	Shiribeshi-Toshibetsu River	SCP, DC
Hokkaido-toho-oki	8.2	1994/10/4	Kushiro River	-
Hyogoken-nanbu	7.3	1995/1/17	Yodo River	DMM, SSP
Tottoriken-seibu	7.3	2000/10/6	Naka-umi Lake	SSP
Miyagiken	6.2	2003/7/26	Naruse River	SCP
Tokachi-oki	8.0	2003/926	Tokachi River	SCP, JG
Niigataken-chuetsu	6.8	2004/1023	Shinano River	SCP, GG

CW: cut-off wall, SCP: sand compaction pile, DC: dynamic compaction, DMM: deep mixing method, SSP: steel sheet pile, JG: jet grouting, GG: geo-grid, -: without special treatment

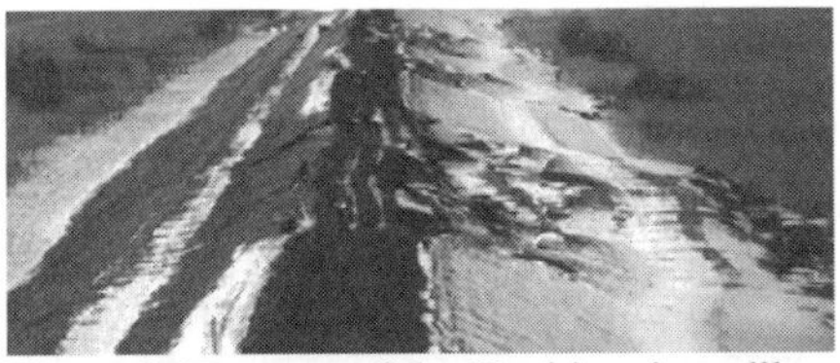

Figure 2. Damage of the Kushiro river dike

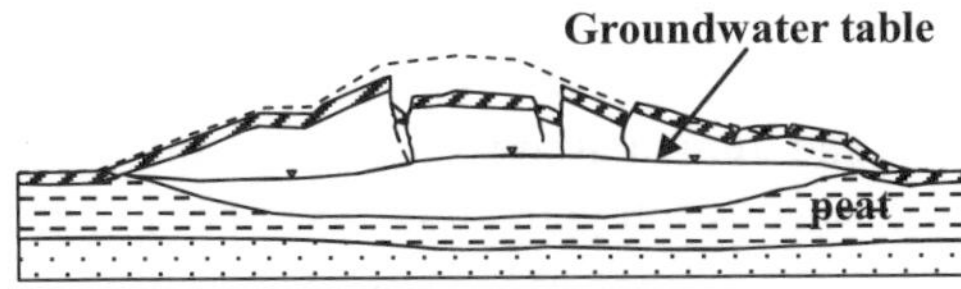

Figure 3. Deformation at the section in Fig. 2

Figures 2 and 3 illustrate the damaged section of the 8-9 m high dike of the Kushiro River. It was noticed first from this case why the liquefaction of bottom part of dike takes place. The consolidation of highly compressible peat layer caused dike settlement, and this settlement had increased saturated zone in the lower part of the dike which was liquefied during the event (Sasaki et. al 1994).

Figure 4 shows the damaged section of the Shiribeshi-Toshibetsu River dike during the Hokkaido-nansei-oki earthquake in 1994. In this case of damage, it was apparent from the traces of sand boiling nearby the toe of the dike that the large deformation was caused from the liquefaction in the subsoil layer. It should be noted that large amount of depression was observed at its crest as shown in Figure 4. Careful observation during restoration work at this site revealed apparent two slip surfaces which ran from the shoulders inside the failed dike as shown in Figure 5.

Figure 4. Damage to the Siribshi-Toshibetsu River dike

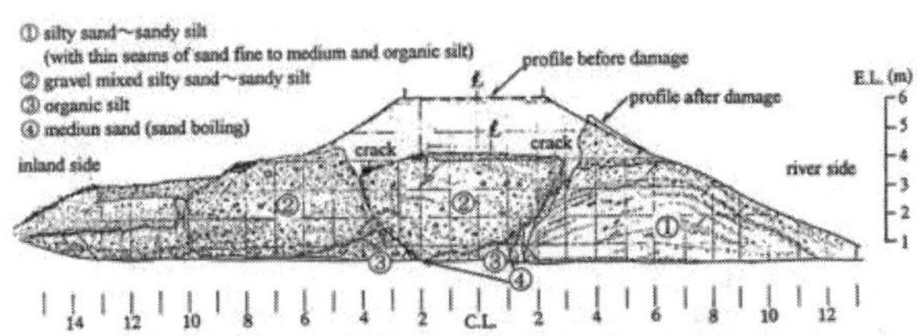

Fig. 4 Cross section of the failed dike at 0 km 865 of Shiribeshi-toshibetsu

Figure 5. Cross sectional sketch at the site in Fig. 4

It was reported elsewhere (Sasaki et. al 1997) that these slip planes were brought by the stress change in the dike associated with the change of stress at its bottom boundary due to the loss of shear strength of the foundation layer by its liquefaction.

This finding implies that it is necessary to take the deformation of dike into account adding to the large deformation of liquefied subsoil layer when to estimate the seismic settlement of dike during earthquakes. Also it should be noted that stretching type of deformation usually causes longitudinal cracks and depressions of crest.

REMEDIAL TREATMENT OF DIKES AGAINST SESMIC DAMAGE

As was seen in previous chapter, the soil liquefaction is the key to whether or not trigger large deformation to dikes during earthquakes. So it is considered that the prevention of the occurrence of soil liquefaction is the fundamental measure to mitigate dike damage. Most of the seriously damaged sections in the past shown in Table 1 were restored by using remedial treatment method to prevent the liquefaction (Sasaki et. al 2004) as shown in this Table. The remedial treatments to improve foundation ground against liquefaction are compiled elsewhere (Japanese Geotechnical Society 1998).

Comprehensive study revealed that the main cause of the damage at the Torishima section of the Yodo River dike shown in Figure 1 was subsoil liquefaction. The deformation and settlement of the dike was considered to be caused along the process shown in Figure 6 (Sasaki et. al 2004). Therefore this section was restored with conducting Deep Mixing Method for improving the liquefiable 10 m thick sand layer beneath the dike as shown in Figure 7.

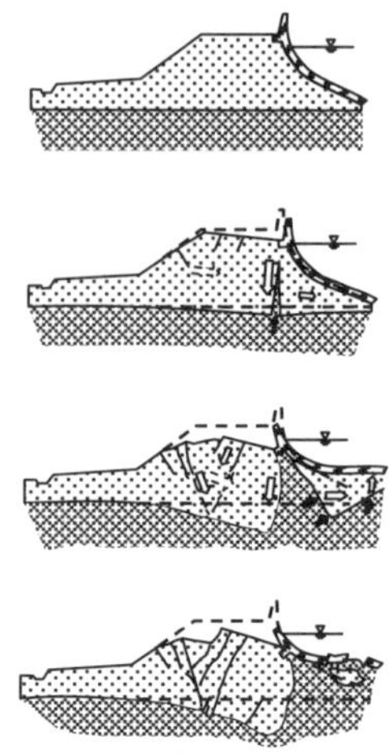

Figure 6. Deformation process of the Yodo River dike

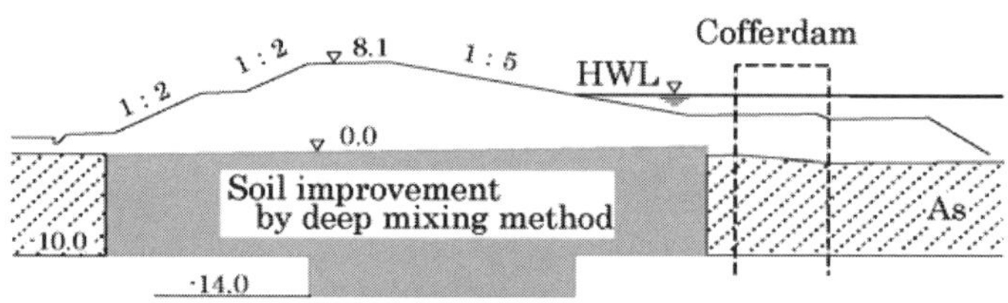

Figure 7. Remediation of the Torishima dike by DMM

The finding gained from the case of the Shiribeshi-Toshibetsu River dike was utilized for the newly constructed Naka-umi dike. Soil profile at this site showed that liquefaction would be easily triggered during an earthquake as shown by F_L value in the Figure 8. However it was hesitated to take soil improvement method to avoid the damage for 3 m high dike in the rural area. So the geo-grid was placed at its bottom to prevent stretching type of failure as shown in the Figure 9. Most of the treated length of the dike section performed well without apparent deformation during the Tottoriken-seibu earthquake in 2000 (Sasaki et. al 2004).

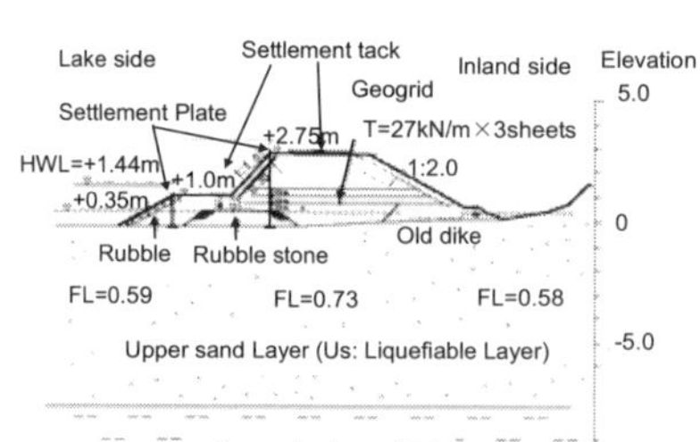

Figure 8. Cross section of the Naka-umi dike

Figure 9. Construction stage of the Naka-umi dike

EVALUATION OF DIKE DAMAGE DURING EARTHQUAKES

As the role of dikes which protect lowland area from flooding becomes more important than the past, so it is essential to raise the resistance of them against causative action of natural hazard. Necessary room for them to be enlarged is limited in urbanized area, therefore existing dikes must be added ductile quality at their locations. In order to achieve this demand effectively, it is necessary to know the weak sections of dike.

To filtering out the weak sections against earthquakes from the long spanned whole dikes, sections which protect so-called zero meter area are firstly to be selected. Then the seismic performance of the dike in the section is to be examined. As the stability of dike is mainly governed by the occurrence of liquefaction, the subsoil conditions of these selected sections are to be examined. If the liquefaction is anticipated, amount of the dike deformation should be estimated. Crest settlement is taken as the measure of dike deformation so that the dike function is to be evaluated. And then remedial measure to reduce the estimated deformation is to be selected.

There are hidden concepts in the background of the selection of zero meter area as a primary step of the procedure, those are based on the past experiences in Japan that the damaged dike can be tentatively rehabilitated by using soils as its construction material in comparatively short period after the earthquake, and that flooding does not occur at the same time with earthquake.

The most important thing from the geotechnical point of view is the estimation of the seismically induced deformation of dikes. For this purpose, conventional stability analysis by circular arc method has been used for a long time. In this empirical method, both safety factor against inertia force without reducing soil strength, and safety factor against reduced soil strength due to the liquefaction induced pore water pressure in the liquefiable layer where the inertia force is not taken into consideration are separately examined. And the obtained safety factor, smaller than the other, is converted to crest settlement by using empirical relation between the crest settlement and the safety factor.

However it is being disclosed that the deformation mode of the liquefied layer is different from what is assumed in the conventional stability analysis. Therefore revision is being made for estimating the seismically induced crest settlement. Adequacy of four numerical analyses was examined. Those numerical analyses are: a computer code named ALID for static analysis using softened soil concept: the Towhata method, a static approach using the viscous liquid concept: a computer code named LIQCA using coupled effective analysis, and a computer code named FLIP using an uncoupled analysis (Japan Institute of Construction Engineering 2002).

Computed deformations were compared with the dike settlements observed during the Hokkaido-nansei-oki earthquake and the Kobe earthquake. Figure 10 shows the result of this comparison. As seen in this figure, the estimated settlements by analytical approach agree fairly well, but the empirical method predicts always larger settlements than the observed ones. This too conservative estimation by the empirical method arises from the used relationship between the safety factor and the settlement. It is also noticed that the static approach by Towhata method can estimate the dike settlement well as the dynamic approaches (LIQCA and FLIP) do.

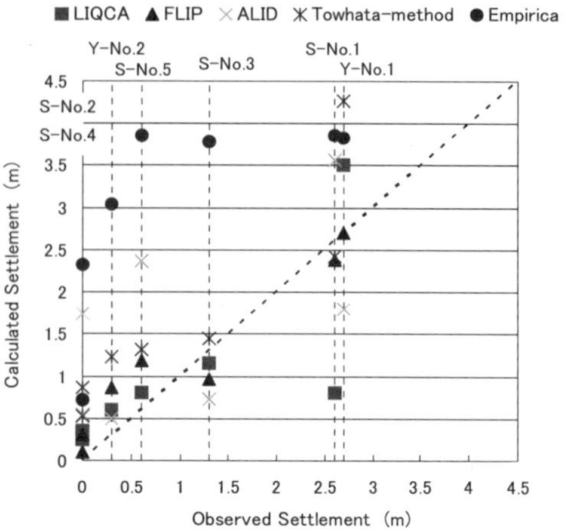

Fig.10 Comparison between calculated and observed settlements

CONCLUSION

The damage to river dikes due to past earthquakes in Japan and their rehabilitation works were summarized. It was shown that the soil liquefaction was the main cause of seismic failure of dikes.

Although the seismic stability of existing dike is already diagnosed, but strengthening is delayed due to limitation of budget. In order to reduce the necessary cost by raising the precision of diagnosis, an adequate method to evaluate the seismic deformation of dike is desired to be established as soon as possible so that the effective strengthening is conducted for the necessary section to mitigate the extensive disaster.

REFERENCES

Japanese Geotechnical Society Ed. (1998). *Remedial Measures against Soil Liquefaction from investigation and design to implementation*, A. A. Balkema.

Japan Institute of Construction Engineering (2002). *Analytical methods to predict seismically induced dike deformation, JICE Report* No. 102001. (in Japanese)

Oike, K. (1995). "*Seismic islands set in an active stage*", Iwanami library of science 33, Iwanami. (in Japanese)

Oshiki, H. and Sasaki, Y. (2001). "Restoration works of seismically damaged river dikes using remedial treatment of liquefiable layer." *Journal of construction management and engineering, Japanese Society of Civil Engineer,* No. 686/VI-52, 15-29. (in Japanese)

Sasaki, Y., Oshiki, H. and Nishikawa, J. (1993), "Embankment failure caused by the Kushiro-oki earthquake of January 15, 1993." *Performance of Ground and Soil Structure during Earthquakes, 13th International Conference on Soil Mechanics and Foundation Engineering, New Delhi, JSSMFE*, 61-68.

Sasaki, Y., Moriwaki, T. and Ohbayashi, J. (1997). "Deformation process of an embankment resting on a liquefiable soil layer." *Deformation and progressive failure in geomechanics, Proc. IS-Nagoya'97*, 553-558.

Sasaki, Y., Kano, S. and Matsuo, O. (2004). "Research and practices on remedial measures for river dikes against soil liquefaction." Journal of Japan Association for Earthquake Engineering, Vol.4, No.3 (Special Issue), 2004.

International Conference on Geotechnical Engineering for Disaster Mitigation & Rehabilitation
Chu, Phoon & Yong (eds) © 2005 World Scientific Publishing Company ISBN 981-256-469-1

ZONING FOR LIQUEFACTION POTENTIAL OF TAIPEI BASIN, TAIWAN

MEEI-LING LIN
Department of Civil Engineering, National Taiwan University
Taipei, 106, Taiwan

JIE-JUNG DAI
Department of Civil Engineering, National Taiwan University
Taipei, 106, Taiwan

In this research, zoning for liquefaction potential of Taipei Basin, Taiwan, was performed using a geotechnical database constructed from geological data and borehole information, and by combining the Grade-2 and Grade-3 zoning results, a more solid liquefaction zoning map could be achieved.

INTRODUCTION

Taiwan is located on the juncture zone of Philippine sea plate and Euroasia continental plate and is frequently struck by earthquakes. During the Chi-Chi earthquake 1999, severe cases of soil liquefaction occurred in central Taiwan, and caused significant properties losses. The Taipei basin is situated on soft sandy soil sediments and with high groundwater table, and thus is susceptible to the liquefaction hazard caused by earthquake. Development of a liquefaction susceptibility zoning map can provide important reference for building design and drafting of hazard mitigation policy. Considering the data requirement of the Grade-3 rigorous zoning method, the Grade-2 zoning method could provide good supplementary results where the data are limited. Thus, in this research, the Grade-2 zoning and results of Grade-3 analysis for liquefaction potential of Taipei Basin was performed to establish the liquefaction zoning map.

GEOTECHNICAL PROPERTIES OF TAIPEI BASIN

The Taipei basin is formed by alluvial deposits of the three rivers flowing into the basin, namely, the Tan-Shui river, the Shin-Tien river, and the Kee-Long river. The engineering properties of soil layer vary with different watershed and the deposition environment. Accordingly, the soil properties of the Taipei basin could be divided into seven subzones based on the three watersheds. The soil formations of the Taipei basin were composed of three different formations, namely; Shin-Juang formation, Jin-Mei formation, and Song-Shan formation from the bottom up. Among all three formations, the Song-Shan formation is the formation closest to the ground surface, and distributes with an average thickness of 40 to 50 m. The Song-Shan formation is composed of alternating layers of silty clays and silty sands, and is the formation where most of the engineering facilities situated. The general distributions of the sublayers in Song-Shan formation are as follows,

(1) The first sublayer is situated on the Jin-Mei gravel formation with a thickness of 0-15 m, and is composed of dense to intermediate silty sand or gravelly sand.
(2) The second sublayer is composed of grey silty clay with low to medium plasticity and a thickness of 0-19 m.
(3) The soil of the third sublayer is non-plastic medium dense silty sand containing shells and with a thickness of 0-19 m.
(4) The fourth sublayer is composed of grey silty clay containing shells; the layer thickness is about 6-29 m.
(5) The fifth sublayer is composed of grey silty sand with coarse sand and gravels and with a layer thickness of 2-20 m.
(6) The sixth sublayer is the top soil layer with a thickness of 3-8 m, and is composed of brown silty clay.

The fifth sublayer of the Song-Shan formation is of silty sand and typically distributes within a depth of 20 m from ground surface. With high ground water level of the Taipei basin, the fifth sublayer is thus susceptible to soil liquefaction. The shape of the basin is as shown in Fig.1, where the flood plane area with elevation lower than 20m is considered as susceptible to liquefaction according to Youd and Perkins (1978).

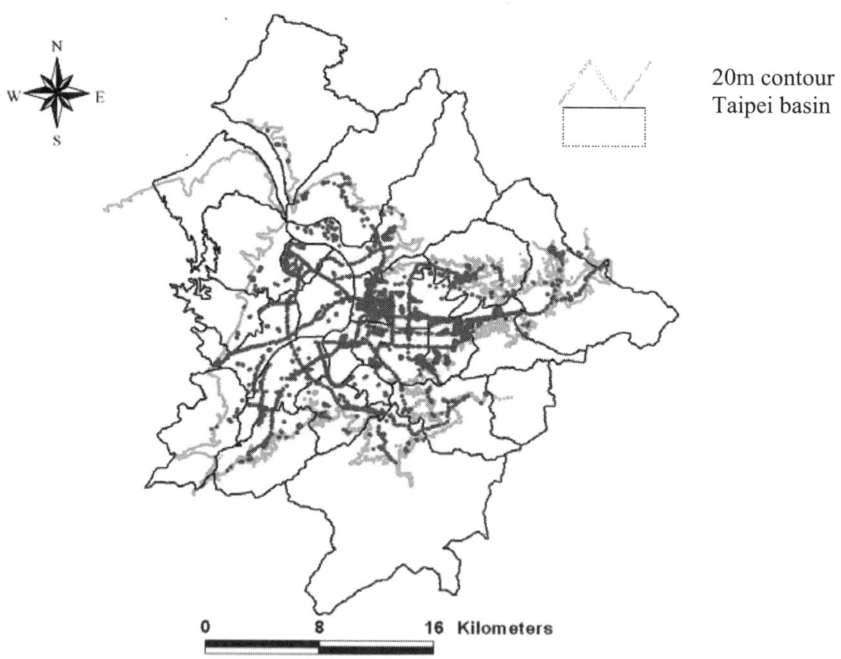

Fig. 1. The area with elevation lower than 20 m and eligible boreholes in Taipei basin.

GRADE-2 ZONING OF LIQUEFACTION POTENTIAL

The Grade-2 assessment of liquefaction susceptibility of the Taipei Basin was made based on the geological and geomorphological criteria proposed by Youd and Perkins (1978) and Wakamatsu (1992). Accordingly, only the alluvial deposited since Holocene age (younger than 6000 years) appeared likely to liquefy. The sedimentary history of the basin was reviewed, and it was found that the basin was flooded by sea water, and sea water receded to expose the basin in late Holocene. By comparing the old maps and geological data, area affected by the late Holocene lake is identified as with medium liquefaction potential, and its area is consistent with the flood plane enclosed within elevation lower than 20m (Pang, 1988).The basin was flooded again about 300 years ago as the sea water level rose, the lake formed was known as the Kang-Shi Taipei Lake (Lin, *et.al.* 1998). The area affected by the Kang-Shi Taipei Lake coincides well with the region with elevation smaller than 3m, and is considered as highly susceptible to liquefaction. Other areas considered as with high susceptibility include recent river deposits and low-land deposited within about 100 years. Combining the areas affected by Kang-Shi Taipei Lake and recent deposits, the zone with high liquefaction potential is determined. The area affected by late Holocene lake is considered as with medium potential, and the rest of the flood plan with elevation lower than 20m is considered as with low potential. The overall Grade-2 zoning of the basin is as shown in Fig. 2.

DATABASE FOR EVALUATION OF LIQUEFACTION POTENTIAL

In order to perform more detailed assessment of liquefaction potential of the soil layers inside the basin, it was necessary to construct the database of the engineering properties of the soil layers. The database based on collected geotechnical site exploration reports was constructed. Considering the data requirement of liquefaction potential analysis, data files are constructed in a standard format with four layers of data sheets. The first layer includes basic borehole information such as: borehole ID, coordinates, elevation, depth of water table, total drilling depth, and name of driller, etc. The second layer contains properties of the specimens of each borehole, which include the standard penetration test blow-count (SPT-N), soil classification, soil description, grain size distribution, specific gravity, water content, and other engineering properties. The third layer contains parameters evolved during liquefaction potential analysis, such as calculated cyclic resistance ratio, cyclic stress ratio, and safety factor at each potentially liquefiable depth for each bore hole. The final results are presented as liquefaction potential index for each borehole based on the weighted depth method proposed by Iwasaki (1982), in addition, the minimum safety factor for each borehole is also included in the fourth layer.

Data from more than 4000 boreholes were collected and most of those were from public construction projects. Considering the possible liquefiable depth, boreholes with depth smaller than 20 m or data not satisfying the logical algorithm were rejected. The boreholes with ground elevation higher than 20m were also excluded. The final distribution of eligible borehole data is as shown in Fig. 1, and the total number of borehole data used in the liquefaction potential analysis is 3329. The area inside the basin with elevation lower than 20m is about 270 km^2, thus the density of the borehole distribution is approximately 12.3 boreholes per km^2 of area.

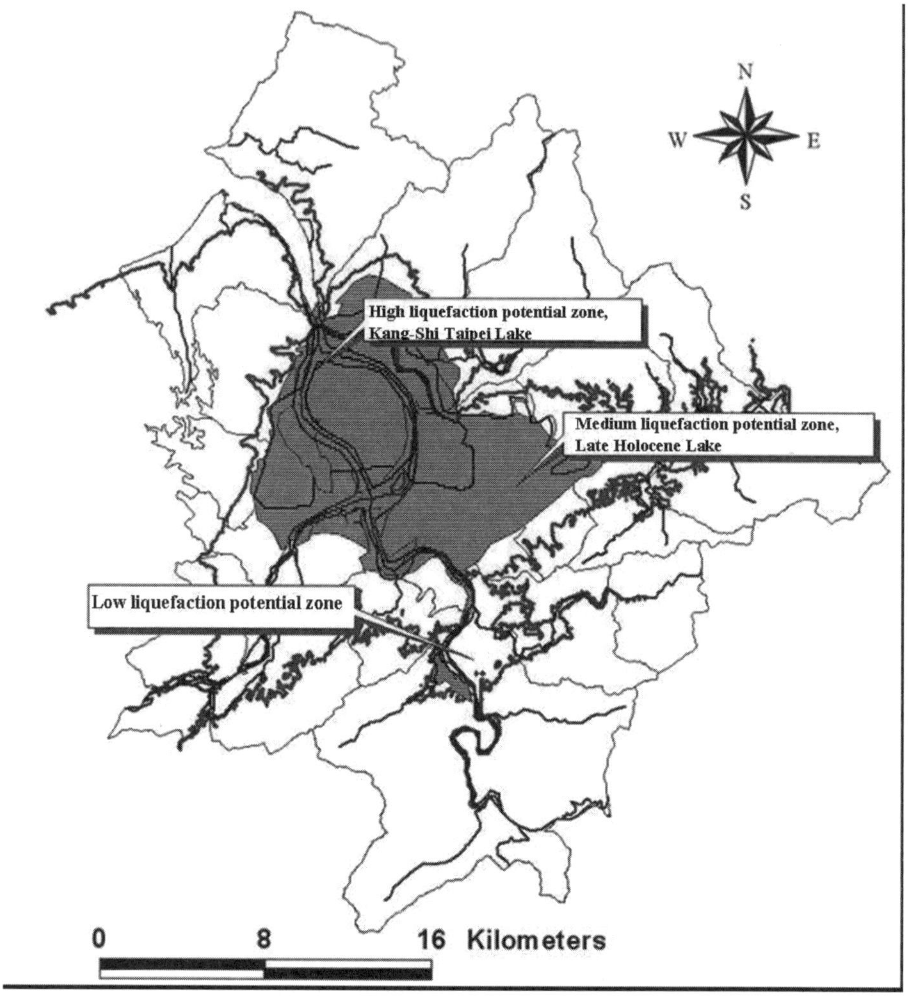

Fig. 2 Grade-2 zoning of liquefaction potential of Taipei Basin

GRADE-3 LIQUEFACTION POTENTIAL ANALYSIS

To evaluate the liquefaction potential of the soil layer, it appeared that the standard penetration test blow count is the readily available parameter, and the simplified procedures proposed by Seed (1972) using blow count number is widely accepted. Different modifications were proposed by many researchers to take into account the effects of the fine contents, and each produced different results. Lin, *et. al.* (2000) used the field liquefaction data of the Yuen-Lin township induced by the Chi-Chi earthquake to conduct comparisons among different methods. It was suggested that the method by Seed *et. al.* (1985) combining with liquefaction potential index proposed by Iwasaki (1982) yielded stable results in good consistency with field observation. Considering the silty sand in Taipei basin with high fine contents resembling the soil in Yuen-Lin, the Seed *et. al.* (1985) method with the potential index by Iwasaki (1982) is used in this study. The safety factor for the liquefaction is defined as the ratio of the cyclic shear resistance of soil to the cyclic shear stress induced by the ground acceleration. When computing the cyclic shear stress, the peak ground acceleration of 0.23g is used for Taipei basin according to the seismic zoning code. For the cyclic shear resistance, the blow count number of given soil data is corrected according to the overburden pressure and energy input. The earthquake magnitude is assumed to be 7.5, which is about the same magnitude as the Chi-Chi earthquake. Computations of the liquefaction potential analysis were performed by implementing the computing procedures using Excel linking to the database and the results were output to the proper data layer. The liquefaction potential analysis were performed on soil layers identified as liquefiable following criteria of : (1) The ground water table is within a depth of 10 m with saturated sand layer within a depth of 20 m. (2) The fine content of soil is smaller than 35% or larger than 35% but the plasticity index is smaller than 15. (3) The mean grain size of the soil is smaller than 10 mm, and the effective grain size is smaller than 1 mm. The liquefaction potential index of each borehole is classified as with low, high, and very high liquefaction potential, as shown in Fig. 3 in green, blue, and red colors, respectively.

ZONING OF LIQUEFACTION POTENTIAL

For the Grade-3 analysis, contours of liquefaction potential index can be plotted, however, care should be taken for areas without sufficient density of borehole data and when data distribution is uneven. Although the data were not evenly distributed, the results provided good reference for evaluation of liquefaction susceptibility of the basin. However, by combining the Grade-2 zoning results, which takes into account the deposition environment of the soil layers, and the Grade-3 results, a more comprehensive liquefaction potential zoning map could be derived as shown in Fig. 3. The distribution of Grade-3 results appeared to be in good consistency with the Grade-2 zoning, as most of the boreholes showing very high liquefaction potential located in the high potential zone in Fig.2, and most of the boreholes showing low liquefaction potential located in the low potential zone.

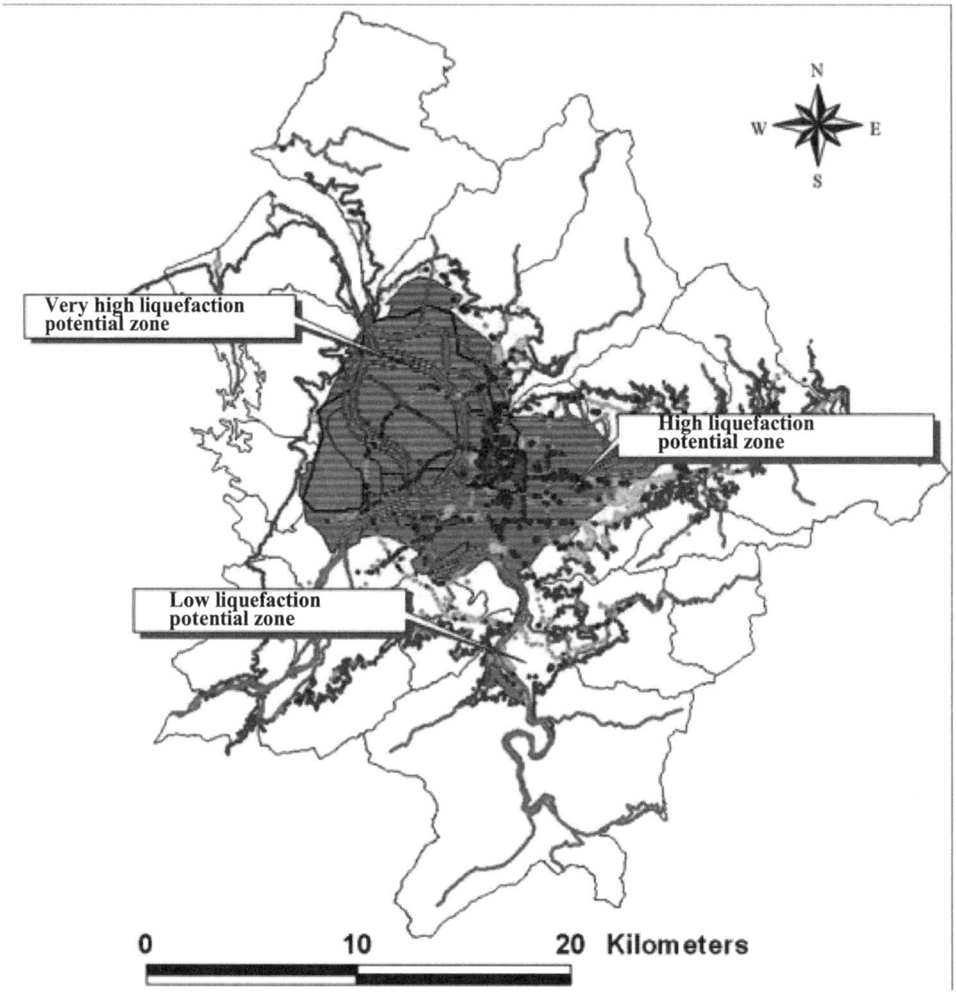

Fig. 3 Liquefaction potential zoning results of the Taipei Basin

CONCLUSIONS

In this research, Grade-2 and Grade-3 zonings for liquefaction potential of Taipei Basin were performed. The Grade-2 zoning was conducted based on the topography, sedimentary history, and geological formations of the basin. The Grade-3 analysis was performed using the database established from borehole data. The resulting liquefaction susceptibility zonation using Grade-2 and Grade-3 methods appeared to be quite consistent. By combining the Grade-2 and Grade-3 zoning map, a more solid liquefaction zoning map could be achieved, which could provide a basic reference for drafting of hazard mitigation policy.

REFERENCES

Iwasaki, T.,Tatsuoka, F.,Tokida, K., and Yasuda, S, (1978), "A Practical Method for Assessing Soil Liquefaction Potential Based on Case Studies at Various Sites in Japan", *Proc., 2nd Int. Conf. Microzonation Safer Construct. Research Apply,* Vol. 2, 885-896

Lin, M.S., Shiou, C.L., Shah, L.M., Chen, H.Z., and Chu, T.H., (1998), "Investigation of the Kang-Shi Taipei Lake", *Proc., the 3rd Taiwan Geography Conference*, 125-143.

Lin, M. L., Ueng, T. S., Chen, M. H. and. Huang, S. C, (2000), "A Preliminary Study on Evaluation of Liquefaction Potential of Yuen-Lin Area", *Proceedings of International Workshop on Annual Commemoration of Chi-Chi Earthquake,* Vol. - Geotechnical Aspect, 74-82

Seed, H. B., (1979), "Soil Liquefaction and Cyclic Mobility Evaluation for Level Ground During Earthquake," *Journal of the Geotechnical Engineering Division*, ASCE, Vol. 105, No. GT2,. 201-255.

Seed, H.B., Tokimatsu, K., Harder, L.F., and Chung, R.M., (1985), "Influence of SPT Procedures in Soil Liquefaction Resistance Evolution", *Report, Earthquake Research Center, University of California, Berkeley*, No. UBC/EERC-84/15.

Wakamatsu, K, (1992), "Evaluation of Liquefaction Susceptibility Based on Detailed Geomorphological Classification", *Proc., Technical Papers of Annual Meeting Architectural Institute of Japan*, Vol. B, 1443-1444.

Youd, T. L. and D. M. Perkins, (1978), "Mapping Liquefaction-Induced Ground Failure Potential," *Journal of the Geotechnical Engineering Division*, ASCE, Vol.104, No.4, 433-466.

International Conference on Geotechnical Engineering for Disaster Mitigation & Rehabilitation
Chu, Phoon & Yong (eds) © 2005 World Scientific Publishing Company ISBN 981-256-469-1

STRATEGY FOR RE-CONSTRUCTION IN THAILAND FOLLOWING THE 26 DECEMBER 2004 TSUNAMI EVENT

KJELL KARLSRUD
Norwegian Geotechnical Institute (NGI), P.O.Box 3930 Ullevaal Stadion, N-0806 Oslo, Norway

HILMAR BUNGUM
NORSAR, P.O.Box 53, N-2027 Kjeller, Norway

CARL B. HARBITZ, FINN LØVHOLT, BJØRN VIDAR VANGELSTEN
Norwegian Geotechnical Institute (NGI), P.O.Box 3930 Ullevaal Stadion, N-0806 Oslo, Norway

SYLFEST GLIMSDAL
Simula Research Laboratory, P.O. Box 134, N-1325 Lysaker, Norway

On December 26 2004 an earthquake, of magnitude 9.3 on the Richter scale and with an epicentre off the NW tip of Sumatra, occurred. In an area covering 300 km by 1200 km the earthquake caused up to approximately 3 meter vertical displacement of the seabed. This movement generated a massive tsunami wave that according to the World Health Organisation caused the loss of between 170,000 and 290,000 human lives. Even though Indonesia suffered the worst consequences, all the countries around the Indian Ocean were affected. Along the west coast of Thailand the inundation levels were +5 and +12 meters above sea level along the most severely affected areas covering the coastline from Phuket Island in south and 200 km further north. The tsunami caused extensive damage in Thailand, and approximately 8000 people were killed. In cooperation with Thai authorities, a project has been initiated that aims at assessing the risk of similar future tsunami events and how the consequences of such events may be reduced to an acceptable level. The most feasible mitigation measures are to establish safe escape routes, elevate the land and to construct protective walls or dikes. It will be a significant challenge to gain acceptance for the recommended mitigation measures and to ensure that they will be implemented.

INTRODUCTION

The earthquake on the 26 December 2004 that triggered the catastrophic tsunami event had its epicentre about 150 km west of the Aceh province on the northern tip of Sumatra, see Figures 1 and 2. Within 10 minutes after initiation, the earthquake slip had propagated 1200 km northwards from the epicentre. The earthquake was caused by stress relief on the subduction zone created by the ongoing NE displacement of the Indian continental plate in under the Burma/Sunda plates. The ongoing plate movement is typically 6.5 cm/year.

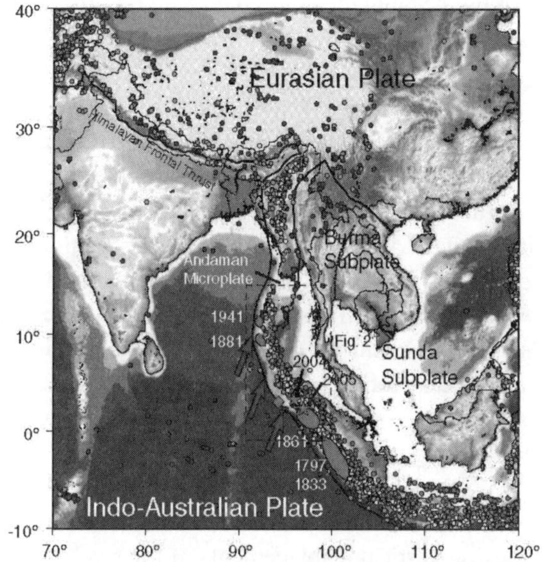

Figure 1. Continental plate movements and recorded earthquakes along the Sunda arch (From Bilham et al. 2005)

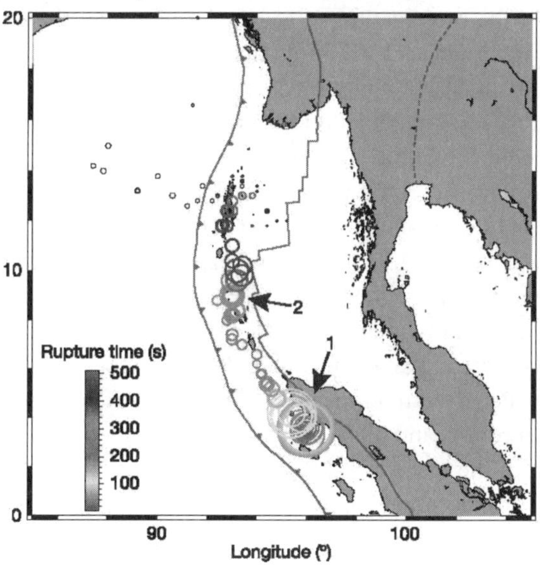

Figure 2. Rupture trace of the Sumatra earthquake, with the size of the circles indicating the amount of energy released and the colours the time. Numbers 1 and 2 indicated the position of the two major seismic energy releases (From Krueger et al. 2005).

The tsunami generated by this earthquake is the most devastating that has ever been registered in historic time. The tsunami affected all countries around the Indian Ocean and Andaman Sea, and even reached the west coast of Africa. About 170,000 people have been reported killed by the tsunami and 120,000 are reported missing (WHO, 2005), whereof about 120,000 killed and 110,000 missing along the west coast of Sumatra.

The Norwegian Geotechnical Institute (NGI) took an early initiative to see how Norwegian expertise could contribute to develop plans for re-construction that account for the potential risk of similar tsunami incidents in the region in the future. This led to the initiation of a project financed by the Royal Norwegian Ministry of Foreign Affairs to assist the authorities in Thailand with establishing re-construction and mitigation plans for Thailand. In this connection it is established close interaction with the Department of Mineral Resources (DMR) under the Ministry of Natural Resources and the Environment in Thailand, which also acts as client for the study. It is an additional objective of the study to serve as an example case for all countries affected by the 2004 tsunami.

The study involves the following main activities:

1. Identification of risks off future earthquakes in the region and their associated potential for causing vertical displacements of the sea bottom, and which in turn will generate tsunamis. Hereunder define specific scenarios that should be considered when plans for re-construction and mitigation are made.
2. Modelling of the propagation of tsunami waves, and their expected inundation levels and consequences if no mitigation measures are implemented.
3. General assessment of alternative mitigation measures that can limit future losses of human life, destruction of buildings and infrastructure, or other direct or indirect socio-economic consequences due to tsunamis.
4. Recommend specific re-construction plans and mitigation measures for three selected areas in Thailand; Patong City, Ban Niang tourist resort area just north of Khao Lak, and Nam Kem fishing village.
5. Establish contact and interaction with local authorities and organisations, property owners and other stakeholders in the affected areas to ensure that proposed plans are realistic and viable. In this connection it has been established an Advisory Panel with 11 members from the affected areas that reviews and comments on results and plans proposed.
6. Presentation of the results of the study at training seminars and workshops with invited participants from all countries affected by the 2004 tsunami.

The study is in its nature very multi-disciplinary. It is therefore established a project team which in addition to NGI staff involves seismologists at the Norwegian Seismic Array (NORSAR) and University of Bergen (UiB), hydrodynamic expertise at the Institute of Applied Mathematics at the University of Oslo (UiO) and at the Coast and Harbour

Research Laboratory at SINTEF in Trondheim, urban and regional planners at the Norwegian Institute of Urban and Regional Research (NIBR), and architect B.Heyerdahl.

The Coordinating Committee for Geoscience Programs in East and Southeast Asia (CCOP) coordinates arrangements of regional seminars and workshops, whereas DMR plays an important role in acquiring all relevant data that are required for the study, including assimilation of data from the 2004 tsunami, and establishing interaction and cooperation with local and central governments.

The study shall be completed in February 2006 (2 months after this conference). This paper therefore limits itself to give a general presentation of the most important aspects, without giving any final and specific recommendations.

FUTURE EARTHQUAKE AND TSUNAMI POTENTIALS

The Sunda trench is a very active fault zone as shown in Figures 1 and 2. Some of the historical earthquakes in the northernmost areas are listed in Table 1. From southeast of Sumatra the trench is passing along all of Sumatra and then off the west side of the Nicobar and Andaman islands in the Indian Ocean. The total length of the subduction zone is about 5000 km.

The dislocations associated with the great Sumatra earthquake extended over a distance of about 1200 km from a little south of the epicentre, west of the Banda Aceh province in Sumatra, to the northern parts of the Andaman islands, as shown in Figure 2.

Table 1. Overview of historical earthquakes along the Sunda trench (mostly based on Bilham et al. 2005).

1681, 1756, 1770, 1897,1818	Tsunami reported for the last three earthquakes.
1833, Nov. 24	M= 8.75. Tsunami along 550 km of the Sumatra coast. Felt also in Singapore and Java.
1847, 31 Oct. [1)]	M=[7.5-7.9]. Five weeks of aftershocks.
1861, Feb. 16	M= 8.25-8.5. Tsunami along 500 km of the Sumatra coast.
1881, 31 Dec. [1)]	M= 7.9 & 7.0. Double rupture. Great tsunami observed in many harbours in the Bay of Bengal.
1941, 26 Jun. [1)]	M= 7.7. Rupture zone 50x150 km, less than 3 m displacement.

1) Along the Andaman-Nicobar segment of the Sunda arch.

Figure 3 illustrates the deformations (dislocations) that are occurring when stress is released in a subduction zone. As the subducting plate is slowly moving down, pulling the overlying plate along in a locked fault zone, the stress is increasing until the frictional strength is exceeded and the overlying plate is bouncing back. The motion is primarily in the fault plane, but since that plane is dipping typically 15-20 degrees (along the Sunda arc, at least) some of the movement will also be vertical, extending all the way to the sea bottom. Most tsunami-generic earthquakes have surface (i.e., sea bottom) ruptures.

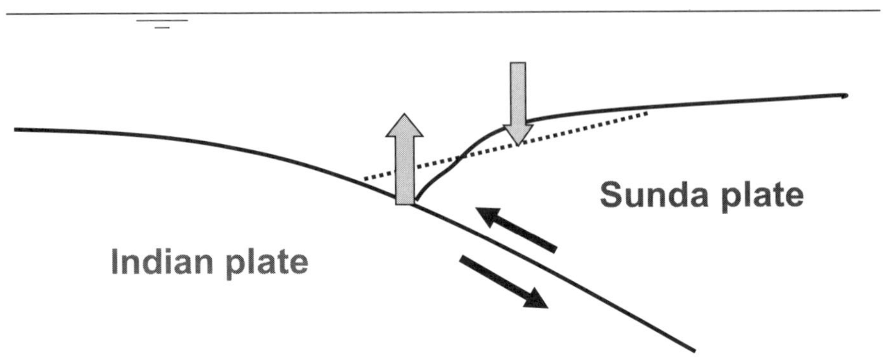

Figure 3. Schematic illustration of stress release in a subduction zone.

In the case of the great Sumatra earthquake on December 26, 2004, the seabed level deformations extended over most of the rupture length, albeit with quite varying dislocations.

Figure 4 shows a dislocation model that has been developed for the purpose of tsunami modelling, based on several published seismological inversions of the in-plane slip distributions and calibrated also against geodetic data. The largest vertical uplift in this model is about 3 m and the largest negative movement is about half of that. The width is typically 200 km. This means that huge volumes of water are set in motion from the dislocations.

Figure 5 illustrates the levels of seismic activity along the northern and southern parts of the Sunda arc, showing the activity is a lot less in the northern areas which has the greatest potential influence in Thailand, Sri Lanka and India. The difference in activity is almost a factor of ten.

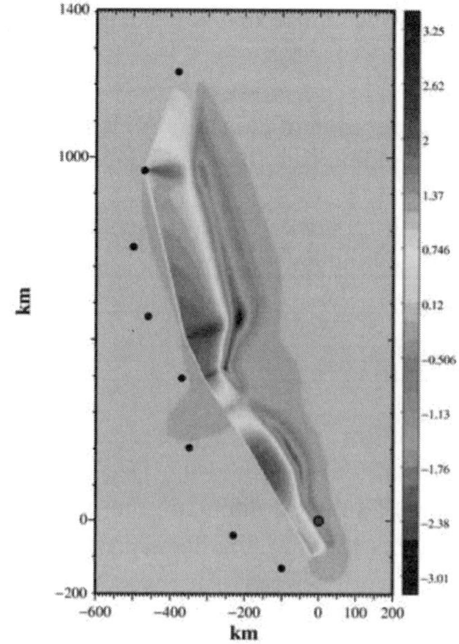

Figure 4. A model for vertical sea-bottom dislocations used to model tsunami effects from the December 26 earthquake.

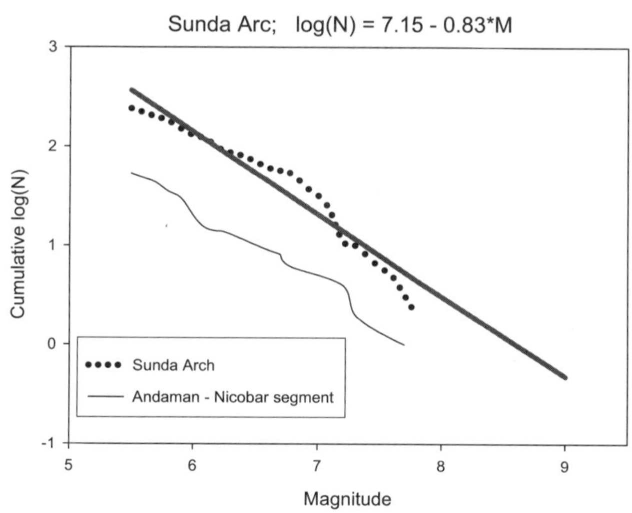

Figure 5. Frequency-magnitude distribution of earthquakes along the northern (Andaman-Nicobar) and southern part of the Sunda arc. (N= number of incidents per 100 year).

Table 2 shows the way in which the magnitude of shallow earthquakes (with potential surface rupture) is controlling the tsunami potentials. In this statistics the size of the tsunami is not considered, however, since that is influenced also by other factors than the magnitude.

Table 2. Probabilities for generation of tsunamis from shallow earthquakes, based on global data (many subduction zones).

Earthquake magnitude (M)	Tsunami probability (%)
7.0	33
7.5	71
8.0	84

It still remains to assess the return periods of large tsunami-generic earthquakes in this region, but it is quite clear that an event like the December 26, 2004, earthquake will require several hundred years of stress accumulation to occur. Even so, tsunamis may be generated also by fault ruptures that are smaller than this megathrust event, and such events of course have smaller return periods.

Which events and associated risks it is reasonable for a society to protect itself against is a very complicated question. It is particularly so for events which have very large return periods and very large potential for loss of human lives and property. However, in today's societies it is clear that it is vital to create more understanding for our responsibility to consider future generations that are going to live in the areas exposed to large natural disasters. This means that in choosing mitigation measures, we should have a strong focus on their long lasting effects.

TSUNAMI BEHAVIOUR AND MODELLING

Tsunami waves generated by earthquake seabed dislocations are normally long. In the case of the 26 December 2004 earthquake, the estimated wavelength is up to 200 km, as can be seen in Figure 4. Even though the water depth in the Indian Ocean is up to 5000 m, the wavelength is still very large compared to the water depth. In addition, the wave height is very small compared to the water depth. In such a case, linear long wave theory is valid. A consequence of this is that the water pressure is approximately hydrostatic, and that the speed of the wave propagation, c, is given by:

$c = (gH)^{0.5}$ where g = gravitational acceleration and H= water depth

Figure 6 shows the water depth in the Indian Ocean and the Andaman Sea, which is located between the Nicobar-Andaman islands and the west coast of Thailand and

Myanmar. For the largest water depth of 5000 m, the tsunami wave propagation speed will be 805 km/h, which is close to the speed of sound.

At a water depth of 1000 m the propagation speed is 360 km/h, at 100 m depth 113 km/h and at 10 m, 36 km/h.

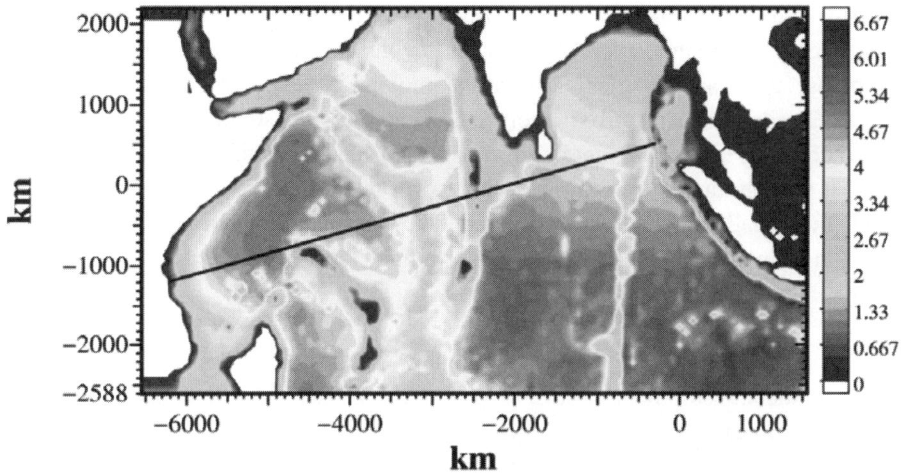

a) Plane view

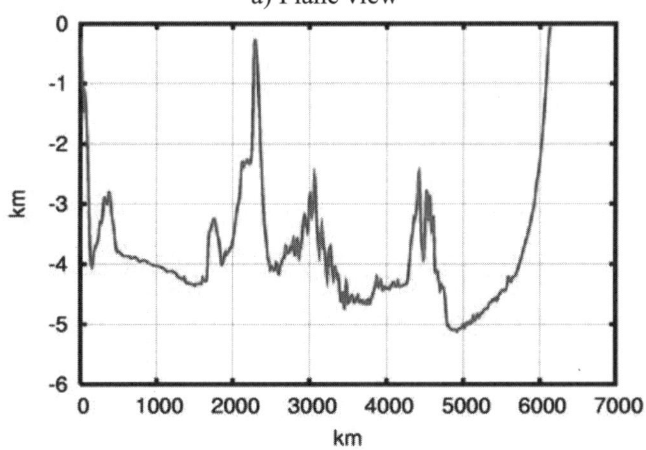

b) Sectional view east to west (seen from the north) from the Nicobar islands to the coast of Africa

Figure 6. Water depths in the Indian ocean

When the tsunami wave moves into shallower water, the wave length will be reduced and the wave height will increase. The wave theory governing this case is more complex, particularly when the wave starts to break and potentially split.

Along the coasts of Thailand and Indonesia, it was observed that the initial effect of the tsunami was a withdrawal of the sea. This agrees with what could be expected on the east side of the Sunda arch because of the depression of the seabed, see Figure 3. Next was the arrival of a wall-like breaking wave front followed by steadily rising water that in some areas flooded the land several kilometres beyond the coastline. It took around half an hour before the wave receded. Maximum inundation level for the tsunami at the coastline of Thailand was approximately +5 meters at the south tip of Phuket and gradually increasing up to +12 meters in the Ban Niang tourist resort area just north of Khao Lak. It then gradually decreased further north towards the border to Myanmar. In the inland, the inundation level was not reduced significantly with distance from the coastline, and the water stopped only as it met hillsides or mountains, that typically are found 1 to 3 kilometres behind the coastline.

The University of Oslo (UiO) and Norwegian Geotechnical Institute (NGI) have cooperated in the development of a 3-D numerical model for simulation of tsunami waves based on linear long wave theory. For the coastal zone, a 2-D model has been applied. These models have been used to back-calculate the 2004 tsunami event. The analyses results presented in the following are based on the dislocation pattern shown in Figure 4.

Figure 7 shows a snapshot of calculated surface elevations 1 and 2 hours after the earthquake was triggered. The wave pattern is significantly influenced by interference with waves that are propagating between Sumatra, Nicobar and the Andaman islands, and the local variations in sea depth. The tsunami wave hit Sri Lanka, approximately 1200 km from the Sunda arch fault, and the west coast of Thailand, 600 km from the fault, almost at the same time. This is caused by the significantly larger sea depths towards Sri Lanka than towards Thailand, see Figure 6a.

Figure 8 shows calculated wave form as the tsunami wave reaches the coastline in Patong City and Ban Niang. The result is obtained by using the result from the 3-D analysis 1 hour after the earthquake as input to the more sophisticated 2-D model that propagates the wave into the shallower water near the coastline.

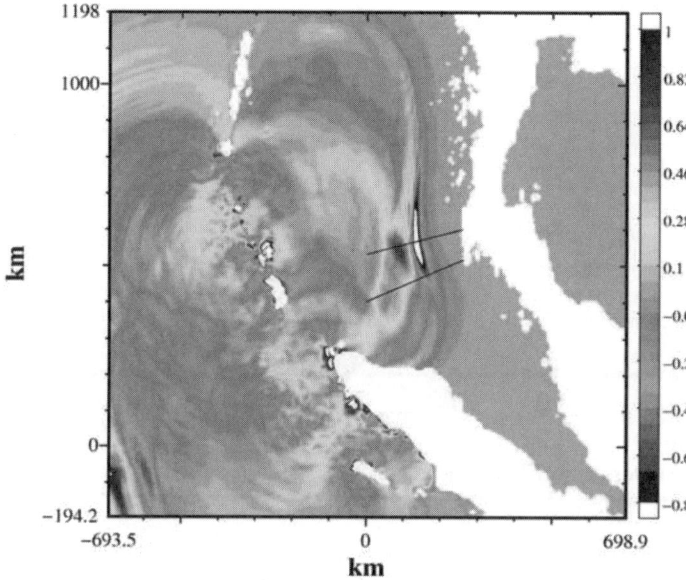

a) After 1 hour

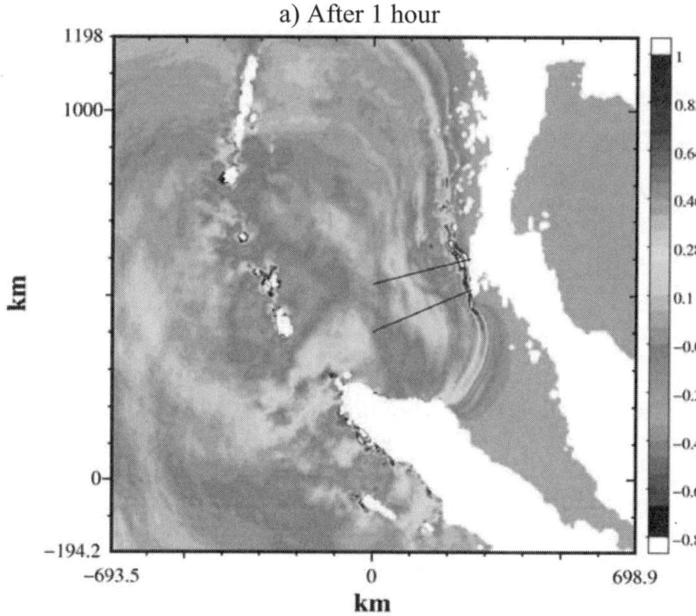

b) After 2 hours

Figure 7. Surface elevation [m] as the tsunami propagates towards the coast of Thailand 1 and 2 hours after the earthquake. Profiles towards Patong (south) and Ban Niang (north) drawn with black lines.

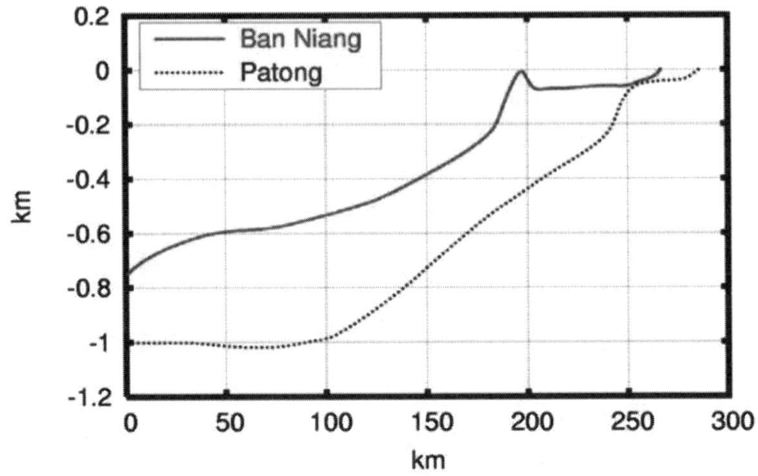

a) Depth profile along the cross sections in Figure 8

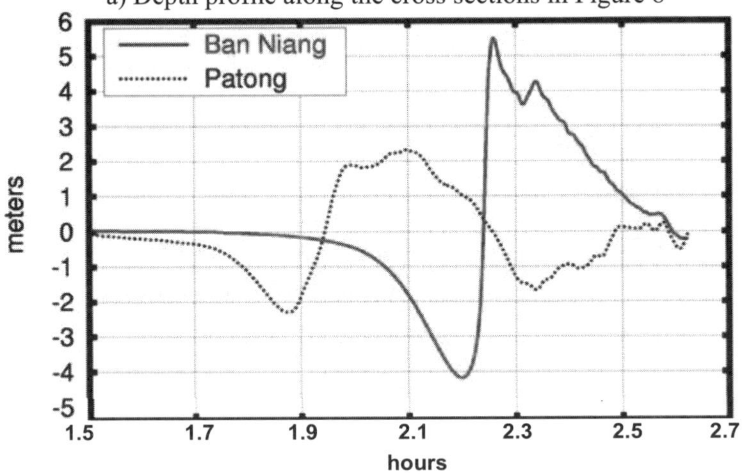

b) Water levels as function of time after the earthquake

Figure 8. Time series of the tsunami wave as it approaches the coast of Thailand for Patong City and Ban Niang along the cross sections in Figure 8

The analyses confirm the observed initial withdrawal of the sea, and that the duration of the tsunami wave is approximately half an hour. There is also good agreement between the wave propagation speed of 40 km/h as it hits the shoreline with estimates of wave speed made from video recordings of the event. The wave propagation speed far exceeds the running speed of an average person. Comparisons, also based on observations from various sites around the Indian Ocean, give the following preliminary conclusions:

- The simulations agree well with observations with regard to arrival time of the tsunami wave for the whole Indian Ocean, and with regard to the general pattern of withdrawal of the sea before the flooding including the duration of the event.

- The analyses confirm the observed relative variation in inundation levels along the coast of Thailand, but absolute levels are at this stage underestimated. This is most likely primarily caused by an underestimation of the seabed movements in Figure 4, and neglect of possible further amplification during inundation.

Further studies of the fault patterns will hopefully bring the analyses even more in agreement with the observations.

The selection of future scenarios to be used as a basis for planning of reconstruction and choice of mitigation measures has not yet been made. Two earthquakes with the same magnitude may cause very different seabed dislocation patterns with regard to both size/position of the dislocated area and direction of the dislocation. Predicting tsunami potential for Thailand based on earthquake magnitude alone is therefore associated with uncertainty. In addition comes the uncertainty related to the return period for a certain earthquake magnitude.

CONSEQUENCES AND POSSIBLE MITIGATION MEASURES

The 26 December 2004 tsunami clearly demonstrated the serious consequences of a tsunami that strikes areas unexpectedly or where there is little or no public awareness of what a tsunami is and how to react to it when the first signs emerge.

Data from Thailand shows as expected a clear correlation between consequences of the tsunami and the maximum inundation level of the tsunami wave as it passed across land. This is illustrated by some examples from the selected study areas in the following.

Patong City, Figure 9, is located at the bottom of a 3-4 km deep and 3-4 km wide bay. The natural ground level is typically + 3 m above sea level. The tsunami inundation level was +5 til +6, which means it reached just up under the second floor level in the buildings. For buildings facing the sea/beachfront located typically 50 m away, most brick walls, plaster walls, windows and doors facing the sea were blown out by the tsunami wave. No buildings collapsed because they had load carrying concrete columns that remained intact. This damage picture also partly applied to the front of buildings facing the streets running perpendicular to the beachfront within about 200 m distance from the sea. Further back from the sea the damage was generally limited to broken windows and damage associated with buildings getting filled with water. The number of casualties in Patong City was limited to 160 out of the approximately 20,000 people that were in the city when the tsunami struck. Most casualties were caused by people who could not get up from the basement floor in a shopping store.

Figure 9. View of part of Patong City and example of damaged building

The Ban Niang tourist resort area stretches about 12 km north of Khao Lak. In this area the tsunami inundation level was generally around +10, but locally up to + 12 near the beachfront. The natural ground level behind the beachfront typically lies at +4 to +5. The resorts consisted mainly of 2 to 3 story buildings. The tsunami wave smashed through both the 1st and 2nd floors of the buildings, and on its way blew out most brick and plaster walls both at front, back and the sides of most buildings within 500 m distance from the sea. The damage gradually decreased with larger distance from the shoreline.

Even in Ban Niang relatively few buildings collapsed completely because they had reinforced concrete columns that were able to carry the loads associated with the walls being blown out by the tsunami impact, see for example Figure 10. There were about 100 resorts in the Ban Niang area with about 15,000 present when the tsunami wave struck, whereof about 3000 lost their lives.

Figure 10. Example of damaged resort building, Ban Niang

In the fishing village Nam Kem located closer to the border against Myanmar, the tsunami inundation level was +8. The ground level in most of the village lies at +3 to +6. The houses consisted for the most part of rather simple 1-2 storey brick structures without solid columns. 6000 people lived in the village whereof 3000, or half the population, lost their lives in the tsunami. This is relatively speaking much more than in the Ban Niang tourist resort area, which may be explained by the following factors:

- Many people were at the fishing harbour at the seafront, and were killed by debris from wharf structures and boats that were thrown around by the tsunami impact wave.
- Many buildings collapsed completely due to poor quality and lack of solid columns to support the structure.
- The main roads in the area were running more or less parallel to the waterfront giving long or difficult escape paths to safer high areas inland.

Implementation of tsunami warning systems has the potential of reducing human casualties in possible future tsunami events. Such systems are expected to be implemented in Thailand as well as in other countries around the Indian Ocean. It will however be a challenge to ensure that warning and evacuation will be effective if a tsunami strikes at night time when people are asleep. If no tsunamis occur within the nearest decade or two, it will also be a challenge to keep a warning system running. It should therefore be seriously considered to implement other measures to reduce the consequence of possible future tsunamis, as outlined in the following.

There is little doubt that knowledge and information to people about how they should respond to the signs of an approaching tsunami could have saved many lives in Thailand and other affected countries in the 2004 tsunami incident. The time elapsed from the sea receded to the tsunami wave hit land with full force was about 10-15 minutes, ref. Figure 8b), which could mean a lot in terms of getting time to reach safe areas. Measures that can enhance public awareness and knowledge are therefore considered an important element in the reduction of human casualties in future tsunamis.

The type of building structures that are or will be in the affected areas will have a limited life span, say 20-30 years for typical hotel and tourist resorts and 20 -50 years for houses, shops, industry buildings etc. It may therefore not be an important factor to introduce measures for the sole purpose of protecting against building damage. The main objective must be to reduce human casualties and suffering. In this context the following mitigation measures are considered in the on-going study:

1. Well marked and easily accessible escape routes to areas safe from inundation by tsunami. For large low lying areas close to the sea this will require construction of "escape pads" that can be reached within a distance of say 200-500 m. Such "escape pads" may be in the form of earth fills or buildings that have easily accessible staircases to higher and safe floor levels accessible to the public in general.
2. A general raising of the ground level, or "vertical land reclamation", where buildings are located as illustrated in Figure 11. Such a measure could be particularly worthwhile to consider for re-construction of the tourist resort areas in Ban Niang, because fill materials are easily accessible, for instance by dredging from the sea. If the 1st storey in the resort buildings are used for general functions only and not for bedrooms, one may allow the water inundation level to reach the roof in the 1st floor, which would reduce the height of the required fill level by 2.5 m. That presupposes that the buildings are designed such that they will not collapse, and that there is easy access to the safe 2nd floor.
3. Construction of vertical seawalls or dikes around populated areas, Figure 12. Such structures may be built so high that they impound the tsunami completely, but one may also consider to allow limited overtopping. That is, their function would be to limit the direct impact of the tsunami and reduce the rate of rise in water level inside the seawall or dike. If overtopping is allowed that would require safe escape routes to high areas outside the seawall or dike structure.
4. Construction of energy absorbing moles or breakwaters in the sea. In the present case this is considered a less viable measure due to high cost and possible consequences for the present equilibrium state in terms of coastal or beach processes. Such coastal processes involving sand transport by waves and currents may easily be disturbed by structures in the sea.

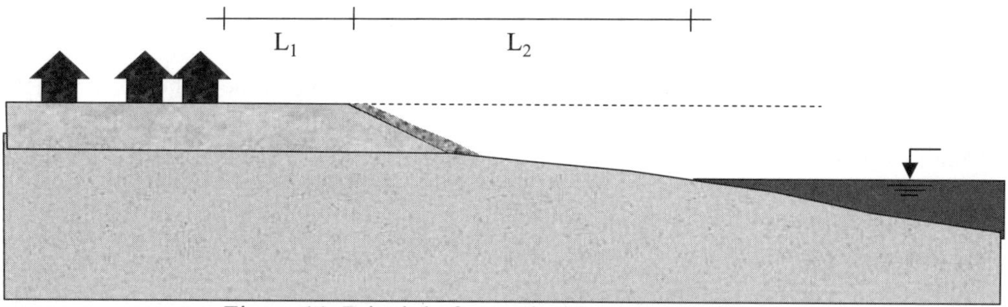

Figure 11. Principle for vertical land reclamation

a) Sea wall

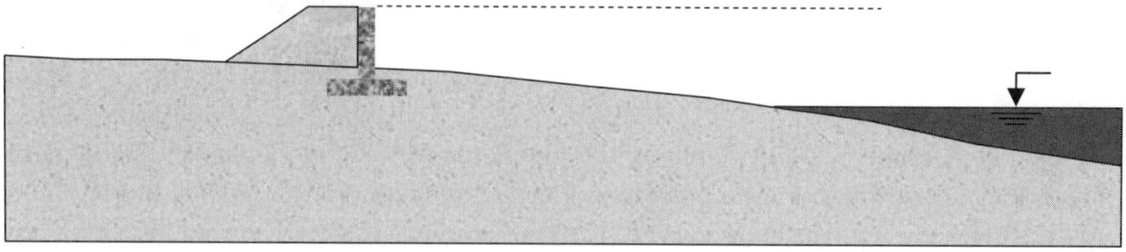

b) Dike

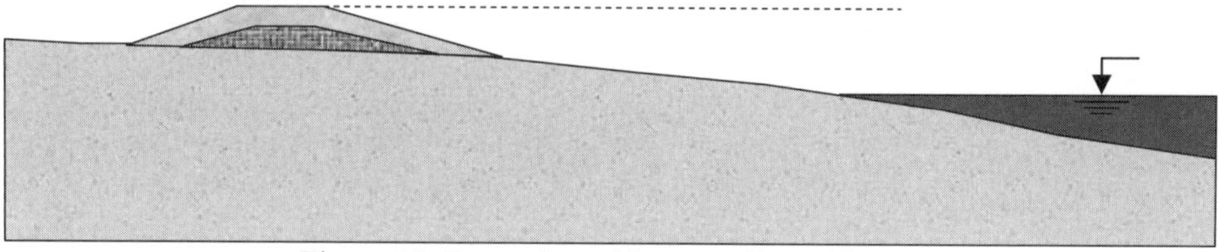

Figure 12. Protection with sea walls or dikes

What measures that will finally be recommended in the different areas will of course depend on the "design level" for the tsunami that the study ends up with. In addition, considerable weight must be given to what it in practice will be realistic and possible to implement considering the national and regional socio-economic framework, property rights etc. In any case it will require a strong political commitment and willingness to implement new laws and regulations that may be required on both national and regional levels. That may be a particular challenge in countries with limited economic resources and with limited experiences and tradition for dealing with such issues. It will therefore be given particular attention to presenting the results of the study at both national and regional government levels, including both elected politicians and bureaucrats.

ACKNOWLEDGEMENTS

The authors would like to give special thanks to other members of the enthusiastic project group which so far have contributed to this study, including Farrokh Nadim (NGI), Conrad Lindholm (NORSAR), Kuvvet Atakan (UiB), Aud Tennøy (NIBR), architect Birger Heyerdahl.

REFERENCES

Bilham, R., Engdahl, E.R., Feldl, N., Satyable, S.P. (2005). "Partial and complete rupture of the Indo-Andaman plate boundary 1847-2004", *Seismological Research Letter*, Vol. 76, No. 3, May/June 2005

Glimsdal, S., Pedersen, G.K., Atakan, K., Harbitz, C.B., Langtangen, H.P. and Løvholt, F. (in press). "Propagation of the Dec. 26 2004 Indian Ocean Tsunami: effects of dispersion and source characteristics" submitted for publication in *Int. J. of Fluid Mech. Research.*

Kruger, F., Ohrnberger, M. "Tracking the rupture of the Mw = 9.3 Sumatra earthquake over 1,150 km at teleseismic distance." *Nature*, Vol. 435 (7044), 937-9, Jun 16, 2005

World Health Organisation (WHO) internet site, http://www.who.int/hac/crises/international/asia_tsunami/3months/en/index.html, 21 October 2005

EARTHQUAKES & SOIL DYNAMICS

International Conference on Geotechnical Engineering for Disaster Mitigation & Rehabilitation
Chu, Phoon & Yong (eds) © 2005 World Scientific Publishing Company ISBN 981-256-469-1

GEOTECHNICAL ISSUES OF THE RECENT TURKISH EARTHQUAKES

KORHAN ADALIER
Department of Civil and Environmental Engineering, Florida State University, Panama City, FL, USA

AHMED PAMUK
Department of Civil Engineering and Engineering Mechanics, Columbia University, New York, NY, USA

Being in the Eastern Mediterranean sector of the Alpine-Himalayan earthquake belt, which is second to only the Pacific-Belt in terms of seismicity, Turkey is frequently exposed to destructive earthquakes. In the twentieth century alone, earthquakes in Turkey have caused about 85,000 deaths and destruction of more than half a million structures. Although the ground motions were solely responsible for majority of damages and casualties during most of these earthquakes, geotechnical aspects such as soil liquefaction and site amplification effects considerably influenced the damage patterns during many as well. This paper briefly discusses the recent 1999-Izmit and 1998-Adana earthquakes where geotechnical issues such as liquefaction and associated ground failures and site amplification effects played particularly significant role in the observed earthquake damages. Observations from such earthquakes provide us with an opportunity to further improve our understanding of the observed phenomena and their effects that can be expected during other future earthquake events around the world.

INTRODUCTION

According to the seismicity map of Turkey, 92% of the Turkish land, 95% of the population, and 98% of the industry lies on the seismically active ground. In the twentieth century alone, earthquakes in Turkey have caused about 85,000 deaths, 200,000 injuries and destruction of 500,000 structures. In many of these earthquakes, geotechnical problems such as soil liquefaction and site amplification considerably influenced the damage patterns. This paper briefly discusses the recent 1999-Izmit and 1998-Adana earthquakes where widespread liquefaction associated ground deformations such as lateral spreading, flow failures, ground fissures and extensional cracking, sand boils, ground subsidence and slope failures were observed. The extent of liquefaction provided a significant amount of data useful for analyzing the effects of ground liquefaction on different type of structures. The extensive nature of ground deformations observed during these two earthquakes, significantly increased the public awareness to earthquake hazards associated with foundations and soils.

GEOTECHTONICS AND SEISMICITY OF TURKEY

Turkey lies within the Mediterranean segment of the Alpine-Himalayan orogenic system, which is the second major earthquake belt following the Pacific-Belt. Figure 1 shows the major tectonic elements of Turkey along with the earthquake epicenters for magnitudes Ms > 5.5 during the 1881-1999 period. The tectonics of Turkey is greatly influenced by the movements of Arabian, Eurasian, and African plates (Fig. 1). The Arabian plate moves in NNE and pushes the Eurasian plate along the Bitlis Thrust and Fault Zone. Due to this continuing movement of the Arabian plate, the Anatolian block shifts westward along the North (NAF) and the Northeast Anatolian faults. On the other hand, African plate moves in NE direction and collides with the Eurasian plate and subducts along the Hellenic-Cyprus Arc, somewhat retarding the westward movement of the Anatolian block and initiating its tendency to rotate to the SW. The interaction of these complex plate motions caused several E-W trending blocks bounded by oblique normal faults in Southwest Turkey. Consequently, this region covering the Cyprus-Hellenic Arc and the Aegean Graben System has very high seismicity. Although the frequency of seismic events within the North and the Northeast Anatolian Faults and the Eastern Anatolian Contractional province has been less, these tectonic belts have produced most destructive earthquakes such as the 1939 Erzincan (M_s= 7.9) and the 1999 Izmit (M_s= 7.8) earthquakes. More detailed information on the tectonics of the region is given by Ambraseys (1976), Jackson and McKenzie (1984), and Saroglu *et al.* (1987).

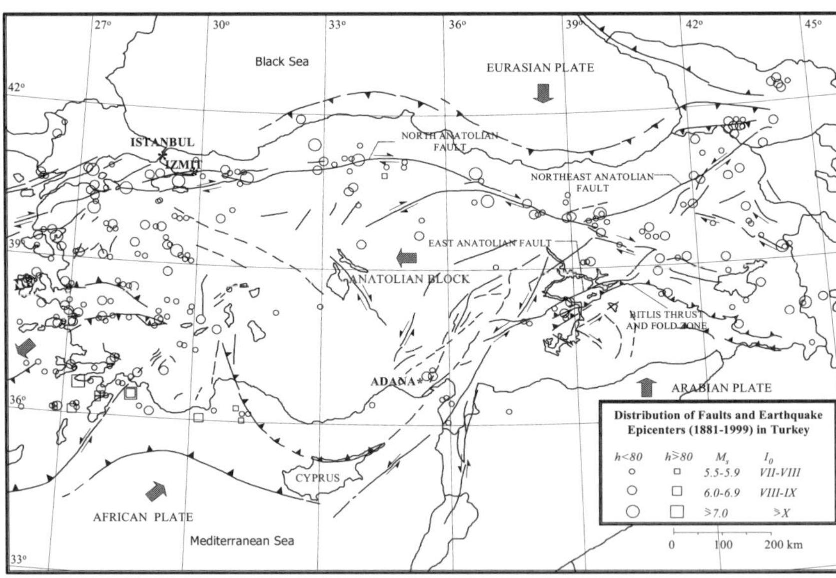

Figure 1. Major tectonic elements and distribution of epicenters for earthquakes with Ms>5.5 (during 1881-1999) in Turkey (Adalier and Aydingun, 2001).

ADANA EARTHQUAKE OF 1998

The Adana-Ceyhan (June 27, 1998) earthquake was caused by the left-lateral strike-slip fault which generated an earthquake with M_L = 5.9 (Richter Magnitude) and M_s= 6.3. It resulted in 145 deaths and a significant damage to several thousand mostly poorly constructed structures. Strong ground motion was about 25 seconds with maximum lateral accelerations of about 0.3g. Although the earthquake was relatively moderate in size, the coincidence of the earthquake epicenter and the fault with a very vulnerable geological surface formation (geologically young thick alluvial deposits of loose fine silty sands overlaid by soft silty-clay) led to widespread geological deformations (mainly due to liquefaction). This provided a unique opportunity to assess the performance of various types of structures, under conditions of loss of bearing capacity and lateral support, subjected to lateral spreading and subsidence. The geological deformations included lateral spreading, flow failures, ground fissures and extensional cracking, sand boils, ground subsidence and slope failures. The two most significant geotechnical aspects of this earthquake has been the widespread soil liquefaction and associated ground failures and ground motion amplification at "soft" soil sites. These two geotechnical phenomena significantly influenced the damage patterns.

Due to severe space limitation, herein only the main lessons learned from this earthquake related to geotechnical aspects will be discussed very briefly. Comprehensive studies on this earthquake include Adalier and Aydingun (1998, 2000, and 2001).

Main Observations and Lessons

i. Local site amplifications, or site effects, can result in significant differences in structural damage within the same general area. This was the reason of observing at several sites a heavily damaged building while a building of similar construction a block or two away may be completely unaffected. Damage patterns at several sites suggested topographic amplifications even on minor hills.

ii. Young sedimentary cohesionless soil deposits with low water table are exceptionally vulnerable to liquefaction. When this liquefaction happens in sloping ground (or ground with initial static shear stress), it is usually accompanied with lateral spreading (Fig. 2), which was found to be the most damaging liquefaction induced phenomena during this earthquake as well as most of the other previous big earthquakes.

iii. At areas away from incised river channels (or any other free face) and any slope where principal effect of liquefaction was vertical ground settlements, surprisingly little damage to man-made structures were observed. However, at areas not far from river channels or sloping ground, where liquefaction generated lateral ground displacements, foundation performance was typically poor. Likewise, ground failure rather than ground shaking, appeared to be the dominant cause of damage to buried utilities such as pipelines and irrigation canals.

iv. Liquefaction of loose sand layers were usually accompanied by the formation of sand boils due to the low permeability and cohesion of an overlying upper silty-clay layer.

In general, the greater was the thickness of the overlying layer, the fewer but larger were the sand boils. Sand boiling by itself caused minor damage only in few cases when it led to localized differential settlements in foundation soils.

v. The buildings with strong continuous mat foundations or thick strong base slab performed noticeably better in the areas subjected to minor to moderate effects of liquefaction, such as lateral spreading, sand boils, and ground subsidence. In these instances, the foundation behaved as a diaphragm, preventing differential ground displacements from propagating upward in to the superstructure.

vi. In the cases where the structures settled evenly in the liquefied foundations, the sunken buildings in many cases were less damaged from shaking than adjacent buildings that were on soils that did not liquefy during the earthquake. This is attributed to the base isolation effect provided by the liquefaction-softened foundation soils. In many cases, the magnitudes of sinking appeared to be proportional to the weight of the buildings.

vii. As in many other seismic disasters such as in Mexico City, the San Francisco Bay area, Kobe-Japan, and Izmit-Turkey (1999), significant damage in the Adana-1998 event was clearly related to soft ground conditions. Microzonation for ground deformations and amplifications at such sites are imperative. Mitigation of ground deformation effects on built environment may involve ground improvement or improved foundation systems. For amplification effects, design on soft ground most importantly must be based on the suppression of resonance by either ground improvement or structural improvement by damping.

Figure 2. Ground shaking triggered liquefaction in a subsurface loose silty sand layer, leading to differential lateral and vertical movement in an overlying unliquefied silty clay layer, which moved towards the incised Ceyhan River channel. This mode of ground failure, termed “lateral spreading” was the most striking and damaging adverse effect of liquefaction caused by the 1998 Adana Earthquake.

IZMIT (KOCAELI) EARTHQUAKE OF 1999

This earthquake hit densely populated areas of Izmit-Kocaeli (Fig. 1) and Sakarya provinces, situated on an alluvial fan at the western part of the NAF with M_s of 7.8. It caused more than 17 thousand deaths and more than $20 billion in economic losses. The earthquake was caused by a right-lateral strike-slip fault. Epicenter was about 7 km southeast of the city of Izmit and at a focal depth of about 16 km. Most of the damage and fatalities were in towns located on the narrow flat shorelines of the Marmara Sea, where mostly soft alluvial soils prevail. Basically of all the geotechnical issues observed during the 1998 Adana earthquake were observed during this event as well, but at a much larger scale. In addition to the observations made for the 1998 Adana earthquake summarized in the last section (i.e., i – vii), the following major points can be offered:

i. Unlike in the Adana earthquake, where no distinct fault related surface fracturing was observed in the field, in the Izmit earthquake a major surface fault rapture was observed. The total length of the fault rupture was about 110 km in the east-west direction. The fault zone was typically a narrow zone of cracks (5 to 25 m wide) that either followed or formed parallel sets of faults oblique to the trend of faulting. The surface expression of faulting included tension cracks and fissures with localized positive relief. Lateral displacements in the range of 3 to 4 m were common with maximum reported displacement of about 5 m immediately east of Arifiye where many structures were ribbed apart by the fault movement. Vertical displacements of as much as 2.4 m were also recorded. Surface faulting destroyed hundreds of buildings, damaged industrial facilities, port facilities, a naval base, pipelines and roads, and two major bridges (Kalkan *et al.,* 2004). Surface faulting in conjunction with earthquake induced densification settlements also damaged the modern Trans European Motorway-TEM, which was built to the highest standards (Pamuk *et al.,* 2004).

ii. Particularly in the areas of delta fans along the coastal margins near the epicentral regions, much of the damages were caused by liquefaction and associated ground deformations. These delta basins are constructed at the discharge points of creeks and rivers that transport sediment from the adjacent elevated areas. The depositional processes within delta fans usually produce loose, saturated sediments that often contain fine sand and silty sand layers that are exceptionally susceptible to liquefaction.

iii. In these coastal basins, much of the damage was a result of: (1) direct surface fault rupture; (2) tectonic subsidence and seawater intrusion; (3) liquefaction associated lateral spreading; and, (4) sea-wave drawdown and inundation. Particularly, where the fault parallels the coast, large areas subsided and extensive parts of Golcuk and Degirmendere were permanently inundated. Regional normal faulting subsidence combined with liquefaction deformations caused a 4-km-long section of the coast near Gölcük to submerge about 2 to 3 m.

iv. Hundreds of buildings founded on the liquefied soils settled due to soil bearing failure associated with the loss of strength caused by liquefaction. Especially in Adapazari and coastal areas many blocks of neighborhoods were affected. Building settlements as large as 3 m was observed. Although most buildings simply sank upright into the soil, many were also overturned, which exposed their foundations. Most of the excessively tilted buildings were of slender types. Liquefaction-induced large lateral ground displacements were observed also observed particularly along some of the coastlines.

CONCLUSIONS

The recent earthquakes of Adana-1998 and Izmit-1999 in Turkey once again emphasized the importance of the geotechical issues such as soil liquefaction, fault rupturing, and site amplification on the performance and damageability of structures. These earthquakes provided an invaluable field laboratory from which to improve our understanding of earthquake effects. Three causes of permanent ground deformation in general were related to: liquefaction, surface faulting, and densification of dry soil. Also, these two events clearly demonstrated that although earthquake shaking typically causes vast majority of damages, earthquake-induced permanent deformation or shifting of the ground can be more devastating to structures locally as few structures are built to withstand large permanent ground movement. As such, zonation studies on local level considering the local geology are imperative to better define the earthquake risks on both natural and man-made built environment.

REFERENCES

Adalier, K. and Aydingun, O. (1998). "*June 27, 1998 Adana-Ceyhan (Turkey) Earthquake: Engineering Reconnaissance Report*", Technical Report No. Geo98-9, Dep. of Civil Engineering, Eastern Mediterranean University, N. Cyprus, 45pp.

Adalier, K. and Aydingun, O. (2000). "Liquefaction during the June 27, 1998 Adana-Ceyhan Earthquake", *Geotechnical and Geological Eng.*, Vol. 18(3), 155-174.

Adalier, K. and Aydingun, O. (2001). "Structural Engineering Aspects of the June 27, 1998 Adana-Ceyhan Earthquake", *Engineering Structures*, Vol. 23(4), 343-355.

Ambraseys, N.N. (1976). "Some characteristic features of the Anatolian Fault Zone", *J. Tectophysics*, Vol. 9, 143-165.

Jackson, J. A. and McKenzie, P. (1984). "Active tectonics of the Alpine-Himalayan belt between western Turkey and Pakistan", *Geophys. J. Astr. Soc.*, Vol. 77, 185-264.

Kalkan, E., Adalier, K. and Pamuk, A. (2004). "Near Source Effects and Engineering Implications of Recent Earthquakes in Turkey", *Proc., 5th International Conference on Case Histories in Geotechnical Engineering*, NY, April 13-17, Paper No. 3.30.

Saroglu, F., Emre, O. and Boray, A. (1987). "Seismicity and active faults of Turkey", *Mineral Research and Exploration Institute of Turkey*, Report No. 8174, 294pp.

Pamuk, A., Ling, H.I., Leshchinsky, D., Kalkan, E. and Adalier, K. (2004). "Behavior of Reinforced Wall System during the 1999 Kocaeli (Izmit), Turkey, Earthquake", *Proc., 5th International Conference on Case Histories in Geotechnical Engineering,* New York, NY, April 13-17, Paper No. 3.45.

International Conference on Geotechnical Engineering for Disaster Mitigation & Rehabilitation
Chu, Phoon & Yong (eds) © 2005 World Scientific Publishing Company ISBN 981-256-469-1

INFLUENCES OF CONFINING PRESSURE AND CYCLIC DEVIATOR STRESS ON LIQUEFACTION OF CLAYEY SAND

PUTERA AGUNG M AGUNG
Civil Engineering of State Polytechnic of Jakarta (PNJ), University of Indonesia-Campus Depok (16425), West Java, Indonesia

A study on the liquefaction behaviour of a reconstituted clayey sand is presented. Clayey sand (CS) sample was made by rivers sand and clay caolinite in a certain composition. This study used instrument of cyclic triaxial test. Some treatments of confining pressure on every sample were 25 kPa, 50 kPa and 100 kPa, respectively. Test results shows that liquefaction would be occurred at 13.20 kPa (CS.A); 31.35 kPa (CS.B); and 65.34 kPa (CS.C) cyclic deviator stresses. Number of cycle required to liquefy the samples would decrease along with increment of confining pressure and cyclic deviator stress. Besides that, pore water pressure would increase due to increment of clay content in each sample. Therefore, more clay content in clayey sand would be more difficult to liquefy sample.

INTRODUCTION

Indonesia is one country having natural hazards potential by earthquake. Coastal zone in Indonesia has soil foundation which consisted of fine sand and in saturated condition and they are relatively susceptible to liquefaction.

In several areas in Indonesia, coastal zone are found layers combination between sand and silt or clay. Coastal zone is alluvial area consisted of sandy, silty or clayey soil deposited coming from river and sea. Typical of soil layers are located at depth variation from 0.00 – 20.00 m. Mean water table depth is located at depth of 0.40 – 1.50 m. All clayey sand layers are in saturated condition.

Seed and Idriss (1976; 1982) concluded that clay soils would be lost their effective stress by liquefaction, if percentage of grain size fraction of 0.005 mm < 15%, liquid limit (LL) < 35% and water content > 0.9 times percentage of LL. Yammamuro and Lade (1998) in their study also concluded that the sand having silt fraction would be occurred static liquefaction process. According to Vallejo (1988) found that combination of sand and silt or clay is not only liquefaction but also tensile cracks around their layers.

This study is addressed to know liquefaction potential on a reconstituted clayey sand by experimental. The study uses cyclic triaxial test instrument and all procedure conducted are based on ASTM D5311 – 92. The study is not only observation to liquefaction phenomena but also influence of confining pressure and cyclic deviator stress on liquefaction of clayey sand. The samples are made from a certain composition between river sand and clay caolinite. Their compositions were 95% river sand and 5% clay caolinite (CS.A); 90% river sand and 10% clay caolinite (CS.B); 85% river sand and 15% clay caolinite (CS.C).

Main Problem

1. Gradation of CS.A; CS.B and CS.C compared with several sand types liquefied.
2. Influence of confining pressure and cyclic deviator stress on liquefaction based on reconstituted samples between natural sand and caolinite clay. General properties of both materials are shown in Table 1.

Table 1. Properties of sandy soil and caolinite clay.

Soil properties	Natural sand	Caolinite clay
Weight volume (γ)	1.65 gr/cm^3	1.83 gr/cm^3
Specific gravity (Gs)	2.84	2.60
Natural water content (w)	3.30 %	49.42%
• **Liquid limit (LL)** • **Plastic limit (PL)** • **Shrinkage limit (SL)** • **Plasticity index (IP)**	- - - -	62.0% 41.0% 29.2% 21.0%
• **USCS** • **AASHTO**	SW A – 3	CH A – 7 – 5 (30)

3. Relationship between confining pressure, number of cyclic stress and cyclic deviator stress parameter on every sample.
4. Susceptibility level from reconstituted sample based on their available composition.

Boundary conditions

1. Natural sand has gradation of grains, 0.20 mm < D_{50} < 0.40 mm. Natural sand in this study is taken from river Kali Ciliwung from Southern Jakarta City, Depok, Indonesia. Sand before mixed with caolinite clay is obtained from natural sand which is separated from fractions of gravel; silt and clay.
2. Caolinite clay used in this study is produced by Bitung-Indonesia Company.
3. Confining pressure used in cyclic triaxial test is limited by 25 kPa, 50 kPa, and 100 kPa on reconstituted sample of clayey sand.

METHODOLOGY

Fractions of natural sand of Kali Ciliwung or sandy soil are consisted of 1.6% gravel; 91.3% sand; 4.4% silt and 2.7% clay. Sand before mixed with caolinite clay is taken from after separated from fractions of gravel; silt and clay. Furthermore, aall procedures in this study are shown in Figure 1.

Table 2. General properties of reconstituted clay sand.

Soil type	CS.A	CS.B	CS.C
Weight volume	1.82 gr/cm^3	1.82 gr/cm^3	1.82 gr/cm^3
Water content	0.42%	0.53%	0.95%

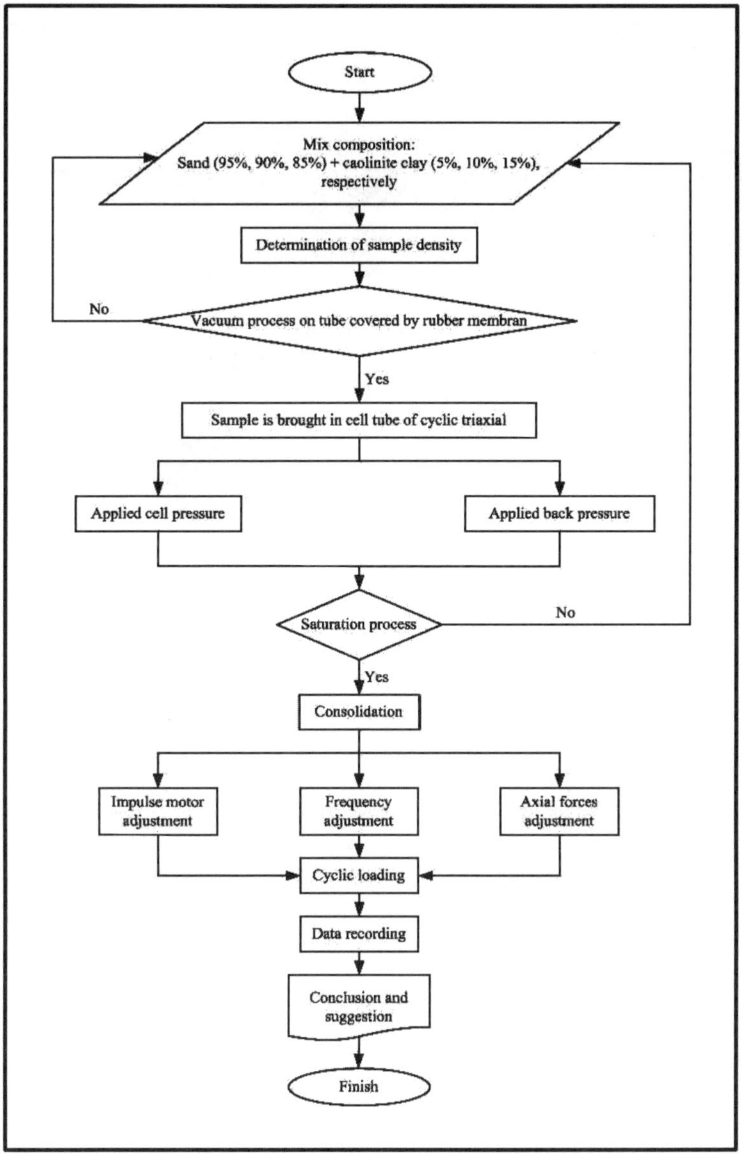

Figure 1. Cyclic triaxial test procedure

RESULT AND DISCUSSION

Based on the previous study, grain distribution analysis of sand is important property in occurrence of liquefaction (Pecker, 1984). Furthermore, all samples are compared with the other sands which have experienced the liquefaction phenomena, such as: Ukai; Tenughat; Obra; Niigata; Solani. The grain distribution analysis for all samples is plotted on Pecker's graphic (1984), as shown as in Figure 2.

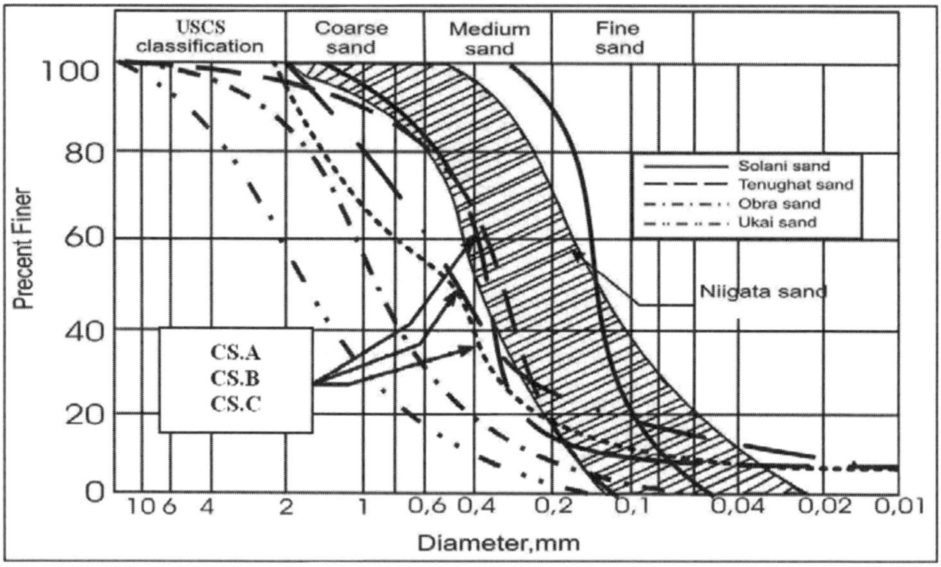

Figure 2. CS.A; CS.B and CS.C compared with several sands liquefied by Pecker.

Generally, Figure 3 and 4 shows number of cycle when clayey sand sample liquefied is very depended on confining pressure; cyclic deviator stress; and clay fraction. Based on Figure 3, more increase confining pressure applied on every sample will enlarge cyclic deviator stress to liquefy the sample. In addition, number of cycles (N-liquefy) required to liquefy the sample is very difference on every confining pressure and cyclic deviator stress parameter. The sample will be liquefied in short time on high confining pressure and cyclic deviator stress, so vice versa.

Figure 4 shows number of clay fraction will influence confining pressure and cyclic deviator stress. More increase clay fraction in reconstituted clayey sand will require increment of confining pressure and cyclic deviator stress to liquefy the sample. Furthermore, more decrease cyclic deviator stress and confining pressure will enlarge liquefaction time of sample. In addition, the level of liquefaction disturbance to reconstituted clayey sand is more influence by number of clay fraction in every composition. The reconstituted clayey sand sample having low clay fraction is very susceptible to liquefy.

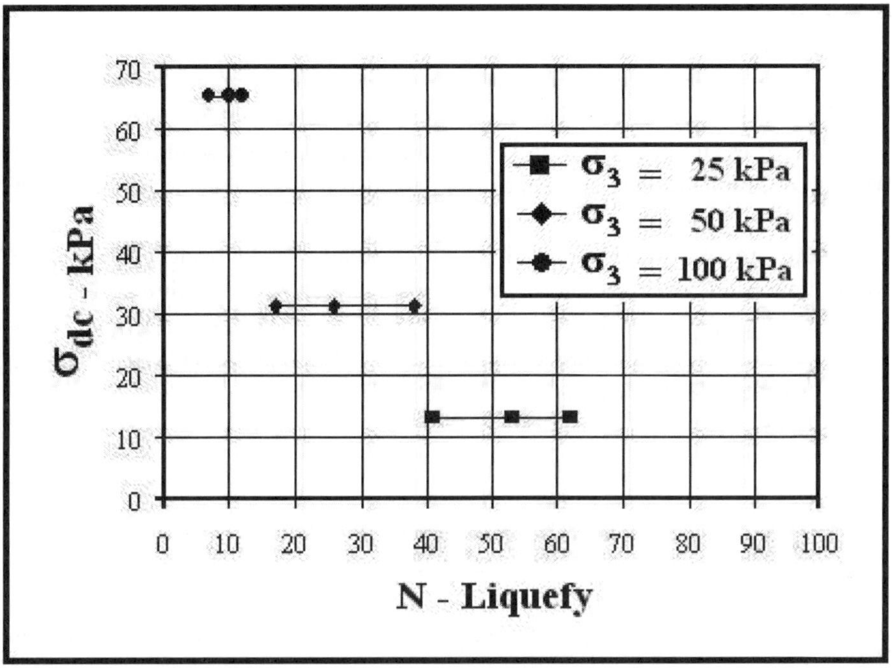

Figure 3. Relationship between cyclic deviator stress and number of cycles (N-liquefy.

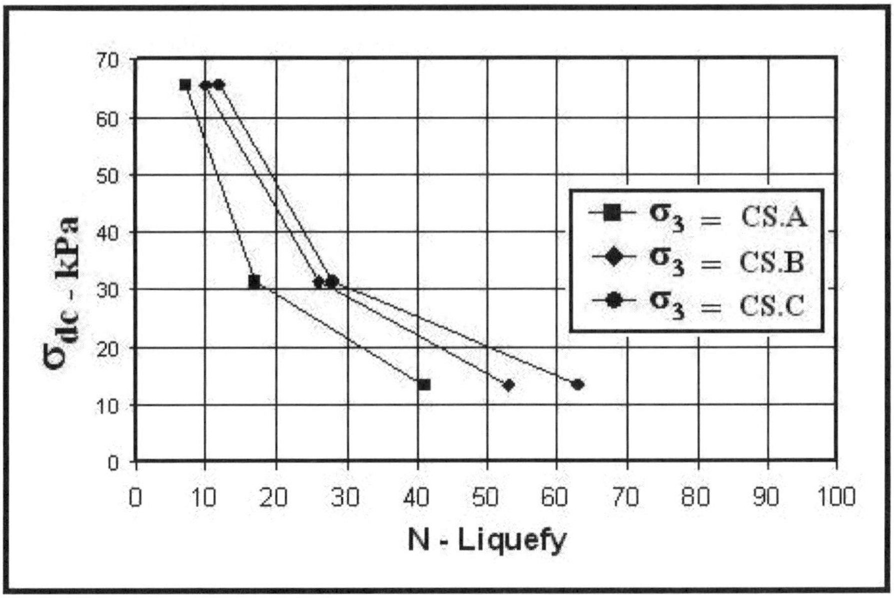

Figure 4. Relationship between cyclic deviator stresses and number of cycles (N-liquefy).

CONCLUSION

1. Gradation of CS.A having value of D_{50} = 0.380 mm easier to liquefy than others (CS.B and CS.C). CS.B and CS.C have value of D_{50} closing to 0.45 mm.
2. Confining pressure applied to 25 kPa, cyclic deviator stress would reach 13.20 kPa. And, if confining pressure applied to 50 kPa and 100 kPa, cyclic deviator stress would reach 31.35 kPa and 65.34 kPa, respectively.
3. Confining pressure at 25 kPa, CS.A would liquefy in 41 cycles, CS.B and CS.C would liquefy in 53 cycles and 62 cycles, respectively. The confining pressure at 50 kPa, CS.A; CS.B and CS.C would liquefy in 17 cycles, 26 cycles and 38 cycles. Finally, the confining pressure at 100 kPa, CS.A; CS.B and CS.C would liquefy in 7 cycles, 10 cycles and 12 cycles.

REFERENCES

Pecker, A. (1984). *"Dynamique des sols",* Presses de I,' ENPC, Paris.

Seed, H.B. & Idriss, I.M. (1976). *"Ground motions and soil liquefaction during earthquakes",* Earthquake Engineering Study Institute (EERI).

Seed, H.B., Idriss I.M., (1982). *"Evaluation of liquefaction potential of sand deposits based on observations of performance in previous earthquakes",* University of California.

Vallejo, L.E., 1997, "Modeling Lateral Sliding of Slope Due to Liquefaction of Sand Layer", *Journal Geotechnical Engineering, ASCE,* Vol. 123, 80-82.

Yammamuro, J.A, Lade P.V, 1988, "Steady-State Concepts and Static Liquefaction of Silty-Sand", *Journal Geotechnical Engineering, ASCE,* Vol. 124, 868-877.

International Conference on Geotechnical Engineering for Disaster Mitigation & Rehabilitation
Chu, Phoon & Yong (eds) © 2005 World Scientific Publishing Company ISBN 981-256-469-1

EFFECTS OF SOIL CROSS-SECTIONAL MODELING ON THE SITE FUNDAMENTAL PERIOD

MUSAID ASSAF
KACST-Science and Technology Unit 2600 Virginia Ave. NW, Suite 800
Washington, DC 20037, U.S.A.

One of the most important tools in earthquake engineering is the site fundamental period. A site fundamental period can be determined by one of three methods; the approximate method, the microtremors' spectra, and the dynamic site response analysis. The dynamic site response analysis and the approximate method require the determination of the site's soil cross-sectional model in the form of the shear wave velocity profile(s). The shear wave velocity model for a certain location can be determined by invasive methods such as cross-hole tests, or by non-invasive methods such as Controlled Source Spectral Analysis of Surface Waves tests, which is also referred to as the CXW test. This paper examines the effects of the different soil cross-sectional modeling methods on the site fundamental period. In addition, the results are also compared with the site fundamental period from microtremors' records. To achieve this objective, a site was selected and different tests as well as microtremors recordings were performed at different locations within the site. The tests included cross-hole as well as CXW tests. The results showed that using different methods to determine a site fundamental period can lead to significantly different results.

INTRODUCTION

A site soil cross-sectional model of its shear wave velocity (SWV) can be determined by invasive tests such the cross-hole (CH) tests, or by the more economical non-invasive tests, such as Controlled Source Spectral Analysis of Surface Waves tests, which is also referred to as the CXW test. The National Earthquake Hazard Reduction Program (NEHRP) Provisions General Procedure (FEMA, 2003), requires the SWV model to be determined for the site's top 30 m, or to the bedrock, whichever is less. For a site with (n) soil layers, the Provisions define the site average SWV ($\overline{v}_s$) as:

$$\overline{v}_s = \frac{\sum_{i=1}^{n} d_i}{\sum_{i=1}^{n} \left(\frac{d_i}{v_{si}} \right)} \tag{1}$$

where (d_i) and (v_{si}) are layer's (i) thickness and SWV, respectively. The site fundamental period (FP), (T) (also called the site characteristic period) can be determined or estimated by one of the following methods:

1. The approximate method (Kramer, 1996):

$$T_s = \frac{4H}{\bar{v}_s} \quad (2)$$

where H is the thickness of the soil cross-sectional model

2. The microtremors' spectral ratios method (Nakamura et al, 1986): This method utilizes records of microtremors at the site in question. The procedure is to select a section of the records to be analyzed. Concurrent sections of all three components of the record are then segmented into 20 to 30 seconds long segments. The segments are tapered, and a fast Fourier transform is performed on each segment. The segments are then smoothed before calculating the $(\frac{H}{V})_m$ as follows:

$$(\frac{H}{V})_m = \frac{\sqrt{S_{EW}^2 + S_{NS}^2}}{S_V} \quad (3)$$

where:
S_{EW} = the smoothed east-west component of the record.
S_{NS} = the smoothed north-south component of the record.
S_V = the smoothed vertical component of the record.
The site FP is the period at which the peaks of the plots occur.

3. The dynamic site response analysis (DSRA) method: This method utilizes the site soil SWV cross-sectional model and its soil properties. These soil properties include the soil unit weight, the modulus reduction curve, the damping curve, and the layers' thickness (Kramer, 1996). The water table depth for the soil cross-sectional model is also required for this type of analysis. There are a number of available computer programs for performing DSRA, including SHAKE and ProShake (EduPro Civil Systems, 1998).

THE TESTING SITE

A seismically active site was selected to conduct the required field tests. The site is located on the northeastern coast of the Gulf of Aqaba, which is the most seismically active area in Saudi Arabia. The general site coordinates are approximately 34° 57.3" east, and 29° 21.5" north. The site has an area of one square kilometer and its elevation above sea level varies from 0 m at the shore line to 30 m the site's east side. Figure (1) shows the site location and an approximate east-west cross-section.

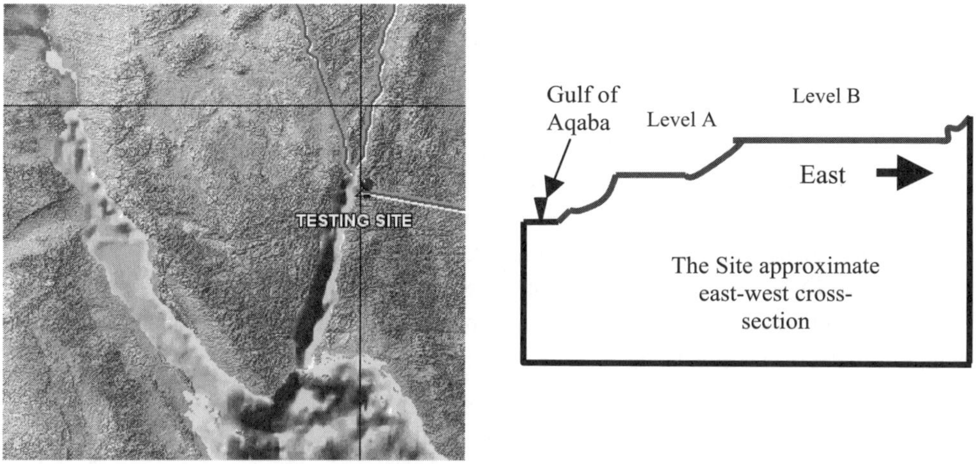

(a) (b)

Figure 1: The testing site; (a) Site location, and (b) Approximate east-west cross-section

METHODOLOGY

DSRA can be performed using one, two, or three-dimensional SWV models. However, a one-dimensional soil SWV model is adequate to capture first-order site response characteristics (FEMA, 2003). In addition to the soil cross-sectional SWV model and the soil properties, the DSRA requires the determination of the water table and bedrock depths (when the bedrock depth is < 30 m). Therefore, additional conventional exploration in the form of Standard Penetration Test (SPT), and vertical electric sounding (VES) were conducted at the site. The SPT was used to estimate the soil properties and determine the depth of the water table at each test's location. Since the SPT boreholes did not reach the bedrock, the VES tests were used to estimate the depth and inclination of the bedrock at the site. The type and number of tests that were conducted at the site were:

i. Twelve SPT
ii. Four CH tests
iii. Four CXW tests
iv. Three Microtremors recording stations
v. East-west and north-south VES to estimate bedrock depth and inclination

The tests were arranged so that at each CH tests location, one SPT and one CXW test were undertaken.

TESTS RESULTS

The soil cross-sectional SWV model was established at four locations within the site. One set of the four models was based on the SWV results of the CH tests, while another set was based on the SWV results of the CXW tests. The results of the CH tests were corrected for the presence of casing. Figure (2) shows an example of the CH and CXW tests results. The two sets were subsequently used to calculate the site approximate FP, and to perform DSRA using the program ProShake. Two randomly generated earthquakes were used in the DSRA. A previous study showed that the expected peak ground acceleration (PGA) for 10% probability of being exceeded in 50 and 100 years for the research area were about 0.2g and 0.3g, respectively (Al-Haddad et al, 1998). The two random earthquakes were based on records of regional earthquakes and were scaled to produce PGA at the site that is compatible with these values. The DSRA was repeated four times for each soil cross-sectional model using the two earthquakes and two damping ratios of 5% and 10%.

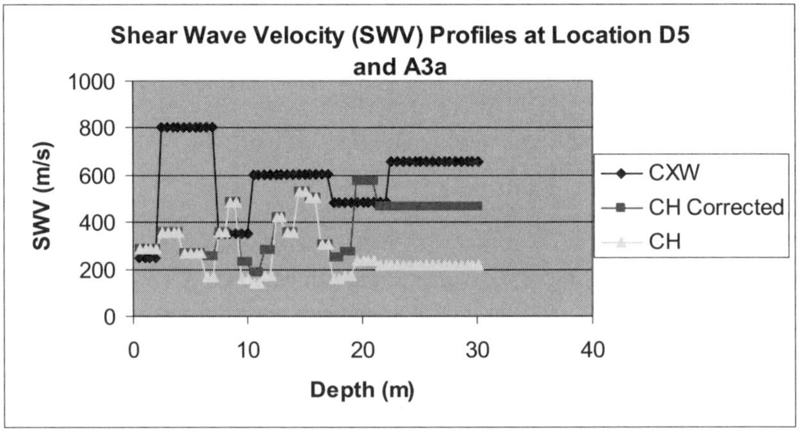

Figure 2: Results of SWV model from CH and CXW tests at location D5

Microtremors recording stations were established at three locations at the site, namely DRA1, DRB1, and DRB3. Figure (3) shows the microtremors spectral ratios plots at station DRB1

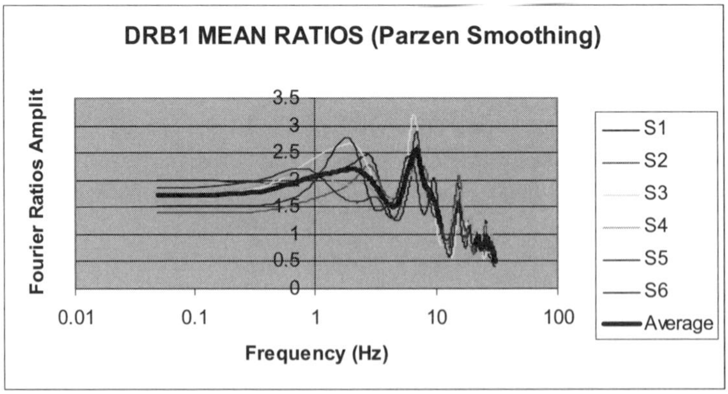

Figure 3: Microtremors spectral ratio plots for station DRB1

Table (1) presents a comparison of the results of the site's approximate FP based on the results of the CH and CXW tests. The results show that the difference between the values of the FPs of the two tests at each location can be as high as 66% (expressed as a percentage of the lower FP value).

Table 1: The site approximate FP from the CH tests and the corresponding CXW tests

Test Location	CH Test Approximate FP (sec)	CXW Test Approximate FP (sec)	FP Difference as %
D5	0.381	0.2294	66.1
D6	0.274	0.3604	31.5
D11	0.315	0.3235	2.7
D12	0.4396	0.3399	29.3
Average	0.3524	0.3133	32.4

Table 2: The site FPs by the dynamic site response analysis

Test Type	Test Location	Site FPs (sec)			
		Earthquake One 5% Damping		Earthquake Two 5% Damping	
CXW Tests	D5	0.064	0.064	0.25	0.064
	D6	0.064	0.054	0.32	0.098
	D11	0.077	0.077	0.098	0.098
	D12	0.064	0.054	0.077	0.054
Average		0.06725	0.06225	0.18625	0.0785
Cross Hole Tests	D5CH	0.064	0.054	0.088	0.088
	D6CH	0.064	0.054	0.098	0.098
	D11CH	0.064	0.064	0.098	0.077
	D12CH	0.077	0.064	0.33	0.064
Average		0.06725	0.059	0.1535	0.08175

Table (2) presents the site's FP from the DSRA based on the soil cross-sectional models of the CH and CXW tests, and using the two earthquakes and two damping ratios. Table (3) presents the percentage of the differences in the value of the FP from the two tests for each location. The percentage of the difference in the values of the FP for each location varies from 0% to 328%. Table (4) presents the results of the microtremors FP of the three stations at the site.

Table 3: Percentage of the difference between the values of the DSRA FPs of the CH and CXW tests

Test Location	Earthquake RAN230 5% Damping		Earthquake RAN330 5% Damping	
D5	0	18.5	184.0	37.5
D6	0	0	226.5	0
D11	20.3	20.3	0	27.2
D12	20.3	18.5	328.5	18.5

Table 4: The site FP s of the microtremors spectral ratios

Recording Location	DRA1	DRB1	DRB3
FP (sec)	0.294	0.159	0.070872

Figure (4) shows the variations of the site FP based on the cross-sectional modeling method, the damping ratio, and the input earthquake.

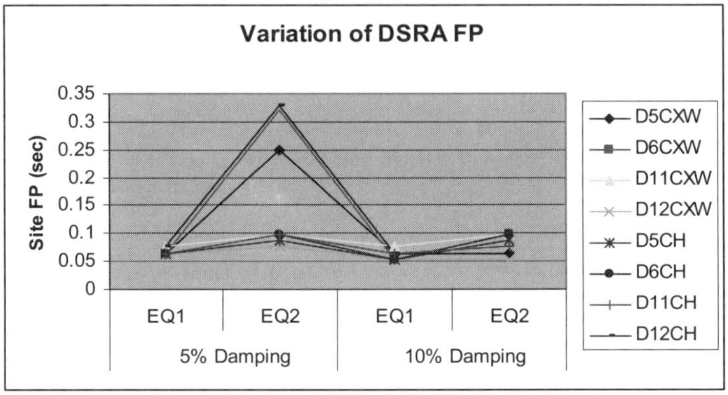

Figure 4: Variation of DSRA FP

CONCLUSIONS

The results show that different soil cross-sectional modeling methods can lead to significantly different FP for the same location. For this site the difference in the FP values for the same location were up to 66% for the approximate method, and 328% for the DSRA method, depending on the input earthquake and the damping ratio. The microtremors FP were not consistent, although the FP of DRB1 and DRB3 showed good agreement with the FP values of the DSRA.

REFERENCES

Al-haddad , etal, "A Study Leading to Preliminary Seismic Design Criteria for the Kingdom", King Abdul Aziz City for Science ant Technology, Riyadh, Saudi Arabia, 1998, p.65.

EduPro Civil Systems, Inc., "ProShake Ground Response Analysis Program, Version 1.1", User Manual, Redmond, Washington

FEMA, "NEHRP Recommended Provisions for Seismic Regulations for New Buildings", 2000 Edition, Part 1-Provisions, Washington, DC, 2001, pp. 39-48

Kramer, Steven L., "Geotechnical Earthquake Engineering", Prentice Hall, Upper Saddle River, NJ, 1996.

Nakamura, Y., and Ueno, M., "A Simple Estimation of dynamic Characteristics of Subsoil", Proceedings of the 7th Japan Earthquake Engineering Symposium, pp. 265-279.

International Conference on Geotechnical Engineering for Disaster Mitigation & Rehabilitation
Chu, Phoon & Yong (eds) © 2005 World Scientific Publishing Company ISBN 981-256-469-1

DYNAMIC RESPONSE OF BRIDGE PILES ON SOFT GROUND USING NUS GEOTECHNICAL CENTRIFUGE

S. BANERJEE

Department of Civil Engineering, National University of Singapore, Block E1A, #07-03, No.1 Engineering Drive 2, Singapore-117576;

D.K. PAUL

Department. of Earthquake Engg. IIT-Roorkee, India -247667;.

F.H. LEE

Department of Civil Engineering, NUS, Singapore-117576;

Pile foundations are widely used for both inland and offshore structures, especially under adverse soil conditions. Of particular interest is the case of bridges which are structurally simple, yet highly response-sensitive to soil-structure interaction effects. The performance of these structural systems under seismic excitations has been the subject of considerable attention in recent years, particularly after the numerous failures of pile-supported bridges during recent major earthquakes. Series of dynamic centrifuge tests were performed on the model piles placed in soft kaolin beds in order to study the interference of the vibration of piles on the free-field ground motion. The experimental investigation had been carried out at Geotechnical Centrifuge Laboratory of Centre for Soft Soil Engineering of National University of Singapore under a joint venture of IIT-Roorkee and National University of Singapore. The investigation reveled the pile-soil interaction under the effect of super-structural load

INTRODUCTION

Pile foundations are widely used for both inland and offshore structures such as bridges, port and harbour structures, tall structures like water tanks, chimney etc. especially under adverse soil conditions. The effect of pile-foundation-soil interaction on the response of such structures under earthquake excitation is still not well understood. The performance of these structural systems under seismic excitations has been the subject of considerable attention in recent years, particularly after the numerous failures of pile-supported bridges during major earthquakes. To develop a better understanding on structure- foundation-soil interaction a series of centrifuge testing on piles in clay under earthquake excitation have been undertaken under a joint collaboration between Indian Institute of Technology, Roorkee, India and National University of Singapore.

Since the early 1980's, many model earthquake tests on the centrifuge have been conducted using stacked ring or laminar containers, which were designed to permit the soil to deform in horizontal shear mode with minimal reflection from the ends. But almost all the researchers concentrated their studies for pile on sand or liquefiable soil.

Nevertheless, few works had also been carried out on pile-clay model but most of them under static condition. In Cambridge Geotechnical Laboratory, number of static analysis had been carried out on integral abutment bridge resting on pile foundations in both clay and sand (Ellis and Springman 2001). Pile research in National University of Singapore has so far focused mainly on the static response of laterally loaded pile foundation in clay to examine the effect of excavation on piles installed behind the retaining walls (Ilyas *et al.*2004). At University of California, Davis model piles in clay were tested in a flexible shear beam (FSB) container at a centrifugal acceleration of 30*g* to evaluate a dynamic beam on a nonlinear Winkler foundation (BNWF) analysis method (Boulanger *et al.*1999, Christina *et al.*2001). These results provide experimental support for the use of dynamic *p-y* analysis methods as well as dynamic BNWF analysis methods in seismic soil-pile-structure interaction problems involving pile-group systems.

The objective of the present study is to perform dynamic centrifuge test on the pile model placed in a kaolin bed in order to study the dynamic response of pile foundation in clay.

TEST SET-UP

The tests were performed on a centrifuge at National University of Singapore at a centrifugal acceleration of 50g. Soft kaolin clay (bulk unit weight=16kN/m^3, W.C=66%, LL=80%, PL=35%) bed models were prepared by pouring clay slurry into the laminar container, followed by episodes of consolidation to re-create the desired strength profile and stress history.

White kaolin powder was mixed with water in a ratio of 1: 1 and the slurry was poured in the rubber packed laminar box in layers and the transducers were placed at the prescribed positions. The laminar box has an inner dimension of 524 mm long, 300 mm wide and 420 mm deep.

The base plate of the box is connected to shaking apparatus consisting of actuator, accumulator and motor pump assembly. Total four accelerometers were placed as shown (Figure 1). Before pouring the slurry to the box a perforated pipe underlying a sand bed of 10 mm thickness had been fixed at the bottom to provide the drainage path and above that a thin layer of geotextile had been placed.

1g consolidation was carried out on all clay bed models to pre-compress the latter so as to reduce the time needed for centrifuge consolidation. The models were connected to the vacuum chamber by perforated pipe in order to get linear effective stress envelope.

After two weeks of 1g consolidation the laminar box was mounted on the Centrifuge along with all the equipments for the shaking. Two tests were conducted; one with steel pile with no load on top and other with the pile with 1kg load on top. The 18mm-diameter and 25 cm-long model piles actually simulate prototype piles of 900 mm-diameter and 12.50 m depth under 50g model 'gravity'. Now models were consolidated at 50g for 10hrs and then three earthquake time histories were fired to the shaking table.

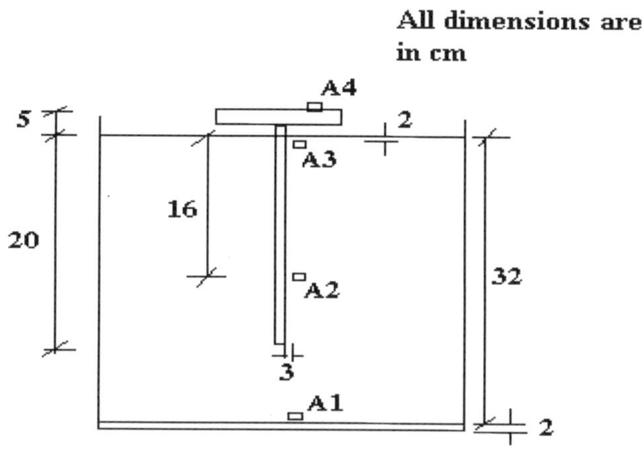

Figure 1. Instrumentation layout

DATA PROCESSING

The output signals from the transducers were stored in .xls format using DAISYLAB version 3.0. The acceleration datasets were detrended using MS EXCEL to remove any dc offset and low frequency drift while retaining the cyclic components. This makes it compatible with an in-house programme which is used to calculate the response spectrum. The programme requires two input files:

i) The wave file contained the detrended acceleration datasets. The maximum acceleration points are limited to 5000

ii) The period file consisted of period points at which response spectrums will be calculated and the structural damping. The maximum numbers of period points are limited to 150 and damping was assumed as 5%.

RESULTS

Predominant time periods obtained from response spectra are given in the Table 1 and Table 2.

Table 1. Predominant time period for pile model with load

Excitations	Earthquakes	Base	Predominant Time Period (sec)		
			Mid depth	Surface	Pile top
First excitation	Small (0.026g)	0.26	1.75	4	0.26
	Medium (0.05g)	0.26	2	4.1	0.26
	Large (0.11g)	0.5	2.6	4.1	0.5
Second excitation	Small (0.026g)	0.3	1.75	4.2	0.15
	Medium (0.05g)	0.26	2.8	4	0.26
	Large (0.11g)	0.5	3	4.1	0.5
Third excitation	Small (0.026g)	0.26	2	4.2	0.26
	Medium (0.05g)	0.3	3	4.1	0.3
	Large (0.11g)	0.26	2.8	4.1	0.26

Table 2. Predominant time period for pile model with load

Excitations	Earthquakes	Predominant Time Period (sec)			
		Base	Mid depth	Surface	Pile top
First excitation	Small (0.026g)	0.28	1	1	1
	Medium (0.05g)	0.26	1.75	1.75	1.75
	Large (0.11g)	0.125	1.75	1.75	1.75
Second excitation	Small (0.026g)	0.26	1.75	1.75	1.75
	Medium (0.05g)	0.3	1.75	2	2
	Large (0.11g)	0.33	2	2	2

The time period is found to be increasing towards top layers for single pile with no load (Figure 2, Table 1).But at the pile top time period is again reduced to that of at the base. On the contrary, for pile with load on top the time periods at the mid depth and surface are found to be same as at the pile top. Although, time period at base is almost identical with the former case (Figure 2).

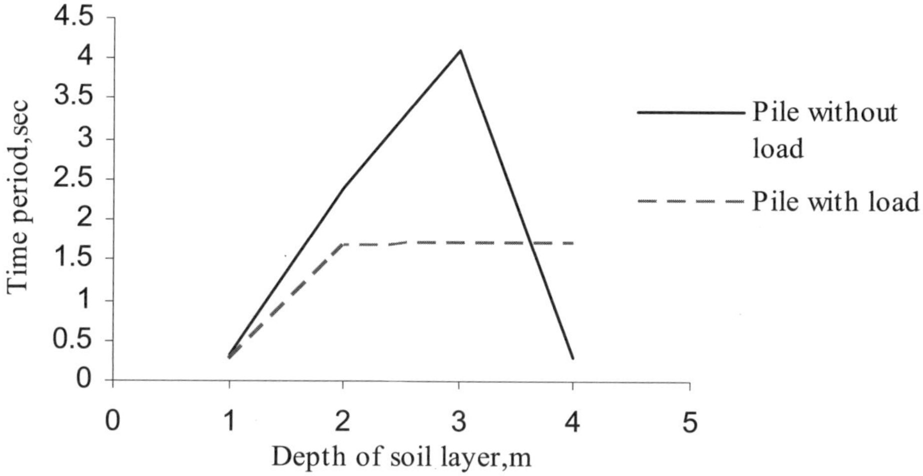

Figure 2. Variation of time period with depth

CONCLUSION

The dynamic centrifuge modeling is, thus, found to be a handful tool for pile dynamics problem. Observations obtained are quite justified with respect to the basic principles of soil dynamics. The single pile with no load actually behaves as a single reinforcing bar in soil attenuating vibration at the pile top. For pile with load, due to strong coupling between soil and pile, whole pile-soil system was behaving as a unit showing unique predominant time period. But time period at base remains almost same in two tests because the accelerometer placed at base is far below from the bottom of pile tip. Hence soil pile interaction can not come into action in either of the tests.

ACKNOWLEDGEMENTS

The authors are grateful to IIT, Roorkee, India and NUS, Singapore for providing all necessary financial and academic support without which this study would have been a distinct dream.

REFERENCES

Brandenberg, Scott J., Singh, P., Boulanger, Ross W., Kutter Bruce L. (2001). “Behavior of piles in laterally spreading ground during earthquakes”, *CALTRAN Workshop, U.C. Davies*.

Boulanger, Ross W, Christina, J. Curras, Kutter Bruce L., Wilson, D. (1999). "Seismic soil-pile-structure interaction experiments and analyses", *Journal of Geotechnical and Geoenvironmental Engg. ASCE, Vol-125, No.9.*

Christina, J. Curras, Boulanger, Ross W., Kutter Bruce L., Wilson, D. (2001). "Dynamic experiments and analyses of a pile-group-supported structure", *Journal of Geotechnical and Geoenvironmental Engg. ASCE, Vol-127, No.7.*

Ellis, E.A. & Springman, S.M. (2001). "Modeling of soil-structure interaction for a piled bridge abutment in plane strain FEM analysis", *Computers and Geotechnics, Vol-28, pp.-79-98.*

Ilyas, T., Leung, C.F., Chow, Y.K. and Budi, S.S. (2004). "Centrifuge model study of laterally loaded pile groups in clay", *Jn. of Geotechnical and Geoenvironmental Engg. ASCE, Vol-130, No.3.*

International Conference on Geotechnical Engineering for Disaster Mitigation & Rehabilitation
Chu, Phoon & Yong (eds) © 2005 World Scientific Publishing Company ISBN 981-256-469-1

THE EXPERIMENTAL STUDY ON DYNAMIC SHEAR MODULUS RATIO AND DAMPING RATIO OF RECENTLY DEPOSITED SOILS IN THE SOUTHERN AREA OF JIANGSU PROVINCE IN CHINA

CHEN GUOXING
Institute of Geotechnical Engineering, Nanjing University of Technology, Nanjing, Jiangsu, China

LIU XUEZHU, ZHU DINGHU, HU QINGXING
Institute of Geotechnical Engineering, Nanjing University of Technology, Nanjing, Jiangsu, China

Site soils in the southern area of Jiangsu province in China belong to washland sediment or limnetic facies sediment. The components of sediment are very complex and the process of sedimentation is in disorder. This paper analyzes the test results of free vibration column apparatus on 155 undisturbed samples of recently deposited soils taken from the southern area of Jiangsu province in China, including mucky silty clay, clay, silty clay, silty clay and silty sand interbedded strata, silt, silty sand, fine sand and medium coarse sand, during last four years. The fitting curves of the average relation between dynamic shear modulus ratio G/G_{max} and damping ratio D versus shear strain amplitude γ and their parameters in the empirical formula of $G/G_{max} \sim \gamma$ and $D \sim \gamma$ curves for 8 types of soils mentioned above are presented. The results in this paper will make us understand more about dynamic characteristics of recently deposited soils in this area.

INTRODUCTION

The southern area of Jiangsu province refers to the south flank of the Yangtze Delta and the Taihu Lake region within Jiangsu province, China, mainly including the cities of Zhengjiang, Changzhou, Wuxi and Suzhou. In recent years, large numbers of major projects in this area, such as Jiangyin Bridge, Sutong Bridge and Runyang Bridge on the Yangtze River, one after another is being constructed or has been completed. In the Tenth Five-Year Plan period (2001-2005), many important projects including lifeline engineering projects are being at the stage of programming or construction in the southern area of Jiangsu province, such as freeways, intercity passenger trains, harbors, docks, electric power facilities, water supply works, and so on. The sites of these important engineering projects are located on washland, delta or limnetic facies recently deposited soils in the southern area of Jiangsu province.

The recently deposited soils were formed in the medium or later term of the Holocene Epoch. Generally speaking, the recently deposited soils belong to underconsolidated soils and its strength is quite lower. With different depositional environments, the strength of recently deposited soils differs greatly. As a special type of soils, the engineering geologic property of recently deposited soils is worse with its special engineering characteristic. The concept and classification method of recently deposited soils are often adopted in the realm of Geotechnical engineering and have already been brought into the China Code for the Investigation of Geotechnical Engineering GB50021-2001.

The southern area of Jiangsu province belongs to the delta plain of the Yangtze River or low mountain mound terrain. Soft soils distribute widely in this area. Especially, the component of sediments is very complicated and asymmetrical greatly, and the underground water level is close to ground surface in this area. Based on the typical geologic section on recently deposited soils of the Yangtze delta plain in the southern area of Jiangsu province, from the bottom up, sediment strata are divided into riverbed facies, washland-firth facies and deltaic facies sediment, including medium sand, fine sand and sandy clay with horizontal bedding and cross bedding. Especially, clayey fine sand widely distributes in the south flank of the Yangtze Delta.

The dynamic characteristic of soils is a major factor affecting ground motion characteristic. Dynamic shear modulus and damping ratio of soils, i.e., G_{max} , $G/G_{max} \sim \gamma$ and $D \sim \gamma$, as two parameters of the soil dynamic characteristic, are utilized in the site seismic response analysis when the equivalent linearization model of soils is adopted. Similarly, dynamic shear modulus and damping of soils are indispensable for the seismic safety evaluation of engineering sites. Whether a selection of soil dynamic characteristic parameters matches a practical site condition, there will be a great influence on the reliability of numerical analysis results (Chen guoxing et al.1995). So, whether these parameters are rational or not, they will directly effect on safety and economical efficiency of important projects.

FOUNDATION OF THE STUDY

According to the analysis of the test results of free-vibration columniation apparatus for soil samples of six sorts including clay, silty clay, silty clay and fine sand interbedded strata of recently deposited soils in the Nanjing and its neighboring cites, the dynamic shear modulus and damping ratio were investigated primarily (Chen guoxing and Liu xuezhu 2004). Based on the research results of Chen guoxing and Liu xuezhu (2004), through the test study and theoretical analysis for seven sorts of recently deposited soils from 15 cities in the lower reaches of the Yangtze River including some cities in the southern area of Jiangsu province, the elementary regularity of dynamic shear modulus ratio G/G_{max} and damping ratio D versus shear strain amplitude γ was presented (Chen guoxing et al. 2005). Similarly, the recommend values of the parameters were also presented and they may be used in more areas. Furthermore, for attempting to provide nationwide curves of dynamic shear modulus ratio G/G_{max} and damping ratio D versus shear strain amplitude γ for various soils, only using some limited test data, its representation and creditability ought to be deliberate (Chen guoxing et al. 2005). As a whole, regional mean curves of $G/G_{max} \sim \gamma$ and $D \sim \gamma$ various soils through large numbers of experimental investigation should be more credible and referenced. Consequently, with further consideration of the regional character of soils and similarity of depositional environment of soils, this paper investigated detailedly the dynamic characteristic of recently deposited soils in the southern area of Jiangsu province, based on the research results of Chen guoxing and Liu xuezhu (2004) and Chen guoxing et al. (2005). Those results have been applied in the site seismic safety evaluation for some important projects in the southern area of Jiangsu province, such as across Yangtze River bridges, orbit transportation and electric establishments etc.

THE EMPIRICAL FORMULA OF DYNAMIC SHEAR MODULUS RATIO AND DAMPING RATIO ON RECENTLY DEPOSITED SOILS IN THE SOUTHERN AREA OF JIANGSU PROVINCE

The relationship of dynamic shear modulus ratio versus the amplitude of shear strain is expressed as (Martin P.P. and Seed H.B 1982):

$$G/G_{max} = 1 - H(\gamma) \tag{1}$$

$$H(\gamma) = \left\{ \frac{(\gamma/\gamma_0)^{2B}}{1 + (\gamma/\gamma_0)^{2B}} \right\}^A \tag{2}$$

where A, B and γ_0 are fitting parameters related with soil behavior.

The damping ratio of soils D varies with shear strain amplitude. The empirical equation of damping ratio is recommended in this paper as follows:

$$D = D_{min} + D_0(1 - G/G_{max})^n \tag{3}$$

where n and D_0 are fitting parameters related with soil behavior. D_{min} is the minimum damping ratio relative to the initial dynamic shear modulus G_{max}. Limited to existing test technique, instrument accuracy and experiment condition, it is difficult to measure accurately minimum damping ratio relative to the initial dynamic shear modulus. So, D_{min} can be treated as fitting parameters related with soil behavior.

These tests were operated with self-developed free vibration column test machine (Chen guoxing et al. 2003). The undisturbed samples of recently deposited soils came from the southern area of Jiangsu province, such as Zhenjiang, Changzhou, Wuxi and Suzhou including mucky silty clay, clay, silty clay, silty clay and silt interbedded strata, silt, silty sand, fine sand and medium coarse sand, and the total number of soil samples is equal to 155. The amount and distribution of soil samples are listed in Table 1.

The test results of $G/G_{max} \sim \gamma$ curves can be fitted well by empirical Eq.(1) and Eq.(2) (Chen guoxing et al. 2005). In this paper, the test results on $G/G_{max} \sim \gamma$ of recently deposited soils are also fitted by empirical Eq.(1) and Eq.(2), and the test results on $D \sim \gamma$ of recently deposited soils are fitted by empirical equation (3). In addition, the fitting parameters on $G/G_{max} \sim \gamma$ and $D \sim \gamma$ curves are listed in Table 2. Figure 1 indicates that the fitting results on $G/G_{max} \sim \gamma$ and $D \sim \gamma$ curves are more appropriate under small strain range.

Table 1. Locations of recently deposited soil samples from the southern area of Jiangsu province

Soils	Quantity of samples in different areas				Amount
	Zhenjiang	Changzhou	Wuxi	Suzhou	
Mucky silty clay	4	4	6	5	19
Clay	2	2	3	9	16
Silty clay	16	4	7	26	49
Silty clay and silt interbedded strata	3	—	4	6	13
Silt	3	3	2	11	16
Silty sand	4	6	3	12	21
Fine sand	—	4	3	7	14
Medium coarse sand	1	3	—	4	8

Table 2. Parameters of eight recently deposited soils in the southern area of Jiangsu province

Soils	Parameters							Amount
	A	B	γ_0 ($\times 10^{-4}$)	n	D_0 (%) Mean	D_0 (%) Range	D_{min} (%)	
Mucky silty clay	1.06	0.47	2.7	1.06	20.9	19.8~23.6	1.06	19
Clay	1.10	0.44	3.4	0.90	19.1	17.4~22.1	1.16	16
Silty clay	1.09	0.44	3.0	0.95	18.4	15.6~23.7	1.25	47
Silty clay and silt interbedded strata	1.06	0.44	2.9	1.09	19.3	18.2~23.3	1.27	12
Silt	1.03	0.43	3.3	1.02	19.1	17.9~23.7	1.09	14
Silty sand	0.93	0.43	5.5	0.91	19.5	17.9~23.3	0.49	19
Fine sand	0.99	0.34	6.9	0.98	19.8	18.2~22.2	0.80	14
Medium sand	0.90	0.45	5.6	0.98	18.4	18.0~21.0	0.63	8

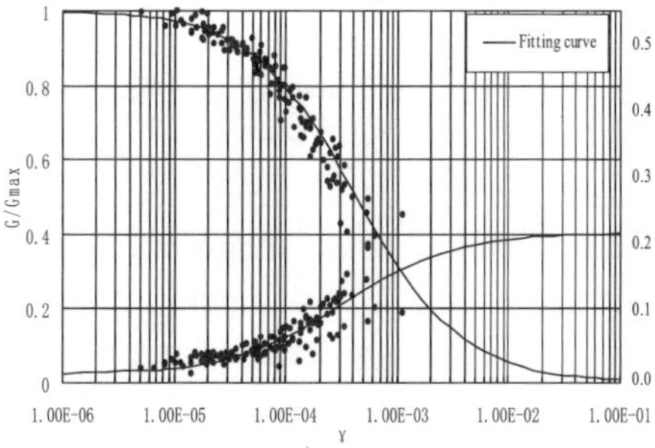

(a) Mucky silty clay

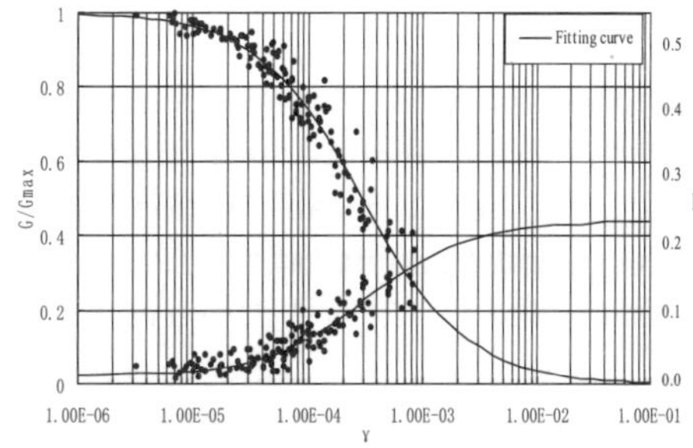

(b) Clay

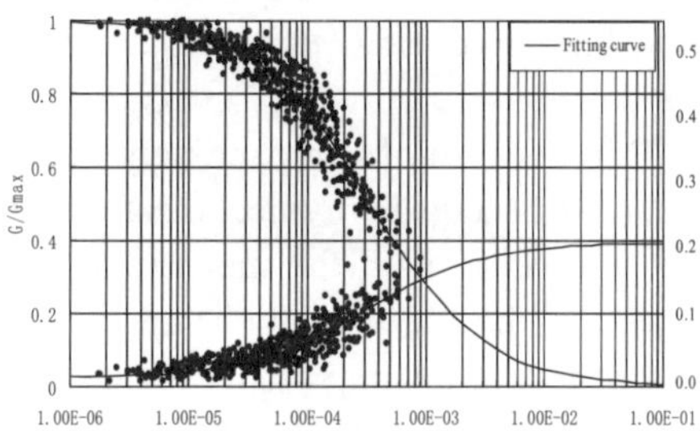

(c) Silty clay

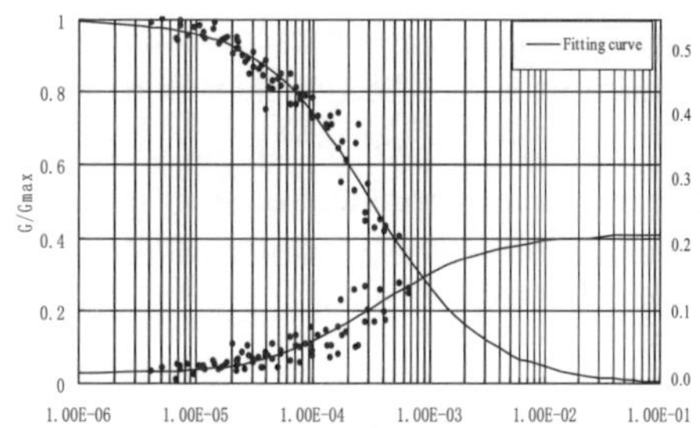

(d) Silty clay and silt interbedded strata

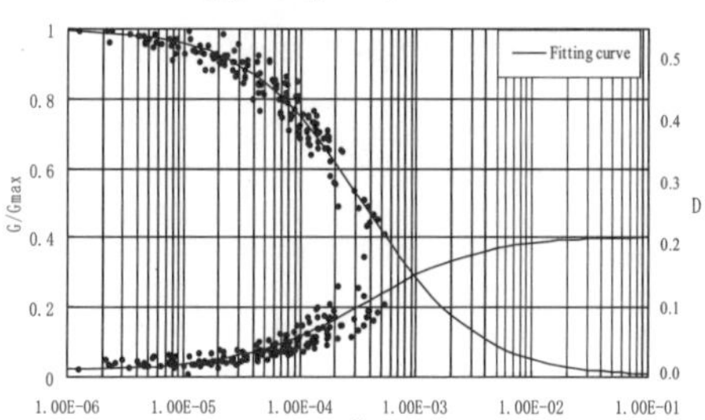

(e) Silt

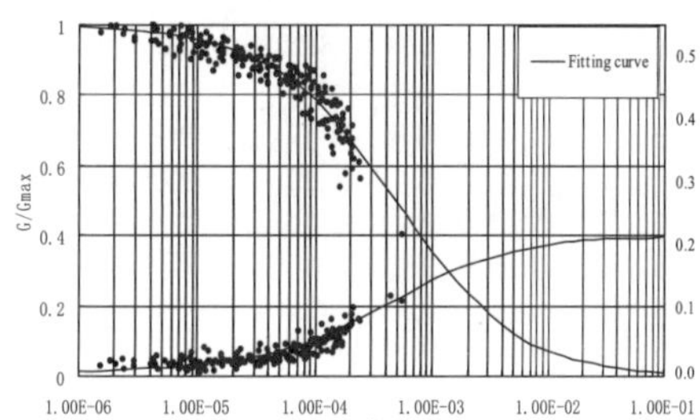

(f) Silty sand

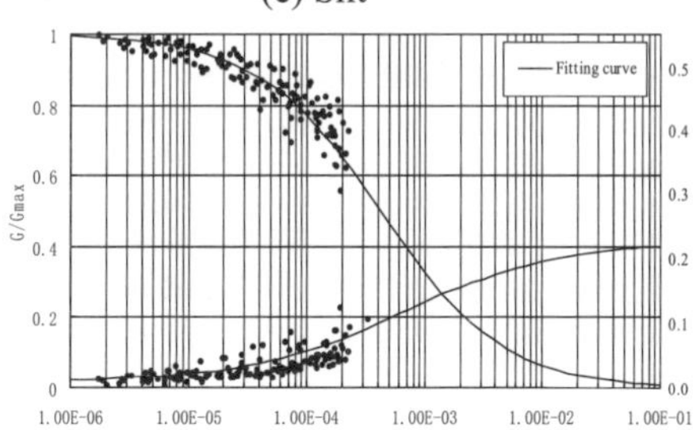

(g) Fine sand

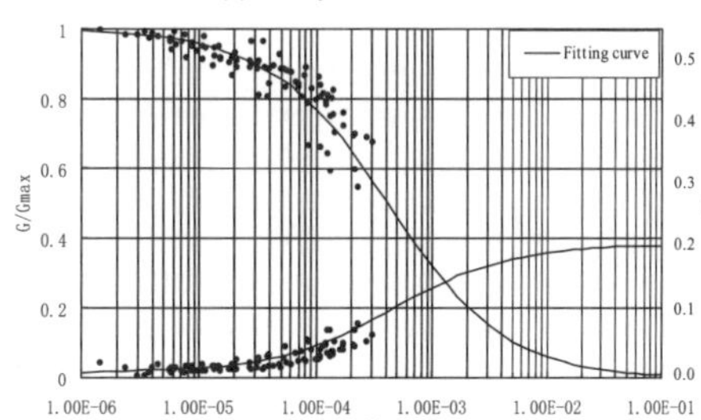

(h) Medium coarse sand

Figure 1. Fitting curves of shear modulus ratio G/G_{max} and damping ratio D versus shear strain amplitude γ of eight types of recently deposited soils in the southern area of Jiangsu province

CONCLUSIONS

Based on the free vibration column test results of 155 undisturbed soil samples of recently deposited soils taken from the southern area of Jiangsu province, the fitting parameters of the mean relationship curves between dynamic shear modulus ratio G/G_{max} and damping ratio D versus shear strain amplitude γ are suggested by means of Eq.(1) and Eq.(2) and the recommended empirical equation of damping ratio in this paper. The results will make us understand more about dynamic characters of recently deposited soils in the southern area of Jiangsu province and provide some essential data for seismic safety evaluation of general engineering sites in this area.

For seismic response analysis of general engineering sites in the southern area of Jiangsu province, when there is a lack of relationship curves of $G/G_{max} \sim \gamma$ and $D \sim \gamma$ test data of soils, the typical values of the mean relationship curves of $G/G_{max} \sim \gamma$ and $D \sim \gamma$ recommended in this paper can be used. But, because of comparatively complex depositional environment of soils in the southern area of Jiangsu province, there are some dispersion in the test results of given site soils comparing with the mean relationship curves of $G/G_{max} \sim \gamma$ and $D \sim \gamma$ curves. So, quoting the mean relationship curves of $G/G_{max} \sim \gamma$ and $D \sim \gamma$ in this paper should be deliberative for important projects in this area. In order to insure the result reliability of site seismic response analysis, model parameters or representative value on the curves of $G/G_{max} \sim \gamma$ and $D \sim \gamma$ for engineering site soils should be obtained by experiment with free vibration column apparatus or dynamic triaxial apparatus.

REFERENCES

Chen guoxing, Xie junfei and Zhang kexu (1995). “The empirical of soil modulus and damping ratio for dynamic analysis”. *Earthquake engineering and engineering vibration*, Vol. 15, No. 1, 73-84.

Chen guoxing and Liu xuezhu (2004). “Testing study on ratio of dynamic shear modulus and ratio of damping for recently deposited soils in Nanjing and its neighboring areas”. *Chinese Journal of Rock Mechanics and Engineering*, Vol.23, No.8, 1403-1410.

Chen guoxing, Chen jihua and Liu xuezhu, et al. (2005). “Experimental study on dynamic shear modulus ratio and damping ratio of recently deposited soils in the lower reaches of the Yangtze river”. *Journal of Disaster Prevention and Mitigation Engineering*. Vol. 1, 49-57.

Martin P.P. and Seed H.B (1982). “ One dimensional dynamic ground response analysis” . *Journal of geotechnical engineering*. ASCE, Vol.108, No.7, 935-954.

Chen guoxing, Zhu dinghua and He qizhi (2003). “Development and property test of GZZ-1 free vibration column test system”. *Earthquake engineering and the engineering vibration*, Vol. 1, 110-114.

International Conference on Geotechnical Engineering for Disaster Mitigation & Rehabilitation
Chu, Phoon & Yong (eds) © 2005 World Scientific Publishing Company ISBN 981-256-469-1

SEISMIC ANALYSIS OF SINGLE END BEARING PILE

DEEPANKAR CHOUDHURY
Assistant Professor, Department of Civil Engineering, IIT Bombay
Powai, Mumbai – 400 076, India

Research in the area of pile foundation under seismic condition are still lacking though it is extremely important topic in the area of geotechnical engineering. In this paper, a comprehensive analysis of end bearing pile under seismic excitation is reported. The case of single end bearing pile without mass at the top of the pile is considered for the analysis. A seismic excitation of harmonic nature is considered at the base of the end bearing pile for the analysis. Based on Winkler model, a differential equation for pile displacement as a function of depth and time has been obtained by considering the equilibrium equations of a proposed mathematical model. Using the boundary conditions, the pile displacement as a function of depth and time is obtained. Also the variations of rotation, shear force and bending moment with pile depth at a particular time are obtained. An illustrative example shows the results of the analysis. The positions of maximum displacement, rotation, shear force and bending moment along the depth of the pile for a given seismic excitation can be found out.

INTRODUCTION

Behaviour of pile under seismic loading is an important topic of research for the geotechnical engineers. It carries more attention in the recent days due to the devastating effects of some recent earthquakes related to the foundations specially pile foundations. Analysis of piles under axial load is extensively discussed by Poulos and Davis (1980). Several researchers like Randolf (1981), Kagawa and Kraft (1981), Flores-Berrones and Whitman (1982), Gazetas and Makris (1991), Leung et al. (2000) and many others have discussed about the pile-soil interaction problem under lateral loading conditions.

But still the analysis of pile-soil interaction problem under horizontal seismic excitation varying with time is important due to the complexity of the problem. The equilibrium equations of the soil-pile system depend on the idealization of the soil medium as well as the boundary conditions e.g. – soil can be considered as homogeneous and elastic medium or it can be considered as a stratified material with non-linear characteristics. There are different analytical methods for studying the soil-pile interaction. Each of them is having some advantages or shortcomings.

METHOD OF ANALYSES

One of the simple methods is based on Winkler's assumption. The basic assumption of Winkler's model is for any applied load on a beam (or pile), there is a displacement at the point of application, which is proportional to the force. Similar to the findings of Flores-Berrones and Whitman (1982), the following theory can be discussed.

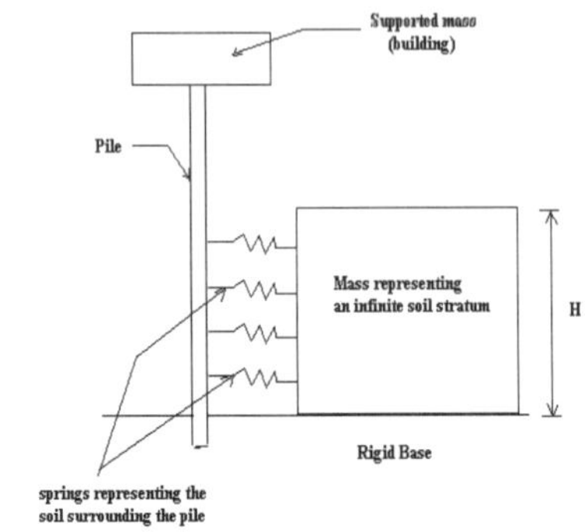

Figure 1. Mass-spring model of the pile-soil interaction problem considered in the present study as per Flores-Berrones and Whitman (1982).

In case of end bearing pile under seismic excitation, the pile base can be represented by a horizontal movement of the type,

$$u_g = u_0 e^{i\omega t} \tag{1}$$

In equation (1),
u_0 = Amplitude of displacement
ω = exciting frequency

This consideration is useful since a seismic motion can be assumed as the superposition of an infinite number of harmonic type of waves, which vary with time in amplitude and frequency of vibration. Based on the assumptions, the equilibrium equation for the forces acting in a pile element such as shown in Fig. 1 is given by,

$$\frac{\partial V}{\partial z}dz + \bar{m}\frac{\partial^2 \bar{y}}{\partial t^2}dz + kD(\bar{y} - \bar{u})dz = 0 \tag{2}$$

where,
D = the pile diameter
$\underline{\mathrm{k}}$ = horizontal subgrade modulus of soil given in force/(length)^3
m = mass density of pile per unit of length in gravity direction
$\underline{\mathrm{u}}$ = the relative soil displacement with respect to base motion
u = the absolute soil displacement
V = the shear force at a depth z

$\underline{y} = y(z,t) =$ relative displacement of pile with respect to the base motion
$\underline{y} =$ the absolute pile displacement
$y - u =$ relative pile displacement with respect to the base motion.

Equation (2) can be re-written in the following form,

$$EI\frac{\partial^4 y}{\partial z^4} + \overline{m}\frac{\partial^2 y}{\partial t^2} + kDy = kDu - \overline{m}\frac{\partial^2 u_g}{\partial t^2} \tag{3}$$

in which, EI = pile stiffness.
The solution of equation (3) will depend on the dynamic soil and pile characteristics (EI, $\overline{m}$, k).

Analysis of single pile without mass at top
First of all, the soil displacement can be obtained from the one-dimensional shear wave equation, which is represented by,

$$\frac{\partial^2 \overline{u}}{\partial t^2} = \mathrm{v}_s \frac{\partial^2 \overline{u}}{\partial z^2} \tag{4}$$

where v_s is the shear wave velocity.

The solution to equation (4) has the form,

$$\overline{u} = (A_1 \sin\frac{\omega z}{c_s} + A_z \cos\frac{\omega z}{c_s})e^{i\omega t} \tag{5}$$

Assuming zero relative displacement between the bed rock and soil stratum base $\left(\overline{u}_{z=0} = u_g\right)$ and that the shear stresses at the soil surface is also $\left[\left(\frac{\partial u}{\partial z}\right)_{z=H} = 0\right]$ zero, solution for equation (5) is,

$$u = u_0 e^{i\omega t}\left[\cos\delta z + \tan\delta H \sin\delta z - 1\right] \tag{6}$$

where, $\delta = \dfrac{\omega}{\mathrm{v}_s}$

Assuming that pile displacement is due to a harmonic base motion, one can write,

$$y = y_0 e^{i\omega t} \tag{7}$$

where, $y_0 = A + B\cos\delta z + C\sin\delta z$ (8)

Now putting the expression of y, equation (3) can be written as,

$$EI(Bd^4\cos\delta z + C\delta^4\sin\delta z) + kD(A + B\cos\delta z + C\sin\delta z) - \overline{m}\omega^2\left(A + B\cos\delta z + C\sin\delta z\right) = kDu_0\left(\cos\delta z + \tan\delta H\sin\delta z - 1\right) + \overline{m}u_0\omega^2 \quad (9)$$

By comparing the constant terms and the coefficients of $\cos\delta z$ and $\sin\delta z$ in both sides, the coefficients A, B and C are obtained and the resulting response due to forced vibration is,

$$y_0 = u_0\left[\xi\left(\cos\delta z + \tan\delta H\sin\delta z\right) - 1\right] \quad (10)$$

where, $\xi = \dfrac{kD}{\left(EI\delta^4 + kD - \overline{m}\omega^2\right)}$ (11)

This last expression represents the amplitude displacement of the pile-soil system due to the soil motion. To obtain the displacements of this system at any time equation (10) is multiplied by $e^{i\omega t}$.

Analyzing the above equation, it can be seen that the difference in soil displacements without pile and soil displacement with pile depends upon factor ξ. This function is a function of the soil stiffness (represented by kD) and the pile rigidity (given by EI).

From equation (11) it can be observed that when $kD \to \infty, \xi \to 1$ and therefore, the tendency of a pile is to follow the soil motion; on the other hand, when $kD \to 0 or EI \to \infty$, the tendency of the pile is to remain still.

Expression for rotation (θ), shear force (V) and bending moment (M) in the pile can be give by,

$$\theta = \frac{\partial y}{\partial z} = u_0\partial\xi e^{i\omega t}\left[\tan\delta H\cos\delta z - \sin\delta z\right] \quad (12)$$

$$M = EIu_0\delta^2\xi e^{i\omega t}\left(\tan\delta H\sin\delta z + \cos\delta z\right) \quad (13)$$

$$V = -EIu_0\delta^3\xi e^{i\omega t}\left(\tan\delta H\cos\delta z - \sin\delta z\right) \quad (14)$$

CASE STUDY

Let us take an example to understand the effect of earthquake on a pile.
Practical data for constants: Pile rigidity, $EI = 4.3x10^6$ Nm^2; Exciting frequency, $\omega = 20$ rad/s; Shear wave velocity, $v_s = 200$ m/s; Mass of the pile in gravity direction per unit length = $5.4x10^{-2}$ Ns/m^2; Length of pile, H = 20m; K = Horizontal subgrade modulus; D = Diameter of the pile; KD = $702x10^3$ N/m^2; Amplitude of displacement, $u_0 = 5x10^{-5}$ m.

Compute the displacement, rotation, shear force and bending moment distribution along the pile depth.

Using the above values, at a particular instant of time we can calculate displacement (y), rotation (θ), shear force (V) and bending moment (M) at different depths using equations (10), (12), (13) and (14). These results are reported in Figs. 2a, 2b and 3a, 3b respectively.

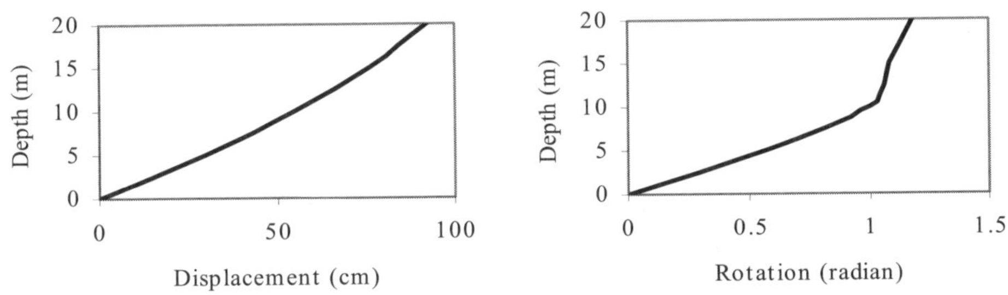

(a) (b)

Figure 2. Result for (a) depth versus displacement (b) depth versus rotation.

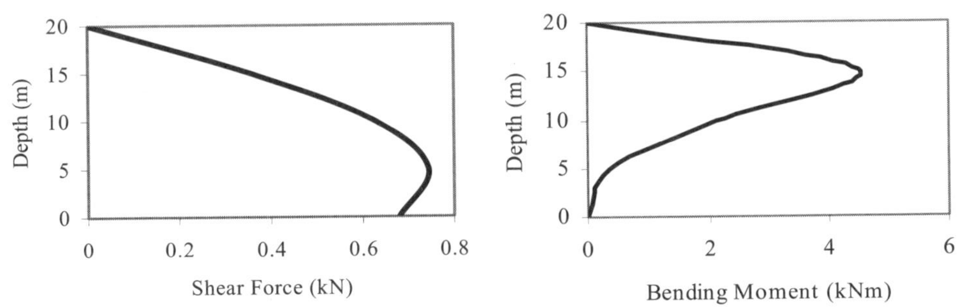

(a) (b)

Figure 3. Result for (a) depth versus shear force (b) depth versus bending moment.

CONCLUSIONS

Based on the analysis presented in the study here, the following conclusions can be derived.

(i) The dynamic response of pile subjected to earthquake motions is a consequence of two effects: (a) The one originated by the soil movements. (b) The one produced by the inertia of the supported structure.

(ii) For practical purposes, most of the foundation elements known in the literature as 'piles' can be considered as flexible elements inside the ground. This means that in general piles follow the ground displacements.

(iii) By a worked out example, the variation of displacement, rotation, shear force and bending moment along the pile depth is shown.

REFERENCES

Flores-Berrones, R., and Whitman, R.V. (1982). "Seismic response of end-bearing piles," *Journal of Geotechnical Engineering Division, ASCE*, Vol. 108, 554-569.

Gazetas, G., and Makris, N. (1991). "Dynamic pile-soil-pile interaction," *Earthquake Engineering and Structural Dynamics*, Vol. 20, 115-132.

Kagawa, T. and Kraft, L.M. (1981). "Lateral pile response during earthquakes," *Journal of Geotechnical Engineering Division, ASCE*, Vol. 107, 1713-1731.

Leung, C. F., Chow, Y. K. and Shen, R. F. (2000). "Behavior of pile subject to excavation-induced soil movement," *Journal of Geotechnical and Geoenvironmental Engineering, ASCE*, Vol. 126, 947-954.

Poulos, H. G. and Davis, E. H. (1980). "*Pile foundation analysis and design*". John Wiley, New York.

Randolf, M.F. (1981). "The response of flexible piles to lateral loading," *Geotechnique*, Vol. 31, 247-259.

International Conference on Geotechnical Engineering for Disaster Mitigation & Rehabilitation
Chu, Phoon & Yong (eds) © 2005 World Scientific Publishing Company ISBN 981-256-469-1

PREDICTION OF PORE PRESSURE DEVELOPMENT IN THE KARKHEH DAM IN IRAN DURING EARTHQUAKE

S.M.MIR MOHAMMAD HOSSEINI
Amirkabir University of Technology,&
Senior Geotechnical Engineer of the Karkheh Dam, Mahab Ghods Consulting Eng. Co.

H. NASROLLAHI
Amirkabir University of Technology-IRAN

Pore pressure within the core of embankment dams plays a decisive role in the performance of earth dams. If these dams experience heavy earthquakes in the short time, the rapid increase of the pore pressure may put the dam at risk in some critical conditions. In this paper the pore pressure development of the Karkheh large embankment Dam, newly constructed in the south-west of Iran, has been investigated during the earthquake loading. The FLAC finite difference Package was used and pore pressures at different levels of the core were estimated and evaluated.

INTRODUCTION

The Karkheh Dam is the largest storage dam in Iran that has been recently constructed in the south-west of the country. This is an earth dam by a mixed clay core. The length of the dam crest is 3030 m. and its maximum height at the river section is 127 m. The width of the dam is 12 m. at the crest and 1100 m. at the widest point at the foundation level. The Karkheh Dam has been built on an alluvium consisting of alternative layers of conglomerates and mudstones. To monitor the performance and behavior of the dam during construction and also in long term utilization, a comprehensive instrumentation system consisting of more than 1000 different devices has been designed and performed (Mahab Ghods Reports).

It is of great importance to control and monitor the stability of the dam at different loading conditions, (ASCE-2000), One of the most critical conditions is when the dam experiences the earthquake loading. This kind of loadings due to uncertainties in magnitude and its occurrence time may cause some serious problems for dams. The exact response of an earth dam during an earthquake loading is not clear. Many factors such as dam characteristics, site conditions, and earthquake loading specifications highly affect the dynamic responses of the dams. The non-elastic and non-linear behavior of the soil materials, extensively influences the dam responses. While the dam usually behaves elastic during small dynamic waves, it may totally moves to the plastic ranges when is subjected to heavy earthquake loadings.

In this study the pore pressure development within the core of the Karkheh dam during the earthquake loading is investigated and evaluated. To verify the analyses results, the dam first has been analyzed under static conditions using the field data measured from different instruments in the site. Then based on the seismicity studies of the region, an appropriate accelerogram has been selected and applied to the model to do the dynamic analyses and to estimate the pore pressure development within the core of the dam during the earthquake loading.

THE ANALYSIS METHOD

In the analyses carried out in this study, the FLAC-2D was used (Detournay & Hart, 1999). It is a finite difference computer software capable of modeling many soil phenomena such as seepage, consolidation, and analyzing different dynamic responses of geotechnical structures under plane-strain conditions. In this model, the motion equations are derived in a continuum media. In this program the motion equations are used to obtain the new velocities and displacements from existing stresses and forces. The strain rates are then calculated according to the new nodal velocities in each element.

To solve a problem, the strain rate, having the velocity gradient, can be calculated from the following equation:

$$e_{ij} = \frac{1}{2}\left[\frac{\partial \dot{u}_i}{\partial x_j} + \frac{\partial \dot{u}_j}{\partial x_i}\right] \qquad (1)$$

where;

e_{ij} = components of the strain rate

$\dot{u}_i$ = component of the velocity

To obtain the stress tensor, the specific mechanical relationships are used which in general are as below:

$$\sigma_{ij} = M(\sigma_{ij}, \dot{e}_{ij}, \kappa) \qquad (2)$$

where;

M = the specific rule of behavior

K = the history parameters (which based on the specific rules may or may not exist)

In this program all equations mentioned so far are solved by means of finite deference.

DYNAMIC ANALYSIS OF THE KARKHEH DAM

To do the analysis of the dam, it is required to select a typical or a critical section of the dam. In this study the critical section has been selected in a way that if the dam stability is met in this section, no other section will be unstable in the same condition. Taking into account all sections of the Karkheh dam from different aspects such as their base elevations, the height of the filling, the depth of the river, the induced settlements, and the pore pressure developments in them, the section 5-5 of the dam which is located at the station 1+230 km. would be the dam critical section. Therefore, this section has been concentrated upon in this study which is shown in Fig. 1.

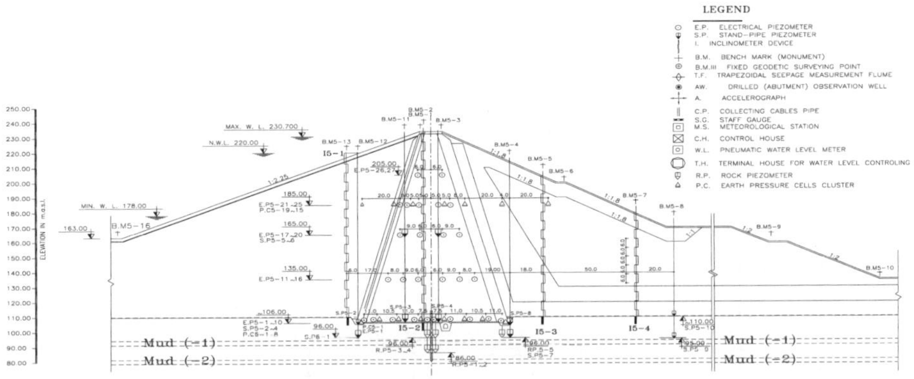

Fig. 1- The critical section of the Karkheh Earth Dam with the ground profile and position of the installed instruments (Sec.5-5 at station 1+230 km)

The modeling zone covers dam body and cofferdam together with a part of the dam foundation down to 70 m. depth and 250 m length on both sides of the dam axis. To model the dam a mesh of 23x80 has been used. To model the stage construction, the dam body has been divided into 5 m. layers . Due to concentration of this study on the pore pressure developments within the dam core, and also the sensitivity of this zone, the selected mesh in the core had smaller grids, so that 20 elements were put in the core width. The general view of the body and foundation is shown in Fig. 2.

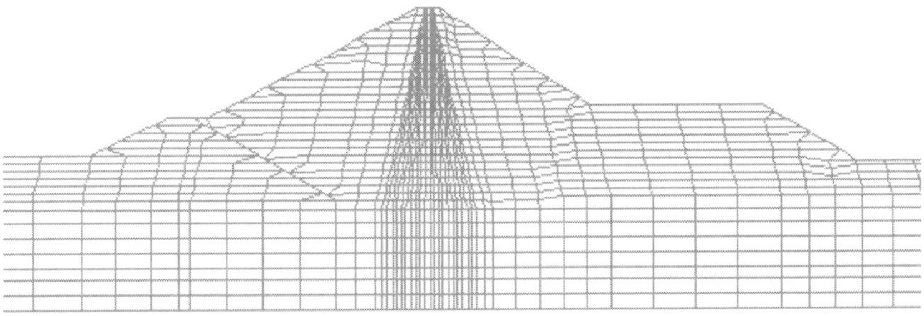

Fig. 2- The general view of the selected zone and the finite difference mesh

THE INPUT ACCELERATION

To evaluate the seismic parameters of the Karkheh site, the seismic risk analyses have been carried out. According to the performed seismic studies on the basis of probabilistic seismic hazard analysis (PSHA), the maximum design acceleration, velocity, and displacement are 0.256g, 23 cm/sec., and 15 cm. respectively for the design based earthquake (DBE) of the Karkheh Dam site (I.W.P.C. ,2001).

Comparing the above parameters with those belong to different recorded earthquake occurred in Iran, the Deyhook accelerogram by the record No. 1082-1 was found to be the most similar to the Karkheh site seismic parameters, thus it was selected for the dynamic analyses in this study. To optimize the analyzing time and prevent the divergence of the numerical results, the first 15 sec. of the accelerogram was used and the frequencies more than 5 Hz. Were omitted. The acceleration was applied to the base of the model which was considered at the depth of 70 m. from the ground surface. The accelerogram is shown in Fig. 3.

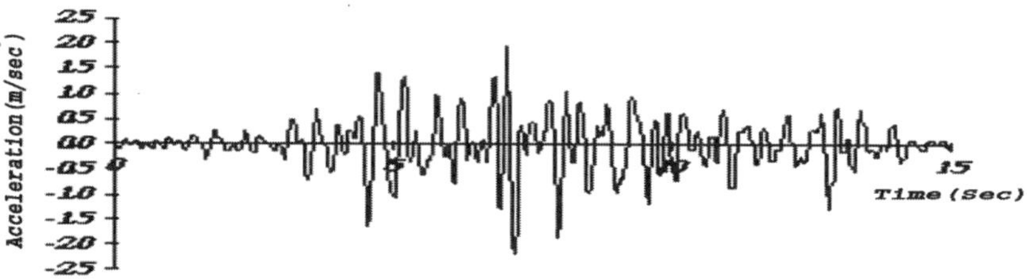

Fig. 3- The input accelerogram used in the dynamic analyses

THE EVALUATIONS OF THE PORE PRESSURE WITHIN THE CORE OF THE KARKHEH DAM DURING EARTHQUAKE

After application of the selected accelerogram to the model and doing the dynamic analysis, even for the case when there is no seepage flow in the core and the flow analysis mode of the model is off, considerable changes happen both in values and in the pattern of the pore pressure developments (Figs. 4 and 5). The maximum pore pressure value in this case reaches to 1043 (kPa) is 26% greater than that of exist in the core before earthquake. It can be seen in figure 5 that the more we move towards the high levels, the more influences of the earthquake loading happen within the core.

Also it is quite evident that at the same level in the core, the more we move towards the sides and filter zones, the more decrease occur in the pore pressure. This may be attributed to the lateral expansion of the core under constant volume and undrained condition due to the vertical displacement and settlement of the core during the earthquake loading. Thus, the pore pressure dissipation happens considerably near the core boundary zones.

The trend of the pore pressure variations during the earthquake loading in the central zones of the core at different levels have been shown in Figs. 6 & 7. The selected points are so that they can show the amount of the pore pressure developed at the position of the electrical piezometers EP5-3, and EP5-12 installed at different levels of the sec. 5-5 of the dam.

SUMMARY AND CONCLUSIONS

The response of the Karkheh large embankment dam with mixed clay core, under earthquake loading was investigated and evaluated numerically. The input parameters were justified and the model was calibrated using the field data obtained from the regular measurements of more than 1000 instruments installed in the dam. According to the obtained results, the main following points can be concluded:

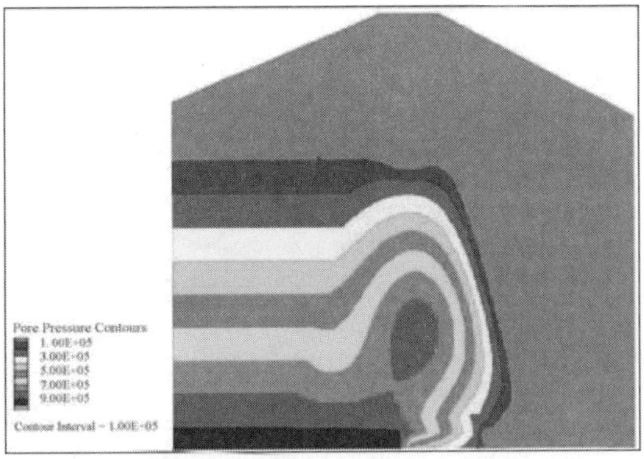

Figure 4- The pattern of pore pressure development before earthquake

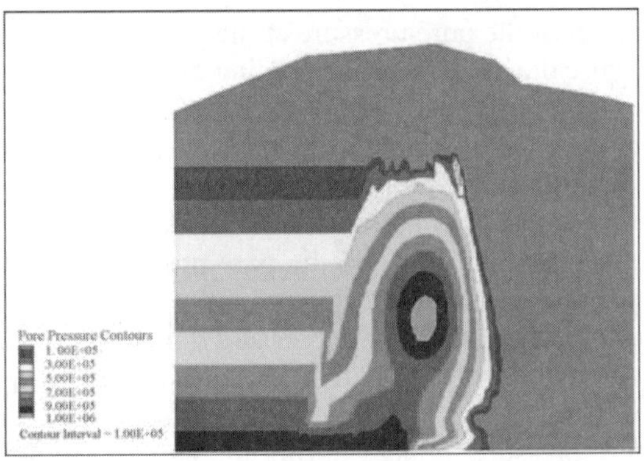

Figure 5- The pattern of pore pressure development after earthquake

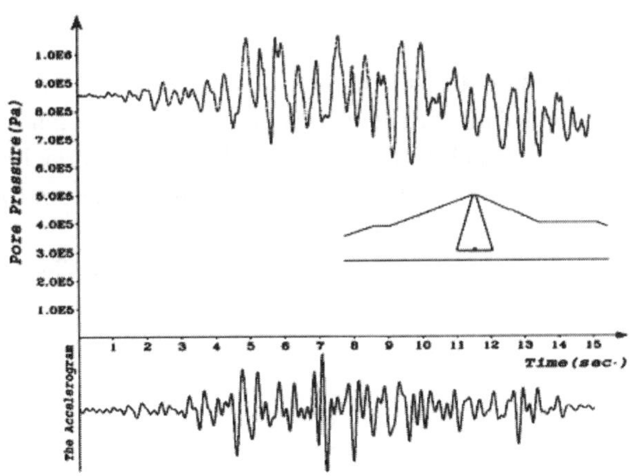

Fig.6- Variation of the pore pressure at the position of the EP5-3 (elevation of 106), in the core due to earthquake loading estimated from numerical analyses.

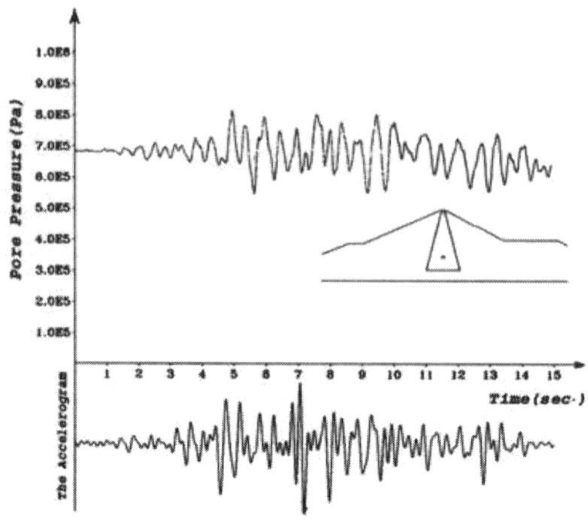

Fig. 7- Variation of the pore pressure at the position of the EP5-12 (elevation of 135), in the core due to earthquake loading estimated from numerical analyses.

- The pore pressure developments move towards the higher levels of the core in comparison with those developed in the lower levels under static conditions. The maximum value of the pore pressure happens at the middle level of the core.
- At the same level within the core, the pore pressure decreases as moving towards the filter zones in the sides of the core.
- The maximum increase in the pore pressure during the earthquake loading is about 26% of the one that developed before the earthquake.
- The pore pressure ratio, Ru, increases to about 0.72 at the point which the maximum pore pressure induces during the earthquake loading. It means that the mixed clay core of the Karkheh large embankment dam shows satisfactory behavior if an earthquake similar to Deyhook occurs in the site.

REFERENCES

ASCE (2000), "Guidelines for instrumentation and measurements for monitoring dam performance." Published at American Society for Civil Engineers.

Detournay, C. and Hart, R., (1999), "FLAC and Numerical Modeling in Geomechanics." *Proc.Int. FLAC Symp. on Numerical Modeling in Geomechanics*, Minneapolis, Minne Sota.

Iran Water and Power Resources Development Company (IWPC) (2001). "The abstract of the first phase studies of the Karkheh project." (in Persian).

Mahab-Ghods Consulting Engineers (2000), "Technical reports (Phase I, II, & III) of KARKHEH project." (in Persian), Tehran, Iran.

Mahab-Ghods Consulting Engs. (1997 to 2003), "Technical reports of instrumentation of Karkheh Dam" (in Persian), Tehran, Iran.

Mahab-Ghods Consulting Engineers (1998), "Trial embankments of Karkheh Dam" (in Persian), Technical Report, Tehran, Iran.

International Conference on Geotechnical Engineering for Disaster Mitigation & Rehabilitation
Chu, Phoon & Yong (eds) © 2005 World Scientific Publishing Company ISBN 981-256-469-1

MODELING THE RESPONSES OF SATURATED SAND UNDER DYNAMIC LOADS

KIM, SOO-IL
Department of Civil Engr., Yonsei University, 134 Shinchon-Dong, Seodaemoon-Gu Seoul, 120-749, Korea

PARK, KEUN-BO
Department of Civil Engr., Yonsei University, 134 Shinchon-Dong, Seodaemoon-Gu Seoul, 120-749, Korea

PARK, SEONG-YONG
Department of Civil Engr., Yonsei University, 134 Shinchon-Dong, Seodaemoon-Gu Seoul, 120-749, Korea

PARK, INN-JOON
Department of Civil Engr., Hanseo University, 360 Daegok-Ri, Haemi-Myeon Chungnam, 356-706, Korea

The main objective of this paper is to develop an improved method for the analysis of liquefaction potential and to verify the modified model that can simulate behavior of saturated sand under dynamic loading conditions. The model is developed through modification of the DSC model. For the development of the model, general formulation for the DSC constitutive model is modified and calibrated for more realistic description of dynamic responses of saturated sand using the energy dissipation approach. The research is focused on evaluation of the disturbance in the DSC which is defined and applied by dissipated energy induced by stress. This paper presents the basic theory and verification of the modified model based on results from triaxial tests for saturated sand with various input motions including sinusoidal, linear increasing, and real earthquake motions.

INTRODUCTION

Over the last 30 years or so, there have been several soil models for describing undrained behaviors of fully saturated sands under dynamic loading conditions. Recently, the energy-based approach (Figueroa et al. 1994) and the disturbed state concept (DSC) (Park and Desai 2000) to evaluate the liquefaction potential have received great interests because these approaches provide more detailed information for dissipated energy degradation and cumulated deformations. While those models have been successfully verified for many geotechnical dynamic problems, they have been primarily for cyclic loading test results. For the liquefaction analysis, it is known that the sinusoidal type of cyclic loading does not realistically represent actual irregular earthquake motions.

Therefore, effects of the actual irregular earthquake motions should also be taken into account for more detailed and advanced liquefaction analysis. In this paper, the general formulation of the DSC constitutive model is modified and calibrated using the energy dissipation approach for better numerical prediction with various loading types. Especially, the disturbance caused by an applied force in the DSC model is defined by dissipated energy. A procedure for back-calculation is developed based on the incremental integration method as well. In order to verify the modified model, dynamic tests are performed and compared with results obtained from the model.

BASIC THEORY

Dissipated Energy Approach

Dissipated energy in dynamic loading is an important measure to define progressive changes of undrained behavior for sands. The dissipated energy can be represented by the area of the hysteresis loop given the stress-strain curve under dynamic loading conditions. The expression for dissipated energy W is given by:

$$W = \int_0^t \sigma d\varepsilon^p \approx \sum_{i=1}^{n} \sigma_i \Delta\varepsilon_i^p \tag{1}$$

where σ = stress vector; $\Delta\varepsilon^p$ = plastic strain increment vector; t = time; and n = total number of increments or cycles.

Disturbed State Concept

According to the DSC proposed by Park and Desai (2000), external forces cause changes and disturbances in the microstructure system of a material. The stress-strain response of the material at a certain loading condition is then determined from a degree of the disturbance caused by the external force. There are two reference states defined in the DSC model: relative intact (RI) and fully adjusted (FA) states. A material in the RI state upon loading modifies continuously through a process of natural self-adjustment, and a part of it approaches the FA state at randomly disturbed locations in the material. As a result, observed responses of the material can be determined from responses of RI and FA states in terms of the disturbance D. The RI state is defined with continuum soil models, such as the elasto-plastic stress-strain models, while the FA state is defined as a response of a material at the ultimate state. In the original DSC model, the RI state is determined by Hierarchical Single Surface model with the isotropic hardening and associated flow rule whereas the FA state is given by the critical state model (Roscoe *et al.* 1958).

In DSC model, the yield surface in the HiSS model for the RI state is given by:

$$F = \frac{J_{2D}}{p_a^2} - \left[-\alpha \left(\frac{J_1}{p_a} \right)^n + \gamma_u \left(\frac{J_1}{p_a} \right)^2 \right] (1 - \beta S_r)^{-0.5} = 0 \tag{2}$$

where J_1 = the first invariant of the stress tensor; J_{2D} = the second invariant of the deviatoric stress tensor; p_a = atmospheric pressure in the same unit as the stress tensor; n, β, and γ_u = model parameters; α = hardening function. The hardening function α can be defined in terms of the dissipated energy [i.e., Eq. (1)] for undrained conditions as follows:

$$\alpha = h_1 / W^{h_2} \tag{3}$$

where h_1 and h_2 = hardening function parameters. The FA state, on the other hand, can be represented by:

$$\sqrt{J_{2D}^c} = mJ_1^c \tag{4}$$

where J_1^c = the first invariant of stress tensor at critical state; J_{2D}^c = the second invariant of deviatoric stress tensor at critical state; and m = slope of the critical state line.

The disturbance D in the DSC model determines the stress-strain response and shear resistance at a certain loading stage. From stress–strain responses of RI, FA, and observed states, the D and effective stresses in the DSC model are given by:

$$D = (\sigma_{ij}^{'i} - \sigma_{ij}^{'a})/(\sigma_{ij}^{'i} - \sigma_{ij}^{'c}) \tag{5}$$

$$\sigma_{ij}^{'a} = (1-D)\sigma_{ij}^{'i} + D\sigma_{ij}^{'c} \tag{6}$$

where D = disturbance at current observed state; and σ'^{i}_{ij}, σ'^{c}_{ij}, and σ'^{a}_{ij} = effective stresses at RI, FA, and observed states, respectively. Differentiation of Eq. (6) leads to incremental equation as follows:

$$d\sigma_{ij}^{'a} = d(1-D)\sigma_{ij}^{'i} + D(d\sigma_{ij}^{'c}) \tag{7}$$

Integrating Eq. (7), the incremental dissipated energy dW is given by:

$$dW = \int \sigma_{ij}^{'a} d\varepsilon_{ij}^{a} dV = \int (1-D)\sigma_{ij}^{'i} d\varepsilon_{ij}^{i} dV + \int D\sigma_{ij}^{'c} d\varepsilon_{ij}^{c} dV \tag{8}$$

where $d\varepsilon_{ij}^i$, $d\varepsilon_{ij}^c$, and $d\varepsilon_{ij}^a$ are increments of the deviatoric plastic strain tensor at RI, FA, and observed states under undrained conditions, respectively. Integrating the Eq. (8), D can be defined as:

$$D = (W_1^{i^*} - W_1^{a^*})/(W_1^i - W_1^c) \tag{9}$$

where $W_1^{i^*}$ = dissipated energy in total RI state; $W_1^{a^*}$ = dissipated energy in total observed response; W_1^i and W_1^c = dissipated energy at RI and critical states, respectively.

ANALYSIS

Experimental Program and Test Results

The saturated Jumunjin sand, a representative silica sand in Korea, was tested using a cyclic triaxial device under the stress-controlled loading condition with three different wave shapes including the sinusoidal, linear increasing, and actual earthquake motions. The Jumunjin sand is a clean quartz sand and defined as SP according to the Unified Soil Classification System. Diameters of the sand particles range from 0.1 to 0.84 mm. The coefficient of uniformity C_u is 1.47, and the mean grain size D_{50} is 0.52 mm. The maximum and minimum void ratios e_{max} and e_{min} are 0.89 and 0.63, respectively. Its specific gravity G_s is 2.61.

Based on test results, various dynamic soil responses, including the deviatoric stresses, pore water pressures (PWP), and deviatoric stress-strain relationship, were analyzed. Figure 1 shows input motions and corresponding mobilization of the PWP with time for different types of dynamic loads. Test results with sinusoidal loading are given in Figure 1(a). As shown in Figure 1(a), the excess pore pressure continuously builds up until a certain number of loading cycles and then a rapid increase and oscillation of the excess pore pressure at a value approximately equal to the initial confining stress are observed. In Figure 1(b) for the linear increasing load, it is observed that the PWP builds up gradually until a certain time [i.e., t = 8 sec in Figure 1(b)] and eventually reaches a value equal to the initially applied confining pressure. For the actual earthquake motion shown in Figure 1(c), the PWP curve with time appears to be quite different from those for sinusoidal and linear increasing loads. As shown in Figure 1(c), it is seen that the pore water pressure does not increase in earlier stages, but suddenly jumps up when the maximum deviatoric stress is reached, whereupon liquefaction sets in.

Comparison of Disturbances

In order to compare disturbances D in earthquake loading types, disturbances D using the original DSC method (method 1) and modified method (method 2) were adopted. Model parameters for calculation of D were obtained from Kim (2004).

Figures 2 and 3 show values of D for sinusoidal load and earthquake motion tests with the plastic strain trajectory ξ and dissipated energy W for method 1 and 2, respectively.

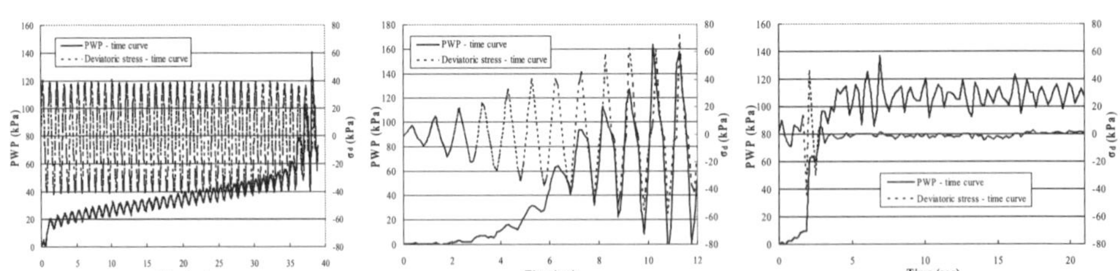

(a) sinusoidal motion (b) linear increasing motion (c) earthquake motion

Figure 1. PWP and deviatoric stress versus time with different types of loads.

Figure 2 shows sinusoidal loading test results. As shown in Figure 2, *D* curves from the two methods show good agreement. Figure 3 shows earthquake motion test results. Since strains are remarkably unsystematic in comparison with sinusoidal loading test results, *D* curve with ξ appears to be of irregular shape. This indicates that *D* function corresponding to the initial liquefaction in DSC cannot be properly estimated. *D* curve from method 2 based on the dissipated energy, on the other hand, is observed to be well defined, and thus superior in the liquefaction analysis for the determination of *D* function.

Comparison of Stress-Strain Response

In order to verify the modified model based on the dissipated energy, dynamic tests were performed. Test results are shown in Figure 4. Figure 4 shows the deviatoric stress versus axial strain curves for the different types of dynamic loading tests. As shown in Figure 4, significant differences in stress-strain curves are observed in the tests for sinusoidal, linear increasing, and earthquake loading tests.

A numerical code of the modified model based on the incremental integral scheme for modified DSC model was developed. The modified model is used to calculate the entire stress-strain responses of sands for cyclic triaxial tests. This is achieved by numerically integrating Eq. (8) based on the procedure given by Park and Desai (2000). Necessary input DSC parameters were re-calculated for static triaxial tests as presented in Kim (2004). Parameters E, v, γ, β, m, n, h_1, h_2, $\overline{m}$, λ, e_o^c, D_u, Z, and A are 210,000 kPa, 0.380, 0.250, 0, -0.5, 2.667, 0.1515, 0.0922, 0.5, 0.045, 0.0636, 0.99, 11.64, and 0.35, respectively.

Figure 4 also compares calculated and tested cyclic stress-strain responses. As shown in Figure 4, stress-strain curves obtained from the modified model represent overall agreement with tested results. As the stress-strain curves arrived at the initial liquefaction stage, it is seen that the difference between calculated and tested results becomes slightly pronounced at compression or extension region. This is because actual soils under dynamic loading conditions behave as a composite material of liquid and solid and thus produce large deformations with significant reductions of elastic modulus. The discrepancy is not considered to be severe and may be seen due to experimental and/or computational errors. From Figure 4, it is demonstrated that the modified model provides highly satisfactory correlations with the tested laboratory behavior of saturated sands under linear increasing loading and earthquake motions.

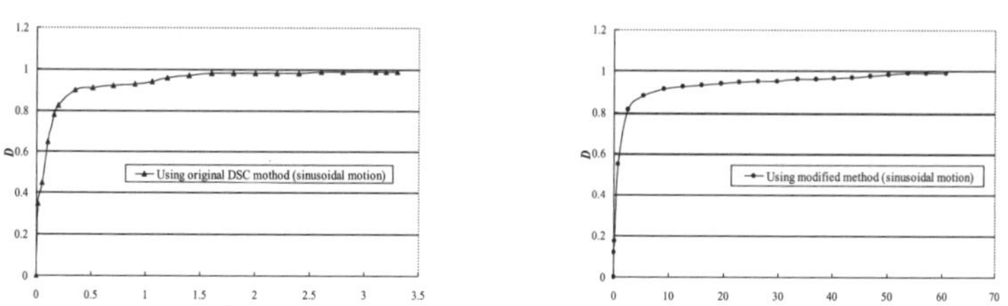

(a) method 1 (*D* versus ξ) (b) method 2 (*D* versus *W*)

Figure 2. Comparison of disturbances for sinusoidal loads.

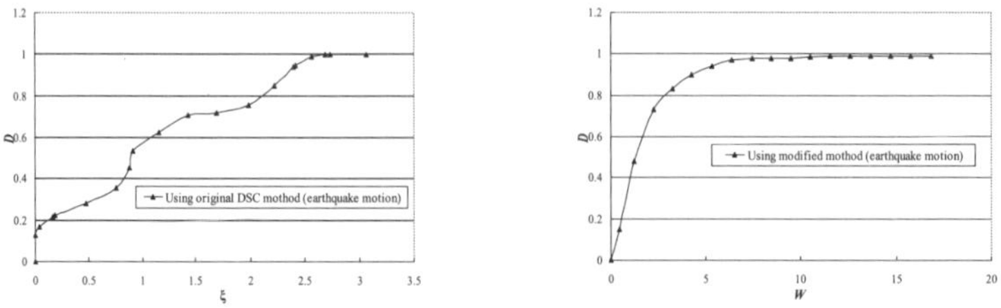

(a) method 1 (D versus ξ) (b) method 2 (D versus W)

Figure 3. Comparison of disturbance for earthquake motion.

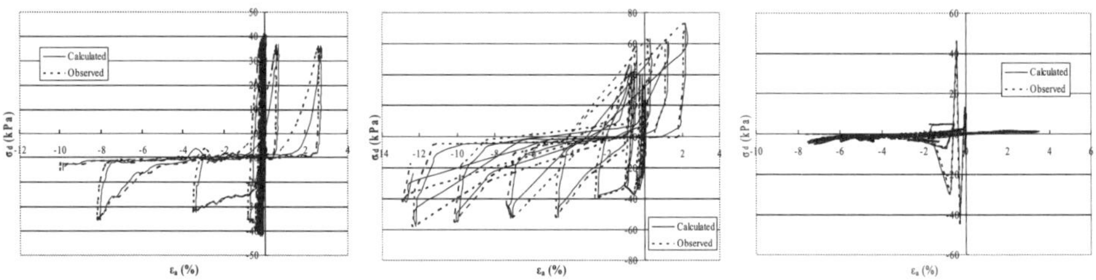

(a) sinusoidal motion (b) linear increasing motion (c) earthquake motion

Figure 4. Comparison between back calculated and tested stress-strain curves.

CONCLUSION

For the analysis of dynamic soil behavior under undrained loading conditions, a model was developed for better simulation of dynamic responses for saturated sands based on the DSC. A constitutive model of the original DSC was modified and calibrated using the dissipated energy approach. In order to verify the modified model, results of the numerical program based on the incremental solution of integral scheme developed in this study were compared with those from experimental tests. Back-calculation results showed good agreement with tested results until the initial liquefaction. From back-calculation results, it is found that the modified model describes reasonably well the stress-strain responses for various loading types of both regular and irregular motions.

REFERENCES

Figueroa, J.L., Saada, A.S., Liang, L., and Dahisaria, M.N. (1994). "Evaluation of soil liquefaction by energy principles", *Journal of geotechnical engineering*, ASCE, Vol. 120, No. 9, September, 1554-1569.

Kim, S.I. (2004). "Liquefaction potential in moderate earthquake regions", *12th Asian Regional Conf. on Soil Mechanics & Geotechnical Engineering*, Leung et al. (eds), World Scientific Publishing Co., Vol. 2, 1109-1138.

Park I.J. and Desai C.S. (2000). "Disturbed state modeling for dynamic and liquefaction analysis", *Journal of Geotechnical and Geoenvironmental Engineering*, ASCE, Vol. 126, No. 9, 834-846.

Roscoe, K.H., Scofield, A., and Wroth, C.P. (1958). "On yielding of soils", *Geotechnique*, Vol. 8, 22-53.

International Conference on Geotechnical Engineering for Disaster Mitigation & Rehabilitation
Chu, Phoon & Yong (eds) © 2005 World Scientific Publishing Company ISBN 981-256-469-1

ZONING FOR LIQUEFACTION AND ACTUAL DAMAGE DURING THE 2005 FUKUOKAKEN-SEIHO-OKI EARTHQUAKE

HIDEO NAGASE
Kyushu Institute of Technology, 1-1 Sensui, Tobata-ku, Kitakyushu, 804-8550, Japan

AKIHIKO HIRO-OKA
Kyushu Institute of Technology, 1-1 Sensui, Tobata-ku, Kitakyushu, 804-8550, Japan

NORIYUKI YASUFUKU
Kyushu University, 6-10-1 Hakozaki, Higashi-ku, Fukuoka, 812-8581, Japan

KOUZOU HIRAMATSU
WESCO Co., Ltd., 2-5-35 Shimadahonmachi, Okayama, 700-0033, Japan

KENJI HASHIMURA
Nihon Chiken Co., Ltd., 5-25-25 Moro-oka, Hakata-ku, Fukuoka, 816-0094, Japan

On March 20, 2005, an earthquake of magnitude 7.0, which was named as the 2005 Fukuokaken-seiho-oki earthquake, occurred at the northwest part of Fukuoka City. During the earthquake, liquefaction took place mainly in the reclaimed lands of the Hakata Bay area and caused structural damage. In this paper, the distribution of the sites where liquefaction occurred and the characteristics of the grain size of sand boils collected at the sites were clarified. Further, the occurrence of liquefaction in the reclaimed land was discussed with reference to some instances, and the structural damage was summarized. Furthermore, liquefaction analysis was performed using a program called "SHAKE," and several soil profiles of the reclaimed land and alluvial ground in order to investigate the relationship between the occurrence of liquefaction and the distribution of acceleration, which was estimated from earthquake response analyses and the mechanical and physical properties of the soils.

DISTRIBUTION OF SITES WHERE LIQUEFACTION OCCURRED

In the survey performed immediately after the earthquake, the occurrence of liquefaction was evident by the existence of sand boils on the ground surface. Figure 1 indicates the distribution of the sites where liquefaction occurred in the Hakata Bay area. It can be seen from this figure that the sites that exhibited liquefaction were distributed mainly in the reclaimed lands of the Hakata Bay area and were studded with reclaimed lands. Therefore, the degree of severity of liquefaction due to the earthquake is not regarded to be high as compared with that observed in past earthquakes, for example, the 2000 Tottoriken-seibu earthquake. On the contrary, the occurrence of liquefaction in alluvial ground was nearly negligible. It may be considered that the occurrence of liquefaction is related to the characteristics of earthquake motion and the magnitude of the liquefaction strength of soil. The grain size distribution curves of sand boils observed in several reclaimed lands in the Hakata Bay area were investigated. The particle size D_{50} is 0.15–1.06 mm and the uniformity coefficient Uc is 1.7–7.1. The boiled sand is classified as a relatively well graded middle sand to poorly graded fine sand, provided certain sand samples are excluded from the sand boils.

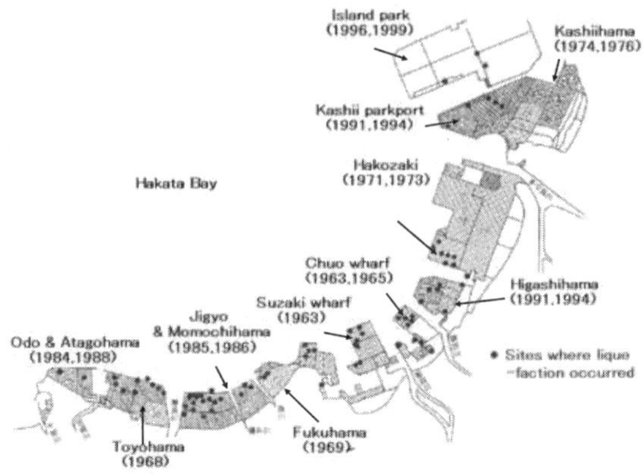

Figure 1. Distribution of sites where liquefaction occurred in reclaimed lands of Hakata Bay area

OCCURRENCE OF LIQUEFACTION IN RECLAIMED LANDS

With regard to the occurrence of liquefaction in the reclaimed lands, the following features are evident upon investigating the reclamation method and properties of the soil used for the reclamation.

1) In general, it is more difficult for liquefaction to occur if the duration of sedimentation is long. However, the trend of liquefaction resistance is not observed because several types of soils and methods of reclamation were used in the construction of the reclaimed lands.
2) In the east side of the reclaimed lands, for example, in Hakozaki wharf, dredged soil composed of sand, silt, and clay, which was supplied from the sea bed of Hakata Bay, was reclaimed by a pumping method, and the fine content and clay content of the sediment soil are variable in the reclaimed land. Therefore, it can be considered that the sites where liquefaction occurred during the earthquake were especially studded with reclaimed lands.
3) In Island Park City, most of the land was not liquefied because dredged soil with a high proportion of fine content was used for the reclamation. However, a section of the temporary embankment of the sea wall and the backfilled ground of the quay wall were liquefied because clean sand with a small proportion of fine content was used for reclamation in these places. This phenomenon was observed in the western parts of the reclaimed lands, such as Momochihama and Atagohama.

STRUCTURAL DAMAGE DUE TO LIQUEFACTION

During the earthquake, shore structures and roads were damaged to a large extent due to liquefaction. In addition, a huge flow phenomenon occurred due to liquefaction in the gently sloping ground in a park. In this chapter, the typical damage incurred by quay wall structures and the flow failure of sloping ground are introduced.

Damages to quay walls

In Tyuou wharf, a quay wall moved around 1 m toward the sea and the backfilled ground sank by around 1.2 m. This represents the maximum extent of the damage to the quay walls. Several sand boils were observed on the surface of the backfilled ground. This quay wall is of an L-shaped concrete block type. In the groin portion of the Tyuou wharf area, the L-shaped type quay wall also moved into the sea by around 20 cm and several cave-ins occurred in the ground surface of the groin. In Hakozaki wharf, the difference between the levels of the quay wall and the backfilled ground was observed to be 30–50 cm. The quay wall is an auxiliary sheet pile type.

Flow failure of sloping ground

At the Uminonakamichi Seaside Park, flow failure occurred in an open space with a gently sloping ground toward a pond named as "Kamo-ike." Photo 1 shows the state of the flow failure immediately after the earthquake, and Figure 2 indicates the locations at which the survey was carried out for measuring the ground displacement. Figure 3 indicates the relationship between the ground displacement due to flow failure and its distance from the waterline of the pond. It can be seen from the figure that the ground displacement reached a maximum of around 10 m, although the maximum displacement in the c-c line was around 1 m. It appears that a stone stairway that was constructed near the waterline had a restraining effect on the ground displacement. Furthermore, a worker who manages tourist boats was interviewed regarding the flow phenomenon. It was clarified from this interview that the flow failure occurred almost immediately after the occurrence of liquefaction during the earthquake.

Photo 1. Flow failure of the ground at the Uminonakamichi Seaside Park

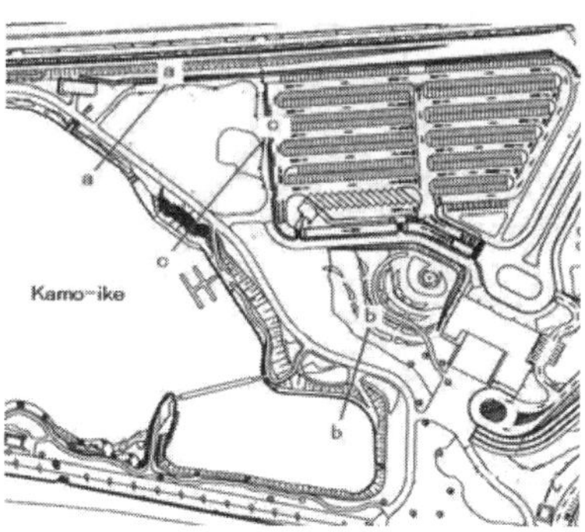

Figure 2. Locations at which the survey of ground displacement due to flow failure was carried out

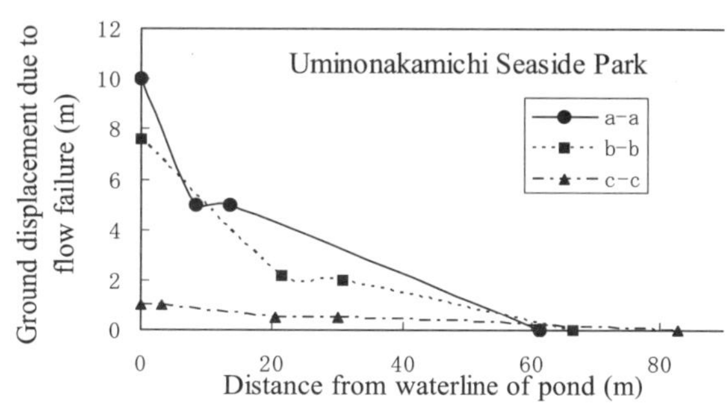

Figure 3. Relationship between ground displacement due to flow failure and distance from the waterline of the pond

ANALYTICAL INVESTIGATION

Analytical procedure

The sites where liquefaction occurred during the earthquake were almost restricted to the reclaimed lands. In order to clarify the reason for this peculiarity in the liquefaction phenomenon, a liquefaction analysis was performed for the reclaimed land and alluvial ground. The method of the liquefaction analysis is explained as follows;

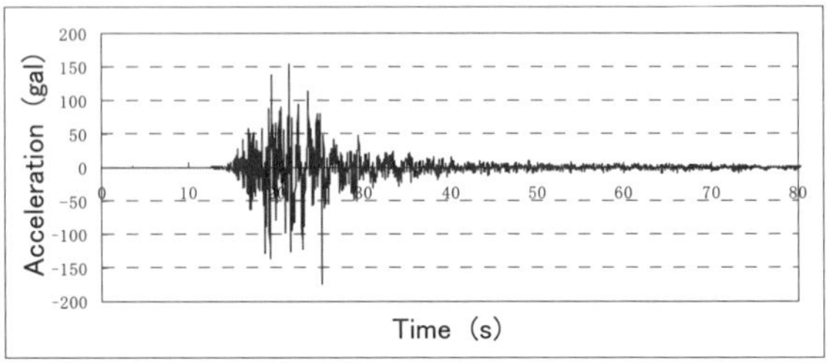

Figure 4. Wave form used for one dimensional seismic response analysis

1) A program for one dimensional seismic response analysis, "SHAKE," was used to obtain the accelerations on the ground surface.
2) The seismic waveform for the analysis is obtained from a modification involving a conversion of the seismic record of the EW component measured on the ground surface at Fukuoka public hall into that for the basement layer using the "SHAKE" program. Figure 4 shows the seismic waveform for the analysis. The maximum acceleration is 174.911 gal.
3) The shear wave velocity was obtained from the following empirical formulae;

$$Vs = 157\ N^{0.180} \quad \text{for clayey soil} \qquad Vs = 144\ N^{0.159} \quad \text{for sandy soil} \tag{1}$$

4) The strain dependencies of the dynamic shear modulus ratio G/G_0 and damping constant h were expressed by empirical formulae proposed by Yasuda et al. (1985).
5) The soil boring logs used in the analysis are shown in Figure 5.
6) The shear stress ratio during the earthquake, L, was calculated by the following equation;

$$L = (1 - 0.015\ Z) \cdot k_{hc} \cdot \sigma_v/\sigma_v' \tag{2}$$

Z and k_{hc} denote the depth from the ground surface and the ratio of the maximum acceleration on the ground surface to the gravity acceleration, respectively. σ_v and σ_v' denote the vertical total and effective stress, respectively.
7) The liquefaction strength ratio R was calculated from two sets of equations, Methods 1 and 2, as shown in Table 1. These methods were referred to the Japanese certifications for highway bridges given in Japan Road Association (1996, 1990). In Method 2, the value of 0.187 was added to the liquefaction strength ratio R, on the basis of the results of Yasuda et al. (1991). This indicates that the modification for the value of N is generally underestimated because alluvial soil contains a relatively higher proportion of fine content. Using these methods, the liquefaction ratio was obtained for the following three cases;

Case 1: Method 1 was used; the fine content ratio Fc of the silty or clayey sand and silt in the alluvial soil layer were 30% and 60%, respectively.

Case 2: Method 1 was used; the Fc value was 40% and 80%, in the same manner as in Case 1.

Case 3: Method 2 was used; the D_{50} value of the sandy soil in the alluvial soil layer was 0.35 mm.

8) The ratio of F_L—a safety factor against liquefaction—was obtained from the following equation.

$$F_L = R/L \tag{3}$$

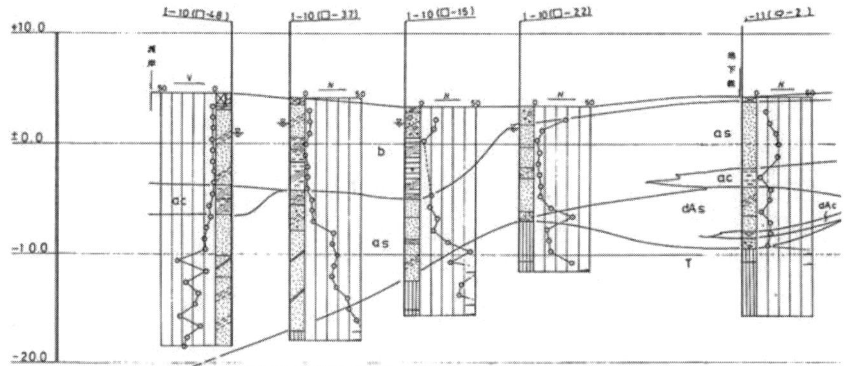

b: Reclaimed ground, embankment	ds: Diluvial sandy soil layer
as: Alluvial sandy soil layer	dc: Diluvial clayey soil layer
ac: Alluvial clayey soil layer	T: Tertiary layer

Figure 5. Soil boring logs for liquefaction analysis

Table 1. Explanation of two sets of liquefaction strength ratios

Method 1	Method 2
$(N_a < 14)$	$(0.02\ \text{mm} \le D_{50} \le 0.05\ \text{mm})$
$R = 0.0882\sqrt{\frac{N_a}{1.7}}$	$R = 0.0882\sqrt{\frac{N}{\sigma_v' + 0.7}} + 0.19 + 0.187$
$(N_a \ge 14)$	$(0.05\ \text{mm} \le D_{50} \le 0.6\ \text{mm})$
$R = 0.0882\sqrt{\frac{N_a}{1.7}} + 1.6\times10^{-6}(N_a - 14)^{4.5}$	$R = 0.0882\sqrt{\frac{N}{\sigma_v' + 0.7}} + 0.225\log_{10}\left(\frac{0.35}{D_{50}}\right) + 0.187$
$N_a = c_1 N_1 + c_2$	$(0.6\ \text{mm} \le D_{50} \le 2.0\ \text{mm})$
$N_1 = \frac{1.7N}{\sigma_v' + 0.7}$	$R = 0.0882\sqrt{\frac{N}{\sigma_v' + 0.7}} - 0.05 + 0.187$
$c_1 = \begin{cases} 1 & (0\% \le F_c < 10\%) \\ (F_c + 40)/50 & (10\% \le F_c < 60\%) \\ F_c/20 - 1 & (60\% \le F_c) \end{cases}$	
$c_2 = \begin{cases} 0 & (0\% \le F_c < 10\%) \\ (F_c - 10)/18 & (10\% \le F_c) \end{cases}$	

Results of the analysis

Figure 6 shows the results of the liquefaction analysis for Cases 1 to 3. The F_L-value for the reclaimed land is lesser than 1.0; this implies that liquefaction occurs in the reclaimed land in Cases 1 to 3, except for (c). However, in the alluvial soil layer, liquefaction occurs apparently in (a), (c), and (d) in Case 1; does not occur in (a) and (c); and its occurrence seems unlikely in (d) in Case 2. On the contrary, liquefaction does not occur in the alluvial soil layer in Case 3. It is evident that the result of the prediction of the occurrence of liquefaction depends on the manner in which the liquefaction strength is modified using the fine content ratio. And, it may also be considered for Cases 1 and 2 that the effect of fine content on the liquefaction strength was not properly accounted for in the analysis. It is noteworthy that the occurrence of liquefaction was suitably predicted in Case 3, although other predictions in this case would be necessary in this regard to confirm the accuracy of the analysis.

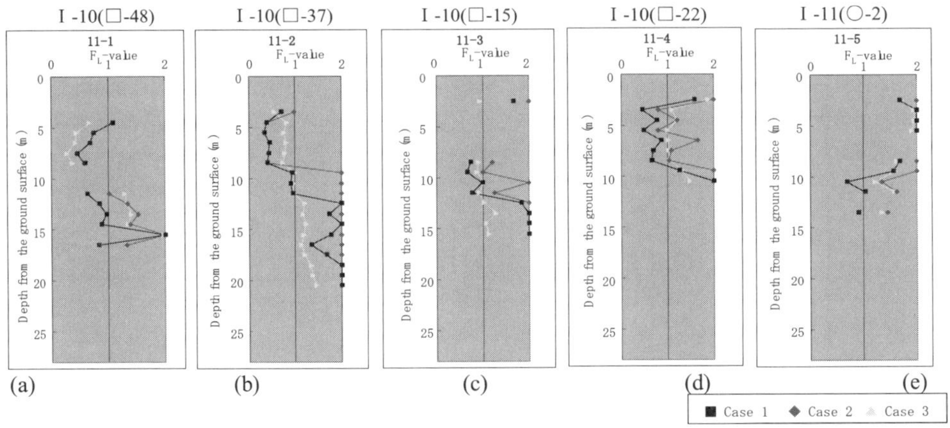

Figure 6. Results of liquefaction analysis

CONCLUSIONS

The 2005 Fukuokaken-seihou-oki earthquake in Fukuoka City was unexpected. During the earthquake, liquefaction occurred in the reclaimed lands of the Hakata Bay area and caused structural damage, although the magnitude of the earthquake was not significantly large. In this paper, the special features of liquefaction phenomena were summarized, and the results of liquefaction analysis were discussed in order to examine the susceptibility of the reclaimed lands of the city to the liquefaction phenomenon.

REFERENCES

Yasuda, S. and Yamaguchi, I. (1985): " Dynamic soil properties of undisturbed samples," Proc. of the 20th Japan National Conference on Soil Mechanics and Foundation Engineering, pp.539–542 (in Japanese).

Yasuda, S. and Matsumura, S. (1991): “Microzonation for liquefaction slope failure and ground response during earthquake in Fukuoka City,” Proc. of the 4th International Conference on Seismic Zonation, Vol.3, pp.703–724.

Japan Road Association (1990): " Specification for highway bridges," Part 5 Earthquake Resistant design (in Japanese).

Japan Road Association (1996): " Specification for highway bridges," Part 5 Earthquake Resistant design (in Japanese).

International Conference on Geotechnical Engineering for Disaster Mitigation & Rehabilitation
Chu, Phoon & Yong (eds) © 2005 World Scientific Publishing Company ISBN 981-256-469-1

THE DISASTER OF RESIDENTIAL LAND FILL CAUSED BY EARTHQUAKES

TAKASHI OKIMURA

Research Center for Urban Safety and Security, Kobe University, 1-1 Rokkodai Nada Kobe, 657-8501, Japan

In Japan, the disaster of residential land fill has been increasing lately due to the frequent earthquakes. This paper describes the features of the housing lots damaged by the earthquakes in Kobe and Niigata. The discrimination of hazardous valley fill is important for future quake-resistant design, and it is enabled by the construction of ground data base. For the development of the data base, it is essential to input more data obtained by ground research.

INTRODUCTION

In Japan, the collapse of buildings has occurred frequently with liquefaction or strong motion as damage caused by earthquakes, and recently the disaster of residential land fill such as failure and displacement has happened repeatedly. The reason of this phenomenon is considered that manmade lands have been increasing for developing residential lands.

Residential lands are individual assets in Japan, so that personal expenses are required in case of their rehabilitation. The collapse of retaining wall occurs at relatively confined places, and the scope of personal costs is clear. On the other hand, deformation is seldom seen at a single site of residential land but involves several sites, which saddles with serious problems that the restore of them can not be carried out individually.

In this paper, residential land disaster in earthquakes in Japan is overviewed, its features are described, and the effects of future earthquakes are predicted.

THE EARTHQUAKE DISASTER OF RESIDENTIAL LAND IN JAPAN

Japan has experienced the frequent occurrence of heavy rainfall in rainy season or typhoons, as well as of earthquakes. Since the heavy rainfalls started to cause more and more residential land disasters, "the Residential Land Development Control Law" was established in 1962. Under the regulation, strict technical standards have been applied to the residential development with 1-meter-or-higher embankment or 30-degree-or-steeper slope. Before the establishment, however, housing lots were developed with little or no attention given to the compaction of embankment or the drain of underground water. Many of the lots remain still now. In addition, the stability analysis in an earthquake was not required on 30-degree-or-below gentle slope at that time.

When the Miyagi Offshore Earthquake occurred in 1978, the residential lands were damaged for the first time in Sendai City. It caused the complete collapse of 26 houses, the half collapse of 142, and the damage of 23 retaining walls. More significant damage was the collapse or crack of 292 housing lots. An apartment complex had the large-scale collapse of its fill, and 28 households of the complex moved together due to the difficult rehabilitation of the fill. After the earthquake, the coseismic risk of fill, particularly of valley fill started to be pointed out.

In 1995, the widely known Kobe Earthquake (officially, the Southern Hyogo Earthquake) deprived more than 6,400 people of their precious lives, and damaged more than 5,000 retaining walls. Regarding 2,300 of the walls, recommendation on their restoration was issued to their owners based on the regulation. On the other hand, 65 housing lots suffered damage, including slope deformation. Most of the damage was inflicted on fills, 78 % of which were valley fill. This elicited that valley fill is vulnerable. Thereafter, only in case in which the damage of retaining walls or building sites effects public facilities such as lifeline, roads, etc., special measures for restoring private retaining walls with public money have been taken. This is the extension of "the Prevention Law against Steep Slope Failure."

In 2001, there was the occurrence of the Geiyo Earthquake whose epicenter was below the Seto Inland Sea in West Japan. This earthquake damaged many residential retaining walls in Kure City, Hiroshima Prefecture. The special measures that had taken in Kobe were adopted here also, and the free rehabilitaion spending public fund was provided for the walls in order to ensure the residential area's safety.

In 2004, the Chuetsu Earthquake hit Niigata Prefecture, East Japan. Its characteristic is that an aftershock which was similar in scale to the main shock repeated as often as 4 times within about 12 hours. Heavy damage was caused by the ground failure in this earthquake, and the damaged roads isolated many communities in the mountain area. A large number of houses were also damaged. The judgement was delivered urgently that 439 housing lots were in danger and 300 were caution- required. About one third of the housing lots were reported to be valley fill.

This year, 2005, the Fukuoka Western Offshore Earthquake hit Kyushu. It gave severe damage to the housing lots that had been developed on the narrow steep slope in Genkai Island.

Herewith, recently in Japan, the series of earthquakes have caused damage to so many residential sites. The research result of the damage in the Kobe Earthquake, 1995, and in the Chuetsu Earthquake, 2004, will be described in the following sections.

THE FEATURES OF THE DAMAGE IN THE KOBE EARTHQUAKE

Figure 1 is the seismic wave that was observed in the Kobe Earthquake, showing that the

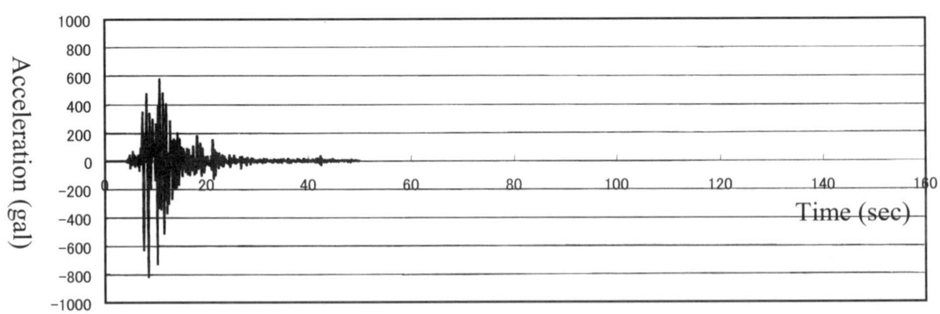

Figure 1. The seismic wave of the main shock measured at Kobe Marine Observatory in the Southern Hyogo Earthquake in 1995.

maximum acceleration was 770 gal, the duration was approximately 20 seconds. Furthermore, the predominant period was approximately 1 second, which is the general characteristic of the inland earthquakes in Japan. In the Miyagi Offshore Earthquake, 1978, the large-scale valley fill collapsed severely, while in the Kobe Earthquake, only 1 site suffered large-scale collapse (approximately 100 m wide, approximately 100 m long, and approximately 10 m deep) and most of the others were damaged in small-scale with 2~3 m displacement, see Figure 2. Most of the displacement occurred on the gentle slopes whose inclination was 10~20 degrees, see Figure 3. Figure 4 shows the relation of

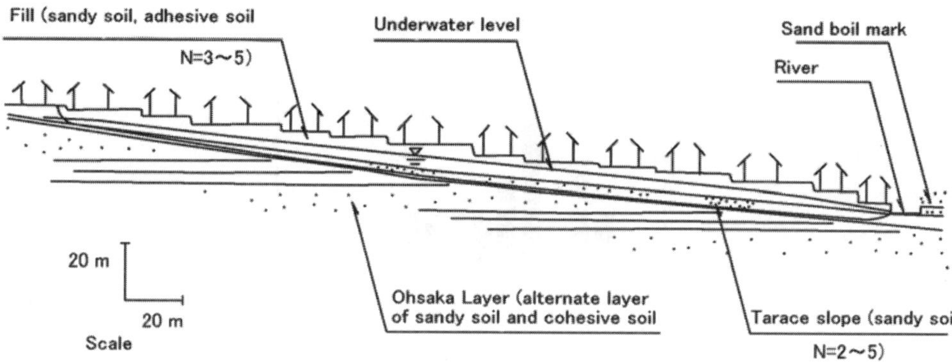

Figure 2. The displacement pattern common in the Southern Hyogo Earthquake

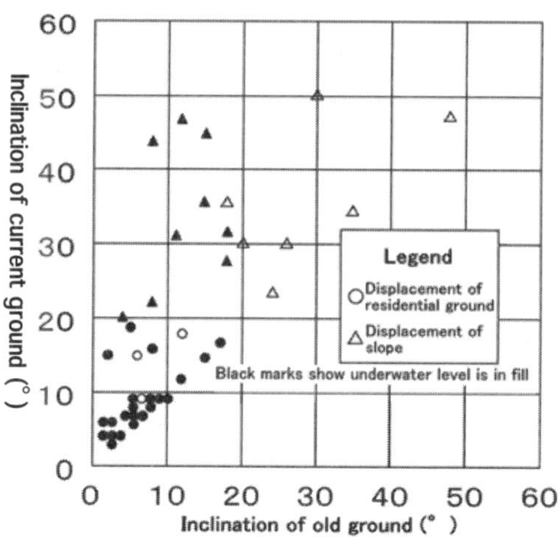

Figure 3. The relation of the inclination between old ground and current ground

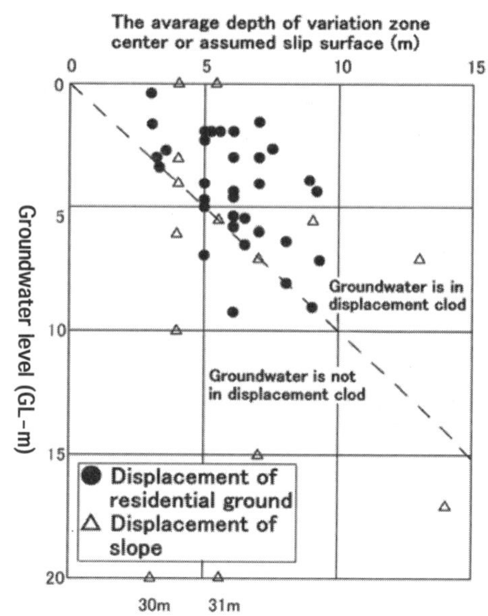

Figure 4. The relation between the depth of assumed slip surface and underwater level

the groundwater level and the depth of estimated slip surface that was read in the post-quake boring data. In this figure, above the dashed line means the appearance of the groundwater level in the fill. Most cases are above the dashed line, and their groundwater levels were found in the fill generally. Figure 5 shows the depth of the fill in the damaged area and of the talus slope that appears below the fill. The average depth of the fill is 5 m, the talus slope layer is 2~4 m, and the addition of them is 6~10 m in most sites. Herefrom, it is considered that a lot of slides were caused in the shallow layers of the residential land in the Kobe Earthquake.

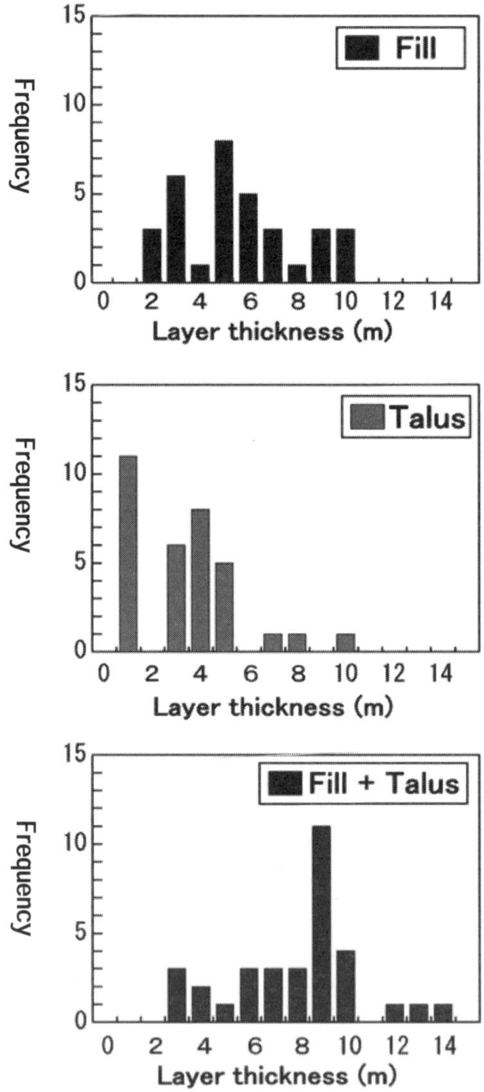

Boring results of 33 residential ground displacement

Figure 5. Layer thickness distribution of fill and talus.

THE FEATURES OF THE DAMAGE IN THE CHUETSU EARTHQUAKE

The seismic waves that were observed in the Chuetsu Earthquake can be seen in Figure 6. The maximum acceleration was 1,713 gal at Tohkamachi and 456 gal at Nagaoka, the duration was 10~15 seconds. The predominant period was 0.2 seconds at Tohkamachi and 0.2~2 seconds at Nagaoka, which were a far smaller value than that in the Kobe Earthquake, but many slope failures took place. This is because the aftershocks hit as often as 4 times as described before. The interview to the residents revealed that these aftershocks were the cause of the slope failures. Frequent collapses of valley fill were triggered in this earthquake also. This paper will report the case of Takamachi apartment block where a great number of valley fill were destructed.

This apartment complex was improved in the late 1970s. In the development, the gentle mountaintop was scraped down, and the scraped soil was used for the valley fill. After the development, the mountaintop was flat like terrace, around which roads were constructed, and they were surrounded by slope as if the feature was terrace scarp. It is well known that during earthquakes the maximum acceleration appears near knick points generally, and this is why large velocities appear on the boundary between the terrace

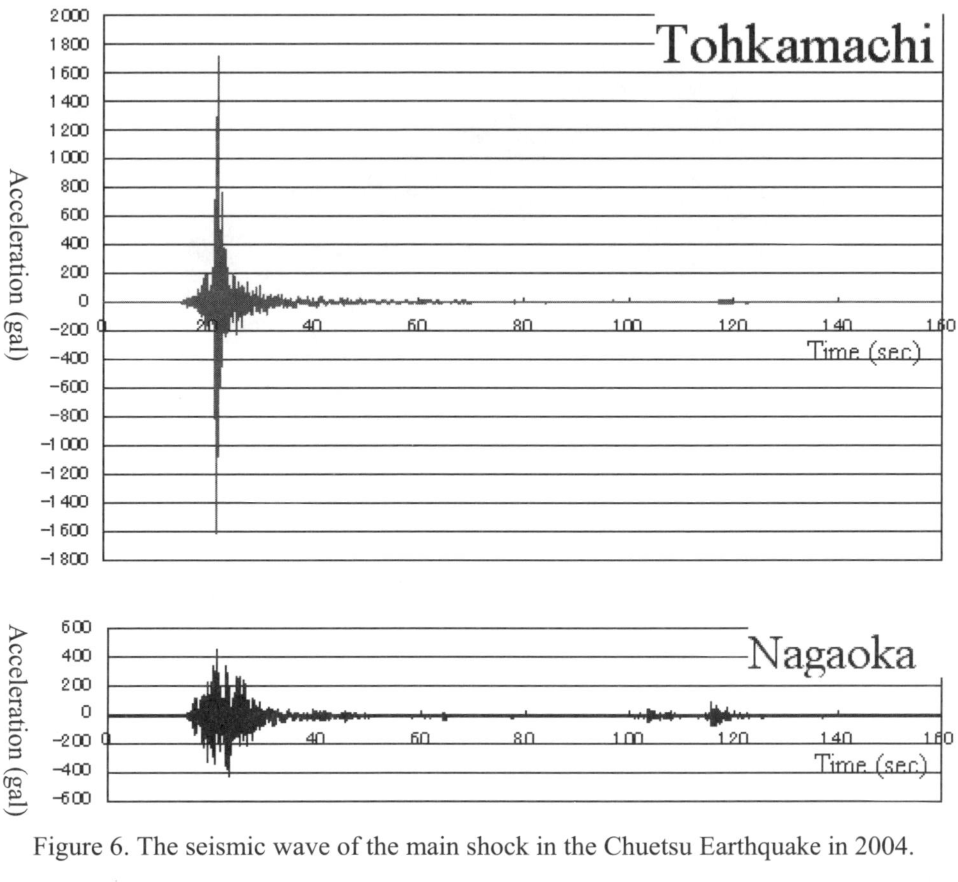

Figure 6. The seismic wave of the main shock in the Chuetsu Earthquake in 2004.

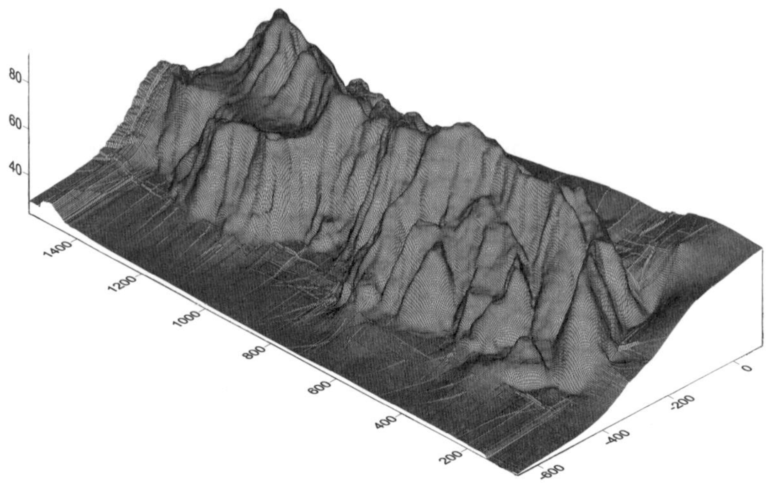

Figure 7. The block diagram before development
(Z-axis (height) is scaled 5.5 times X and Y-axis)

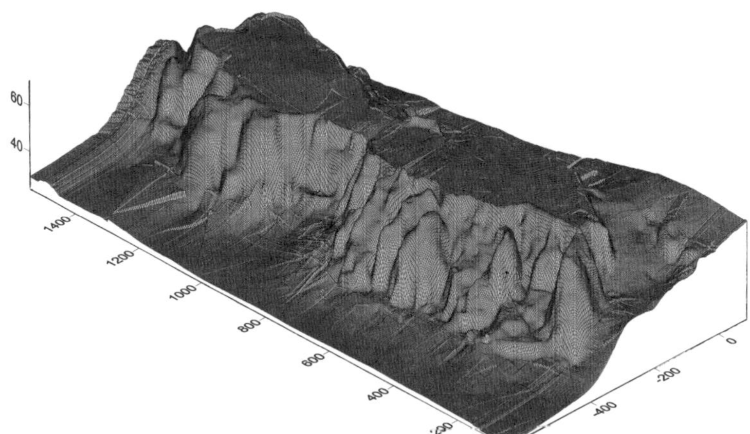

Figure 8. The block diagram after development
(Z-axis (height) is scaled 5.5 times X and Y-axis

surface and the terrace scarp. Accordingly, the valley fill on such area collapsed severely, and the outer roads as well as the residential sites adjacent to it were damaged. Figure 7 and 8 show the topographies before and after the development, by which it is found easily that the mountaintop was planalized. Figure 9 shows the damaged building sites and roads. It is understood that many areas close to the outer roads, which were knick points, were affected and heavy collapse occurred on the valley fill. It is desirable, hereafter, to emphasize the need for paying attention to the location (or landform) of valley fill, with earthquake-resistant design using dynamic analysis.

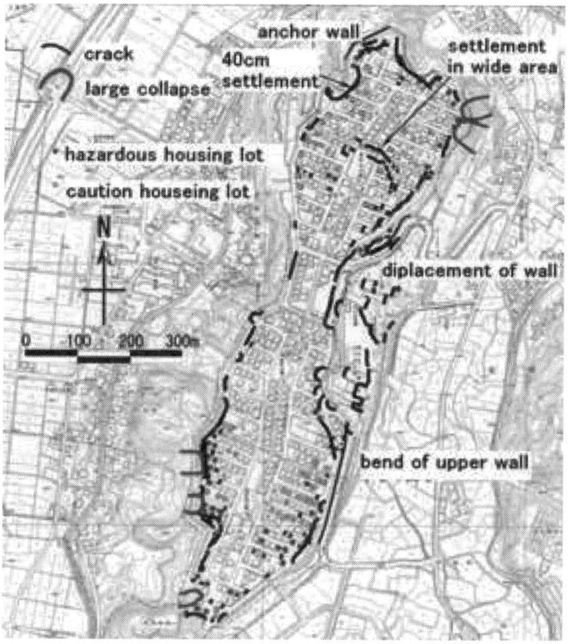

Figure 9. Damage distribution of Takamachi apartment block

THE METHOD FOR DISCRIMINATING HAZARDOUS HOUSING LOTS BASED ON THE FACTS OBTAINED IN THE KOBE EARTHQUAKE

In Kobe Earthquake, there was the rash of collapse and displacement at the valley fill. Whether a current housing lot is cut land or fill can be found out easily with the cut and fill drawings of the time, or topographic maps or aerial photographs before development. Without these sources, exploration on the scene makes the judgement possible. In case the onsite exploration is unhelpful, the depth and area can be figured out by simple research: Swedish sounding, static or dynamic cone penetration test, etc.

The damage patterns of the housing lots on valley fill in the Kobe Earthquake are categorized into 3 groups:

(1) Slumping collapse occurs at the ground.
(2) The ground does not go far enough to collapse but deforms.
(3) The residential retaining walls are damaged.

The method for discriminating residential sites at risk of these damages is described:

(1) The fills at risk of slumping collapse in its ground are located where:
 1) the relevant building site does not have the application of "the Residential Land Development Control Law," 2) the depth of the fill slope is more than 15 m, 3) the groundwater level is high, 4) the intensity of the fill material is low, and 5) the soft talus slope sediment exists along the valley.
(2) The fills whose ground do not go far enough to collapse but at risk of displacement are located where:

1) the relevant building site does not have the application of "the Residential Land Development Control Law," 2) the groundwater level is high, 3) sandy soil is distributed, and 4) the ground is not solidified enough.

(3) The fills on which the residential retaining wall is at risk of damage are located where:

1) the retaining wall does not have the application of "the Residential Land Development Control Law," 2) the retaining wall is not up to the standard of "the Residential Land Development Control Law," and 3) the bearing power of the foundation ground below the retaining wall is not sufficient.

In the housing lots that have these characteristics, the high risk of earthquake damage can be pointed out.

THE NEED FOR THE INTRODUCTION OF THE GROUND DATA BASE

The hazardous housing lots described above are at high risk of earthquake damage, but if their locations are found out beforehand, the enforcement of quake-resistant construction is possible. Under the present situation, however, the locations of fills are not identified and site investigation is the only way for it in most cases. On the other hand, when ground condition is known by data base or something in advance, valley fill can be located and hazardous residential lots can be also identified. Moreover, using the boundary condition and input condition obtained by the ground data base makes dynamic analysis available. For this reason, it is highly important to construct ground data base in the future.

In Kobe City, the wooden houses damaged by the Kobe Earthquake were distributed in the shape of a belt. In order to clarify the reason, the digitized ground data base "Kobe JIBANKUN" was constructed with collaboration from Kobe City, using the data of about 4,000 borings and the result obtained at 5 cross sections by Physical Reflective Prospecting Method. Planar GIS and Section GIS are utilized for it. The dynamic analysis by this ground data base made it clear that the boundary condition of the deep underground had effected the belt-shaped emerging of the damage distribution.

Once the ground data base is constructed, it enables not only the simulation for investigating what caused the past facts but the prediction of the damage in future earthquakes. On the Japanese government's own showing, the Toh-Nankai Earthquake and the Nankai Earthquake will occur with probability 60 % and 50 %, respectively, within the next 30 years. They also announced these earthquakes' predictive time history waveforms, with which the predictive velocity in future earthquakes can be calculated. In this kind of prediction method in Japan, ground condition is estimated from land-use condition, and then dynamic analysis using the ground condition is made. Regarding precision criteria, however, what is obtained by the ground data base is far more reliable. Other researchers have conducted the prediction of lateral flow in case of seaside structures also. It is important to make quake-resistant design through such analysis in future, and therefore the input of comprehended ground conditions into the data base is greatly expected hereafter.

The presentation will introduce the high-density ground data base constructed in Kobe, the dynamic analysis of future earthquakes using it, and the dynamic analysis that was conducted at Nagoya Port.

International Conference on Geotechnical Engineering for Disaster Mitigation & Rehabilitation
Chu, Phoon & Yong (eds) © 2005 World Scientific Publishing Company ISBN 981-256-469-1

SEISMIC GROUND MOTION SITE REPONSE EFFECTS FOR REGIONS OF MODERATE SEISMICITY

JACK PAPPIN
Ove Arup and Partners Hong Kong Ltd., Level 5, Festival Walk, 80 Tat Chee Ave., Kowloon Tong, Hong Kong

MATTHEW FREE & JULIET BIRD
Arup, 13 Fitzroy Street, London, W1T 4BQ

The methodology used for the classification and zonation of the ground conditions in a region of low to moderate seismicity and the development of the region-specific seismic ground motion site amplification factors is presented. Following NEHRP guidelines, and using published geological maps and a large database of geotechnical information, the Hong Kong region was sub-divided into five site classes. One-dimensional site response analyses were then undertaken for representative ground profiles in each site class to investigate expected amplification at different hazard levels, and the potential variability. The results indicate that the NEHRP recommended short and long period site amplification factors do not provide an optimum representation of the potential amplification characteristics for the moderate seismicity appropriate to Hong Kong. The incorporation of site amplification and its variability into the seismic demand curves for a regional loss model are described.

INTRODUCTION

The input to a regional seismic loss estimation model can be divided into two parts: the *demand* i.e. the seismic hazard data, and the building and infrastructure *inventory* data, including performance and occupancy classifications. Seismic hazard demand comprises ground shaking hazard and secondary collateral hazards (i.e. tsunami, liquefaction, landslide, fire, inundation and release of hazardous materials). Since these secondary hazards (apart from the tsunami hazard) are dependent on the strength of ground shaking, it is clear that the initial correct assessment of the strength of surface ground shaking including its potential variability, is of fundamental importance to understanding and interpreting the results of the study. This paper presents a case study of the methodology used for classification, zonation and definition of surface ground shaking in the low to moderate seismicity region of Hong Kong. The particular features of a regional loss estimation model are taken into account, whereby the constraints of model size versus time and cost restraints mean that the available data are likely to be significantly less detailed than for a site specific study.

SITE AMPLIFICATION ACCORDING TO THE NEHRP GUIDELINES

Most seismic design codes include site classification schemes, with factors to apply to the design rock spectra. The site classifications and soil amplification factors proposed in the NEHRP Provisions for the seismic rehabilitation of buildings (FEMA, 1997) are widely used both in the United States and elsewhere. Amplification factors are presented for both the short and long period ranges of the spectra, as opposed to many other codes which adopt only a single factor and thus fail to capture either period dependence or non-linearity in the site amplification.

Dobry *et al.* (2000) describe the derivation of the NEHRP amplification factors, which originate from a combination of empirical data (response spectral ratios of soil to rock sites from the 1989 Loma Prieta earthquake) and numerical site response analyses.

The use of site classification schemes such as the NEHRP one is well suited to regional hazard studies, since the zoning can be undertaken using a manageable level of input data and the amplification factors may be applied to the bedrock spectra using simple algorithms in database or GIS environments. However, given the potential importance of site effects, there are some inconsistencies in this approach. The NEHRP scheme is based upon the properties of the top 30m of the soil only and therefore can overlook particular features of the deep geology. Furthermore, these amplification factors have a very significant scatter associated with them, due to the complex behaviour of soils under cyclic loading, and the need to group many sites within a single classification.

SITE AMPLIFICATION FOR A SEISMIC RISK MODEL

There is a growing awareness that regions of low to moderate earthquake hazard may face large potential losses. For example, the Hong Kong Special Administrative Region of China (Hong Kong), is in a region of low to moderate seismicity but has a potentially very high exposure to losses due to its large population density (6.7 million inhabitants), its exceptional concentration of high rise buildings and its important economic status.

Seismic hazard

A recent probabilistic seismic hazard assessment (PSHA) for Hong Kong has been published by Free *et al.* (2004) and the results of this evaluation are used as the basis for the analyses presented in this paper. The results of the PSHA are presented in terms of uniform hazard bedrock response spectra, where bedrock was defined as NEHRP Site Class A (FEMA, 1997) or 'Hard Rock', with a shear wave velocity of 1500 m/s, in accordance with measured data from Hong Kong. Three probabilities of being exceeded were considered, namely 50% in 50 years, 10% in 50 years and 2% in 50 years.

Site classification

The subsoil conditions in Hong Kong have been categorized according to the NEHRP Site Classes A to E, as presented by Pappin *et al.* (2004). Surface geology in Hong Kong typically comprises either igneous or sedimentary rocks, in varying states of weathering, Quaternary age alluvial or colluvial deposits, or reclamation areas. A database of over 1,200 boreholes was used, in conjunction with geological maps, to develop a site classification map for the territory of Hong Kong.

The main requirements of a site classification system are to provide a logical set of criteria from which to zone a study area, and to group similar soil profiles, such that they can be represented by single amplification factors. In this case, the NEHRP system was adopted, and the uncertainty and variability related site amplification, including the deep geology, was incorporated into the site amplification factors, as described below.

As noted by Pappin *et al.* (2004), site class boundaries were found to coincide with the geological boundaries across the territory. For example, the boundary between Site Classes D and C broadly agrees with the boundary between Quaternary and Pre-Quaternary deposits. The boundary between Site Classes C and B was found to be related to the presence of a layer of residual soil over bedrock. This layer is generally present on shallower slopes, and absent where the gradient is steeper. Site Class E, as would be expected, coincides with areas of thick reclamation fill or alluvial deposits.

Soil amplification

Table 1 compares the seismicity, in terms of spectral accelerations, used in the NEHRP guidelines with that of Hong Kong presented by Free *et al.* (2004). The NEHRP guidance presents a single amplification factor for all accelerations below the minimum value, and in most cases in Hong Kong, the spectral accelerations fall below this value. The amplification at low accelerations is therefore an important consideration for Hong Kong with respect to reducing uncertainties.

Table 1: Comparison of Hong Kong short and long period bedrock spectral accelerations, at 0.2 s and 1.0 s periods (S_{AS} and S_{AL}) respectively, with the ranges used for the NEHRP amplification factors.

Probability of Exceedance in 50 years	Spectral Accelerations (g)			
	S_{AS} Hong Kong	S_{AS} NEHRP	S_{AL} Hong Kong	S_{AL} NEHRP
50%	0.08	≤ 0.25	0.02	≤ 0.1
10%	0.23	≤ 0.25	0.04	≤ 0.1
2%	0.48	0.25 – 0.50	0.08	≤ 0.1

Rather than use the NEHRP factors, a number of 1-dimensional site response analyses have been undertaken for a range of typical soil profiles and input parameters for each site class. The results were used to determine ground-motion amplitude and period dependent factors for each site class. These factors have the advantage of being specific to Hong Kong in terms of the local soil conditions and the amplitude of the reference input motion. Furthermore, by developing new regional site amplification factors, the uncertainty associated with them was able to be evaluated.

Site response analyses

A suite of representative soil profiles was compiled to represent the range of ground conditions within each site class. In each case the profile extended to bedrock, defined as moderately decomposed rock. Earthquake strong-motion records were used that agreed with Magnitude-distance combinations obtained from disaggregation of the uniform hazard response spectra for short and long period regions of the UHRS. These were then artificially adjusted to match the shape of the appropriate target spectra.

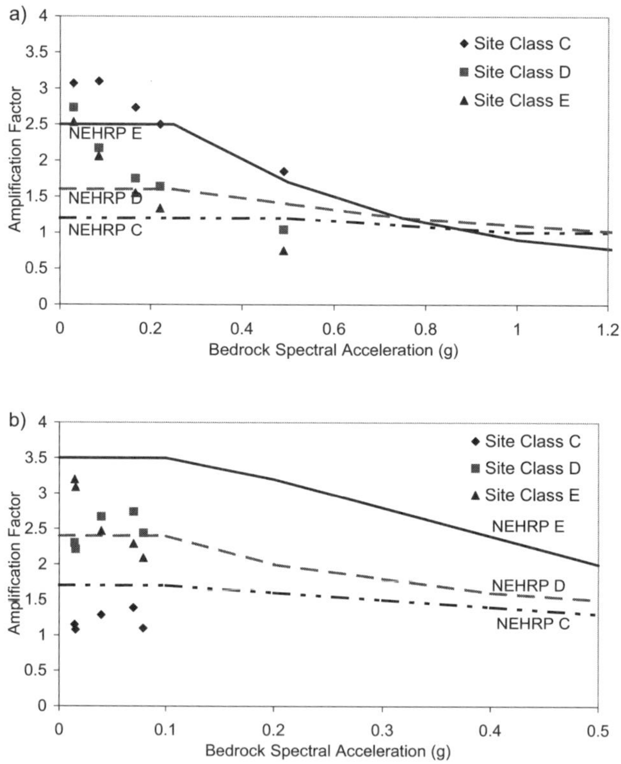

Figure 1. Comparison of computed average site amplification factors for Site Classes C, D and E for Hong Kong soil profiles and seismic hazard levels, compared with the NEHRP factors, (a) short-period, S_{AS}, (b) long-period, S_{AL}

One dimensional site response analyses were carried out using the Arup in-house software SIREN (Heidebrecht *et al.*, 1990). The computed surface response spectra (5%

damped) were used to obtain spectral ratios for each soil profile. Uncertainty in soil parameters, primarily the shear modulus degradation curves, was incorporated by including sensitivity studies within the suite of analyses.

Site amplification factors

Figure 1 compares the average computed site amplification factors to those presented in the NEHRP guidelines. At short periods, the Hong Kong Site Class C factors were predicted to significantly exceed the NEHRP factors whereas the Hong Kong factors are less for Site Class E. At long periods, the NEHRP factors are larger than the calculated Hong Kong factors, with the exception of Site Class D which shows reasonably good agreement. In general, the agreement between NEHRP and the Hong Kong specific factors is poor, justifying the decision to derive new factors. The uncertainty associated with the use of the NEHRP factors, which were found to overestimate amplification by a factor of more than 2 in some cases, could overshadow many of the other uncertainties in this type of regional risk study.

The differences shown in Figure 1 are attributed to a number of factors, including the fundamental differences in the characteristics of the input rock ground motion (particularly the frequency content and the amplitude) and the differences in the near surface geology and associated non-linear site effects.

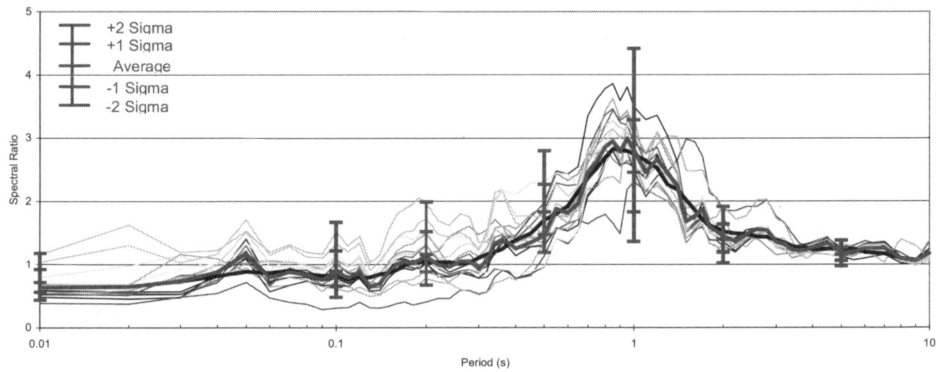

Figure 2. Computed spectral ratios for selected Site Class D soil profiles, showing mean and standard deviations (σ). The input motion in this case was for a Mw 6.3 at 50 km, scenario, scaled to match the appropriate 2% in 50 year spectrum at short periods.

Variability in site amplification

There was a significant scatter in the computed site response spectral ratios, of which the mean is shown in Figure 1. This scatter is associated with the large number of representative soil profiles used for each site class, and is therefore typical of any regional study where there is a need to group soil profiles into a limited number of classes, when, in fact, each profile will respond differently to ground shaking. This variability was incorporated into the regional seismic hazard model for Hong Kong by considering the mean and 1 and 2 standard deviations above and below the mean for the site amplification factors, as illustrated in Figure 2.

DISCUSSION

In many regional seismic loss estimations, site amplification is represented using the NEHRP classifications and factors. The HAZUS methodology (FEMA, 2003) uses this approach, and this is used as the basis for many studies both within and outside the US.

However, as has been demonstrated in this paper, in many regions with different soil characteristics and seismicity to those used as the basis for the NEHRP guidelines, there is potential for significant errors to arise from this approach. This is related to the differences in the characteristics of the input ground motion and in the near-surface geology. Furthermore, non-linear site response can be highly variable between different sites, and the deeper geology can also influence this.

The methodology presented herein, presents significant advantages in terms of reducing the uncertainty associated with the definition of the demand and also in allowing an improved representation of the ground-motion uncertainties that will inevitably be present in any regional model of this kind.

REFERENCES

Dobry, R., Borcherdt, R.D., Crouse, C.B., Idriss, I.M., Joyner, W.B., Martin, G.R., Power, M.S., Rinne, E.E., & Seed, R.B. (2000). New site coefficients and site classification system used in recent building seismic code provisions. *Earthquake Spectra* 16 (1): 41-67.

FEMA (1997) FEMA-273 & FEMA-274. *NEHRP guidelines for the seismic rehabilitation of buildings*. Federal Emergency Management Agency, Washington, D.C.

FEMA (2003) *HAZUS-MH Technical Manuals*. Federal Emergency Management Agency, Washington, D.C.

Free, M.W., Pappin, J.W. & Koo, R. (2004). Seismic hazard assessment in a moderate seismicity region, Hong Kong. *13th World Conference on Earthquake Engineering*, Vancouver, Canada. Paper No. 1659.

Heidebrecht, A.C., Henderson, P., Naumoski, N. & Pappin, J.W. (1990). Seismic response and design for structures located on soft clay sites. *Canadian Geotechnical Journal*. 27 (3): 330-341.

Nordensen, G.J.P., Deodatis, G., Jacob, K.H. & Tantala, M.W. (2000). Earthquake loss estimation for the New York City area. *12th World Conference on Earthquake Engineering* Auckland, New Zealand, Paper No. 0605.

Pappin, J.W., Free, M., Bird, J. & Koo, R. (2004). Evaluation of site effects in a moderate seismicity region, Hong Kong. *13th World Conference on Earthquake Engineering*, Vancouver, Canada. Paper No. 1662.

International Conference on Geotechnical Engineering for Disaster Mitigation & Rehabilitation
Chu, Phoon & Yong (eds) © 2005 World Scientific Publishing Company ISBN 981-256-469-1

EVALUATION OF TRIGGERING ACCELERATION REQUIRED FOR COMMENCEMENT OF LIQUEFACTION

SNEHAL R. PATHAK
Department of Civil Engineering, Govt. College of Engineering
Pune, 411005, Maharashtra State, INDIA

DHANANJAY R. PHATAK (Retd.)
Department of Civil Engineering, Govt. College of Engineering
Pune, 411005, Maharashtra State, INDIA

Earthquake induced liquefaction is one of the most severe seismic hazards that can damage structures founded on both shallow and deep foundations. Significant amounts of damages have been directly or indirectly attributed to the liquefaction phenomenon in several earthquakes. Assessment of liquefaction potential of sandy soils has attracted considerable attention over the past few decades. Number of techniques have been developed to estimate the likelihood that liquefaction may or may not occur at a given site under specified conditions. Richards et. al. (1999) have proposed a new concept of triggering acceleration for commencement of liquefaction. Authors of this paper have carried out the work to investigate a method for evaluating the triggering acceleration for the initiation of liquefaction in fully saturated sands. Once the initial fluidization occurs, the liquefaction is inevitable which ultimately results into considerable damages. The present work deals with the evaluation of triggering acceleration indicating initiation of liquefaction. Authors have developed a method to compute this acceleration based on model already established by them (1999); which has been used to demarcate "yes" and "no" zones of liquefaction in terms of earthquake magnitude, standard penetration values, initial overburden stress on level ground and epicentral distance.

INTRODUCTION

Liquefaction is one of the most severe seismic hazards that can damage structures founded on both shallow and deep foundations and disrupt buried lifelines with potentially catastrophic consequences for water supply and fire following earthquakes. Assessment of liquefaction potential for sandy soils has attracted considerable attention over the past 30 years. A number of techniques have been developed to estimate the liklihood that liquefaction may or may not occur at a given site under specified conditions. Number of mathematical/numerical models are developed by Davis and Berill(1982), Trifunac (1995), Zhang (1998) and Phatak and Pathak(1999), using the field data for prediction of liquefaction.

Phatak and Pathak (1999) have developed a model for the determination of sites prone to liquefaction based on the seismic energy methodology using Gutenberg-Richter (1956) relation for energy released from an earthquake attenuated for geometric spreading (Kayen and Mitchell,1999). According to this methodology, for an earthquake of magnitude, M, the total radiated energy E is, $E = 10^{1.5M + 1.8}$

Only a fraction of this energy will be contained in seismic waves reaching the site. Some energy will be dissipated by an elastic attenuation along ray paths, and further attenuation occurs due to geometric spreading. The most simple energy attenuation model for spherically expanding waves is an inverse relationship based on the square of the epicentral distance. Pore pressure increase Δ u, will be determined from the dissipated energy density within the soil. The factors responsible to cause liquefaction can be considered to be earthquake magnitude, M, source to site distance, R, effective overburden stress, $\bar{\sigma}_o$ and standard penetration values, N.

Richards proposed a concept of triggering acceleration for commencement of liquefaction. The equation proposed indicates that, acceleration required for initial fluidisation is a sole function of the friction angle of the soil. However, authors of this paper have shown that this need not necessarily be true. Triggering acceleration which initiates liquefaction will be a function of the parameters responsible to cause liquefaction as stated above.The purpose of this work is to investigate a method for evaluating the triggering acceleration required for the commencement of liquefaction in fully saturated sands, in terms of these parameters.

METHOD

The model "A" separating the "yes" and "no" zones of liquefaction developed by the authors has been used for estimating triggering acceleration. For this purpose, 57 case histories of liquefaction and no liquefaction were collected from the data of Davis and Berrill (1982).

The following steps are carried out:

1. The relation between Liquefaction Potential $\left(10^{1.5M} / \bar{\sigma}_o^{3/2} . r^2\right)$ and corrected SPT blow count ($\bar{N}$) as obtained in model "A" is given by the equation of the straight line separating the 'yes' and 'no' zones of liquefaction as,

$$\log_{10} \frac{10^{1.5M}}{\bar{\sigma}_o^{3/2} . r^2} = 7.3825 \log_{10} \bar{N} - 9.6375 \qquad \text{---(1)}$$

where,

M	=	earthquake magnitude in Richter scale
r	=	epicentral distance (km)
$\bar{\sigma}_o$	=	effective overburden stress (kPa)
$\bar{N}$	=	corrected SPT blow count
	=	C_N . N

where, N = raw SPT blow count

C_N =overburden correction factor given by Seed et al

$$C_N = 1 - 1.25 \log_{10} \left\{ \frac{\bar{\sigma}_o\,(kPa)}{107.6} \right\}$$

2. Rearranging the terms of equation (1) and obtaining the expression for M on left hand side,

$$1.5\,M = 7.3825 \log_{10} \bar{N} + 1.5 \log_{10} \bar{\sigma}_o + 2 \log_{10} r - 9.6375 \qquad \text{---(2)}$$

3. The values of M calculated from equation (2) for validated cases from 57 case histories are compared with actual M in the database (Fig. 1).
4. Similarly, corrected SPT blow count N can be obtained by substituting the values of M

calculated from equation (2) and $\bar{\sigma}o$ and R from the database as,

$$7.3825 \log_{10} \bar{N} = 1.5 M - 1.5 \log_{10} \bar{\sigma}_o - 2 \log_{10} r + 9.6375 \qquad \text{---(3)}$$

5. From these corrected SPT blow count ($\bar{N}$), the raw SPT blow count (N) values are calculated using the relationship given by Seed et al (1979).
6. The raw SPT blow count (N) values thus calculated are observed to be the same as those in the database (Fig. 2) thus satisfying the equation (1).

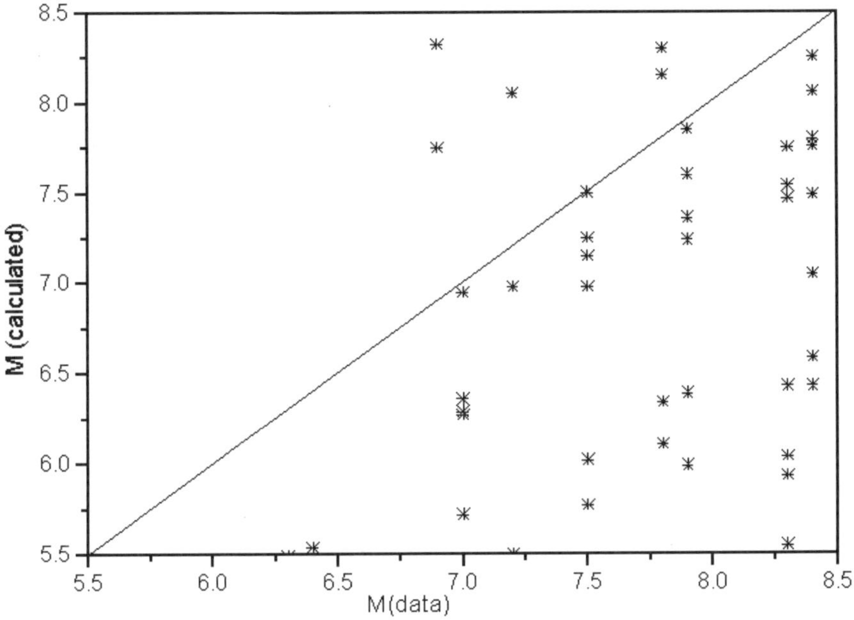

Fig. 1. Graph showing M(calculated) versus M(data)

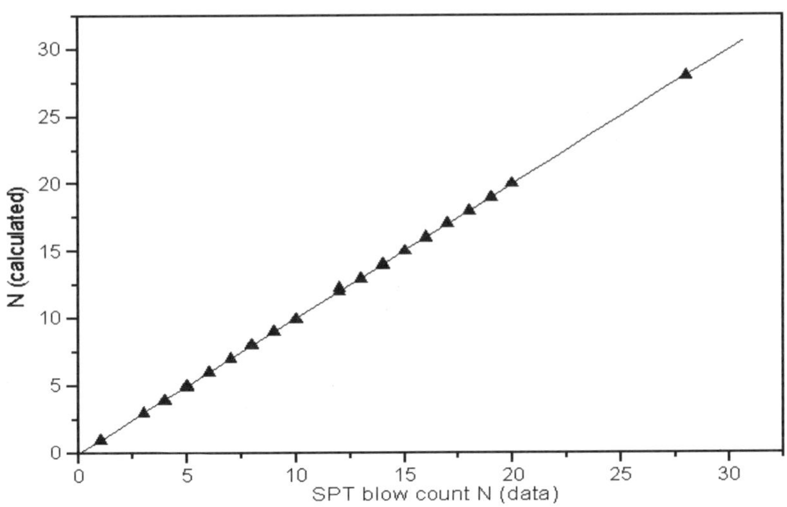

Fig. 2. Graph showing N(calculated) versus N (data)

7. A plot of Log $\left(10^{1.5M} / \overline{\sigma}_o^{3/2} . r^2\right)$ and Log N (Fig 3) for the 57 case histories not only separates the "yes"and "no"zones of liquefaction but the points on the line indicate triggering acceleration.

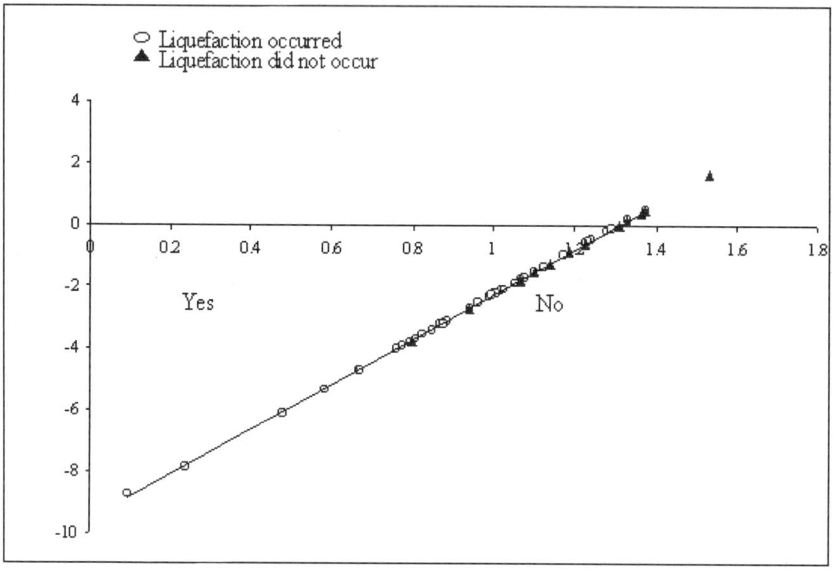

Fig. 3. Plot showing processed data for "yes" and "no" liquefaction cases

8. Thus the boundary separating the liquefaction and no liquefaction is defined by the equation,

$$\log_{10} \frac{10^{1.5M}}{\overline{\sigma}_o^{\;3/2} . r^2} = 7.3825 \log_{10} \overline{N} - 9.6375 \qquad \text{--- (4)}$$

where,

M = Earthquake magnitude calculated from equation (3)
$\overline{\sigma o}$ = Effective overburden stress (kPa)
r = Epicentral distance in km
$\overline{N}$ = Corrected SPT blow count

9. The values of PGA are obtained using the Gutenberg-Richter (1956) peak ground acceleration (PGA) attenuation relation with actual M values as,

$$\log a\,(g) = \log F_a - 2.1 + 0.81\,M - 0.027\,M^2 \qquad \text{--- (5)}$$

where, F_a is given by following relations

$\log F_a = -\,1.029 - 1.725 \log R;$ $\quad R \leq 75$ km
$= -\,3.115 - 0.634 \log R;$ $\quad R > 75$ km

where, M = Earthquake Magnitude in Richter scale
R = Epicentral distance (Km)

10. Thus, substituting the M calculated from equation (4) which is required for initiation of liquefaction, equation (5) will give the acceleration which will trigger liquefaction.
It has been observed that the values of this acceleration are less than the PGA values as expected and for the cases where liquefaction has not occurred, the triggering acceleration value is more than the PGA value thus validating the method developed.

PROOF CHECKING OF THE METHOD

The method developed was checked on totally new data.collected from Niigata-1964, Tokachi-Ken Oki-1968, and Loma-Prieta-1989 earthquakes for the sites where liquefaction has occurred. The data contains earthquake magnitude, M, epicentral distance, r, effective overburden stress, $\bar{\sigma}_o$, and corrected standard penetration values, $\bar{N}$.
For processing of data, following steps are carried out :
1. The liquefaction potential values (L.P) are obtained using model `A' with actual M in the database.
2. The corrected SPT blow count values ($\bar{N}$) are obtained using overburden correction factor given by Seed 1979.
3. Based on the method developed herein, values of M are calculated.
4. The values of PGA are obtained using the Gutenberg-Richter relation with actual M values.
5. .Substituting the M calculated in the Gutenberg-Richter PGA attenuation relation, triggering acceleration is obtained which is required for initiation of liquefaction.
Table 1 shows the processed data. From the table it can be seen that for 'yes' liquefaction cases the triggering acceleration value is less than the PGA value, thus proves reliability of the method.

Table 1. Earthquake cases with 'Yes' liquefaction

Sr. No.	Earthquake	Site	M	a_{max}	Trigging acceleration model A	N	Liquefaction
1.	Niigata (1964)	kwagashicho	7.5	0.0170	0.0029	7.24	Yes
2.	Niigata (1964)	kwagashicho	7.5	0.0170	0.0059	9.0	Yes
3.	Niigata (1964)	kwagashicho	7.5	0.0170	0.0079	9.3	Yes
4.	Niigata (1964)	kwagashicho	7.5	0.0170	0.0085	10.17	Yes
5.	Tokachiken-Oki (1968)	Hachinoe	7.8	0.0581	0.0058	9.5	Yes
6.	Loma-Prieta (1989)	San Francisco Oka - Bay	7.5	0.0690	0.0059	10.0	Yes

CONCLUSION

In order to assess the possibility of liquefaction hazards, it is necessary to determine triggering acceleration which is a function of earthquake magnitude, M, epicentral distance, r, initial effective overburden stress, $\bar{\sigma}_o$, and corrected Standard Penetration values, $\bar{N}$. Authors have developed a method to predict this acceleration which could trigger the liquefaction. Thus to minimize the liquefaction hazard, this acceleration should be obtained for the sites prone to liquefaction, in addition to PGA.

REFERENCES

Davis, R.O. and Berrill, J.B. (1982). "Energy Dissipation and Seismic Liquefaction in Sands", *J. Earthquake Engg. and Structural Dynamics,* 10, 59-68.

Phatak, D.R. and Pathak, S.R. (1999). "Assessment of Liquefaction Potential During Earthquakes by Arias Intensity : *Discussion, J. Geotech. Engg., ASCE*, 125(7),626-627.

Phatak, D.R. and Pathak, S.R. (2000). "Seismic Earth Pressure on Retaining Structures", *Discussion, J. Geotechnical Engineering, ASCE,* 126(12), 1216-1217.

Richard, Rowland Jr., Huang, C. and Fishman, K. L. (1999). "Seismic Earth Pressure on Retaining Structures", *J. Geotechnical Engineering, ASCE*, 125(9), 771-778.

Seed, H. B.(1979). "Soil Liquefaction and Cyclic Mobility Evaluation for level ground during earthquakes*", J. Geotech. Engg., ASCE* ,105 (2), 201-255.

Trifunac, M. D. (1995). "Empirical Criteria for Liquefaction in Sands Via Standard Penetration Tests and Seismic Wave Energy", *J. Soil Dynamics and Earthquake Engineering*, 14(6), 419-426.

Zhang, Liayang (1998). "Assessment of Liquefaction Potential Using Optimum Seeking Method", *J. Geotechnical Engineering, ASCE,* 124(8), 739-748.

International Conference on Geotechnical Engineering for Disaster Mitigation & Rehabilitation
Chu, Phoon & Yong (eds) © 2005 World Scientific Publishing Company ISBN 981-256-469-1

SEISMIC MICROZONATION OF DELHI REGION

K.S. RAO
Department of Civil Engineering,
Indian Institute of Technology Delhi, New Delhi, INDIA

D.NEELIMA SATYAM
Department of Civil Engineering,
Indian Institute of Technology Delhi, New Delhi, INDIA

The national capital region of Delhi, India attracted major attention of several scientific studies after the 2001 Bhuj earthquake. Since Delhi falls in the zone IV (IS 1893:2002) with high seismic activity, there is a great need for site characterization and seismic microzonation of the area. Multi Channel Analysis of Surface Wave (MASW) tests were carriedout in Delhi at 118 sites in predefined grids of 2kmX3km each to estimate the shear-wave velocities. These tests were carried out using 48 channel digital seismograph and the collected data was analyzed using SeisImager/SW software. Also, the average shear wave velocity up to 30m (V_{S30}) is estimated which is used for site characterization. Based on the shear wave velocity and density of the sub strata, soil amplification is estimated using DEGTRA software and the microzonation map is generated. Delhi is divided into four zones e.g., ZA, ZB, ZC1 and ZC2 based on V_S and soil amplification.

INTRODUCTION

Earthquake is one of the most feared natural disasters because of uncertainties involved in predicting it. Seismic microzonation, which is a subdivision of an area into smaller zones depending upon site-specific seismic response, is an effective mitigation method. In most general terms, seismic microzonation is the process of estimating the response of soil layers under earthquake loading. It is a well-known fact that each soil type responds differently, when subjected to the ground motions due to an earthquake. A qualitative and quantitative estimation of site effects is often expressed by the amplification factor and resonance/fundamental frequency, which depends upon the properties of the soil and depth of the bedrock. The site response parameters can be used to distinguish regions where the seismic hazard will be high due to an earthquake of particular magnitude. It is observed that large concentration of damage in specific areas during an earthquake is due to site dependent factors related to surface geological conditions and local soil altering seismic motion. Various studies have demonstrated the ability of geological and geotechnical conditions in altering the seismic motion. Amplification of ground motion refers to the increase in the intensity of ground shaking that can occur due to local geological conditions, such as the presence of soft soils (Kramer, 1996). Surficial materials overlying the bedrock surface can either amplify or deamplify the bedrock ground motions by the time they reach the ground surface. Therefore, to better estimate the shaking and damage that may occur to super structures during an earthquake, it is necessary to modify the bedrock motions with soil amplification values. Amplification of seismic motions under different soil deposit can be significant, especially at sites with soft soil conditions.

Shear-wave velocity, V_S is an important parameter for evaluating dynamic behavior of soil. Multi channel analysis of surface waves (Park et al., 1999, Xia et al., 1999), is a non-invasive method developed to estimate shear wave velocity profile from surface wave energy. Measurements of phase velocity of Rayleigh waves of different frequencies can be used to determine a velocity depth profile. This MASW test is carried out in Delhi at various locations by spreading multiple geophones for estimating 2D Velocity profiles. Shear wave velocity of surface layers is a useful index property for evaluating site amplification. Investigations based on the observations and the analyses of ground motion have revealed that average shear wave velocity of surface soils to a certain depth shows strong correlation with the relative amplification. In this study soil amplification is estimated using DEGTRA software and a microzonation map with respect to soil amplification is generated which can be used for seismic hazard assessment.

MULTI CHANNEL ANALYSIS OF SURFACE WAVE (MASW) TESTING

MASW technique has been developed in response to the shortcomings of SASW. The simultaneous recording of 12 or more receivers at short (1-2m) to long (50-100m) distances from an impulsive or vibratory source gives statistical redundancy to measurements of phase velocities. Multichannel data displays in a time variable frequency format also allow identification and elimination of non fundamental mode Rayleigh waves and other coherent noise during the analysis process. The field setup, acquisition process and analysis procedure are explained below in detail.

Field Study

This MASW test is done in Delhi at various locations by spreading multiple geophones for estimating 2D Velocity profiles. Figure 1 shows the field set up of the MASW test. In this survey spacing between the geophones had been made equal to 6 meters. All the geophones are connected to the seismograph and the source is shifted with 6 m increment. An additional geophone is used as a trigger geophone to intialize the recording. The generated Rayleigh wave data are recorded at all shot points.

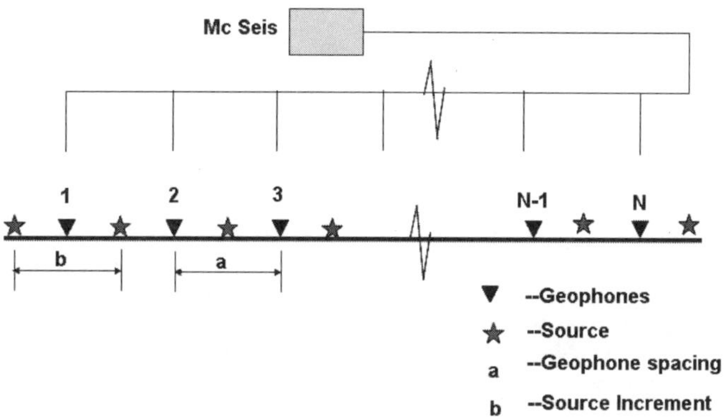

Figure 1. Sketch of MASW Test Spread

Data Acquisition

The 48 channel signal enhancement seismograph is used for acquisition of the seismic data. Geophones of 4.5 Hz are used and they are connected to the seismograph with connecting cables as shown in Fig. 1. A wooden hammer of 11 kg weight manually impacted on 165cm^2 aluminum plate is used for creating the energy. Fixed receiver configuration is used and the source is placed between each receiver and at both ends of the survey line. Around 13 shots are created at each test location in the present study. Data for each shot is digitally recorded and saved in the equipment. Figure 2 shows the multichannel records (shot gathers) of the geophones at JNU campus with source at mid point of the survey line. The acquired data is then transferred for the analysis. The following are the recording parameters (Table 1) during the acquisition process.

Table 1. Seismic Data Recording Parameters

Recording System:	McSEIS-SX 48
Sampling Interval:	500 μs
Memory:	2kb
Recording Format:	SEG-2
Pre Trigger:	ON
Stack Mode:	Summation
Geophones:	12 Geophones of 4.5 Hz frequency
Geophones Array:	Linear with geophone spacing of 6m
Source:	11kg sledge hammer on 165cm^2 metal plate
Source Array:	Source is shifted with 6m interval
Offset Distance :	3m

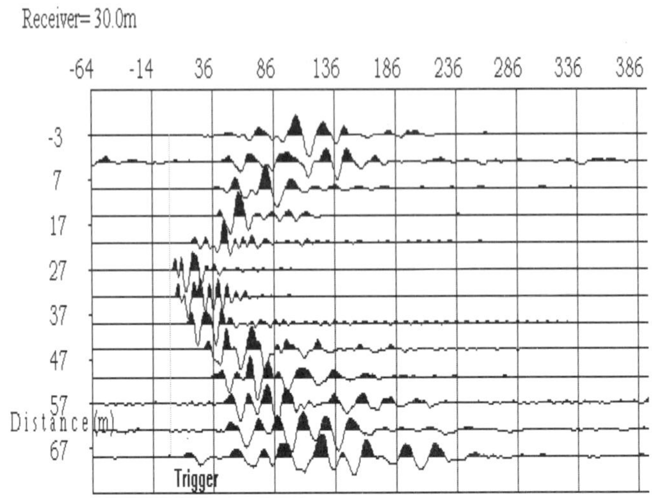

Figure 2. Multichannel Records (Shot Gathers) at JNU Campus with Source at Mid Point of the Survey Length

Analysis of the Data

Acquired surface wave seismic data is transferred to the Seisimager/SW software developed by OYO, Japan through spectral inversion to obtain 1D and 2D MASW shear wave velocity profiles. This SeisImager/SW software has three more softwares i.e., Pickwin95, WaveEq and GeoPlot. Figure 3 gives the flowchart with the detailed steps involved in the analysis of surface wave data for getting two dimensional shear wave velocity model.

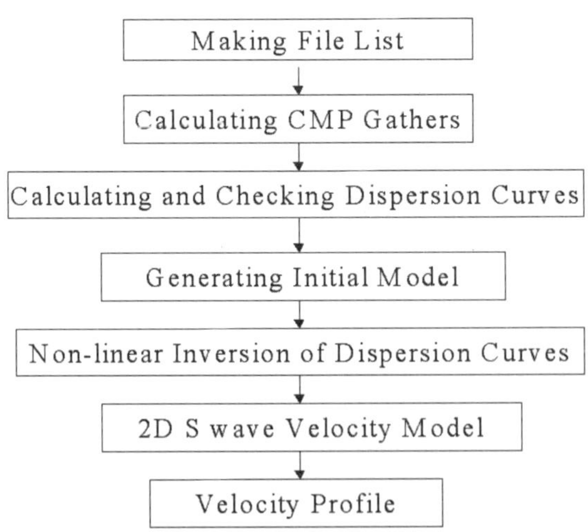

Figure 3. Flowchart for the Analysis of Surface Wave Data Using SeisImager/SW

The first step in the analysis is making the file list in which all waveform files and source receiver configuration are mentioned and then cross correlation CMP gather is calculated. Dispersion curves are calculated by converting it into frequency domain and then checked. Generation of a dispersion curve is one of the most critical steps for generating an accurate shear wave velocity profile. Dispersion curves are generally displayed as phase velocity versus frequency. This relation can be established by calculating the phase velocity from the linear slope of each component of the swept frequency record. Figure 4 shows the phase velocity curves of JNU campus. The 1D shear wave velocity profiles are calculated using non linear least square method using the dispersion data. Finally 2D velocity model can be generated in Geoplot. According to NEHRP guidelines, V_{S30} for a layered structure is determined for site classification.

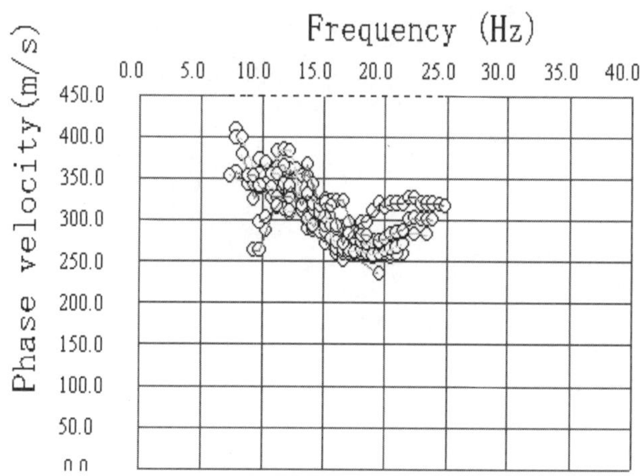

Figure 4. Phase Velocity Curves for JNU Site

SOIL AMPLIFICATION FACTOR

The most important geotechnical parameter which can be used to estimate the amplification factor is the average shear wave velocity. Shima (1978) found that the analytically calculated amplification factor is linearly related with the ratio of shear wave velocity of the surface layer to that of bedrock. When the bedrock shear wave velocity is found to be relatively constant over a wide area, the relative amplification in each locality can be obtained from the shear wave velocity of the surface layer. Various researchers proposed relations between the average shear wave velocity of surface layers and the relative amplification.

Estimation of soil amplification factor is very important in any seismic microzonation process. In this study the shear wave velocities are used to estimate the soil amplification factor (SA) using DEGTRA software. It is windows based computer program designed to process ground motion time histories in several applications. It has been developed at the Institute of Engineering, UNAM, Mexico. This program accepts the ground motion recordings, shear wave velocities and density of the soil strata up to bedrock depth as input parameters. In this study an earthquake input of 6.5 magnitude recorded on the hard rock site is used for the analysis. Amplified site response at the soft soil sites can be obtained using the input shear wave velocities. The ratio between the actual time histories at the hard rock site to the amplified site response at the soft soil site gives the amplification factor. This is estimated at various soil sites in Delhi and an amplification map is generated as shown in Fig. 5. The soil amplification factor (SA) is ranging from 1.03 (rocky sites) to 4.8 (soft soil sites) for Delhi region.

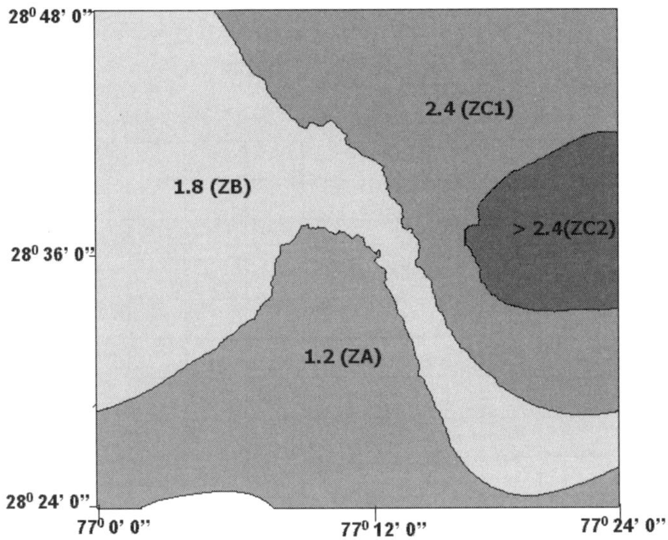

Figure 5. Seismic Microzonation Map of Delhi with Respect to Soil Amplification

RESULTS AND DISCUSSIONS

The two-dimensional velocity structure at all the 118 locations is generated and it is observed that the value of Vs is ranging from 400 to 480 m/s in the rocky sites, it is 120 to 280m/s in trans Yamuna region and 250 to 370m/s in western side of the area. Average shear wave velocity at 30m depth i.e., V_{S30} is also calculated and is ranging from 185 to 495 m/s for Delhi. Soil amplification factor is estimated from the measured shear wave velocities and density of the soil structure. Based on the SA and V_{S30}, microzonation of the Delhi is done by dividing the area into four zones i.e., ZA, ZB, ZC1 and ZC2. The areas in the south and south central part of Delhi (ZA) with weathered quartzite and gravelly deposits are characterized by very low amplification factor (≤ 1.2). The average shear wave velocity in this zone is grater than 350m/s. The western side of the Delhi with older alluvium has an amplification factor ranging from 1.2 to 1.8 (moderate amplification) and the V_{S30} for this zone falls between 250 t0 350 m/s. The soils in this zone are sandy silts and silty sands with clay seams. The north and north eastern side of Delhi has very loose sandy silts and silty sands (newer alluvium) and high water table depths. The amplification factor in these zones (ZC1 and ZC2) is quiet high due to high sedimentary thickness also the V_{S30} is less than 250m/s for these zones. It is clear that the amplification factor is high in the trans Yamuna region which can experience severe damage due to an earthquake of considerable magnitude. Similarly, the areas falling in ZA and ZB can have low to moderate damage respectively. A correlation between V_{S30} and SA for Delhi region is also developed as shown in Fig.6.

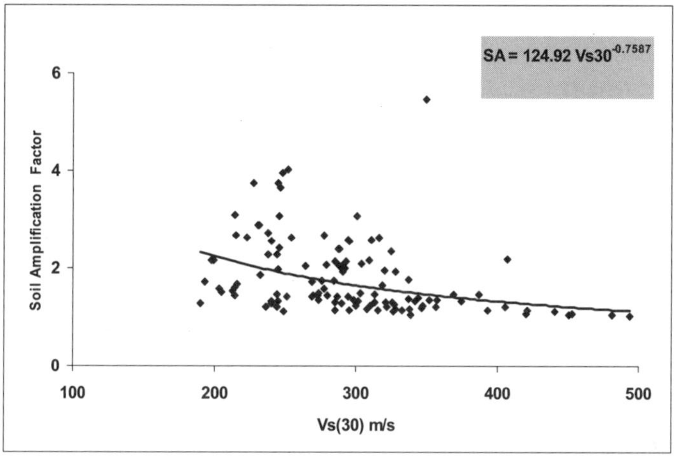

Figure 6. Correlation Between Soil Amplification Factor and V_{S30}

REFERENCES

IS: 1893-2002. "Criteria for Earthquake Resistant Design of Structures"

Building Seismic Safety Council (2000). "NEHRP Recommended Provisions for Seismic Regulations for new Buildings and Other Structures", Part 1: Provisions, *FEMA 38*, Washington, D.C.

Kramer, S.L. (1996) *"Geotechnical Earthquake Engineering",* Prentice Hall, New York

Park, C.B., R.D.Miller, and J.Xia, (1999). "Multi-channel Analysis of Surface Waves", *Geophysics,* V.64, No.3, 800-808.

Shima, E. (1978) "Seismic Microzoning Map of Tokyo", *Proc. 2nd International Conf. on Microzonation*, Vol. I, 433-443.

Xia, J., R.D.Miller, and C.B.Park, (1999). "Estimation of Near surface Shear Wave Velocity by Inversion of Rayleigh Waves", *Geophysics*, V.64, No. 3, 691-700.

International Conference on Geotechnical Engineering for Disaster Mitigation & Rehabilitation
Chu, Phoon & Yong (eds) © 2005 World Scientific Publishing Company ISBN 981-256-469-1

FORMATION OF THE MAP OF SHAKING OF THE TERRITORY IN ALMATY REGION

AITALYEV SHMIDT, BAIMAKHAN RYSBEK, SEINASINOVA ASIMA
Institute of Mechanics, Almaty city, Kazakhstan

DASHDORJ SARANDULAM
Hokkaido University, Sapporo city, Japan

The amplified map of shaking of the Almaty region with the center in Almaty city is formed. The map was based on the recent amplified data of the breaking geological structure. The space photographs and frequency of free vibrations of the territory of city Almaty relative to the Paleozoic foundation are used.

INTRODUCTION

Almaty is the largest city of Kazakhstan with its population of more than 1,300000 people. It was the capital of Kazakstan in 1928-1999. The city still remains the cultural, financial and business center of the country. The biggest tourist programs such as Winter Olympic games take a place in the city. At present an intensive construction is in progress, a stores of buildings goes up to 26, and also railway in metropolitan area is under construction.

The region of Northern Tien Shan where Almaty is in, is situated in high-seismic zone on the planet. Past three geological catastrophic earthquakes held in 1887, 1889 and 1991 took a place in Tien Shan region. The first and the third of those earthquakes have been outlined in the book of J.Aibi "Earthquakes" (Eiby, 1980). An estimation of earthquake level by different parts of the city territory is necessary for the construction with seismic stability against earthquake. Earlier, the map of shaking for a vast territory of the Northern Tine Shan was formed based on the improved method (Aitalyev, Baimakhan, Masanov, 2001; Baimakhan, 2002). The seismic stability of the underground structures in metropolitan area of Almaty has been examined (Aitalyev, Ukshebaev, Baimakhan, 2003).

In this paper the above mentioned method for the determination of shaking is used quite directly to the territory of the Almaty region. The new data of tectonic breaking of the city territory taken from the space are used also. The maps of shaking for different intensity are formed.

The frequency of free vibrations of precipitation thickness concerning about Paleozoic foundation is defined taking into account breaking and block structure of the earth crust based on finite element method.

MAIN BREAKS ON THE TERRITORY OF CITY ALMATY

The space print of the northern slope of Zailisky Alatau and a part of the territory of the city Almaty are shown in Figure 1.

Figure 2a show the main breaks which pass through the territory of Almaty, and geographic coordinates, main blocks of the earth crust, location of the three mountain rivers and the 2 great historic earthquake centers: Vernon of 1887 and Keminsk of 1911. Figure 2b shows the central part of the city which is crossed by the thick net of breaks (breaks are shown with bold lines, thin lines correspond to streets). Some main streets of the city and districts of expected intensity of vibrations of the earth crust in big scale earthquakes are also shown here. White color of the figure corresponds to the intensity of earthquake-8, gray shows intensity-9 where earthquake level is measured in scale of MSK-64. The symbol of a small plane indicates a territory of the international airport.

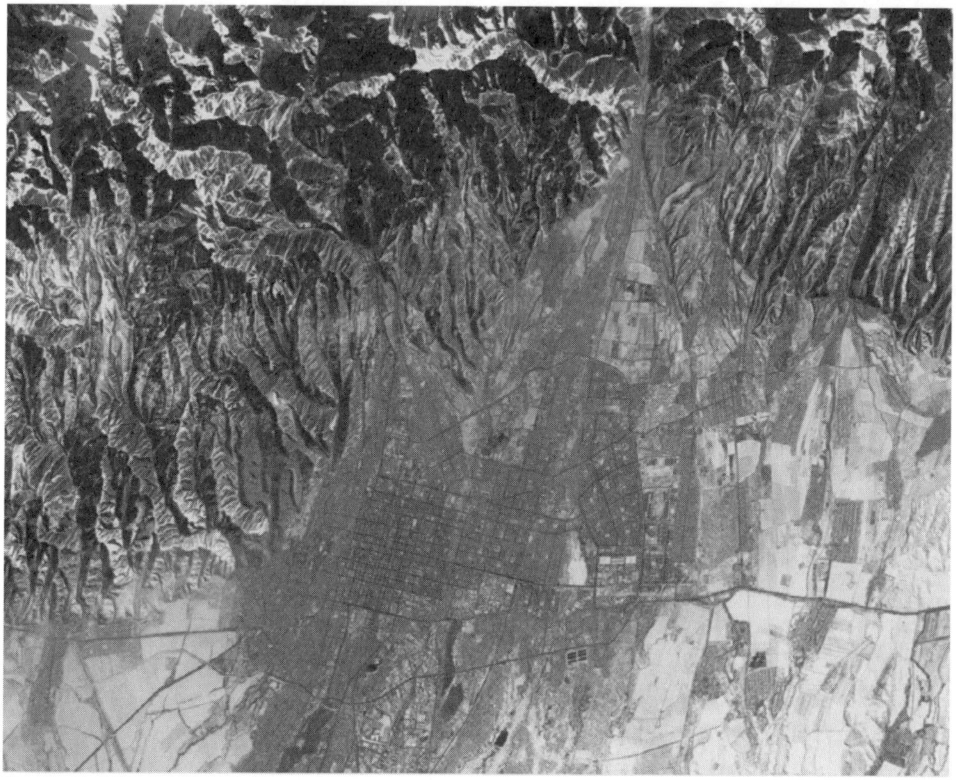

Figure 1. The space print of the Northern side of the mountain system of Zailisky Alatau and the territory of city Almaty.

FORMATION OF THE MAP OF SHAKING FOR STRONG EARTHQUAKES IN ALMATY REGION

For the formation of a map of shaking, the vast territory of the Northern Zhan has been considered. The territory of North-West, Southern Tien Shan and Western Dzhungar are partially taken. The earthquakes of these centers had an effect to territory of Almaty with considerable extent. According to the geographic coordinates the region between 74-82^0 on the east longitude and 40-45^0 on the north latitude is considered.

This huge region is divided into 8000 sub-districts with the area of 6 km x 8 km each. The meanings K_{max} for each sub-district are determined. The shaking was calculated by the following formula of Yu.Znichenko . (Baimakhan, 2002).

$$B_{\sum_I} = \sum \left\{ [10^{q-\frac{vp}{b}} r^{\frac{vp}{b}} 10^{\frac{p_I}{b}}]^{-\gamma} - (10^{K_{\max}})^{-\gamma} \right\} A \frac{10^{\gamma K_0}}{10^{0.5\gamma} - 10^{-0.5\gamma}} \Delta S \qquad (1)$$

Here A –center activity of earthquake; ΔS-area of elements of sub-district, γ-slope angle on recurrence graph; I – earthquake intensity; K_0-10 the fixed meaning; r-hypocentral distance;

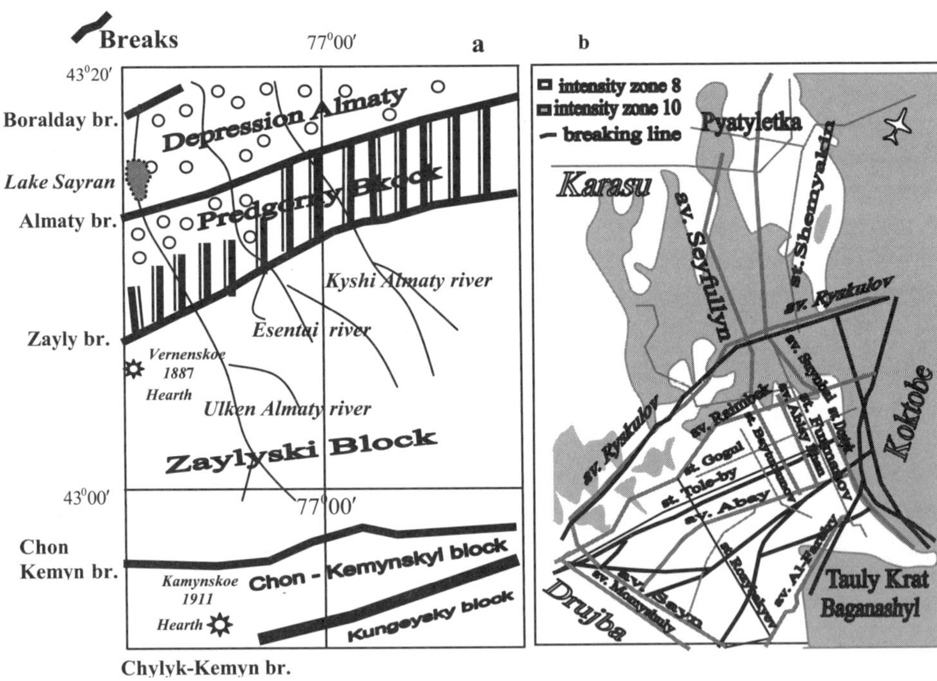

Figure 2. Breaks on the territory of Almaty city: a – with centers of the past earthquakes, b-with zones of different intensity of earthquakes.

p, b, v, q –constants determined based on microseismic data. The reverse magnitude of shaking is the period of earthquakes in years. If the information of a center activity A is not available, A can be determined using K_{max} for Middle Asia by formula: $1gA=2.84+0.21 (K_{max} -15)$. Based on some data obtained, the average meaning γ is equal 0.44 for North Tien Zhan, for Kungei and Alatau : $\gamma=0.45$ and $\gamma=0.30$ respectively.

The results are shown in the scale MSK-64 for intensity $I_{int}=6$-10. Those could be presented in magnitude scale. But there is no principal meaning, as if they are connected in the equation of macroseismic field:

$I_{int} = bM - \nu \lg\sqrt{\Delta^2 + h^2} + c$. Here M-magnitude of earthquake; Δ, h - hypocentral distance and depth of shaking from the ground surface; b, ν and c - coefficients for different regions and zones which vary depending on geological conditions.

Based on the equation (1) calculations for various A and γ are carried out. Average year of periods for case of strong earthquakes in the region were found: for I_{int} =8, 47 years; for I_{int} =9, 87 years and for I_{int} =10, 300 years. And for a territory of city Almaty and its neighborhoods - 872 years. The correspondings to this period, isoclines are shown in the figure 3.

FREQUENCY AND FORMS OF VIBRATIONS OF SEDIMENTARY THICKHESS OF THE BLOCK STRUCTURE

The central part of the city is primarily situated on a site of cones of the rivers of Ulken Almaty, Yesentai and Kyshy Almaty, and adjoining the inclined plain.

The surface deposits of territories from Medeu till Burundai mainly refer to the average ternary and upper ternary. They are covered by fundamental rocks of Paleozoic and tertiary age (Figure 4).

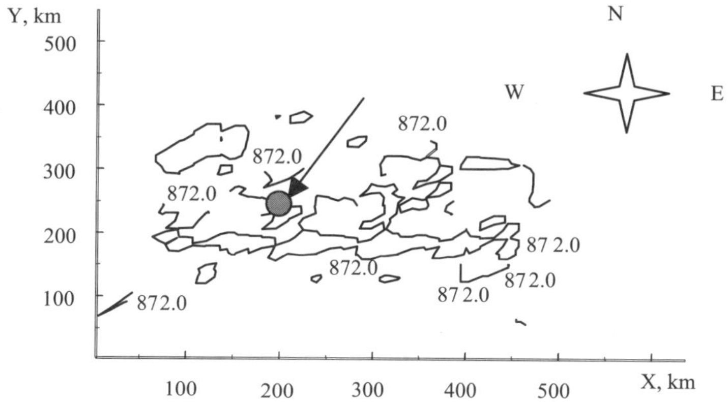

Figure 3. The periods when intensity of earthquakes is 10 in Almaty region and on the territory of city Almaty.

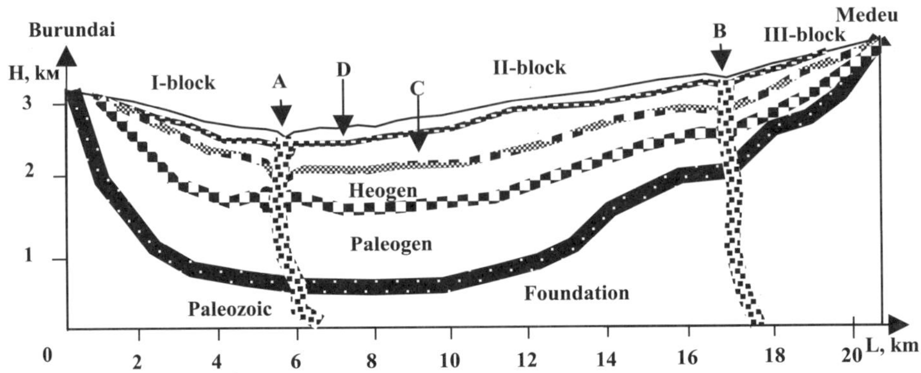

Figure 4. The structure of the soil thickness of city Almaty, Breakings: A – Almaty, B – Zailisky Alatau, C – quaternary deposits, D – contemporary deposits.

The earth crust consists of four horizontal layers (paleogen, neogen quaternary and contemporary deposits) and three vertical blocks I,II and III (divided by Almaty and Zailisky breakings) as shown in the figure. This section is modeled by finite element method.

Over the fundamental rock of Paleozoic foundation, the most powerful paleogen deposit of the ternary period is situated – E=10^4 MPa; v=0.23; $\gamma = 2{,}7 \cdot 10^{-2} tf/m^3$. Above that neogen deposit of the same period is situated with considerable power: – $E \approx 10^3$ MPa; v=0.22; $\gamma = 2{,}5 \cdot 10^{-2} tf/m^3$. Above the mentioned deposit the quaternary layer is situated: – $E \approx 10^2$ MPa; v=0.21; $\gamma = 2{,}6 \cdot 10^{-2} tf/m^3$. On the day-time surface of the territory contemporary deposits with variable power – $E \approx 10^2$ MPa; v=0.20; $\gamma = 2{,}2 \cdot 10^{-2} tf/m^3$. For materials of breakings are accepted: $E \approx 5 \cdot 10^2$ MPa; v=0.23; $\gamma = 2{,}0 \cdot 10^{-2} tf/m^3$.

The system of equation of free vibrations can be written in the form of $[M]\{\ddot{U}\}+[R]\{U\}=0$. Here $[M]$, $[R]$ – matrixes of mass and stiffness, $\{U\}, \{\ddot{U}\}$ – vectors of displacement and acceleration. Thc main lowest frequency are found estimating the system of equations which are shown in the Table 1.

The corresponding form of vibrations of the earth surface is shown in the figure 5. The biggest amplitude of vibrations is observed in the Zones of breaking as shown in figure 5.

CONCLUSION

The analysis based the investigations shows that shaking is based on the account of the intensive energy on the past earthquakes $K_{\max}$ and activity of those centers in present time. The periods of strong shakings estimated here approximately coincides to the periods of registered earthquakes in this region. It is necessary to investigate the forced vibrations of the cart – crust relatively to Paleozoic foundation for more detailed classifications of regions by intensity of vibrations furthermore.

Table 1. The main frequency of free vibrations of Almaty relatively Paleozoic foundation.

number of the frequency	$\omega, rad/sek$	T, sek	f, Hz
1	0.61	10.27	0.10
2	0.67	9.36	0.11
3	0.77	8.14	0.12
4	0.86	7.27	0.13
5	0.92	6.80	0.14

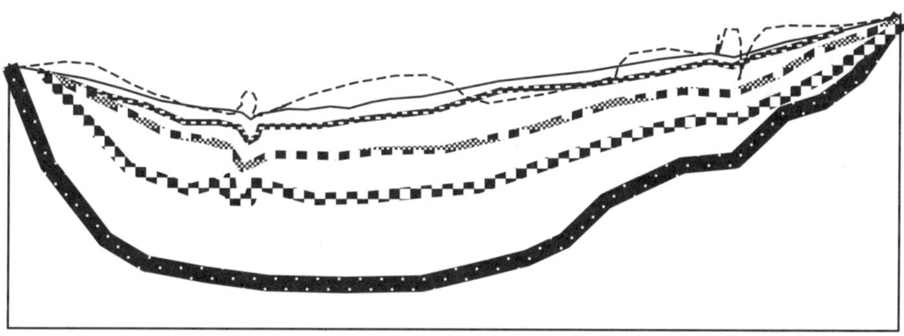

Figure 5. The first form of the vibrations to earth surface on territory of city Almaty.

REFERENCES

Eiby G.A. Earthquakes. Heineman. 1980.

Aitalyev Sh.M., Baimakhan R.B., Masanov Zh.K. (2001). "Formation of the map of shaking by the improved method". Vestnik of Engineering academy. Republic Kazakhstan, № 1(6), p. 62-65.

Baimakhan R.B. (2002). Calculation of seismic stress-state of underground structures in the un-isotropic thickness by finite element method (under edition of Aitalyev Sh.M.). Almaty, 230 p.

Aitalyev Sh.M., Ukshebayev M.T., Baimakhan R.B. (2003). "Geotechnical problems at during time of constraction of Almaty subway".*Abstracts of Japan-Kazakstan joint geotechnical seminar*. Japan, Tokyo, 2003, p.7.

LANDSLIDES & SLOPES

International Conference on Geotechnical Engineering for Disaster Mitigation & Rehabilitation
Chu, Phoon & Yong (eds) © 2005 World Scientific Publishing Company ISBN 981-256-469-1

EXPERT SYSTEM FOR SLOPE STABILIZATION

CHEARNKIATPRADAB. B
Kasetsart University, Bangkok, Thailand.
Civil Engineering Department , South East Asia University, 19/1 Petchkasem rd. Nongkhem Bangkok 10160 Thailand

Soil slope failure is one of the common disasters in Thailand. Several modes of failure are found ranging from debris flow from a mountainous area to a circular deep-seated failure in very soft marine clay deposit in the central plan. Although various methods of the slope stabilization were attempted in the past, but safety and economics of the system are questionable. Some failure slopes are still re-occurred at the same spot or the better alternatives can be proposed. This research is aiming to search for the systematic guidelines to reach the most effective soil slope stabilization for the particular area. The basic knowledge to create this system are from the expert opinions, the past failure cases and the theoretical considerations. After the knowledge acquisition from several sources, the rules are formed to input to the expert system shell program. The GIS techniques are employed to advise the soil data derived from the existing soil boring logs and the agricultural soil map from the Land Development Department. The user can enter the slope location with the pertinent information then the program will suggest the top three suitable methods for slope stabilization. However the final decision has to be the responsibility of the engineer who designs the slope protection himself. The program also links to the stability analysis program (KUslope) in order to provide the useful tool for analysis and design. Finally this program is hoping to reduce the numbers of slope failure in Thailand.

INTRODUCTION

This expert system program is designed to help civil engineers or related users for selection of the possible soil slope stabilizations. The C^{++} program was used as expert shell because of the interfacing capability with the other programs such as stability analysis and GIS programs. The main task of program is consisted of 3 database modules to form knowledge bases and decision making steps as follows.

1. The database of possibly failure modes of soil slopes. In which the data are derived from the technical papers and interviewing the experts in this field.
2. The database for slope stabilization methods. It derived from the expert opinions, the past failure cases and the theoretical considerations.
3. The database of construction cost. The data are from Department of Internal Trades, Ministry of commerce Thailand and other relevant government organizations such as Department of Highways, etc.

The classification of modes of failure is as suggested by VARNES (1978).

The program's flow chart is shown in fig 1. The user can enter the program by the site location of 4 regions of the country according to the Ministry of Interior, as Northern, Southern, North eastern and Central Area regions. Each region is then separated into Provinces, Amphoes and Tambons. The specific locations are entered by the surveying coordinates of the slope such as river bank, embankment, mountainous slope, cut and fill slope. The program proceeds as follows: selection of the region, location, mode of failure, and causes of failure. Then the program will show 3 suitable stabilization methods with safety factor and cost of construction per unit.

EXPERT SYSTEM RULES AND INFERENCE PROCEDURES

The rules of the expert system were derived by analyzing the knowledgebase of stabilization methods from expert opinions, the past failures cases and theoretical considerations. The 3 suitable stabilization methods were derived from weighted scores by the analytic hierarchy process (AHP). The weighted scores were weighted by sequence of construction, difficulty/facility of construction's maintenance, social's reasons, etc. The schematic diagram for analyzing the suitable inference rule of stabilization method by the failure case is shown in fig 2. The inference procedure of this expert's program is based on forward chaining procedures. The chaining procedures start from a set of conditions and move towards a conclusion.

RELATIONSHIP OF ENGINEERING SOIL PROPERTIES AND AGRICULTURAL SOIL MAPS

Because the database of engineering soil does not cover all of the country, GIS agricultural soil's map of the Land Development Department (LDD) is used to find the relationship of engineering and agriculture soil. LDD has distinguished soil of Thailand into 62 groups, classified soil about 2-3 meters from ground surface according to soil taxonomy. Each group of agricultural soil consists of established soil series that has specific properties for each location. The soil series is named according to the location. The agricultural GIS map of LDD is separated into tambons, amphoes, and provinces.

The engineering soil's database used in this research are supported by the mobile phone company, DTAC about 95%, soil boring company, STS Engineering about 5%. The soil's bore holes have about 583 bore holes all over Thailand. The Northern area has 135 bore holes. The Southern area has 65 bore holes, The North eastern area has 176 bore holes, and The Central area has 207 bore holes. Almost all soil databases from DTAC have only SPT and depths, no soil classification and other engineering property data. But the database from STS has soil classification and other engineering property data; these data filled engineering soil data completely. Plotting the location of soil bore holes which have the true longitude and latitude position on the agricultural GIS map, the position of bore hole showed the number of soil's group , as shown in fig 3 . Then the relationship between engineering soil property and agricultural soil property of that bore hole can be known .

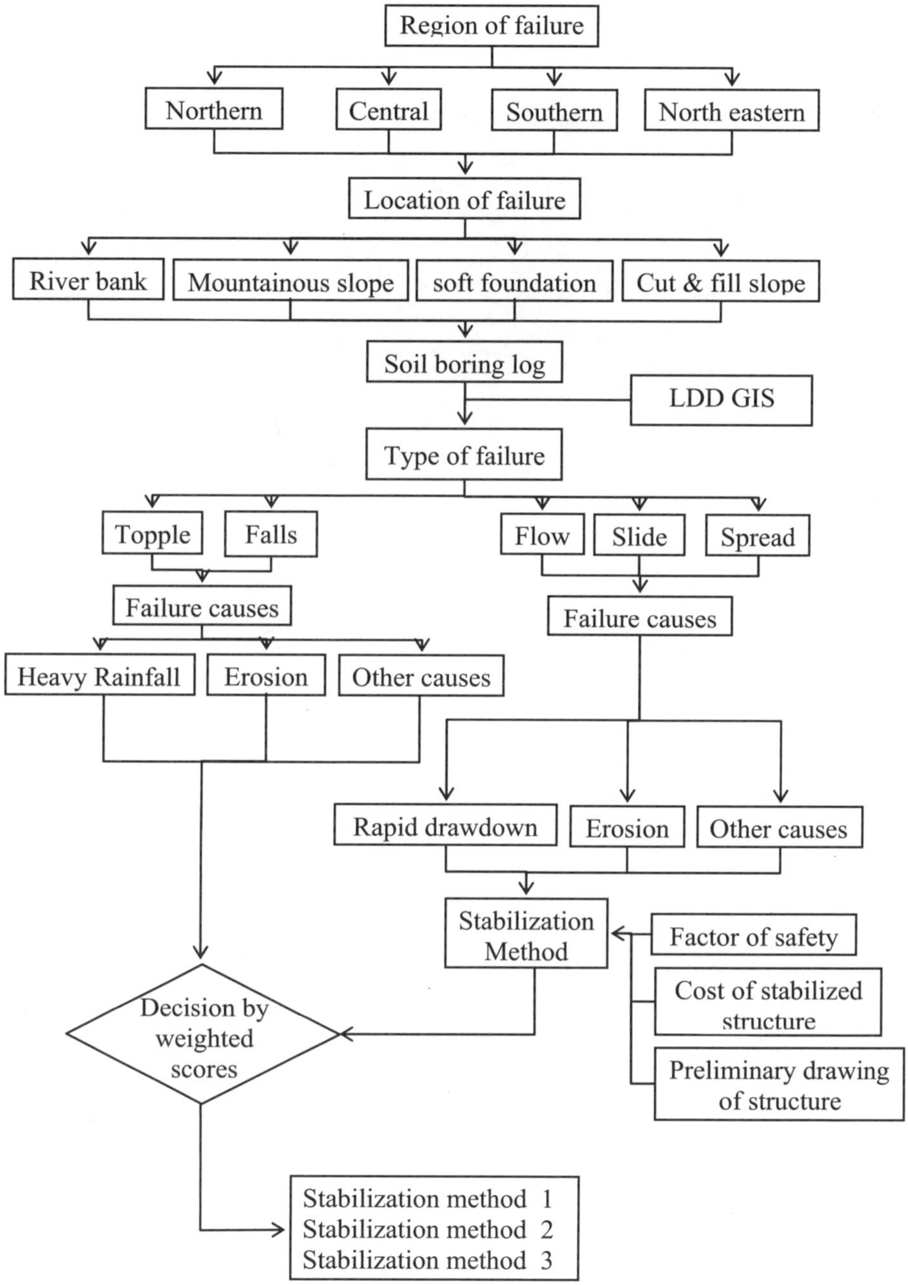

Fig 1. Flow chart of expert system program

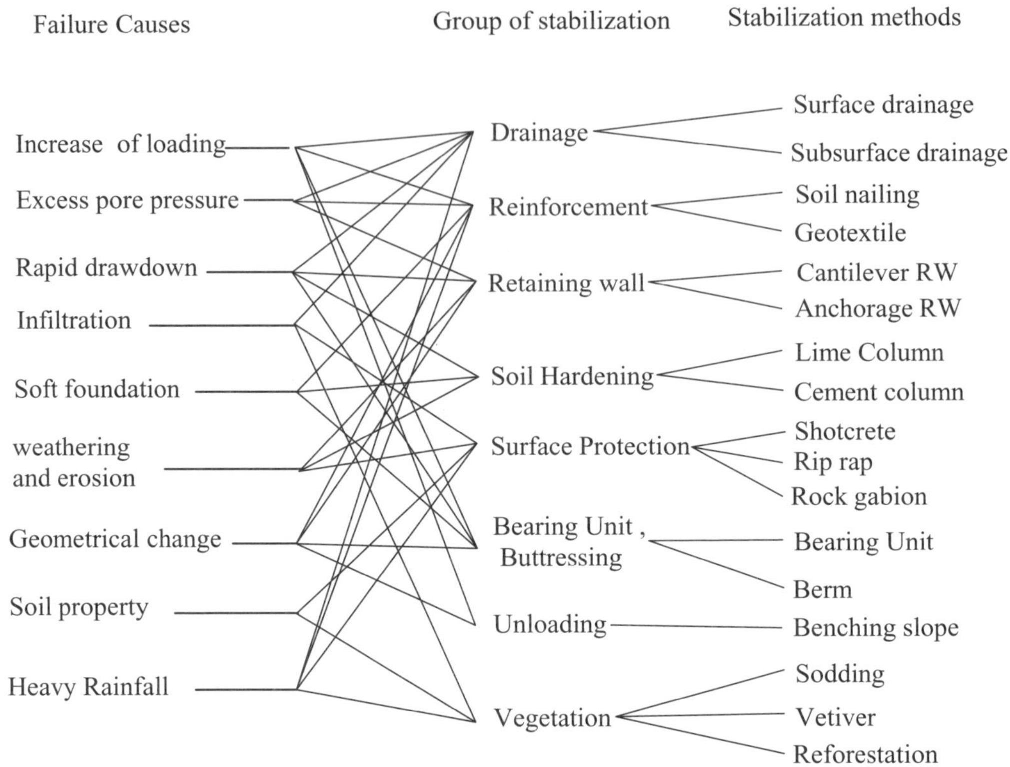

Fig 2 The schematic diagram for analyzed the suitable rules

The essential correlations between the two soils concern soil shear strength and the depth for each soil group. The demonstration of correlations between soil depth and SPT (blows/ft) of soil group No 40 in the North eastern area is shown in fig 4. The strength envelope's equations were fitted by linear equation. By using the lower envelope for calculating the safety factor of stabilization, the stabilized structures were safer than the actual sample. Safety factor of the stabilization method were calculated by the stability program, KUslope. The cost of stabilized structure is derived from estimates the cost of that structure.

Construction's materials costs were derived from the Department of Internal Trades, Ministry of commerce Thailand and other relevant government organizations such as the Department of Highways or the construction company . The costs of the structures are presented in Baht per unit (meter, square meter, etc).

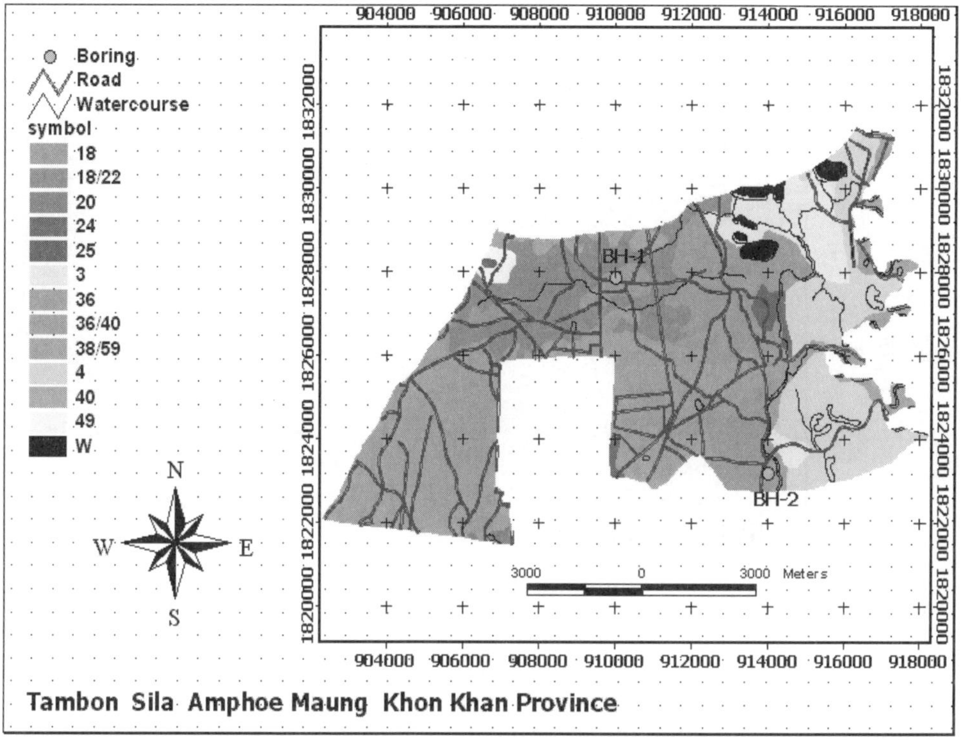

Figure 3 Agricultural GIS map with soil bore hole

CONCLUSION

In Thailand, the growing numbers of landslides caused by human activity and the climatic influence have increased every year. The suitability of stabilization methods depend on the causes of failures. So, the expert system for slope stabilization is a program for providing guidance for selecting a preliminary method for stabilization. Because the soil's shear strength for calculated slope stability of structure uses soil engineering data and soil's agricultural correlations, the user should rechecks soil profiles by boring and recalculated the slope stability of the structure.

The correlations between two soils are rather scattered, the agricultural soil classified by 2-3 meters from ground surface can be used for agricultural purposes only. LDD did not examine the deeper soils and distinguished only 62 groups, so the properties of soils were insufficient to identify any correlations. The correlations between engineering's soil and geological soil may be better than agricultural and engineering's soil because the similarity of processes of soil formation.

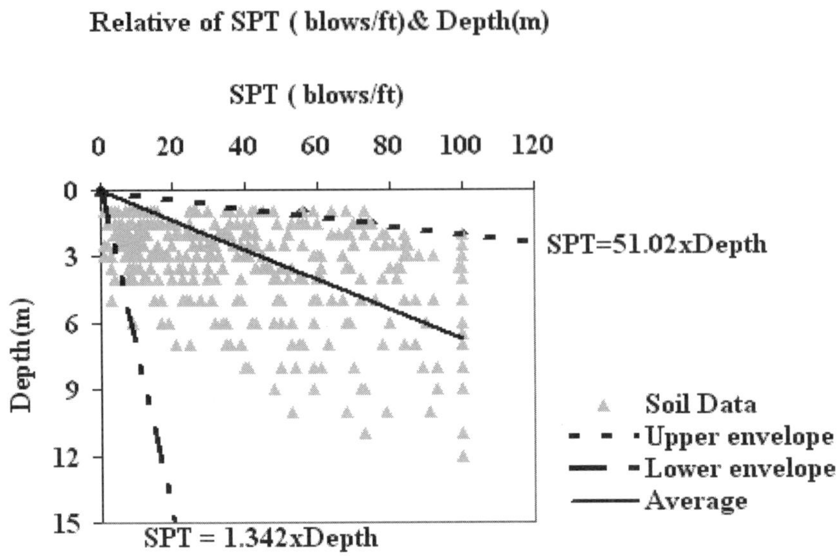

Figure 4 The correlation of SPT & Depth of soil group No 40 on North Eastern area

ACKNOWLEDGEMENT

The Thailand Research Fund for funding of this research. Assoc Prof Warakorn Mairaing W. for the precious advice. Mr. Dale Kvalheim , English lecturer South East Asia University.

REFERENCES

Chearnkiatpradab..B ang Mairaing. W (2002) " Evaluation of soil slope stabilization case studies : Bangpakong river bank and Namkor watered area" , *Proceeding of the eighth National Convention on civil engineering* , , Khon Khan ,Thailand , vol 2 (In Thai language)

Lee W. Abramson "*Slope Stability and Stabilization*" John Wiley and Sons Inc

Mairaing. W (2003) " *Development of Master Plan for Management of Natural Disasters :Landslides* " TRF research , Thailand. (In Thai language)

Varnes,D.J. 1978. "*Slope movement types and processes*", pp.11-33. In R.L. Schuster and R.J. Krizek(eds). Landslide Analysis and Control, Transportation Special Report No 176. National Academy faculty of sciences, Washington D.C.

Tansirimongkol W (1999) "*AHP the most popular decision method in the word* ", 1st edition , Graphic and printing , Bangkok (In Thai language).

International Conference on Geotechnical Engineering for Disaster Mitigation & Rehabilitation
Chu, Phoon & Yong (eds) © 2005 World Scientific Publishing Company ISBN 981-256-469-1

TRIGGERING AND PREVENTION OF LANDSLIDE DISASTERS TO CAVE DWELLINGS IN LOESS REGION OF CHINA

AN DENG
Geotechnical Research Institute, Hohai University, 1 Xikang Rd
Nanjing 210098, China

A cave dwelling is a special but long-standing style of dwelling which is constructed into or against loess physiognomy in north and north-west China. Currently in China over 40 million residents live in cave dwellings. However, landslide disasters to cave dwellings become more frequent and disastrous in loess regions. Numerous casualties and property damages were reported. Loess landslide disasters pose vast threats to residents in cave dwellings. To investigate triggering factors and preventing scenarios regarding loess landslide disasters to cave dwellings is a pressing research topic facing geotechnical engineers and related professionals. This state-of-the-art study is launched to address this issue. Work scope covers three aspects: geological requirements of constructing cave dwellings, prerequisites and triggering factors of landslide disasters, and preventing scenarios of loess landslides. Summarization of triggering factors and disaster prevention scenarios establish a platform mitigating landslide disasters to cave dwellings.

INTRODUCTION

In China, currently over 40 million residents live in cave dwellings which distribute over areas of around 630,000 km^2 surrounding drainage basin of Huanghe river in the north and north-west China (Hou *et al.* 1994, Wang *et al.* 2001 and 2004). Cave dwelling, as a special and long-standing style of dwelling, was initiated 2000 years ago by Chinese ancestors who lived in loess physiognomy, where limited living lands and scarce construction resources enforce residents to resolve their living concerns using local featured geological resources: loess. Self-standing behavior of loess deposits prompts the concept of digging cave dwellings in loess deposit, where the inner of dwellings is kept warm in winter and cool in summer. More advantages of constructing cave dwellings include reduced construction budget and occupation, simplified construction procedures, natural resource and energy reservation, orientation to local geological physiognomy and harmonious living environments (Liao *et al.* 2000).

In recent years, disasters to cave dwellings, particularly caused by landslides, became more frequent and disastrous in China. One-third landslides occurred in loess regions (Qu *et al.* 1999). In loess regions, loess landslide accounts for over 70%. In addition, many large-scale and extremely destructive disasters, such as Bailu plateau landslide, Haiyuan earthquake landslide, and Jiangliu landslide, also happened in loess regions. As a result of loess landslide disasters, numerous casualties and property damages were reported. According to statistical documents, around 1000 persons per year dies from landslide disasters to cave dwelling, which occupy 80% of total disaster to cave dwellings. From 1949 to 1990, at least 1025 people directly died from landslide disasters merely in part areas of three provinces of Shanxi, Gansu, and Qinhai where loess widely exists. Not only residents living in cave dwellings risk their lives and properties each moment, but also the landslide disaster becomes a major obstacle hindering the social and economic development of loess regions. Utilization of abundant resources of mineral, water, electricity, land, light, and heat are delayed, and considerable developing potential of loess regions is depressed. How to prevent and mitigate landslide disasters to cave residents, dwellings, and related facilities is a pressing research topic facing geotechnical engineers and related professionals. A state-of-the-art study is launched to address this issue. Work scope covers three aspects: geological requirements of constructing cave dwellings, prerequisites and triggering factors of loess landslide disasters, warning and preventing scenarios of loess landslide to mitigate disasters.

GEOLOGICAL REQUIREMENTS OF CAVE DWELLINGS

Surrounding drainage basin of Huanghe river in north China, loess deposit is greatly developed, where extensive and consistent loess layers of up to 100 to 200 m thick are uniformly distributed over ground surface (Fig. 1). In this region, the local climate is featured by being few raining and arid most of the seasons, numerous well developed loess precipices. Loess particles have characteristics of being dry, calcic and porous. The shear strength and compressive strength of loess mass are high. Physical properties of two typical loess samples are presented in Table 1. Main ingredients of loess particles consist of clayey and sandy particles. Characteristics of loess particles contribute to its engineering behaviors of self-standing, highly-developed vertical joints, easy excavation. It was reported (Liao *et al.* 2000) that a loess precipice is able to stand with its height up to 10 to 20 m without external supporting structures. Accordingly, local residents call loess deposits as "standing soil" or "lying soil". These geological conditions become technically constructive advantages for digging durable and large-scale cave dwellings in loess deposit. The life time of cave dwellings is generally 40-50 years, with a few cave dwellings still in service aging up to 100-200 years (Chen *et al.* 1995).

Cave dwellings built into or against loess deposits are broadly classified into three types depending upon their site selection and embedding manners: precipice cave dwellings, courtyard cave dwellings, and pit cave dwellings. Cave dwellings are popularly built along the slopes of various scales of loess precipices, which are called precipice cave dwellings. This cave dwelling is horizontally aligned against loess slope, mostly facing south to absorb sunshine during daytime. In circumstances where loess plateaus dominate, cave dwellings are frequently subsided vertically into loess plateau to form a courtyard, which is called courtyard cave dwelling. Obviously, the disadvantages of courtyard cave dwellings are less sunshine absorption and worse drainage system. Some cave dwellings, which are similar to precipice cave dwellings, are partially dug into

Figure 1. Loess plateau in north China.

Table 1. Physical properties of two loess samples (Guo *et al.* 2003).

Sample	Water content, %	Unit weight, kN/m^3	Specific gravity	Void ratio	Saturation degree, %	Vertical permeability, cm/sec
I	7.0	13.92	2.74	1.065	18.5	7.7×10^{-5}
II	3.1	12.94	2.70	1.109	7.5	7.2×10^{-5}
Sample	Compression modus, MPa	Cohesion, MPa	Internal friction angle, °	Liquid limit, %	Plastic limit, %	Plastic index
I	41.61	0.091	24	26.6	17.0	9.6
II	31.92	-	-	26.3	17.3	9.0

loess deposit vertically or horizontally and covered by extra loess mass. This dwelling is called pit cave dwellings. Normally, 3 to 5 cave dwellings cluster together to form a living community for a family. The dimension of a cave dwelling is generally 7 to 10 meters in depth, 3 to 4 meters in width, and around 3.3 m in height, which is competitive against a conventional style of dwelling. An arch style ceiling is adopted to conform to theory of structure mechanics.

FACTORS TRIGGERING LOESS LANDSLIDE DISASTERS

Most cave dwellings are built along the slopes of various scales of loess precipices or plateaus, which means that safety of cave dwellings is closely correlated to the slope stability of loess masses. Once loess slopes fail to stand so as to cause landslides, the cave dwellings in the vicinity of landslides may possibly be buried, overturned, or destroyed. Multiple factors are relevant to triggering landslides, mainly including precipitation, earthquake, geological stratum, freezing-and-thawing cycles, and human activities.

About 90 percent of landslides in loess region are related to precipitation and earthquake in the aspect of external triggering factors. In the other aspect of internal factors, occurrence of loess landslides is closely relevant to collapsibility and liquidation of loess mass. External and internal factors interact during loess landslides. Loess mass has engineering behaviors of collapsibility and liquidation if exposed to infiltration of external water and shaken force incurred by earthquakes. Due to the effect of gravitation and existence of huge pores, loess particles rapidly collapse and subside when wet or soaked by water. If an additional external shaken force, such as an earthquake, is applied, the loess deposit would further collapse. If the water content is high enough to reach a point greater than its plastic limit, the pore water pressure increases significantly with the effect of shaken force. Residual strain of loess deposit under shaken force increases rapidly and loess liquidation occurs which is closely similar to sandy soil liquidation. Besides infiltration and earthquake, Qu *et al.* (2001) found that loess collapsibility is controlled by the content of clay. Over areas where clay content is less than 10%, collapsibility occurs frequently. Over areas with clay content between 10% and 20%, collapsibility occurs sparsely. Over areas with clay content more than 20%, no collapsibility is observed.

Supplemental evidences supporting precipitation as a loess landslide triggering factor is from statistical investigations by Wang *et al.* (2004) that landslides occur sparsely in the northern loess plateau with less than 400 mm of annual precipitation, but frequently in the other area with more than 400 mm of annual precipitation. Loess landslides mostly happen in rainy season of July, August and September, particularly during moderate and heavy raining periods, but relatively less in other rarely raining seasons. It is suggested that loess mass is soaked during raining and post-raining period, which not only leads to the increase of loess slope weight but also reduce the coherent strength of loess mass. When a great number of rainfalls reach confining layer to form phreatic water layer, the cohesive force of soil body and frictional force of interlayer greatly decrease, and thus is favorable to landslide occurrence. Because the loess slopes can be cut down and eroded continuously by the surface running water, the slopes become steeper and unstable, and the landslides may occur readily. The existence and variation of ground water flow can directly affect gravitational condition of slopes and mechanical characteristic of loess mass, which are main factors determining slope stability.

Affected by earthquake, loess landslides are developed densely along earthquake strips. According to relevant data, every more than 6-magnitude earthquake may incur loess landslides, and higher the magnitude, more destructive the landslide disasters. For example, during the periods of Tianshui earthquake of 7.5-magnitude in 1654, Tongwei earthquake of 7.5-magnitude in 1920 and Haiyuan earthquake of 8.5-magnitude in 1920, landslides occurred in all the earthquake areas mentioned. In recent years, with social and economic development, large-scale cities have been emerged over loess areas where earthquakes are relatively active. From the seismic ground motion parameter zonation of China with 10% probability of exceedance in 50 years, around a half of loess region has seismic acceleration greater than 0.2 g, up to 80% has seismic acceleration greater than or equal to 0.15 g. Four provincial capitals and six middle-level cities are surrounded by regions of 0.2 g seismic acceleration. Three provincial capitals and two middle-level cities are surrounded by regions of 0.15 g seismic acceleration. And it is acknowledged that a seismic activity with seismic acceleration greater than 0.15 g is able to cause loess disasters, including seismic landslides, seismic collapsibility, and liquidation of saturated loess mass. Hence, loess landslide disasters incurred by earthquakes are a pressing concern.

Geological stratums of loess deposits may influence occurrence of loess landslides. Loess deposits have been accumulated since Quaternary, and are further divided into ancient loess formed in early Pleistocene (Q1), old loess formed in mid-Pleistocene (Q2), new loess formed in late Pleistocene (Q3) and recent loess formed in Holocene (Q4) according to its geologic depositing time. Owing to long time depositing, Q1 loess and Q2 loess have characteristics of low void ratio, few large pores, and high strength, and accordingly are not readily to incur earthquake landslides. Q3 loess and Q4 loess have abundant large pores, loose structures, developed vertical joints and strong collapsibility, and thus are relatively easy to incur earthquake landslides. Besides the identification of loess stratums, boundaries between loess mass and underlying bedrocks influence loess landslides as well. Sliding failures relatively readily follow those existed structure surfaces of poor shear strength.

Freezing-and-thawing cycles due to climate alternation may induce loess landslides. In the northwest China, during the alternation from winter to spring, freezing-and thawing cycles of loess masses occur. As a result, frozen loess mass may be intenerated or even liquefied. Strength of loess deposit decreases. Stability of loess slopes reduces correspondingly. Landslide disasters to cave dwellings are readily triggered. Thus, cave residents should be aware of landslide disasters during the alternation from winter to spring, or an ice-melting period from March to May. Typical landslides of this type include Saleshan landslide, New Saleshan landslide, and Wolongshi landslide.

In the catastrophic loess landslides happened in recent years, a majority of landslides are related to human activities. In recent twenty years, population in loess regions increases so rapidly that it has reached 81.49 million, accounting for 7.8% of the national population. With the rapid increase of population and the development of capital construction, irrational human activities, such as excavating foothill, filling hillsides and over-cultivating, frequency and intensity of loess landslide disasters are raised. One of the changes related to human activities is the change of loess slope. Slopes of loess deposits become steeper as a result of human activities, such as cultivation, engineering exploitation and excavation, and other abused uses of loess deposit or related natural environments. After removal of original soil portions, stable stress distribution is broken. With the increase of slope angle, the angle of possible failure surface increases as well, which accordingly results in the reduction of slope stability. The worst case, which would finally approach due to human activities on natural environments, is the loss of adequate anti-sliding resistance and occurrence of landslides. From the GIS analysis performed by Wang *et al.* (2004), over loess regions in China, slopes of most loess deposits have angles of 20-30 °. Some loess slopes have angles greater than 40 °, which may readily incur loess landslides.

Landslides in loess regions, affected by geological stratums, climate, earthquake activities and human activities, are distributed with apparent rules of time and space. Loess landslides occur relatively frequently over areas of loose and recent depositing loess stratums including Q3 and Q4 loess deposits, over areas and during seasons of higher precipitation and freezing-thawing cycles, over areas and during periods of active earthquakes, and over locations with dense human activities.

MEASURES AGAINST LOESS LANDSLIDE DISASTERS

Prevention

Measures preventing loess landslide disasters to cave dwellings are established taking triggering factors of loess landslides into account. First of all, construction site selection of cave dwelling is a vital step preventing potential landslides. It is decisive to identify solid and durable Q1 and Q2 loess stratums which are preferred sites to accommodate or support cave dwellings. Newly formed Q3 and Q4 loess deposits and sites are not suitable for cave dwellings. Cave dwellings are better selected along stable loess slopes or precipices where potential of torrents or running streams rarely occur. The worst case is the site close to or against steep precipices which possesses rather poor stability. Cave dwellings should also avoid sites of exploiting areas, mining areas, water reservoirs, and irrigation facilities.

In some circumstances, the steep slopes close to cave dwellings are flatted moderately by excavating sliding, collapsing or foreign parts. The excavation should start from top, then gradually towards bottom. Vegetable growing is proposed over the slopes at the back of cave dwellings. To prevent loess mass from being intenerated by water infiltration and thus shear strength decrease along potential sliding surface, drainage system is suggested to channel water penetrating into loess mass, particularly for courtyard cave dwellings where raining waters are readily accumulated. Drainage system mainly includes two parts: hidden cutting channels in the ground to cut underground infiltration into potential landslide areas, and hidden drainage channels to discharge penetrated water in potential landslide areas. Sometimes, surface drainage system is applied to channel or prevent surface water approaching potential sliding areas.

Geotechnical works and soil treatment techniques are sometimes applied to improve stability of potential sliding mass. Mostly used treatments include anti-sliding piles, retaining walls, baking of soils, blasting and grouting. It is common and effective to combine multiple measures to prevent loess landslide disasters, such as combination of drainage system, retaining walls, and grouting.

Forecast

Monitoring is a popular measure to forecast landslide, which includes displacement monitoring of ground surface and dwellings, crack monitoring of ground surface and dwellings, deep displacement monitoring, dynamic variation monitoring of groundwater and stress monitoring of loess mass. In theory, forecast of loess landslide includes long-term forecast, short-term forecast, and urgent forecast. Creep theory is usually used to obtain the prediction of landslides.

However, due to limited availability of monitoring instruments for most possible loess landslides, professional monitoring techniques and computing theory cannot be used. Hence, forecast based on signals and experiences are favorably adopted. For instance, before, during or even immediately after raining, it is worthwhile and urgent to exam potentially sliding regions. Possible negative signals indicating landslides include water seepage in the back of cave dwellings, heave or swell of ground surface, sliding of dwellings, cracks on sidewalls of dwellings, fallings of soil masses, deformed doors and windows and thus hard to open. A large-scale of steep slope over cave dwellings is also a negative signal indicting a landslide disaster. It was suggested that the height of a slope where a cave dwelling locates is not expected to exceed 8 to 10 m, and the angle of the slope is properly less than 75 to 80 ° (Mou 1995). When the slope is rather high, cutting slope into a ladder style is applied. To train residents living in cave dwellings about how to identify disaster signals and handle urgent events helps mitigate loess landslide disasters. They should have the concept in their mind that raining infiltration may cause rather adverse consequences to their dwellings, and that being aware of any negative signal, such as slope activity, is greatly helpful for their life-savings.

CONCLUDING REMARKS

Loess landslide disasters are identified as vast threats to cave residents living in loess regions due to the special behaviors of loess mass: collapsibility and liquidation given external effects, such as water infiltration and earthquake shaking force. Occurrence of loess landsides is spatio-temporal. Loess landslides occur relatively frequently over areas of loose and recent depositing loess stratums including Q3 and Q4 loess deposits, over areas and during seasons of high precipitation and freezing-thawing cycles, over areas and during periods of active earthquakes, and over locations with dense human activities. Prevention of loess landslide disaster is mainly accomplished by eliminating its triggering factors, such as construction site selection, human impact reduction, supplemental slope treatment scenarios, etc. Due to the unavailability of advanced instruments, judgment on negative signals and experiences help forecast occurrence of loess landslides

REFERENCES

Chen, G., Zhang, K. and Xie, J. (1995). "A seismic performance analysis of the cave dwelling on the loess precipice", *Journal of Harbin Architecture and Civil Engineering Institute*, Vol. 28, No. 1, 15-22, (in Chinese).

Guo, F. and Jiang, C. (2003). "The type and formation condition of geohazard along Dabanliang-Qiangyang section of S202", *Gansu Science*, Vol. 15, 80-83, (in Chinese).

Hou, J., Li, Y. and Li Z. (1994). "Application and development of new type earth sheltered construction in China", *Underground Space*, Vol. 14, No. 3, 185-192, (in Chinese).

Liao, H., Zhao, S., Gao X. and Su, L. (2000). "Environmental engineering problems in developing loess plateau cave dwellings in western areas", *Journal of Xi'an Jiaotong University*, Vol. 20, No. 3, 7-10, 23, (in Chinese).

Mou, W. (1995). "Analysis and prevention of loess cave construction accidents", *Development of Small Cities & Towns*, Vol. 13, No. 1, 26, (in Chinese).

Qu, Y., Zhang Y. and Qin, Z. (1999). "Hipparion laterite and landslide hazards on loess plateau of northwestern China", *Journal of Engineering Geology*, Vol. 7, No. 3, 257-265, (in Chinese).

Qu, Y., Zhang, Y. and Chen, Q. (2001). "Preliminary study on loess slumping in the area between northern Shanxi and western Shanxi — Taking the pipeling for transporting gas from west to east in China", *Journal of Engineering Geology*, Vol. 9, No. 3, 233-240, (in Chinese).

Wang, Y., Zhao Q., He, M., Yang, L. and Liu, J. (2001). "The study on indoor environment of old and new Yaodong dwellings", *Journal of Xi'an University of Architecture & Technology*, Vol. 33, No. 4, 309-312, (in Chinese).

Wang, Y., Wang L., and Zhang X. (2004). "GIS based seismic landslide zonation of the Loess Plateau", *Scientia Geographica Sinica*, Vol. 24, No. 2, 170-176, (in Chinese).

International Conference on Geotechnical Engineering for Disaster Mitigation & Rehabilitation
Chu, Phoon & Yong (eds) © 2005 World Scientific Publishing Company ISBN 981-256-469-1

LANDSLIDE HAZARD ZONATION IN INDIA: A REVIEW

S. SARKAR
Central Building Research Institute
Roorkee – 247667, India

D. P. KANUNGO
Central Building Research Institute
Roorkee – 247667, India

The assessment of landslide hazard and preparation of landslide hazard zonation map is an imperative task in the area of disaster management. There are many methods used by several workers in different parts of the globe. In most of the techniques though input parameters are mostly the same, they differ in ranking the factors. In general the methods are either based on the qualitative approach which dictates the weight assignment to the factors based on the experience and expert knowledge or the statistical approach which involve relationship between existing landslides and the factors. In India several hazard zonation mapping have been carried out in different parts of Himalaya. The paper presents few of the techniques used by the authors.

INTRODUCTION

In recent years, a number of major disasters have made the global community aware of the immense losses of human lives and properties. Although an individual slope failure is, in general, not so spectacular or devastating as an earthquake, a volcanic eruption or a flood, yet, being much more frequent and wide spread over the years, landslides have caused considerable loss of property and life. In many countries, economic losses due to landslides are great and apparently are growing as development expands into unstable hill areas under the pressure of expanding populations.

The landslide hazards, in general cannot be completely prevented, however, the intensity and severity of their impacts can be minimized if the problem is recognized before the development activity or deforestation begins. Hence, there is a dire need for identification of unstable slopes, which can be fulfilled by landslide hazard zonation mapping. The landslide hazard zonation (LHZ) of an area aims at identifying the landslide potential zones and ranking them in order of the degree of hazard from landslides. Several workers using different methodologies have carried out landslide hazard zonation mapping in different parts of the Himalaya. It is not possible to describe all these works; however, a brief outlines of commonly used methods are reported here.

LANDSLIDE PROBLEMS IN HIMALAYA

The occurrence of landslides is a common phenomenon in the Himalaya. In the recent past there were few landslide disasters in Uttaranchal Himalaya. The landslide disaster at Malpa in the Kali valley occurred on 18th August 1998, which completely destroyed the habitation and wiped out temporary shelters of pilgrims going to Kailash-Mansarovar causing a death toll of more than 200 peoples. The Mandakini Valley of Rudraprayag district has been struck by several landslides as a result of heavy precipitation in the 3rd week of July, 2001. The important among them are Phata and Byung Gad landslides which killed 20 lives along with several injured. The landslide occurred in Uttarkashi on 24th September, 2003 has deposited a huge quantity of debris at the foot hill which caused a huge loss of property (Fig.1). It was fortunate to have no casualties as the district administration and scientific organizations took prompt action to evacuate the families as soon as the landslide was triggered.

Figure 1. Uttarkashi landslide (a) Panoramic view, (b) Landslide seen in satellite image

FACTORS AND THEMATIC DATA LAYERS

Landslide can be triggered by both natural and man induced changes in the environment. The most common factors considered for landslide hazard zonation in India are lithology, soil, geomorphology, slope, drainage, lineaments, proximity to fault and landuse which are essentially the preparatory factors. The adverse nature of any of these factors affects the existing equilibrium of stability. These factors can be judiciously studied and the pertinent data can be collected from the existing field conditions. Over the past few years there has been a significant contribution of remote sensing and Geographic Information System (GIS) for preparation of thematic data layers representing the causative factors used for landslide hazard zonation mapping. In India remote sensing and GIS are now become an essential tools for landslide hazard zonation mapping (Gupta *et al.*, 1999; Saha *et al.*, 2005; Sarkar and Kanungo, 2004; NRSA, 2001; Nagarajan *et al.*, 1998).

LANDSLIDE HAZARD ZONATION IN HIMALAYA, INDIA

Landslide hazard zonation mapping in various parts of Himalaya has been attempted by several workers by applying different techniques. The techniques such as qualitative map combination, statistical, fuzzy and ANN based methods are most commonly used. The selection of appropriate method depends on nature of data, expertise and mapping scale. In India most commonly used scale for LHZ mapping is 1:50,000 to 1:25,000. Efforts are being made for preparation of the Indian Standard code for landslide hazard zonation mapping for macro (1:50,000 to 1:25,000), meso (1:25,000 to 1:15,000) and micro (1:15,000 to 1:5000) level scales.

Qualitative Map Overlay Approach

In qualitative map combination method the Land Hazard Evaluation Factor (LHEF) rating scheme of Anbalagan (1992) uses the facet concept for data collection. In this scheme, ratings of the classes of factors are based on personal judgment of their relative contribution to landslide occurrence. The total estimated hazard (TEHD) which is cumulative ratings of the factor classes are suitably classified into different hazard classes. Gupta *et al.*, (1999) applied the ordinal weighting and rating system based on the relative importance of factors derived from field knowledge. The landslide hazard indices for grids are obtained by multiplying ratings of the classes with the corresponding weight of the factor and then classified into various landslide hazard classes. The weights and ratings for factors and their classes respectively can be also computed based on their pair-wise comparison using Analytical Hierarchy Process (AHP). National Remote Sensing Agency, Hyderabad has developed an Atlas on Landslide Hazard Zonation for parts of Uttaranchal and Himachal Himalayas using the concept of AHP (NRSA, 2001).

Sarkar and Kanungo (2004) developed a rating scheme for factors and their classes based on the associated causative factors for landslides surveyed in parts of Darjeeling Himalaya. In this scheme, the factors were assigned numerical ranking on a 1-9 scale in order of importance. Weights were also assigned to the classes of the factors on 0-9 ordinal scale where higher weight indicates more influence towards landslide occurrence. Undertaking several iterations using different combinations of weights suitably modified the scheme. The integration of numerical data layers in GIS produced the landslide susceptibility map of the area (Fig.2).

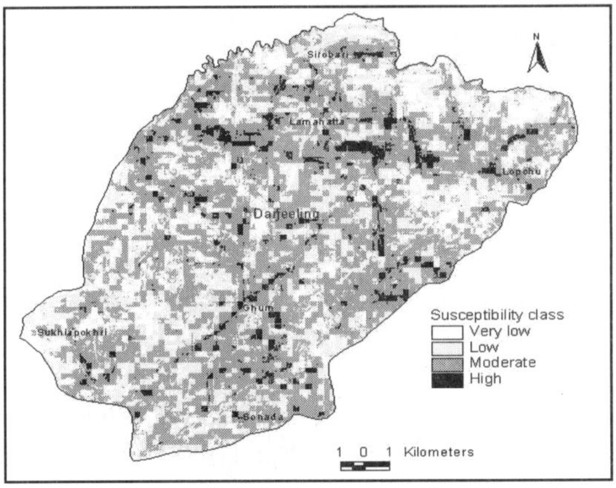

Figure 2. Landslide susceptibility map of parts of Darjeeling Himalaya (Sarkar & Kanungo, 2004)

Statistical Approach

In statistical methods the existing landslides are correlated with the causative factors to predict the landslide potential zones. Gupta and Joshi (1990) adopted an empirical approach wherein statistical relationships of the factors with landslide occurrences are converted to landslide nominal risk factor (LNRF) based on the ratio of landslide incidence in a particular category to the average landslide incidence in various categories of that factor. These LNRF values were integrated to provide cumulative risk factor based on which landslide hazard zonation map was prepared.

The numerical weights of the factor categories derived from the frequency distribution of landslides, termed as Landslide Susceptibility Grades (LSG) was defined by Sarkar and Gupta (2005) for landslide hazard zonation mapping of Srinagar-Rudraprayag area of Garhwal Himalaya. The LSG of a category belonging to a factor is the ratio of number of grids of that category having landslides and the total number of grids with the category. This LSG was considered as the contribution of a particular category in promoting landslides. The landslide potential score for each individual grid was computed by adding the LSGs of different factor categories lying in each grid. These scores were then contoured to produce the landslide hazard zonation map (Fig.3).

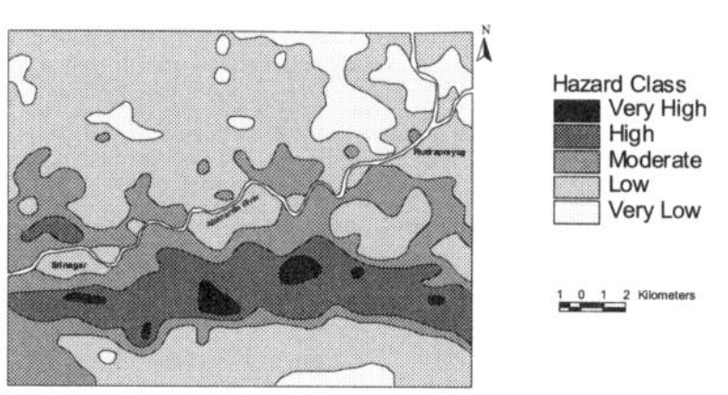

Figure 3. Landslide hazard zonation map of Srinagar-Rudraprayag area, Garhwal Himalaya (Sarkar and Gupta, 2005)

A very popular statistical technique is the Information Model which was first used for landslide susceptibility mapping by Yin and Yan (1988). Information model was used in different parts of Himalaya by Jade and Sarkar, (1993); Saha *et al.,* (2005). In this method the information value, which is the weightage for the category (variable) of a factor, supplied to landslide by variable i is expressed as:

$$I_i = \log \frac{S_i / N_i}{S / N} \tag{1}$$

where, N = total number of grids, S = number of grids with landslide, S_i = number of grids with landslide having variable i, N_i = number of grids having variable i.

Total information value in grids then can be obtained by integrating the information value layers.

The statistical relationship between factors and landslides can also be achieved by the Certainty Factor (CF) model and can thus be used for hazard zonation mapping (Lan *et al.*, 2004). The CF, defined as a function of probability is:

$$CF = \frac{pp_a - pp_s}{pp_a(1 - pp_s)} \quad \text{if } pp_a \geq pp_s \quad \text{OR} \quad \frac{pp_a - pp_s}{pp_s(1 - pp_a)} \quad \text{if } pp_a < pp_s \tag{2}$$

where pp_a is the conditional probability of having a number of landslide event occurring in class a and pp_s is the prior probability of having the total number of landslide events in the area.

Fuzzy Model and ANN Approach

Recently approaches based on fuzzy logic and Artificial Neural Network (ANN) is getting popular for landslide hazard mapping. In fuzzy model, factors are assigned membership grade according to their contribution to landslide occurrences. ANN, a useful technique for regression and classification problems, has several advantages for LHZ mapping, as these have the capability to analyse complex data patterns. Also, ANN can process data at varied measurement scales such as continuous, near-continuous and categorical data, which are often encountered in LHZ mapping. An ANN involves training and testing processes. Based on training and testing accuracies of different neural network architectures, the most suitable network is selected and evaluated for LHZ mapping. ANN technique has been used by Arora *et al.*, (2004) for LHZ mapping of parts of Bhagirathi valley in Garhwal Himalya. A combined application of ANN and fuzzy logic has been used by Kanungo *et a*l., (2005) to calculate the weights and ratings of the factors and the categories respectively.

MAP VALIDATION

The hazard zonation maps prepared should be evaluated for its validity with reference to the existing slope instability conditions of the area. The direct method to check the quality of a map is by field verification of signs of instability present in hazard zones. Another way to evaluate the map is by some statistical means in which the frequency of existing landslide in each hazard class is determined. The map having maximum landslide frequency in high hazard class and minimum in low hazard class can be considered as well representative of existing field condition. Further, statistical significance test such as chi-square test can also be performed to see the effectiveness of the hazard zonation map. Such validation has been attempted by Sarkar and Kanungo (2004) and Sarkar and Gupta (2005).

CONCLUSIONS

Spatial prediction of landslide potential slopes through landslide hazard zonation mapping is an important issue for disaster management. The different methods have their own merits and demerits. Since the landslide contributing factors vary from region to region, uniform rating criteria can not be applied to different geo-environmental condition. The statistical methods or approaches based on ANN and fuzzy logic seems to be more reasonable due to the objectivity in determining approach.

REFERENCES

Anbalagan, R., (1992). "Landslide hazard evaluation and zonation mapping in mountainous terrain", *Engineering Geology*, Vol.32, 269-277.

Arora, M.K., Das Gupta, A.S., Gupta, R.P. (2004). "An Artificial Neural Network Approach for Landslide Hazard Zonation in the Bhagirathi (Ganga) Valley, Himalayas", *International Journal of Remote Sensing*, Vol. 25 (3), 559-572.

Gupta, R.P. and Joshi, B.C. (1990). "Landslide hazard zonation using the GIS approach – A cases study from the Ramganga catchment, Himalayas", *Engineering Geology*, Vol. 28, 119-131.

Gupta, R.P., Saha, A.K., Arora, M.K. and Kumar, A. (1999). "Landslide hazard zonation in a part of Bhagirathi Valley, Garhwal Himalayas, using integrated remote sensing – GIS", *Journal of Himalayan Geology*, Vol. 20 (2), 71-85.

Jade, S. and Sarkar, S. (1993). "Statistical models for slope stability classification", *Engineering Geology*, Vol.36, 91-98.

Kanungo, D.P., Arora, M.K., Gupta, R.P. and Sarkar, S. (2005). "GIS based landslide hazard zonation using neuro-fuzzy weighting", *India International Conference on Artificial Intelligence"*, Pune, India (in press).

Nagarajan, R., Mukherjee, R.A., Roy, A. and Khire, M.V. (1998). "Temporal remote sensing data and GIS application in landslide hazard zonation of part of Western ghat, India", *International Journal of Remote Sensing*, Vol. 19, 573-585.

NRSA. (2001). "Atlas on Landslide Hazard Zonation mapping in the Himalayas of Uttaranchal and Himachal Pradesh States using remote sensing and GIS" National Remote Sensing Agency, Hyderabad, India.

Sarkar,S. and Kanungo, D.P. (2004). "An integrated approach for landslide susceptibility mapping using remote sensing and GIS", *Photogrammetric Engineering & Remote Sensing*, Vol. 70, 617-625.

Sarkar, S. and Gupta, P.K. (2005). "Techniques for landslide hazard zonation – Application to Srinagar-Rudraprayag area of Garhwal Himalaya", *Geological Society of India*, Vol.65, 217-230.

Saha, A.K., Gupta, R.P., Sarkar, I., Arora, M.K., and Csaplovics, E. (2005). "An approach for GIS-based Statistical Landslide Susceptibility Zonation – with a case Study in the Himalayas", *Landslides*, Vol. 2, 61-69.

Lan, H.X., Zhou, C.H., Wang, L.J., Zhang, Li, R.H. (2004). "Landslide hazard spatial analysis and prediction using GIS in the Xiaojiang watershed, Yunnan, China", *Engineering Geology*, Vol.76, 109-128.

Yin, K.L. and Yan, T.Z. (1988). "Statistical prediction models for slope instability of metamorphosed rocks", *5th International Symposium on Landslides*, Lausanne, Switzerland.

International Conference on Geotechnical Engineering for Disaster Mitigation & Rehabilitation
Chu, Phoon & Yong (eds) © 2005 World Scientific Publishing Company ISBN 981-256-469-1

GEOTECHNICAL INVESTIGATION OF AMIYAN LANDSLIDE HAZARD ZONE IN HIMALAYAN REGION, UTTARANCHAL, INDIA

T.N. Singh & Kripamoy Sarkar
Department of Earth Sciences, Indian Institute of Technology, Powai, Mumbai – 400076, India

Himalayas witness each year a series of landslides both small and large, specially in the post monsoon period. These natural disasters have occurred in all the sections of the Himalayas from outer to Higher Himalayas. The southern hills of Kumaun in the proximity of the Main Boundary Thrust (MBT) are very prone to recurrent landslides. The slope failures are related to steeper angle of hill slope, extent and type of vegetation, structural condition of bedrock, high rainfall intensity and neotectonism of the MBT. The Amiya landslide on 11 July 1999, which has claimed 22 lives, including 2 persons in a place 20 km northeast of Kathgodam in Kumaun, Uttaranchal state, India. The flow of debris mainly contains different types of sandstone. These sandstone rocks belongs to the Lower Siwalik Formation, dominantly constituting the Amiyan Landslide area have been investigated to determine the different Physico-mechanical properties, dynamic properties and its chemical analysis.

In this paper, an attempt has been made to determine the all physico-mechanical properties of sandstone and establish correlation among this parameter. The study reveals that the p-wave velocity and uniaxial compressive strength (UCS) decreased with increasing moisture content. Regression analysis demonstrated that wave velocity, UCS decreased with increased quartz content in the area, which is mainly due to the presence of mica and clay. The different geo-mechanical parameters show good correlation with P-wave velocity.

INTRODUCTION

Landslide includes all varieties of mass movements of hill slopes and can be defined as the downward and outward movement of slope forming materials composed of rocks, soils, artificial fills or combination of all these materials along surfaces of separation by falling, sliding and flowing, either slowly or quickly from one place to another. It is a common engineering phenomenon in hilly regions, mainly in North – Eastern India. Recently a huge landslide has been occurred in Amiyan village (Figure 1), Nainital district, Uttaranchal. The frequency of landslides and rock falls in the outer Kumaun Himalaya varies from area to area, depending on underlying rock structure and neotectonic activities along faults and thrusts. The consequence is large landslides and rock falls, particularly in the zones of neotectonically active faults and boundary thrusts (Pant & Luirei, 1999). Rock is the most widely used natural material for construction inspite of its non-homogenous, anisotropic and inelastic nature. The occurrence of many catastrophic failures of the natural and man made structures; in the last few decades have displayed the complexity of the material. This has also time and again proved the lack of understanding of the rock material's response behavior. The problems due to this are blockade of roads and rivers, destruction of property and even loss of lives. There are number of factors such as geological, hydrological, environmental, topographical, weathering etc. affecting the stability of slope. Therefore, there is a need to study the geological and physico - mechanical property of the rock, which will help to minimize the landslide in that area. This will facilitate the design and construction of structures with safety and economy (Singh et.al, 1994). Several factors affecting the properties of rock include poor cementation, weathering, tectonic disturbances and large pore spaces (Olivera, 1993). In addition, it is equally important to know the mineral composition, p-wave velocity as well as the compressive strength of the rock (Bell & Culshaw, 1993). Chemical analysis was conducted for determination of the mineral present in the rock. Different researchers have reported that UCS has a strong positive correlation with quartz content (Shakoor & Bonelli 1991), however Ulusay et al. (1994) and Singh et al. (2005) on the other hand they have found a negative correlation.

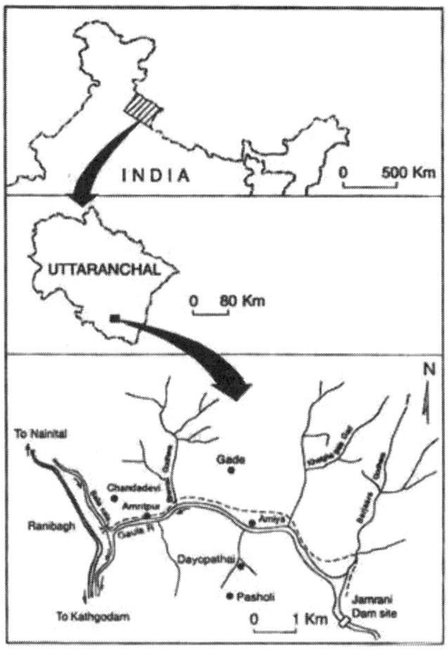

Figure.1 Location Map of Amiyan

Figure 2: Landslide prone area in Amiyan

GEOLOGICAL CONDITIONS

The Amiyan Landslide area falling in the SOI toposheets No. 53 O/11 is located in between the Longitudes $29^0 16'$- $29^0 19'$ and Latitudes $79^0 33'$ - $79^0 38'$. It is almost 12 Kms north of the Kathgodam rail head of Uttaranchal. Geologically the slide lies along to the Main Boundary Fault (MBF) separating the Outer Himalaya (Siwaliks) from the Lesser Himalaya. The slide material is dominantly constituted of the blocks and debris (Figure 2) derived out of the Lower Siwalik sandstones. The area experiences subtropical to temperate climate with temperature ranging from 10.6 °C to 26.7 °C in summers and 2.8 °C to 15.6 °C in winters. The slopes are mostly steep and highly erodable due to heavy rainfall and mineral constituents of the slope forming material. A huge landslide was reported in the area in the year 1992 and later in 1998 with minor sliding in almost all rainy seasons. The heavy sliding along the Amiyan landslide in these years resulted in the blockade and ponding of Gaula River upstream.

EXPERIMENTAL WORK

To understand the geotechnical properties, the various tests of sandstone of study area were performed as per ISRM methods (1980).

Physical properties

The average moisture of the representative rock sample of the area varied from 0.641 to 3.09%. The rock samples were first dried in the oven at 104°C for two hours to remove the existing moisture content. Then, samples were weighted. These samples were put into the water for 72 hrs to make it under supersaturated stage. The average density of the sandstone is found to be 1.62 gm / cc.

Dynamic Wave velocity

The velocity of ultrasonic pulses traveling in a solid material depends on the density and elastic properties of that material. Rafavich *et.al. (*1984) have established the integrated studies of seismic, petrophysical and petrological properties and established useful relationships between these properties.

Determination of Strength Properties

The strength of the 40 samples was determined and five samples were tested for each specimen. Strength of the rock is strongly dependent on the grain size and interlocking material as well as strain rate (Singh and Naidu, 2000).

Chemical Analysis

The geochemical analysis of the rock samples was carried to know the major elemental composition, using the wet chemical analysis (Shappiro & Brannock, 1962).

DISCUSSIONS

Interrelationships between physical properties

The relationship between moisture and UCS is shown in Figure 3, both dry and wet condition.

Dry rock has higher compressive strength as compared to the wet rock. Moisture has significant effect on the strength of rocks (Rehbinder, *et.al,* 1948). The correlation between moisture and UCS shows linear relation with high correlation coefficient.

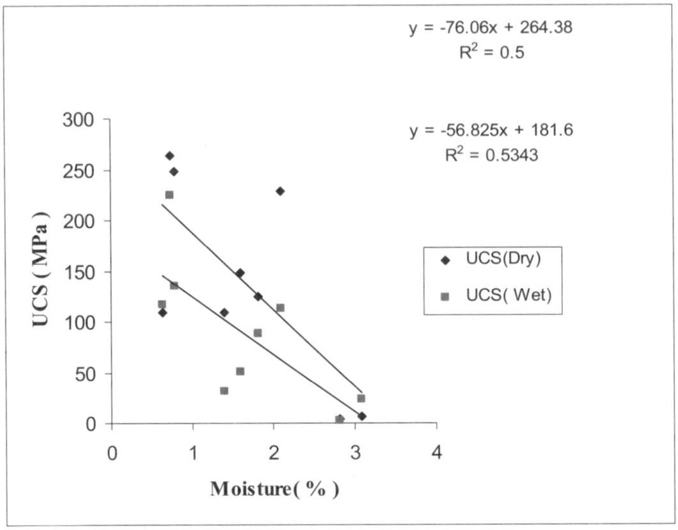

Figure 3: Relationship between U.C.S and moisture content.

Interrelationship with mineral composition

The UCS is correlated with mineral composition and illustrated in the Figures. 4 a-b. The strength of the rock decreases as the silica content increasing. The strength is influence by orientations of minerals also. However, these cases, most of the sample have parallel orientation with the loading direction.

Interrelationships with P- Wave Velocity

A linear relation exists between Vp and moisture (Figure.5). It marks inversely proportional relation with the moisture content. Ultrasonic wave propagation in rocks is controlled by the magnitude, distribution and nature of the porosity within the rock mass. The introduction of water into pore space decreases the shear moduli of the rocks by about 2 GPa, suggesting that there exists a fluid/matrix interaction at grain contacts, which reduces the rigidity (Assefa *et.al,* 2003).

P-wave velocity show linear correlation between the UCS in both dry and wet condition (Fig.6), however, wet rock shows higher values as compared to dry rock. V_p does not show good correlation with density (Figure. 7) due to higher porosity and void ratio, however other reported positive correlation. P- wave velocity decreases in both the dry and wet conditions (Figure. 8) as the silica content increases, it may be due to the presence of high mica content in rock.

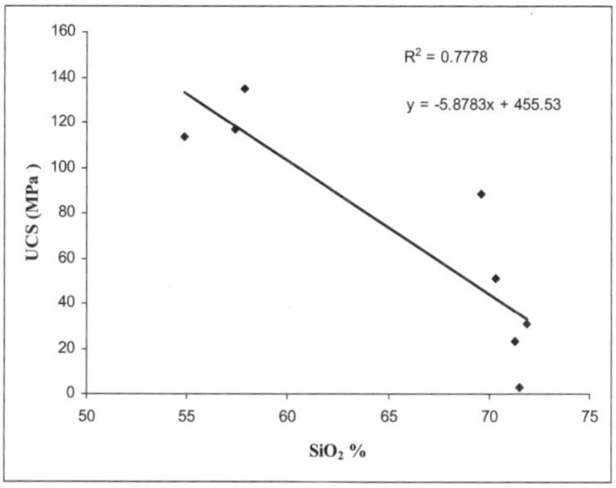

Figure.4a: U.C.S Vs Silica content.

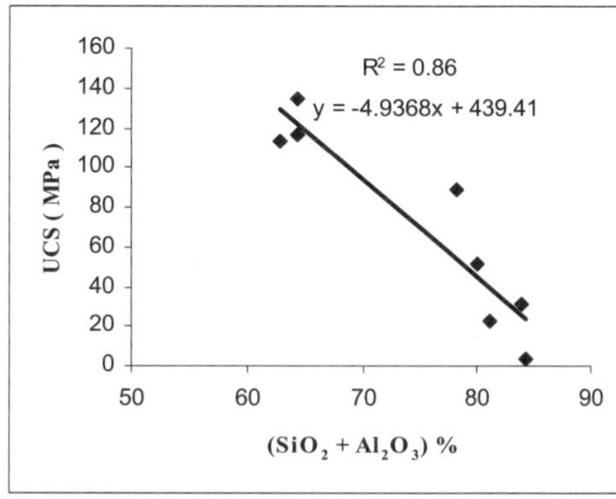

Figure.4b: U.C.S Vs (SiO_2+Al_2O_3) content.

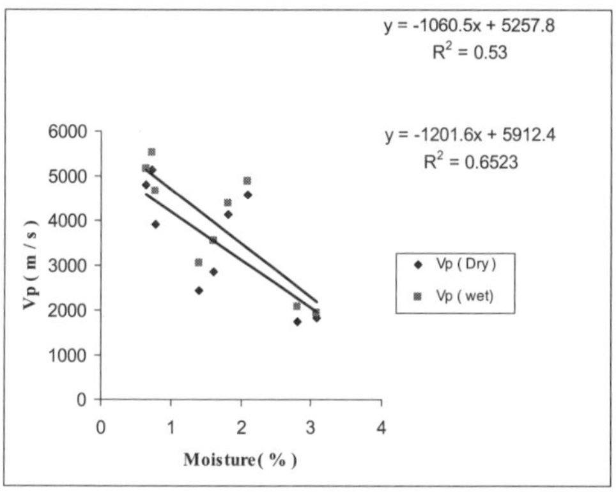

Figure.5: Vp Vs Moisture content.

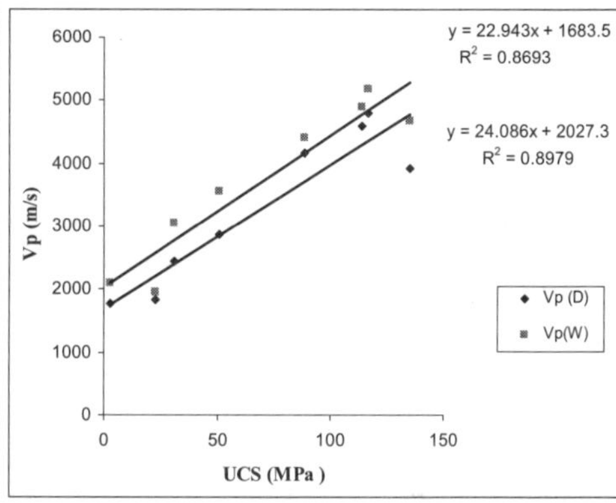

Figure.6: Relationship between Vp Vs U.C.S.

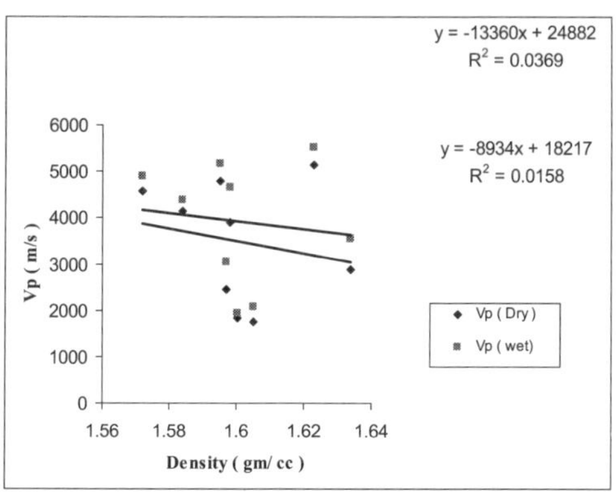

Figure.7: Relationship between Vp Vs Density.

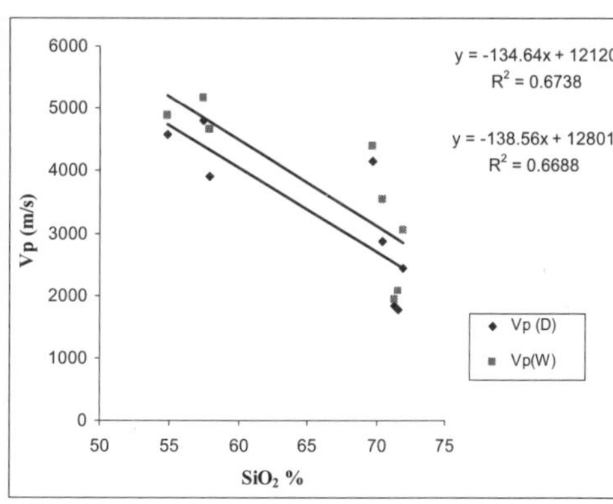

Figure.8: Vp Vs Silica content.

CONCLUSIONS

The present study provides certain useful inferences about the sandstone of the study area which is very susceptible to failure. The physico-mechanical properties have been well correlated with chemical composition of the sandstone. It mainly contains quartz, feldspar and mica (Biotite as well as muscovite). The relation between U.C.S and silica content shows high degree of correlation coefficient 0.77. At higher silica content P-wave velocity decreases linearly. Reduction in strength due to wet condition is mainly responsible for failure in the area particularly in the rainy seasons. The P-wave velocity can be used for understanding of strength properties of rock particularly in Siwalik region.

REFERENCES

Assefa S., McCann C and Sothcott J. (2003). Velocities of compressional and shear waves in limestones.Geophyscial prospecting. V. 51, pp. 1-13.

Bell, F. G. and Culshaw, M. G., (1993). A survey of geotechnical properties of some relatively weak Triassic sandstone. In: Cripps, J. C., Coulthard, J. M., Culshaw, M. G. and Forster, A. (eds) The engineering geology of weak rock. Eng. Group Geol. Soc. Spec. Publ.V. 8, pp. 139 – 148.

ISRM, (1980). Suggested methods for determination of compressive strength. Int. Soc. Rock Mech. Comm. Lab. Tests DOC.1.

Olivera, R., (1993). The engineering geology of weak rock. In: Cripps, J. C., Coulthard, J.M., Rafavich F., Kendall C. and Todd T. (1984). The relationship between acoustic properties and the petrographic character of carbonate rocks. Geophysics. V. 49, pp. 1622 - 1636.

Pant, P. D. and Luirei, K., (1999). Malpa rockfalls of 18th august 1998 in the northeastern Kumaun Himalaya. Jour. Geol. Soc. India, V. 54, pp. 415-420.

Rehbinder, P. A., Schreiner, L. A. and Zhigach, K. F., (1948). Hardness reducers in drilling. Trans. CSIRO Report.

Shakoor, A. and Bonelli, R. E., (1991). Relationship between petrographic characteristics, engineering index properties and mechanics properties of selected sandstone. Bull. Assoc. Eng. Geol., V. 28, pp. 55 – 71.

Shappiro, L. and Brannock, W.W., (1962). Rapid analysis of Silicate, Carbonate and Phosphate rocks. US Geol. Sur. Bull., V. 1144-A, pp.56.

Singh, T.N, and Naidu, S., (2000). Influence of strain rate on mechanical behaviour of rocks. Ind. J. Engg. and Mat. Sci. V.59, pp. 482-486.

Singh, T.N, Ray, S.K. and Singh, D.P., (1994). Effect of Uniaxial cyclic Compression on the mechanical behaviour of rock. Ind. J. Engg and Material Sc. V. 2, pp. 110 – 120.

Singh, T.N, Sarkar, K. &.Bali, R. (, 2005). A Geotechnical Investigation Of Rocks Of Amiyan Landslide Area, Kumaun Himalaya, Uttaranchal. Mining Engineers Journal, V.6 (7), pp 21-26.

Ulusay, R., Tureli, K. and Ider, M. H., (1994). Prediction of engineering properties of selected litharenite sandstone from its petrographic characteristics using correlation and multivariate statistical techniques. Eng. Geol. V. 37, pp. 135 – 157.

International Conference on Geotechnical Engineering for Disaster Mitigation & Rehabilitation
Chu, Phoon & Yong (eds) © 2005 World Scientific Publishing Company ISBN 981-256-469-1

NUMERICAL ANALYSIS OF SYNCRUDE TAILINGS DAM WITH CONSIDERATION OF STRENGTH ANISOTROPY OF FOUNDATION MATERIAL

HOSSEIN SOLTANI-JIGEH
Civil and Environmental Engineering Department, Amirkabir University of Technology, 424, Hafez Ave, Tehran, Iran

ABBAS SOROUSH
Civil and Environmental Engineering Department, Amirkabir University of Technology, 424, Hafez Ave, Tehran, Iran

This paper studies the effect of anisotropy of steady state strength of loose Syncrude sand, obtained from Syncrude mine tailings in Canada, on the stress-strain behavior of a tailings dam constructed over the sand. The tool for this study is the finite element technique. Primarily, numerical analyses are performed with isotropic steady state shear strength of the sand, resulted from triaxial compression loading. Then, the analyses are repeated with anisotropic strength by introducing an anisotropic ratio (A_S). A_S is the ratio of the shear strength from triaxial extension loading divided by the shear strength from triaxial compression loading. The results of anisotropic analyses show that the tailings dam and foundation are comparatively in a more critical condition, with respect to yield zones, displacements and strains.

INTRODUCTION

Inherent and stress-induced anisotropy cause the behavior of soils to be different for various loading paths. It is well known to geotechnical engineers that values of shear strength parameters of geomaterials depend on the direction of the major principal stress and the magnitude of intermediate principal stress, compared to the magnitude of the minor principal stress. In saturated soils another factor that can induce anisotropic behaviour is the undrained loading-induced excess pore water pressure. For a given soil, the value of the excess pore water pressure depends on the stress path, the level of stress, and the principal stress directions.

Recently, deformation behavior of loose sands under monotonic and cyclic loading has been lionized. The anisotropic behaviour of saturated sands is also important from the liquefaction point of view. Therefore, it is necessary to consider the anisotropy of the shear strength parameters in evaluating the behavior of earth structures constructed with or on these types of soils.

In this paper, the finite element technique is employed to evaluate the effect of shear strength anisotropy of geomaterials on the stability and behavior of earth structures. For this purpose, stresses and deformations in the Syncrude tailings dam and foundation subjected to an undrained loading condition are evaluated by means of numerical analysis. First, experimental and theoretical background of the research is briefly presented.

UNDRAINED BEHAVIOUR OF SATURATED SANDS

A considerable number of laboratory tests have shown that the undrained shear strengths of saturated sands decrease with increasing of the angle between the major principal stress and the vertical direction (i.e. α_σ) (Bjerrum, 1973; Vaid et al., 1995a,b; Uthayakumar and Vaid, 1998; Nakata et al., 1998).

Vaid et al. (1995a) carried out a number of undrained triaxial tests on saturated samples of the Syncrude tailings sand. They showed that the undrained response of the loose sand is strongly influenced by the principal stress direction, as shown in Figure 1. Triaxial compression shearing (i.e., α_σ = 0°) resulted in a strain hardening behavior, whereas in triaxial extension shearing (i.e., α_σ = 90°) the sand behavior was strongly strain softening. The behavior of the sand during simple shearing, in which α_σ is about 45°, was intermediate.

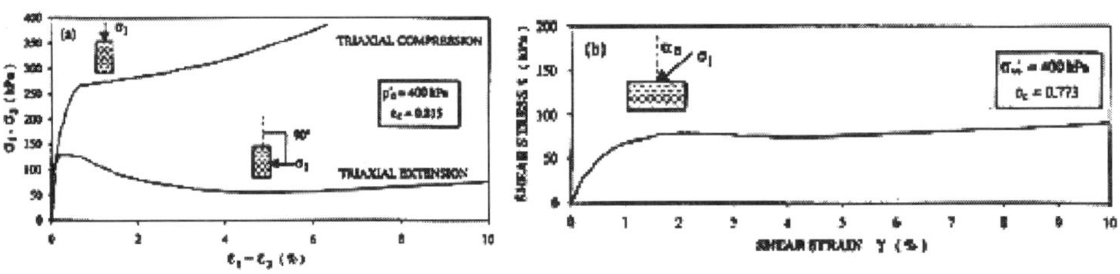

Figure 1. Stress-strain undrained behavior of the Syncrude tailings sand: a) compression and extension triaxial tests, b) simple shear test (Vaid et al., 1995a).

Vaid and Uthayakumar (1998) carried out a number of triaxial tests in undrained conditions and showed that with the change in principal stress direction from vertical to horizontal, the behavior is changed from hardening to softening for the Fraser River sand and from gentle softening to strongly softening for the Syncrude sand. They related this high difference in the behavior to the anisotropic fabric of the sands, which is induced by rearrangement of particles during the sedimentation. Variations of the undrained peak shear strength and the steady state shear strength of the Syncrude sand versus maximum principal stress directions are presented in Figure 2. It is observed that both of the strengths decrease with the increase of α_σ from 0° to 90°.

As shown in Figure 2, the undrained shear strength of the Syncrude tailings sand can be defined as a function of α_σ. For simplicity, this relation can be considered linear;

therefore, the undrained steady state shear strength of the sand under triaxial loading is defined, as follows:

$$\frac{S_{u,A}(\alpha_\sigma)}{S_{uc}} = \left[\left(\frac{A_S - 1}{90}\right)\alpha_\sigma + 1\right] \tag{1}$$

Where:

$S_{u,A}(\alpha_\sigma)$ = anisotropic undrained steady state shear strength

A_s = anisotropy ratio of shear strength, defined as $A_S = S_{ue}/ S_{uc}$.

S_{uc} and S_{ue} = steady state shear strength of sand, obtained from triaxial compression and extension, respectively

α_σ = angle between the major principal stress and the deposition plane

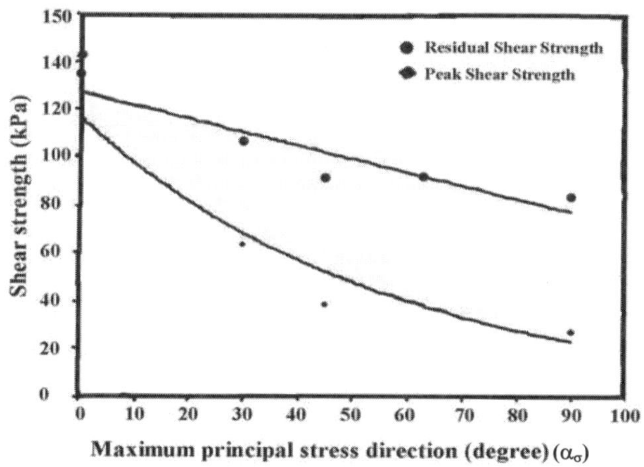

Figure 2. Peak and steady state shear strengths of Syncrude sand versus maximum principal stress direction (adopted from Uthayakumar et al., 1998).

SYNCRUDE TAILINGS DAM

Figure 3 shows a cross section of the Syncrude tailings dam and its foundation, constructed in the Syncrude area in the vicinity of Fort McMurrary, Alberta, Canada in 1995. The structure includes 11 m of submerged tailings sands and a 10 m clay embankment over the tailings. About 1.5 m of soft clay exists between the tailings sand and the in-situ foundation materials. The contained tailings are piped behind the clay embankment in order to exert an undrained loading on the tailings sands of the foundation. The stages of the dam construction and the rapid filling of the dam reservoir by tailings were components of a field event (prototype) as a part of the CANLEX (Canadian Liquefaction Experiment) project. The mechanical properties of different materials used in the analysis are summarized in Table 1.

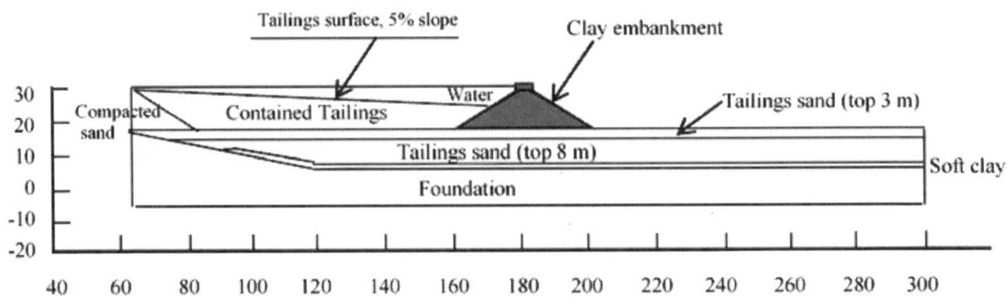

Figure 3. A cross section of Syncrude tailings dam and foundation (Soroush et al. 1996).

Table 1. Material properties used in analyses of Syncrude tailings dam.

	γ (kN/m^3)	E (kN/m^2)	Possion ratio, ν′	Internal friction angle, ϕ′ (°)	Cohesion, c′ (kPa)
Foundation	19	70000	0.3	38	25
Soft clay layer	19	10000	0.47	8	-
Tailings (lower 8.0 m)	18.75	14000	0.3	34	-
Tailings (upper 8.0 m)	19.2	18000	0.3	36	-
Compacted sand	19.25	100000	0.3	45	-
Clayey embankment	19	15000	0.37	25	25
Contained tailings	18.75	10000	0.3	28	-

The finite element program PISA was employed for the analysis. A drained static analysis was performed to simulate the very slow construction rate of the clay embankment and compacted sand embankment. An undrained analysis was performed at the stage of filling of the contained tailings. This analysis was repeated by considering isotropic (A_s=1.0) and anisotropic (A_s=0.60) strengths of the tailings, separately. In the analyses, the undrained steady state shear strength (S_u) of the tailings sand is introduced as a function of mean effective stress (p'_0). Based on the laboratory test results, S_u/p'_0 was assumed between 0.16-0.20.

ANALYSES RESULTS

Results of the analyses at the end of the stage of filling of contained tailings for both isotropic and anisotropic steady state shear strength of the Syncrude sand are presented in Figures 4 to 6. Figure 4 shows that for the analysis with anisotropic strength (A_s=0.60), the yield zone is comparatively wider.

Displacement vectors resulted from the analyses are shown in Figures 5a and 5b. A comparison of these figures indicates that the values of the displacements considering shear strength anisotropy are larger than the displacements resulted from the analysis with isotropic shear strength.

Figure 6 presents shear strains for six points on a given slip surface through the embankment and the tailings foundation. It is evident that the shear strains resulted from the analysis with the anisotropic shear strength of the tailings sands are relatively higher.

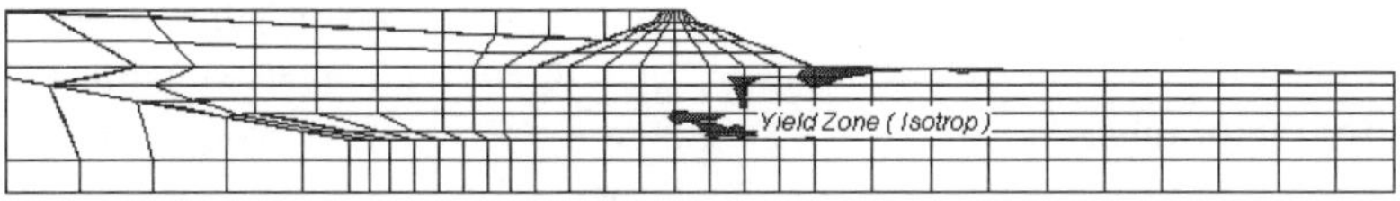

(a)

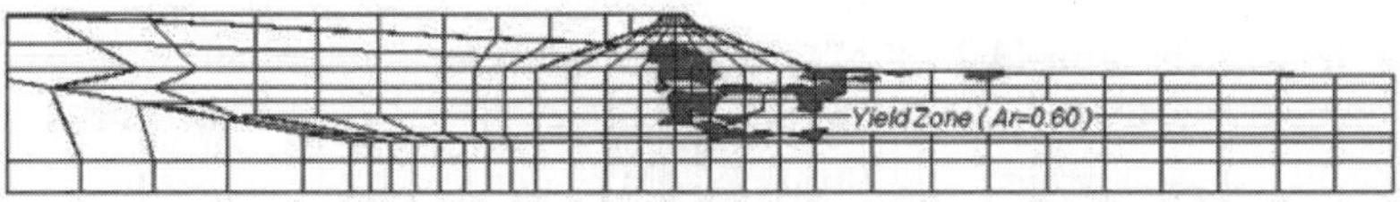

(b)

Figure 4. Yield zones: a) isotropic analysis (A_s=1.0), b) anisotropic analysis (A_s=0.60).

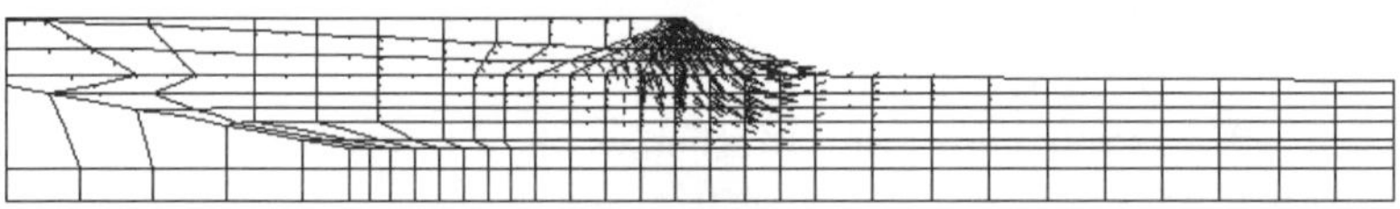

(a)

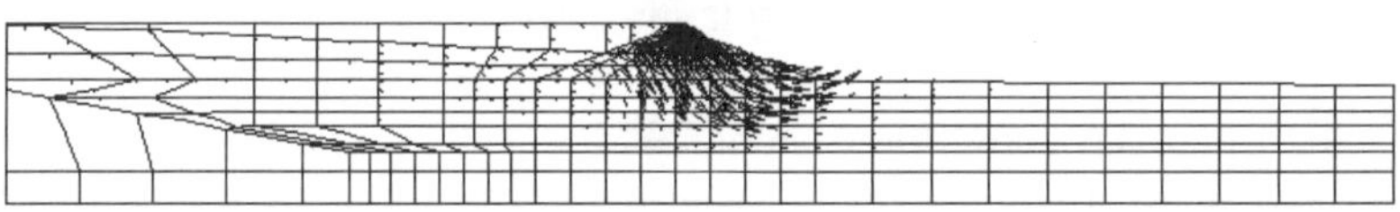

(b)

Figure 5. Displacement vectors: a) isotropic analysis (A_s=1.0), b) anisotropic analysis (A_s=0.60).

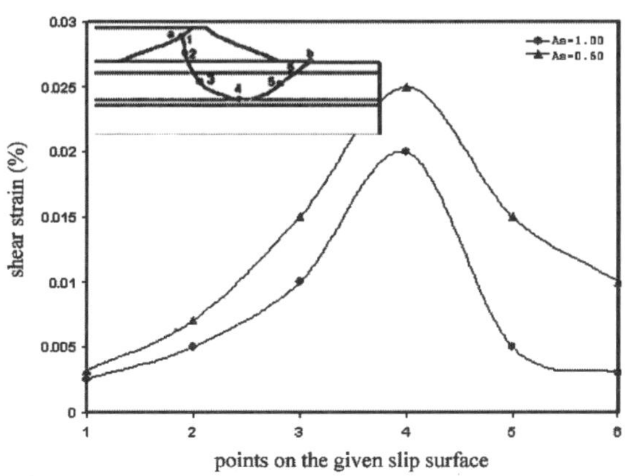

Figure 6. Shear strains of six points at a given slip surface for isotropic analysis (A_s=1.0) and anisotropic analysis (A_s=0.60).

SUMMARY

Laboratory studies have shown that the undrained mechanical behavior of geomaterials is anisotropic. The results of these studies show that the shear strength decreases as the angle between the principal stress and the vertical direction increases. In this paper, effect of the above factor on shear strength of saturated sands was reviewed. Then, the effect of anisotropy of steady state strength of the Syncrude tailings sand on the stress-strain behaviour of Syncrude tailings dam and its foundation was evaluated by employing finite element method. An undrained analysis was performed at the stage of filling of the contained tailings by considering isotropic strength (A_s=1.0). Again, this analysis was repeated by anisotropic strength (A_s=0.60) of tailings. The results of analyses showed that with using the shear strength anisotropy of the Syncrude sand, the tailings dam and its foundation are in comparatively more critical conditions with respect to yield zones, displacements and shear strains.

REFERENCES

Bjerrum, L. (1973). "Problems of Soil Mechanics and Construction on Soft Clays", *Proc. 8th ICSMFE*, Moscow, Vol. 3, 109-159.

Nakata, Y., Hyodo, M., Murata, H., and Yasufuku, N. (1998). "Flow deformation of sands subjected to principal stress rotation", *Soils and Foundations*, Vol. 38, No. 2, 115-128.

Soroush, A., Chan, D. and Morgenstern, N. R. (1996). "Numerical Analysis of the CANLEX Phase III field Event", *Proc. 49th Canadian Geotechnical Conference*, 605-614

Uthayakumar, M., and Vaid, Y. P. (1998). "Static Liquefaction of Sands under Multiaxial Loading", *Canadian Geotechnical Journal*, Vol. 35, 273-283.

Vaid, Y.P., Uthayakumar, M. E, Sivathayalan, S., Roberston, P.K., and Hofman, B. (1995a). "Laboratory Testing of Syncrude Sand", *Proc. 48th Canadian Geotechnical Conference*, 223-232.

Vaid, Y. P., Sivathayalan, S., and Uthayakumar, M., and Eliadorani, A. (1995b). "Liquefaction Potential of Reconstituted Syncrude Sand", *Proc. 48th Canadian Geotechnical Conference*, 223-232.

International Conference on Geotechnical Engineering for Disaster Mitigation & Rehabilitation
Chu, Phoon &Yong (eds) © 2005 World Scientific Publishing Company ISBN 981-256-469-1

SEISMICITY IN SLOPE STABILITY PRACTICE IN MALAYSIA

W. H. TING
Zaidun Leeng Sdn Bhd , Tingkat 5, Jalan Bukit Nanas,
50250 Kuala Lumpur, Malaysia.

T.A. OOI
TAO Consult Sdn Bhd, 17A, Jln Awan Hijau,Taman Overseas Union
Batu5,Off Jln Klang Lama,58200 Kuala Lumpur, Malaysia

AMINUDDIN AHMAD
Zaidun Leeng Sdn Bhd, Tingkat 5, Jalan Bukit Nanas
50250 Kuala Lumpur, Malaysia

B. K. TAN
Universiti Kebangsaan Malaysia,
43600 Bangi, Selangor Darul Ehsan, Malaysia

Recent earthquake experiences have revealed that Malaysia is vulnerable to the earthquake originating from active plate boundaries in terms of tremors. Bridges, high-rise buildings and dams are greatly exposed to these seismic activities during their lifetime. This paper presents the seismic design requirements in terms of different g values used in various dam slope stability designs in Malaysia. The design g value applied in each dam site is dependent on the information, advice and guidance given by the Malaysian Meteorological Service (MMS) as the national information centre for seismology.

INTRODUCTION

Malaysia, in particular Peninsular Malaysia, has been considered as a non-seismic risk country until the tsunami incidence on December 26, 2004. In 1982, the Public Works Department, Malaysia, cautioned in the Keynote address to the Conference on Tall Buildings in Kuala Lumpur that Malaysia is actually located in the 'ring of fire' that marks the areas affected or likely to be affected by earthquakes and predicted that Malaysia will experience in the future severe earthquakes arising from the Sumatra and Andaman fault that runs the length of Sumatra and the Andaman Sea to the north of the Straits of Malacca (Yunus, 1982).The tsunami that occurred on 26 December 2004 can be said to be the prediction that came true. The interest in earthquake design amongst the Malaysian engineers in fact already started then. The Komtar Towers in Penang and the Penang Bridge both built in the 1980s in fact were designed for seismic forces.

TECTONIC SETTING OF MALAYSIA

Malaysia is surrounded by the two most seismically active plate boundaries namely, the plate boundary between the India Plate (part of the huge Indian-Australian Plate that underlies the Indian Ocean and the Bay of Bengal) and the Burma Plate (part of the Eurasian Plate) on the west at the Sunda (Java) Trench off Sumatra, and (b) the plate boundary between Eurasian and Philippine Plates on the east. Large earthquakes originating from in and around these boundaries resulted in tremors being felt in Malaysia (Tibballs, 2005). The maximum observed intensities of these tremors on the Modified Mercalli (MM) scale are VI for Peninsular Malaysia and VII for East Malaysia (Mohd Rosaidi bin Che Abas, 2001). Figure 1 shows the major tectonic plates and plate boundaries around Malaysia.

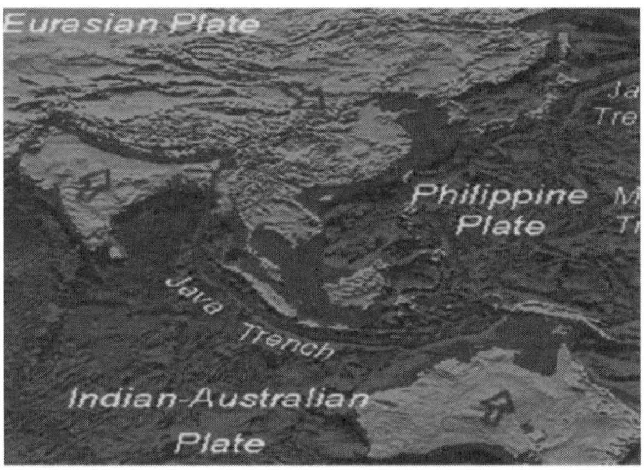

Figure 1. Major tectonic plates and plate boundaries around Malaysia

Two recent major earthquake events occurred in Sumatra on the 4 June 2000 and 26 December 2004 with measured magnitudes of 7.9 and 9.3 respectively on the Richter scale. These events had resulted in people having to evacuate from their homes and office buildings in panic and most significantly the loss of life during the tsunami that struck Kuala Kedah and Penang Island on 26 December 2004. These events clearly indicate that although Peninsular Malaysia is classified as a seismically stable area, we are still vulnerable to the effect of earthquakes that originated from the Sumatran active fault. Large earthquakes from this fault have created considerable ground motions over the western part of Peninsular Malaysia. The locations of epicentres of earthquakes near Peninsular Malaysia and in Sabah and Sarawak are shown in Figures 2 to 4. Earthquakes in plate boundary areas usually cluster in place and time. At a particular earthquake source location, clusters may recur after a long period of time perhaps thousands or hundreds of thousands of years and large amount of energy will have to be released (Seismology Research Centre Victoria Australia, 2000).

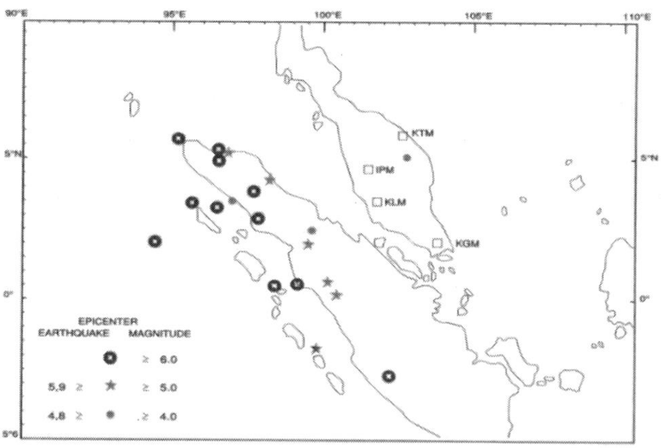

Figure 2. Location of Epicentre of Sumatra Earthquakes felt in Peninsular Malaysia since 1967 (Tungah Surat, 2001).

THEORY AND PRINCIPLE OF SEISMIC FORCES ON DAM SLOPES

In most earthquake hazard studies, the main concern is the response of the structure to earthquake vibrations. There are ground motion hazards, surface rupture, landslip, liquefaction or tsunami. East Malaysia is affected by low magnitude earthquakes that measured below 6.0 on the Richter scale, which is a measure of the earthquake magnitude (M) that includes the logarithm of the amplitude of the motion recorded on seismograph. The earthquake is assumed to produce ground acceleration intensity below 0.15g recorded by the accelero-graph. These intensity measurements are essential in determining the risk to structural damage (Azlan et al. , 2001). The seismic effects are then used in the design of bridges, high-rise buildings and dams in Malaysia since these structures could be exposed to important seismic effect during their service lifetime.It is determining the risk to structural damage (Azlan et al. , 2001). The seismic effects are then used in the design of bridges, high-rise buildings and dams in Malaysia since these structures could be exposed to important seismic effect during their service lifetime.It is also noted that the liquefaction of poor soil deposits is negligible since in Malaysia we do not have high waste dump and thick natural soft clay deposit as in Sweden and other parts of the world. However, the highrise condominiums located in the alluvium over the cavernous Limestone formation in Kuala Lumpur have experienced the effect of earth tremors. During earthquake, landslides in dam slopes and slumping of dam embankment may occur and some failures may be of major consequences. These failures are attributed to increase in shear stresses caused by seismic loading. Decrease or loss of strength during cyclic loading due to ground motion may have caused major and catastrophic

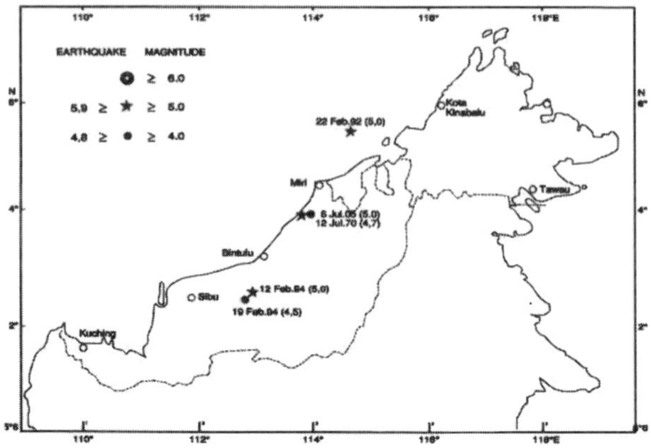

Figure 3. Local Earthquakes with known epicentral position and magnitude in Sarawak (Tungah Surat, 2001)

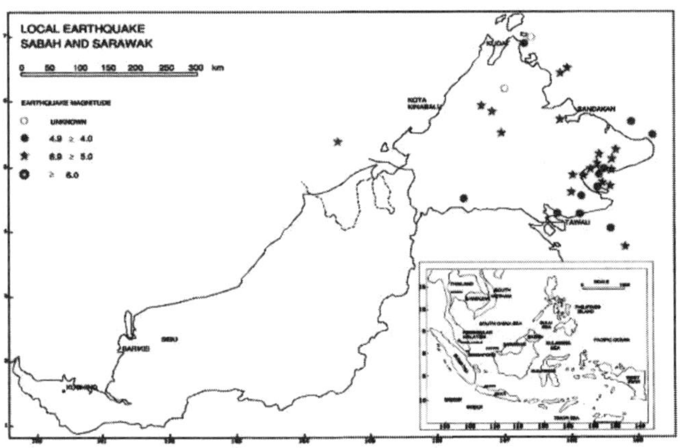

Figure 4. Location of Epicentre of Local Earthquakes in Sabah (Tungah Surat, 2001)

failures. Figure 5 shows the seismic force acting on a potential sliding mass. The scismic force is equal to the weight of the potential sliding mass times a seismic coefficient in terms of g (ground acceleration intensity). Safe slope gradients for dam slopes are determined using slope stability analysis program using the limit equilibrium method to check for slip failure conditions for the undrained, drained and rapid draw-down conditions. The input for g values is normally shown as earthquake loading coefficients.

DESIGN OF SLOPE STABILITY FOR DAMS

It is common practice in Malaysia to meet seismic design requirements by using different g values for dam slope stability design. Table 1 shows the different g values used. The design g value applied in each dam site is dependent on the information, advice and guidance given by the Malaysian Meteorological Service (MMS) as the national

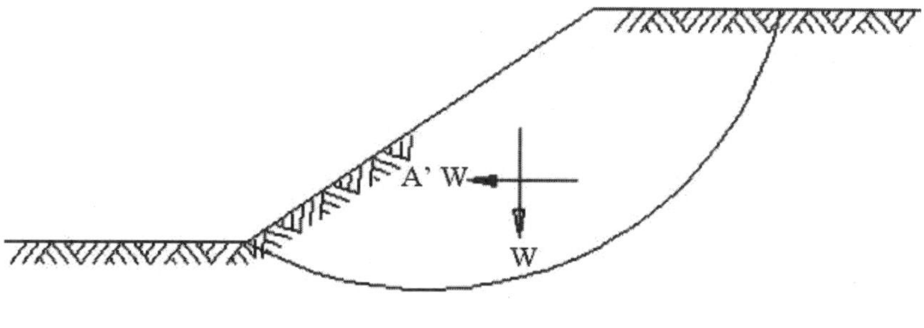

Maximum ground acceleration g

A' : Seismic coefficient
W : Weight of potential sliding mass
A'W : Siesmic Force which is equal to the weight of the potential sliding mass time a seismic coefficient A'.

Figure 5. Seismic force upon potential sliding mass.

Table 1 Different g values used in dam slope stability design in Malaysia

	Dam Locations	Design g values
1.	Gemencheh, Negeri Sembilan	0.10g
2.	Sungai Selangor, Selangor	0.10g
3.	Sg. Kinta, Perak	0.15g
4.	Bakun, Sarawak	0.20g
5.	Kelakong, Sarawak	0.30g

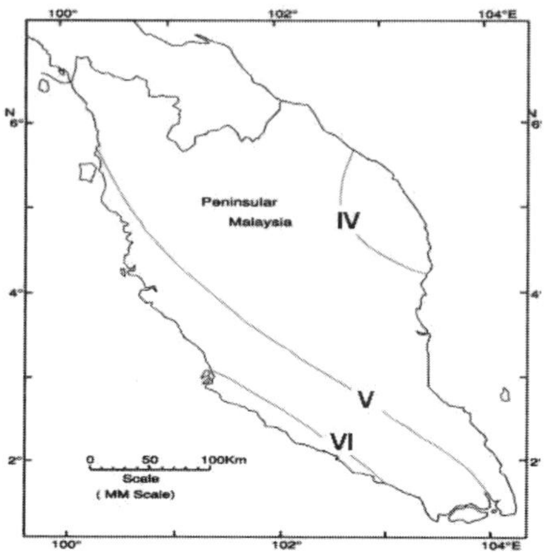

Figure 6. Maximum Observed Earthquake Intensity in MM scale for Peninsular Malaysia.

information centre for seismology. The iso-seismal maps from the Malaysian Meteorological Service (MMS) are used as an indicator for zones that received different levels of recorded intensity on the Modified Mercalli (MM) scale. The maximum observed intensity was VI for Peninsular Malaysia and VII for East Malaysia on MM scale (Mohd Rosaidi bin Che Abas, 2001). Figure 6 shows the iso-seismal map for Peninsular Malaysia.

CONCLUSIONS

Seismic factors have been considered in the design of dam slope stability in Malaysia. The design requirements in terms of different g values (ground acceleration intensity) used in slope stability analyses for dams are dependent on the information, advice and guidance given by the Malaysian Meteorological Service (MMS) as the national information centre for seismology. Based on distinct differences in seismicity, higher design g values are used foe East Malaysia when compared to Peninsular Malaysia. As for the Peninsular Malaysia, the west coast has a higher seismic intensity in MM scale when compared to that of the east coast, it is therefore expected to experience greater effect of the tremors. In view of recent earthquake activities that have affected Malaysia, it is hoped that Malaysia will have its own seismic hazard assessment map for various parts of Malaysia on structures such as bridges, high-rise buildings, dams etc. Only then will it be able to apply its own seismic regulations on all structures in order to safeguard the general public in terms of loss of lives and properties, mitigation measures in disaster prevention and environmental protection.

REFERENCES

Azlan Adnan et al. (2001). "Seismic Hazard Assessment of Malaysia", *Seismic Risk Seminar 2001*, Malaysia.

Mohd Rosaidi bin Che Abas. (2001). "Earthquake Monitoring in Malaysia", *Seismic Risk Seminar 2001*, Malaysia.

Seismology Research Centre Victoria Australia (2000). "*Review of Seismicity, Kelalong Dam for GHD* ", June 2000. Technical Study Report , 25.

Tibballs, G. (2005). "*Tsunami: The World's Most Terrifying Natural Disaster*", Carlton Books Ltd., London.

Tungah Surat. (2001). "Case History of Earth Tremors in Malaysia", *Seismic Risk Seminar 2001*, Malaysia.

Yunus, M. M. Y. (1982). "Keynote Address- Building High", *Proceedings Asian Regional Conf. on Tall Buildings and Urban Habitat, Kuala Lumpur*, Institution of Engineers, Malaysia, K-1 to K-8

International Conference on Geotechnical Engineering for Disaster Mitigation & Rehabilitation
Chu, Phoon & Yong (eds) © 2005 World Scientific Publishing Company ISBN 981-256-469-1

A COMPREHENSIVE ASSESSMENT ON THE DISASTRIOUS CHARACTERISTICS OF LANDSLIDE

JEFFREY WANG
Tritech Consultants Pte Ltd, Singapore

KONG GUANGYA, YANG WENYI
Tiandi Science and Technology Co. Ltd, China

This paper has presented a method by using the tool of weight functions to assess the disastrous characteristics of landslides. This method allows us to consider the key factors which affect landslides to make a comprehensive assessment quantitatively the possibility of disastrous landslides. The method has been used in the feasibility study of the Three Gorges dam feasibility study to assess the slope stability of the dam and the disastrous characteristics of potential landslides.

GENERAL

One of the problems associated with the assessment of landslides in hydraulic dam area is the prediction of disastrous characteristics of landslides during the process of water level rising. Upon the landslide occurs under any circumstance, the sliding speed of landslide is one of the key effects on the impact to the safety of dam and surroundings. The current method on prediction of landslide speed is very primitive, which are only based on site observation and preliminary judgment. The actual prediction on the possibility of high speed landslide is actually impossible, as the factors which affect the movement of landslide is too many to judge if without a systemically assessment.

SALIENT FACTORS AFFECTING LANDSLIDE SPEED

A landslide could be disastrous to surrounding if it is located in urban area, or near a dam, or other sensitive buildings. To classify the disastrous level of a landslide, one sensible way is to use its sliding speed including maximum speed and average speed of landslide (Table 1).

It is prudent to consider a landslide as a system rather than an isolated case. There are many factors, which affects the occurring, sliding of a landslide. It is hardly to combine all those factors to conclude the disastrous level of a landslide. However, if we analysis all factors including geology, geometry, environment, rain, human factor etc., it is possible to summaries the salient factors which affect the sliding speed of a landslide as the followings: 1) volume of landslide (v); 2) slope of slide path (α) ; 3) relative height of shear-out position (h); 4) transitional angle of sliding path (β); 5) magnitude of earthquake (a); 6) rain factor (w); 7) deformation of landslide body (u); 8) ratio of length of a landslide to length of path (K_L).

Table 1. Classification of landslide according to sliding speed.

	Average sliding speed	Maximum sliding speed	Disastrous level	Engineering solutions
High speed landslide	>10 m/sec.	>20 m/sec.	High	Serious precautions, close monitoring. Strengthen if possible
Medium speed landslide	5 to 10 m/sec.	10 to 20 m/sec.	Average	Close monitoring. Strengthen if possible.
Low speed landslide	< 5 m/sec.	>20 m/sec.	Possibly low	Monitoring.

Where the rain factor (w), transitional angle of sliding path (β) and ratio of length of a landslide to length of path (K_L) are defined as (Fig.1):

(1) $w = \dfrac{W}{W_b}$ (where W is water content and W_b is the saturated water content)

(2) $\beta = (\alpha_1 - \alpha_2)$

(3) $K_L = \dfrac{L_2 + L_3}{L_1}$

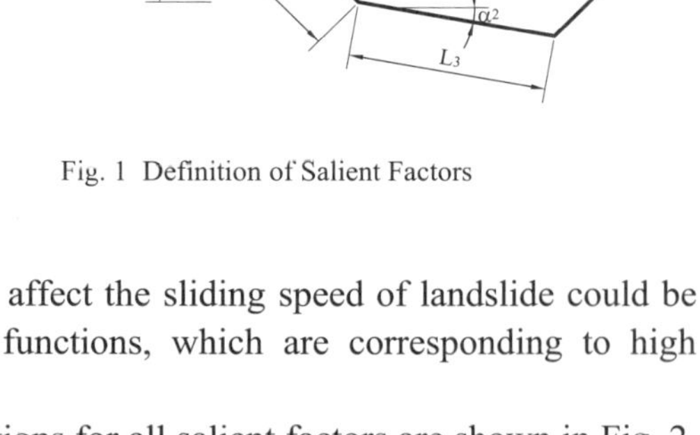

Fig. 1 Definition of Salient Factors

WEIGHT FUNCTION OF SALIENT FACTORS

The contribution of the salient factors which affect the sliding speed of landslide could be quantitatively represented by three weight functions, which are corresponding to high speed, medium and low speed, respectively.

The standard trends of three weight functions for all salient factors are shown in Fig. 2, in which Fig. 2 (a) shows the weight function of the particular salient factor attributive to high speed, whereas Figs.2 (b) and (c) to medium and low speed landslides, respectively. However, the weight functions of transitional angle of sliding path (β) are different from others, which are shown in Fig. 3.

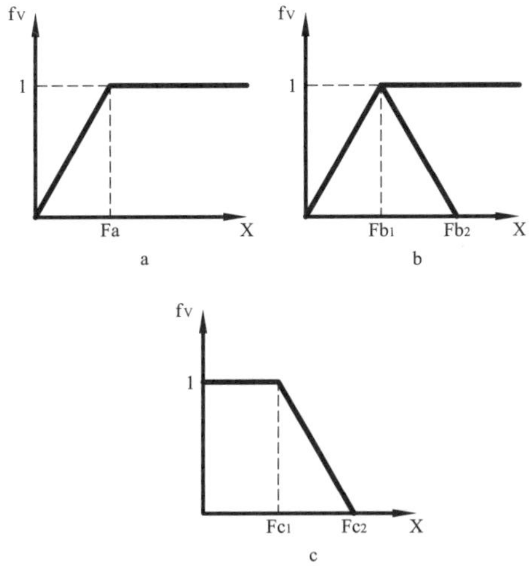

Fig. 2 Typical Fc1 Form of Weight Functions

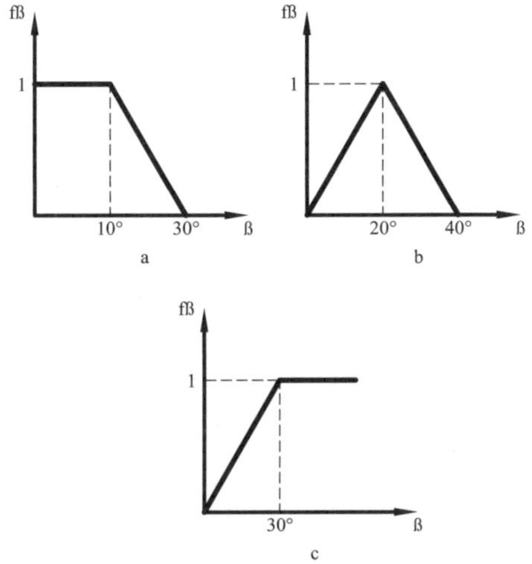

Fig. 3 Weight Function of transitional angle ß

The characteristic values F_{ij} of each salient factor could be obtained from overall analysis of the historical records of different types of landslides (Table 2). On the basis of the above weight function, it is possible to calculate the value of combined coefficient η_{ij}, which reflects the possibility of each case, i.e. high sliding speed, medium sliding speed or low sliding speed.

(4). $\sigma_{ij} = \sum_{k=1}^{8} W_{kj}(d_{ik})\eta_{kj}$

Where, i =1,2,3, representing the cases, i.e. high sliding speed, medium sliding speed or low sliding speed; Wkj is value of white function of each salient factor, j=1,2,3 …8, representing each salient factor; d_{ik} is actual values of salient factors of the particular landslide to be assessed.

$\eta_{kj} = \frac{F_{kj}}{\sum_{i=1}^{8} Fij}$, Where j=1,2,3, representing the cases, i.e. high sliding speed, medium sliding speed or low sliding speed; F_{ij} refer to the characteristic values of each salient as shown in Fig. 2.

ANALYSIS OF OCCURRED LANDSLIDS BY WEIGHT FUNCTION METHOD

A few of large scale landslide which occurred in China were assessed by using the above described method. Table 3 shows the values of all salient factors, which affect the sliding speed, and Table 4 shows the results of assessment and the comparison between the assessed results and actual conditions. As can be seen, the results of assessment of each occurred landslides show a good agreement with the actual conditions, which in return show the validity of the proposed method. It should take a note that when calculating the combined coefficient η_{ij}, the values of slope of slide path (α) and transitional angle of sliding path (β) should be converted to 10 times of radian to minimize the magnitude effect.

Table 2. Index values of weigh function of each salient factors affecting sliding speed.

	Factors	F_{a1}	F_{a2}	F_{b1}	F_{b2}	F_{c1}	F_{c2}
1	volume of landslide (v)	7	-	6	12	5	12
2	slope of slide path (α)	30°	-	20°	40°	10°	30°
3	relative height of shear-out position (h)	20m	-	5m	10m	0.2m	5m
4	transitional angle of sliding path (β)	10°	30°	20°	40°	30°	-
5	magnitude of earthquake (a)	7°	-	6°	12°	5°	12°
6	rain factor (w);	1	-	0.8	1.6	0.5	1
7	deformation of landslide body (u);	1m	-	0.5m	1m	0.2m	1m
8	ratio of length of a landslide to length of path (K_L).	1.5	-	1	2	0.5	1.5

Table 3. Values of all salient factors which affect the sliding speed of some occurred typical landslides in China.

	Landslides	v	α	H	β	a	w	u	K_L
		(x 1000 m3)	°	M	°	°		m	
1	No. 2 landslide	3800	30-45	0	<10	7	0.5	<1	<0.5
2	Longyang	80000	30	0	3-5	7	1	>1	>1.5
3	Mefang	80000	24	0	0	7	0.5	0	1
4	Chana	160000	40	0	0	7	0.5	>1	>1.5
6	Chaxi	46800	30- 45	0	0	7	0.5	>1	>1.5
7	Shaleshan	60000	23	0	0	7	0.5	1.5	>1.5

APPLICATION OF WEIGHT FUNCTION METHOD TO THREE GORGES DAM PROJECTS

Three Gorges dam project has about 686km long slope from Miao River to Jiangjin. It has been found that there are about total 277 number of landslides or potential landslide along the slope, with average frequency of one per 2 to 3km long river slope. As the catastrophic landslide (high speed landslide) could induce overflow of dam, or damage of the dam structure, it is substantial to investigate the sliding mode of each landslide and potential landslide, which would provide rational measures to prevent any possible high speed landslide along the river after the built-up of the dam.

Table 4. Assessment results of occurred landslide with weight functions and comparison to actual conditions.

	Landslides	Assessed results	Actual conditions
1	No. 2 slide	Low sliding speed	Low sliding speed, less catastrophic
2	Longyang	High sliding speed	High sliding speed, catastrophic
3	Mefang	Low to Medium sliding speed	Low sliding speed, less catastrophic
4	Chana	High sliding speed	High sliding speed, catastrophic
6	Chaxi	High sliding speed	High sliding speed, catastrophic
7	Shaleshan	High sliding speed	High sliding speed, catastrophic

Table 5. Values of all salient factors of typical large-scale landslides in Three Gorges.

	Landslides	v	α	H	β	a	w	u	K_L
		(x 1000 m3)	°	M	°	°		m	
1	Baiyiyan	21220	20	0	20	6	>1	0	0.7
2	Qiancaotuo	16250	20	0	20	6	>1	0	0.7
3	Baihuanpin	91120	50	0	50	6	>1	0	0.7
4	Sandenzi	15000	40	0	40	6	>1	0	0.7
5	Xinwu	46800	24	0	5	6	>1	0	0.7
6	Oh-Tang	27000	40	0	15	6	>1	0	0.7
7	Gaojiazui	27000	14	0	0	6	>1	0	0.7
8	Baota	27000	10	0	0	6	>1	0	0.7
9	Jipazi	27000	10	0	0	6	>1	0	0.7
10	Xichen	27000	10	0	0	6	>1	0	0.7
11	Jiuxianpin	27000	15	0	0	6	>1	0	0.7
12	Huangnashi	27000	30	0	0	6	>1	0	0.7
13	Fanjiaping	27000	19.4	0	0	6	>1	0	0.7

The above described weight function method has been adopted for a total 13 number of large scale landslides (volume more than 10million cubic meters). The values of all salient factors, which affect the sliding speed, are surveyed and shown in Table 5, and the results of assessment and the comparison between the assessed results and actual condition are shown in Table 6. As indicated in Table 6, most of landslide or potential landslide, if reactivated after the built-up of the dam, the slide speed is medium to low except Huangnashi landslide, which could be strengthened before the dam construction stage. This conclusion is of substantial to the overall evaluation of slope stability along the river for the feasibility study of Three Gorge dam project.

Table 6. Assessment results of occurred landslide with weight functions and comparison to actual conditions.

j	Landslides	Estimated values of combined coefficient η_{ij}			Assessment results of estimated sliding speed
		η_{1j} (high speed)	η_{2j} (medium speed)	η_{3j} (low speed)	
1	Baiyiyan	0.4496	0.652	0.6232	medium
2	Qiancaotuo	0.4496	0.653	0.6232	medium
3	Baihuanpin	0.462	0.490	0.6371	low
4	Sandenzi	0.461	0.5064	0.6359	low
5	Xinwu	0.460	0.5759	0.5153	medium
6	Oh-Tang	0.4781	0.5569	0.561	low
7	Gaojiazui	0.4467	0.5573	0.5233	medium
8	Baota	0.438	0.4534	0.518	low
9	Jipazi	0.438	0.542	0.535	medium
10	Xichen	0.438	0.538	0.531	medium
11	Jiuxianpin	0.449	0.5484	0.5060	medium
12	Huangnashi	0.615	0.546	0.410	high
13	Fanjiaping	0.459	0.562	0.491	medium

REFERENCES

Hsu, K.J. (1969). "Role of cohesive strength in the mechanics of overthrust faulting and landsliding", *Geol. Soc. Amer. Bull*, Vol. 80, 927-957.

Cruden, D.M. (1980). "Anatomy of landslide", *Canada Geotech.*, Vol. 17,. No.2.295-299

International Conference on Geotechnical Engineering for Disaster Mitigation & Rehabilitation
Chu, Phoon & Yong (eds) © 2005 World Scientific Publishing Company ISBN 981-256-469-1

STABILITY ANALYSIS OF A HIGH EXCAVATED SLOPE

WEN HAI-JIA, ZHANG JIA-LAN, ZHANG YONG-XIN
Faculty of Civil Engineering, Chongqing University, Chongqing 400045, P.R. China

This work aimed to study on stability-assessment way of a high excavated-slope. A stable slope was the key to insure the security of building site adjacent to the high slope, which frequently encountered in town construction at mountain areas by terrain limit. Some crucial problems were summarized for stability assessment of an excavated high slope based on several typical engineering cases in Chongqing. A synthetical-analysis way was explored to assess stability of high slope from its natural geological conditions and its man-destroyed degree, engineering environment, potential failure pattern of a high slope, calculation parameters and analysis methods. As a result, it is indicated that the conclusion of stability assessment can be determined according to the aspects above-mentioned, and the stability assessment is one of the fundamental data to insure the safety of the related construction, site and buildings.
Key words: high slope; geological environment; synthetical-analysis; stability assessment

INTRODUCTION

Mountains and highlands cover about one quarter of the world's landmass and are home to about 10% of the world's population. Building development of mountain areas, has received international attention, and the 2002-year was the United Nations International Year of the Mountains (U.N. IYM 2002).

Figure 1. The site of Wulong landslide, Chongqing

In China, mountains and highlands cover 69% of the country's land surface and are home to 56% of the Chinese population. So the 'All-out Development Strategy for Western Regions', brought into effect by the central government of P.R. China in 2000, is also a great movement towards developing mountainous areas in western China.

Increased development of mountain areas has appeared specific engineering problems associated with hill-side construction. Typically, the safety problem of building sites adjacent to high excavated slopes, widely encountered in west China, is an issue.

This problem is widespread in Chongqing, a city representative of mountainous regions in west China. To ensure the safety of buildings and construction sites, an assessment of stability of the excavated slope is required to determine whether advance reinforcement is necessary. Unstable excavated slopes may lead to landslides, such as the incident at Wulong, Chongqing, (2001), shown in Figure 1, where a 9 storey building collapsed and 86 people were killed/injured.

Therefore, stability assessment is very important for building sites adjacent to excavated high slopes. However, it is difficult to obtain accurate assessments because of the many complicated factors determining the *in-situ* slope stability. Crucial factors include: natural geological conditions, degree of anthropogenic disturbance, engineering environment, potential failure pattern, calculation parameters and analysis methods.

GEOLOGICAL CONDITIONS

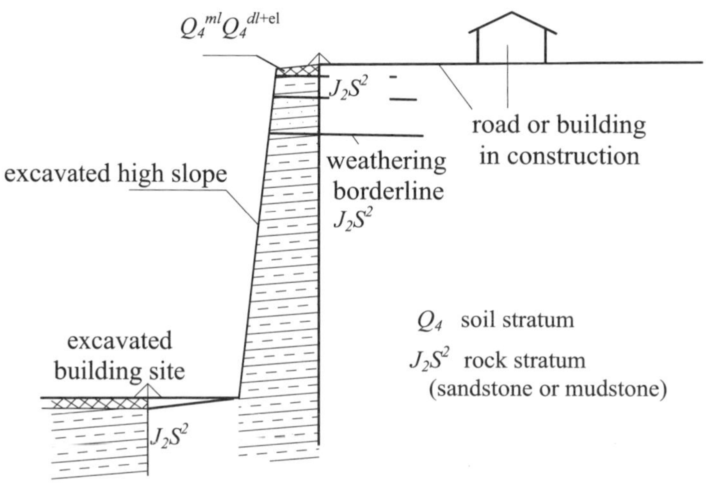

Figure 2. A typical stratum profile of excavated high slope in Chongqing

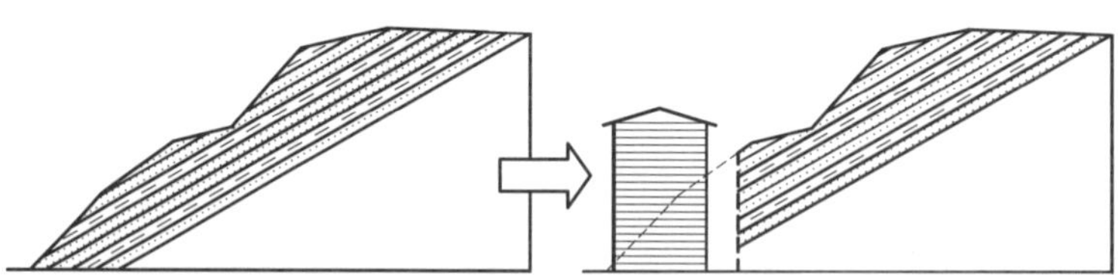

Figure 3. An unstable high slope by manual excavated

The natural geological condition is essential in determining the stability of high excavated slopes. The main factors of influence are: the properties of rock and soil mass, stratum attitude, associated dip situation of the stratum and excavated slope, geological structure (such as the growth degree of fissures, faults and groundwater regime). Building site excavation can also disturb geological conditions with deletesions effects.

Combinations of weak rock and/or soil, widely encountered in slope engineering, usually leads to depressed slope stability. In Chongqing, strata including incompact soil, sandstone, and mudstone are widespread. In the Wulong landslide example (Figure 1.), the main strata are Q^4 soil, T_{3xj} sandstone and mudstone weak interbed. The combined strata of Q^4 soil, J_2S^2 sandstone and mudstone is typical at many building sites in the Chongqing urban district, as shown in Figure 2.

The slope may be stable state if the stratum tilt direction is opposite to the excavated slope's dip. However, many excavated slopes are unstable because of the identical tilt direction of the stratum and slope, following a disadvantageous excavation, without retaining structures, as shown in Figure 2. and Figure 3. For instance, during 319 state railway and building construction, at the site of the Wulong landslide, the foot of the slope has been badly excavated since 1989.

Generally, the integrity of natural rock and soil is destroyed and the mechanics properties are weakened by growing geological structures such as fissures. The Wulong landslide had a slight tilt angle stratum, however the rock mass was split into fractional clump layers mainly by two joint set parallel and vertical to slope surface respectively. This rock slope was unstable because the fractional clump was split by fissures.

According to surveys, over 90% of high slope landslides are attributed to groundwater. Groundwater may increase the unit weight and decrease the mechanical strength of the rock-soil mass, causes static and dynamic water pressure, scours or corrodes the slope foot and results in flotage of rock-soil. During the rainy season, slope stability is greatly reduced due to the groundwater, and additionally, soil moisture and surface water. For instance, the Wulong landslide was preceded by a heavy rainfall.

ENGINEERING ENVIRONMENT

Engineering environment refers to construction-engineering movements at the building site adjacent to the high excavated slope. Generally, natural slopes are stable, but

Figure 4. A drainpipe crossing the upper side of a high excavated slope

Figure 5. The heavily trafficked road at the top of a high excavated slope

slopes become unstable once the equilibrium condition has been disturbed by anthropogenic engineering movement. The Wulong landslide is a typical example.

Other engineering factors of slope stability include unsuitable landfill materials, construction materials, deposited at the top of the slope, engineering blasts construction, traffic vibration and seepage from an engineering service pipe or drainpipe. Such factors are typically observed in Chongqing city, as shown in Figure 4. and Figure 5.: a drainpipe crossing the upper side of an high excavated slope and heavy traffic at the top of an high excavated slope respectively.

STABILITY ASSESSMENT

A stability assessment will provide important data for the design and construction for slope retaining structures. An accurate and precise assessment of the excavated slope stability is essential to ensure the safety of the site and buildings. In Chongqing, it is compulsory to assess the following classes of slope prior to construction: classification slope and its height over 25 m, classification and H≥15m, classification and H≥8m.

An accurate assessment of a high excavated slope may be based on the clear investigation of the afore mentioned natural geological conditions, and the degree of anthropogenic/engineering disturbance to the building site. To obtain an exact assessment conclusion, the following aspects also need to be thoroughly studied, in terms of the related geological survey datum and technical standard.

Potential failure pattern

The potential failure pattern of high excavated slopes may be determined during the stability assessment.

The potential failure pattern of the case shown in Figure3 may be linear shear failure with a single slip surface. Rotary failure with an arc slip surface can occur in homogeneous soil slopes, landfill slopes and rock slopes with a prosperous fissure. Shear failure with multi-replicate slip surfaces may occur in rock masses with identical dips multi-discontinuity or landfill on the replicate surface of bedrock. Excavated slopes with steep gradients and ground fissuration at the top of slope are prone to toppling collapse. Fractionary collapse with complex slip surface may occur at slopes with slight dip angle stratum, layer on layer with crashed part by joint set, as observed at the Wulong landslide, Figure 6.

Assessment method

The potential failure pattern of high-excavated slopes is an important criterion in the selection of stability assessment method.

The ordinary method is appropriate for linear shear failure with single slip surface. Integrative C value method is usually adopted for rotary failure with arc slip surface. The transfer coefficient method may be suitable for shear failure with multi-replicate slip surfaces. The SARMA method is generally used for arbitrary boundary of slip surface and the rock mass separated by distinct discontinuity. Graphic methods include stereographic projection and sphere projection method. The former is suitable for rock masses with less distinct discontinuity and the latter is used for failure controlled by a large dimension wedge. More complex failures need to be analyzed by numeric methods such as FEM (finite element method).

Figure 6. The fractionary collapse pattern of the landslide, Wulong

It is difficult to obtain a convincing assessment conclusion using a single assessment method, therefore the local technical standard in Chongqing recommends that: "... engineering geological analogy and limit equilibrium method should be comprehensively used.... Numeric method ought to be combined to analyze the stability for the slope with complex deformation and failure mechanism."

Calculation parameter

An accurate calculation parameter for rock and soil mass, specific to the site, must to be used or the stability assessment will be inaccurate, regardless of the good computational method used.

The calculation parameter of rock-soil mass can be determined from the geotechnical investigation report of the specific slope plus the related technical standard. The parameters described by the site investigation report may not accurately represent the rock-soil mass *in-situ* because the experiment conditions to test these parameters are often simplified. Therefore Technical Standards recommend that the final calculation parameter should additionally consider factors such as service condition of high slope with the length of building service. In the absence of experimental values, the related technical standard may be used to offer these calculation parameters.

Engineering significance

The conclusions of stability assessment are important to help ensure slope stability and therefore building safety. Moreover, it is an important factor in determining the need for structures and the methods of excavation.

The excavation of natural rock-soil is disadvantageous to the stability of high slopes due to the intrinsic destruction of the equilibrium and the mechanical properties of rock mass, which becomes weakened by weathering over time. Retaining structures should be used when the stability assessment shows that the excavation results in slope failure. Specific measures are necessary to deal with the sensitive factors forward to reduce high slope stability. The disjunction and topdown construction method ought to be used in slope excavation construction if it is shown that a large degree of distortion would occur if a "one-off" excavation procedure was adopted. Some other aspects, including the use of engineering blast methods and the quantity of explosive used in excavation construction, may be decided according to the stability dynamic assessment.

CONCLUSIONS

1) Accurate stability assessment of high excavated slopes is fundamental to ensure the safety of the related excavation construction site and buildings.

2) The stability assessment is influenced by several crucial factors: the natural geological conditions and anthropogenic disturbance of these conditions, the engineering environment, potential slope failure pattern, assessment calculation parameters and analysis method.

3) The conclusion of the stability assessment helps estimate the safety of building sites adjacent to high slope and also provides important data used to determine the strategy for excavated slope-retaining structures.

REFERENCES

Wen, H.J. (2001). "*Stability Assessment Report of Building Site adjacent to J2&K2 Commercial Building of Chongqing Linjiang Plaza*", 10~12

Wen, H.J. (2001). "*Stability Assessment Report of Building Site adjacent to Excavated High Slope, Yaoyong Shock Absorber Workshop*", 10~11

Chongqing Nanjiang Eng. Geo. Inv. Co.. (2001). "*Engineering Geological Investigation Report of Bayangshan Landslide, Wulong County Urban. Chongqing*", 11~15

Wen, H.J (2002). "*Crucial problems on security assessment of a building site adjacent to an excavated high slope*". J of Chongqing University, Vol.1 No.2, 38~41

Construction Supervision Bureau. (1998). "*Chongqing Local Technique Standard. Code for Engineering Geological Investigation (DB50/5005-1998)*", 31~37.

Construction Supervision Bureau. (2001). "*Chongqing Local Technique Standard. Technical Specification for Retaining Structure in Excavation (DB50/5018-2001)*", 7~20

Construction Supervision Bureau. (1997). "*Chongqing Local Technique Standard. Design Code for Building Foundation of Chongqing (Db50/5001-1997)*", 42~55*

International Conference on Geotechnical Engineering for Disaster Mitigation & Rehabilitation
Chu, Phoon & Yong (eds) © 2005 World Scientific Publishing Company ISBN 981-256-469-1

SLOPE MONITORING INFORMATION AND INSTABILITY INDICATION ANALYSIS SYSTEM

YUAN BAOYUAN
Geological and Geotechnical Engineering Department, Hohai University, Xikanglu 1# Nanjing, 210098, China

CHEN YANGUO
Geological and Geotechnical Engineering Department, Hohai University, Xikanglu 1# Nanjing, 210098, China

For handling with monitor information of slope, analyzing and predicting the stable state of a large scale slope timely, authors established the Slope Monitoring Information and Instability Indication Analysis System which has functions of monitor information database management, monitor information visual inquiry and analysis, and slope unstable sign analysis. The system has the oriented-object style, visualization characteristic, and the feature of analysis and predication participated by experts. It has applied in the left bank ship-lock slope of the Wuqiangxi Hydropower station.

INTRODUCTION

In large scale slopes which have great influence on the safety of important buildings, large scale deformation monitor system was established, for discovering unstable signs and taking measures timely, safeguarding the stability of slope. After establishment of a deformation monitoring system, handling and analyzing deformation monitor information, controlling stable state of slope timely is an important work of engineering technicians.

A Slope Monitoring Information and Instability Indication Analysis System could centralize management of monitor information. This makes analysis of monitor information becomes rapid. Authors advanced the train of thought about slope unstable sign analysis. This makes analysis and prediction of slope stability participated by experts, and makes analysis became dynamic and prediction became rapid. This accords with demands of modern design thinking which emphasizes on dynamic design and rapid handling.

Authors developed the Slope Monitoring Information and Instability Indication Analysis System which has oriented-object style and visualization characteristic, based on Access, the oriented-object database developing system, and Visual Basic, the oriented-object programming language. The system is composed by the deformation monitor information management branch system, the deformation monitor information visual inquiry and analysis branch system, and the slope unstable sign analysis branch system.

SLOPE MONITOR INFORMATION CALCULATION

Monitor system has many monitor items. Monitor items of slope mainly are the Surface Observation, the Static Leveling Instrument, the Borehole Inclinometer, the Multipoint Extensometer, the Waterpipe Inclinometer, the Extensimeter, the Bar Rangemeter, the Ditch-covering Extensometer, and the Convergence Indicator. Calculations of information of these items are as follows.

(1) The Static Leveling Instrument

Measure relative displacements A, B, C, D, E between two neighboring mounds, calculate mound displacement related to zero datum point.

$$S_A = A + B + C + D + E \tag{1}$$
$$S_B = B + C + D + E \tag{2}$$
$$S_C = C + D + E \tag{3}$$
$$S_D = D + E \tag{4}$$
$$S_E = E \tag{5}$$

where S_A: displacement of mound A, S_B: displacement of mound B, S_C: displacement of mound C, S_D: displacement of mound D, S_E: displacement of mound E,

(2) The Surface Observation

Measure coordinates X, Y, and Z of a monitor point, calculate displacement value and displacement direction of monitor point.

$$S_X = X - X_0 \tag{6}$$
$$S_Y = Y - Y_0 \tag{7}$$
$$S_H = H - H_0 \tag{8}$$
$$S_{XY} = \sqrt{S_X^{\ 2} + S_Y^{\ 2}} \tag{9}$$
$$S = \sqrt{S_{XY}^{\ 2} + S_H^{\ 2}} \tag{10}$$
$$N = arctg(S_Y / S_X) \times 180/\pi \tag{11}$$
$$F_0 = \begin{cases} 90 + \alpha + N, S_X > 0, S_Y > 0, N < 0 \\ 0 + \alpha + N, S_X > 0, S_Y > 0, N > 0 \\ 180 + \alpha + N, S_X < 0, S_Y > 0, N < 0 \\ 90 + \alpha + N, S_X < 0, S_Y > 0, N > 0 \\ 270 + \alpha + N, S_X < 0, S_Y < 0, N < 0 \\ 180 + \alpha + N, S_X < 0, S_Y < 0, N > 0 \\ 0 + \alpha + N, S_X > 0, S_Y < 0, N < 0 \\ 270 + \alpha + N, S_{Xi} > 0, S_Y < 0, N > 0 \end{cases} \tag{12}$$
$$F = \begin{cases} F_0, F_0 < 360 \\ F_0 - 360, F_0 > 360 \end{cases} \tag{13}$$

where X: measuring value of coordinate X, X_0: initial value of coordinate X, S_X: displacement of direction X, Y: measuring value of coordinate Y, Y_0: initial value of coordinate Y, S_Y: displacement of direction Y, Z: measuring value of coordinate Z, Z_0: initial value of coordinate Z, S_Z: displacement of direction Z, S_{XY}: horizontal displacement, S: whole displacement, α: angle between construction coordinate X and earth coordinate X, F: direction angle of horizontal displacement, N: middle variable, F_0: middle variable.

(3) The Borehole Inclinometer

Measure reading A^+, A^-, B^+, B^- of direction A and B of each point in borehole, calculate displacement value and direction of each point.

$$A = (A_i^+ - A_i^-) - (A_{i0}^+ - A_{i0}^-) \tag{14}$$

$$A_{zi} = \sum_{K=0}^{i} A_i \tag{15}$$

$$B_i = (B_i^+ - B_i^-) - (B_{i0}^+ - B_{i0}^-) \tag{16}$$

$$B_{zi} = \sum_{K=0}^{i} B_i \tag{17}$$

$$S_i = \sqrt{A_{Zi}^{\ 2} + B_{Zi}^{\ 2}} \tag{18}$$

where i: point number from bottom to measuring point; A_i^+, A_i^-, B_i^+, B_i^-: reading of A^+, A^-, B^+, B^- of point i; A_{i0}^+, A_{i0}^-, B_{i0}^+, B_{i0}^-: initial reading of A^+, A^-, B^+, B^- of point i; A_i: variation of direction A of point i; A_{si}: accumulating variation of direction A of point i; B_i: variation of direction B of point i; B_{si}: accumulating variation of direction B of point i; S_i: horizontal displacement of point i.

(4) Items of Extensometer Type

Items of the Multipoint Extensometer, the Waterpipe Inclinometer, the Bar Rangemeter, the Ditch-covering Extensometer and the Convergence Indicator have similar monitor principle, their calculations are same, the basic calculation is as follows.

$$S = (C-C_0)\times f \tag{19}$$

where C: measuring value, C0: initial value, f: a coefficient, S: displacement value.

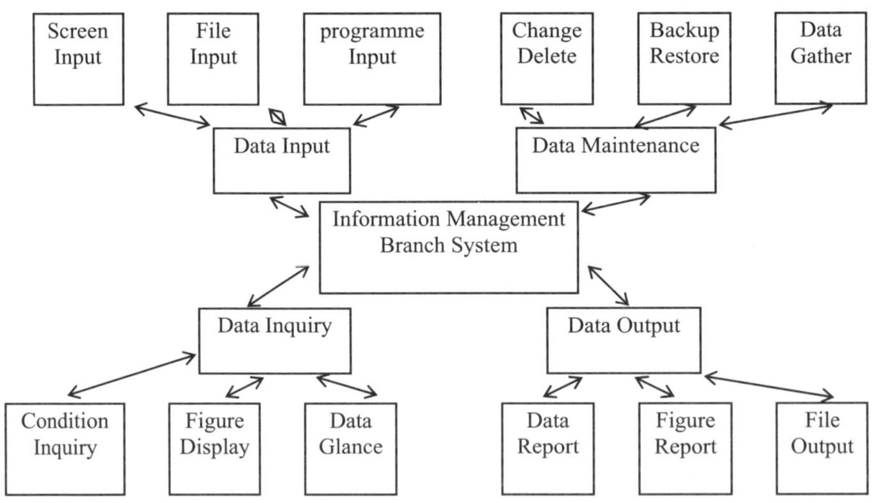

Fig1. Basic Structure of Monitor Information Management Branch System

INFORMATION MANAGEMENT BRANCH SYSTEM

The assignment of monitor information management branch system is in charge of collecting and managing a vast amount of monitor data. Authors developed the monitor information management branch system which has the oriented-object style and visualization characteristic, based on Access, the oriented-object database developing system, and Visual Basic, the oriented-object programming language, under Windows operating system. The structure of the branch system is as Fig. 1.

MONITOR INFORMATION VISUAL INQUIRY AND ANALYSIS BRANCH SYSTEM

This branch system is in charge of calculation and analysis of monitor information. Authors designed the function of monitor point visualization, so technicians could directly perceive through the screen during inquiring and analyzing. User could choose a monitor point by clicking the icon of monitor point on the screen. The distribution of icons of monitor point on screen is same as the distribution of monitor point on slope, so technicians could find a monitor point very easily. The whole structure of the branch system is as Fig2.

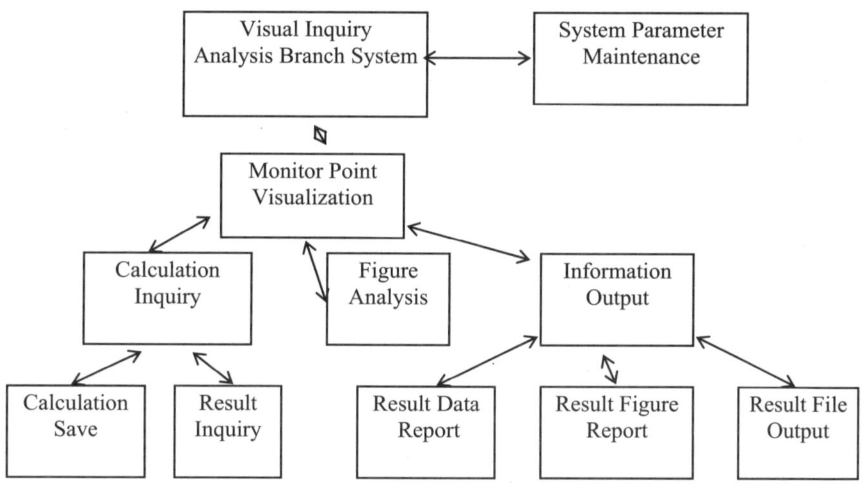

Fig2. Structure of Information Visual Inquiry and Analysis Branch System

SLOPE UNSABLE SIGN ANALYSIS BRANCH SYSTEM

Principle of Unstable Sign Analysis

The stable degree could be graded. Each grade could be distinguished by monitor information. All desirable distinguishing method of sign grade for each item was programmed. For each actual situation, according to inquiry and analyze of monitor information, combined with expert experience, especially experience on-the-spot experts, the sign analysis method and criterion of the research slope were determined. Then, on-the-spot technicians could analyze slope stable state dynamically, using the slope sign analysis branch system. With the promotion of understanding degree of experts for the slope, criterion could be change dynamically.

Authors differentiated stable degree of slope into five grades.

(1) Quite Stable (code is A): Monitor data is quite good.

(2) Stable (code is B): Monitor data is within normal range.

(3) Slight Unstable (code is C): Monitor data exceed permitted limits a little.

(4) General Unstable (code is D): Monitor data exceed permitted limits and has growth tendency.

(5) Serious Unstable (code is E): Monitor data exceed permitted limits enormously, the stability will be worsen.

If stable degree belongs to A and B, the stable state is considered as good; Belongs to C, the stable state is in middle, its tendency should be paid attention to; Belongs to D, slope has a little unstable sign, the stable state is relatively poor, and some measurement should be taken to modify its stability; Belongs to E, slope has obvious unstable sign, the slope should be handled urgently.

The steps of slope unstable sign analysis are as follows. Firstly, analyze the stable state of each monitor point; Then, analyze unstable sign type of each item according the stable state of monitor points; Lastly, analyze the stable state of the whole slope according the stable state of monitor point and the unstable sign type of items.

The basis of unstable sign analysis of monitor point is that each monitor point is established for measuring some deformation feature values. If these feature values reach limit values, the monitor point was considered to reach an unstable sign level. So, judging whether or not deformation feature values reach some limit values is the sign analysis method of a monitor point. The deformation feature values of a monitor point are accumulation displacement, period displacement, displacement speed, displacement period increment, and deformation step value.

The methods of item sign analysis is that the sign type of an item is determined by percentage or number of monitor points reaching a sign type.

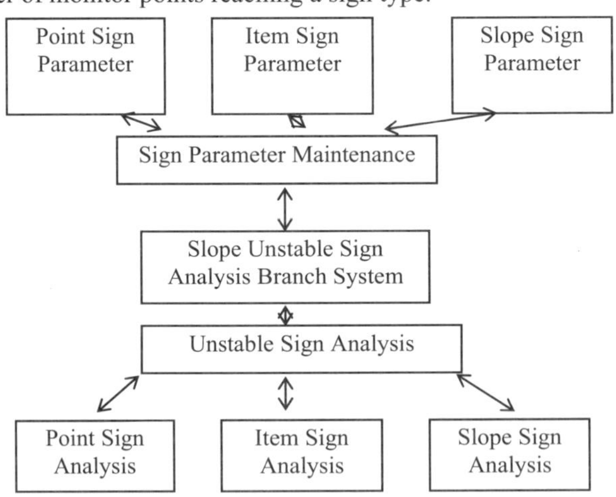

Fig. 3. Structure of Slope Unstable Sign Analysis Branch System

The sign type of the whole slope is determined by sign type of items and general features of unstable monitor points.

The structure of the slope unstable sign analysis branch system is shown in Fig. 3.

APPLICATION IN LEFT BANK SLIP-LOCK SLOPE OF WUQIANGXI HYDROPOWER STATION

The Wuqiangxi Hydropower Station is a large hydropower station in Yuanshui River in the western of Hunan Province, which installed capacity is 1,200,000 Kw, and the height of dam is 87.5m. The navigable building is a three-stage ship-lock, arranged in the left bank. The height of the left bank ship-lock slope is 165m. It's geologic structure is very complex.

For the safety of the ship-lock, a large monitor system which relies mainly on the deformation monitor is established on the ship-lock slope. The system formed an integrated monitor net based on the thought of combination of point, line and plane.

Authors analyzed the stability of the Wuqiangxi Hydropower Station left bank ship-lock slope based on Slope Monitoring Information and Instability Indication Analysis System. The result of unstable sign analysis is as follows. The first point of the Multipoint Extensometer, the first point of the Borehole Inclinometer, and the whole displacement of the Bar Rangemeter, belonged to stable type. The Ditch-covering Extensometer, the shape of the Borehole Inclinometer curves belonged to slight unstable type.

CONCLUSION

1) The slope monitor information management branch system could handle the monitor data timely.
2) The slope monitor information visual inquiry and analysis branch system is helpful for on-the-spot engineers to analyze the feature and trend of a slope rapidly.
3) The slope unstable sign analysis branch system could discover unstable signs, predict stable state of slope. As the sign analysis is participated by experts, and operated in site, the prediction are rapid and reliable.
4) The actual application shows that the Slope Monitoring Information and Instability Indication Analysis System has strong practicality and reliability.

REFERENCE

Yang Zhifa etc. (1997) "The Establishment of Slope Monitor System and Analysis Method of Monitor Information". *Proceedings of National Mountain Area Foundation Symposium*, Chongqing University Press, 1-13.

International Conference on Geotechnical Engineering for Disaster Mitigation & Rehabilitation
Chu, Phoon & Yong (eds) © 2005 World Scientific Publishing Company ISBN 981-256-469-1

STABILITY ANALYSIS AND RISK ASSESSMENT OF SHILIPU DAM DURING THE NINGJIANGBAO LANDSLIDE

FAMING ZHANG
Geology Department, Hohai University, No.1 Xikang Road
Nanjing, Jiangsu Province 210098, China

JING LI
Geology Department, Hohai University, No.1 Xikang Road
Nanjing, Jiangsu Province 210098, China

ZENGDE ZHU
Geotechnical Research Institute, Hohai University, No.1 Xikang Road
Nanjing, Jiangsu Province 210098, China

Shilipu dam project is situated in Sichuan province in south-west China. According to geological works, there are more than 20 landslides distributed in the reservoir area. The Ningjiangbao landslide is the most important landslide since it is near 1400~1600m^3 and adjacent to the selected dam foundation. To analysis the stability and evaluate the risk of the landslide are the most important works before dam construction. Based on field investigation of the property of the landslide, a simplified three dimensional limit equilibrium method which considering the sliding value and sliding vector was used to calculate the stability factor. Finally, by using the law of momentum, the risk of the surge height and velocity induced by the landslide to the dam construction was evaluated.

INTRODUCTION

Ningjiangbao landslide was located at the right bank of upper Mingjiang River, China. It is a rockslide, which the length along the river is 1000m, the height is 500 m, the area is 3500 m^2 and the volume of the landslide is more than $(1400\sim1600)\times10^5$ m^3. The average degree of the slope is 30°, the landslide can be divided into three parts by tension fissures. The vegetation of the area has been destroyed by farmers in order to increase the cultivation. The sliding-body is consisted of layered phyllite where the attitude is N60~70°E/NW70 ~75° and colluviation. The thickness of the accumulation is 10~30 m, in profile of the landslide, the gradient is gently to steeply from the button to top of the landslide, the condition of the slope is easy to form a high-speed and long run-out landslide. So, it is necessary to evaluate the sliding hazards for constructing a dam where has been sit at the lower 9 km from the landslide. In this paper, based on analyzing the geology and geography condition of the Ningjiangbao landslide, the simplified three dimensional limit equilibrium methods was used to analyzing the static stability, furthermore, the sliding speed and surge height was evaluated.

MECHANISM OF THE LANDSLIDE SLIDING

Geological background

Ningjiangbao landslide is distributed in a dip slope, the slip surface is dip to the Mingjiang river, and the sliding body is consisted of layer phyllite, most of the phyllite was lightly weathered. The rock mass structure belongs to the layer rock structure, there is no faults distributed in the sliding area, but, the joints are well-developed. The field working shows that the sliding surface was consisted of joints and layer face.

Properties of the landslide

Ningjiangbao landslide deformation shows that there are many tension fissures appeared on the sliding body surface, the width of the fissure is about 10m and depth 20m, some of the rock filled the fissures. According to the deformation of the slide body based on geological exploration, the process of sliding can be divided into four phases:

(1) When Mingjiang river incised, the gulf becomes higher and higher, so, the stress was liberated. In this stage, the slope is from stability to limit equilibrium;

(2) Since the slope becomes stepper, shear stress was distributed in some areas, the tension fissures were cut the body, sliding surface was not continued;

(3) The sliding surface formed, the slope becomes instability;

(4) Sliding stage. When the reservoir water becomes higher and higher, the weights of sliding body increased, the shear strength of the sliding surface will be decreased, so, the landslide will be occur in high speed since the high potential energy, the hazards will be taken place.

From above discussion, Ningjiangbao landslide is a high speed land slide, the factors forced the land slide is the reservoir water table of Shilipu dam project where 9 km downstream of Mingjiang river. It is necessary to analysis the stability and to evaluate the hazards for dam construction.

STABILITY ANALYSIS OF NINGJIANGBAO LANDSLIDE

Parameters of landslide

Table 1 Calculation parameters of Ningjiangbao landslide

Type	Parameters			
The highest elevation (m)	2198			
The lowest elevation (m)	1605			
Length of slope (m)	1000			
Height difference (m)	500			
Sliding mass thickness (m)	20～80			
Property of the sliding face	Discontinue			
Unit weight(kN/m^3)	Normal	25	Saturated	22

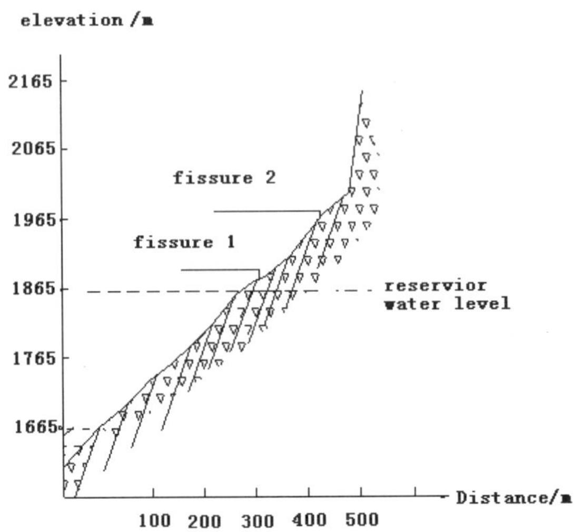

Figure 1 Geology profile of the landslide

According to the geological exploration, the parameters of the landslide mass have been identified (Table 1). The Geology profile of the landslide is shown in Figure 1.

Calculation case

Factors affected on the stability of Ningjiangbao landslide and calculation conditions can be described as follows.

(1) Naturally condition: In this condition, the water table is supposed as the normal level of nature.

(2) Original topography +earthquake: In this condition, the earthquake will be considered as static, the earthquake intensity is supposed VI degree.

(3) Considering the reservoir water level as 1861 m.

(4) Considering the reservoir water level as 1861 m and earthquake.

Sliding surface searching based on annealing simulation method

The failure mass is divided by 3 cross sections. The geotechnical property parameters are shown in Table 1. Although most of the slip surface will coincide with the linking joints, optimization is needed to determine the critical locations of the points defining the interception of the slip surface. In this paper, the simulation annealing method was used to simulate the sliding surface because the shortcoming of geological exploration.

Generally, for the jointed rock mass, the sliding surface is consisted of rock-bridge and joints, and the failure accordance with Mohr-Coulomb criterion. In simulation annealing method for searching the sliding surface, the function can be described as follows:

$$T_i \cos(\varphi_i) = c_i l_i \cos(\varphi_i) + N_i \tan(\varphi_i) \cos(\varphi_i) \tag{1}$$

Where: T_i is the shear strength of sliding surface; c is the cohesion; l_i is the length of sliding surface; N_i is total normal stress and φ_i is friction angle of the sliding surface.

So, the shear strength of sliding surface can be closely calculated as function (2).

$$f_i \approx \sum_{i=1}^{n} (c_i l_i \cos(\varphi_i)) \tag{2}$$

Where: c_i is the cohesion of one joint; l_i is the length of one joint; φ_i is the dip angle of one joint; n is the total number of joints in one sliding surface.

Suppose f(x) →0 the sliding surface can be simulated. The procedure discussed in this paper is:

Firstly, based on field joint geometry data collecting, 3D joint map can be obtained, then, according to the slope direction, cut a two-dimensional map from 3D joint map.

Secondly, give an initial point and end point which the sliding surface may be go through, calculate the shear strength in one path which consisted of joints and rock bridges, in this way, the lowest shear strength route (i.e. the critical failure surface) will be found by using the simulated annealing method.

Calculation methods and results

The procedures of the 3D upper bound approaches proposed by Chen et al. (2001a) can be summarized as follows:

(1)Define a 'neutral plane' in the middle of the sliding mass, on which every point is assumed to move within it, or in a direction inclined at a constant angle to this plane. The xoy co-ordinate system is then set in the neutral plane. The failure mass is divided into a number of prisms with inclined interfaces.

(2)Calculate the velocity for each prism.

(3)Determine the factor of safety by the energy-work balance equation.

$$\sum D^*_{i\leftrightarrow j,e} + \sum D^*_{i\beta\, j,e} + \sum D^*_{i,j,e} = \boldsymbol{W}\boldsymbol{V}^* + \boldsymbol{T}^*\boldsymbol{V}^* \tag{3}$$

By using the 3D upper bound approaches to calculate the stability of Ningjiangbao landslide, the result is shown in Table 2.

Table 2 Evaluation of factors of safety upon various operational cases

Case	r_u	FOS by various methods		
		Upper bound	Limit equilibrium	Weighted - average
Original topography	0.15	0.952	0.938	0.942
Original topography + earthquake	0.05	0.931	0.916	0.923
Water level 1861	0.05	1.041	1.052	1.047
Water level 1861+earthquake	0.05	0.846	0.880	0.868

NOTE: r_u is pore pressure coefficient.

SLIDING RISK EVALUATION OF THE LANDSLIDE

Sliding velocity prediction

Since the volume of Ningjiangbao landslide is more than (1400~1600)×10^5 m^3, the depth of the sliding surface is more than 80m, the slope angle is more than 30°and the length along the Mingjiang river is 550m, so, it is necessary to evaluate the velocity for Shilipu dam construction. The velocity of the landslide can be calculated as equation (4).

$$V_i = \sqrt{V_0^2 + 2gh_0(1 - f \bullet ctg\alpha)} \tag{4}$$

Where h_0 is the height of center of the landslide mass; f is the friction coefficient of sliding surface V_0 is the initial velocity.

The landslide velocity in initial sliding times is more than 125m/s, it will induce the wave to effect on the Shilipu rockfill dam.

The risk for Shilipu dam project

Shilipu dam project is situated in Sichuan province in south-west China, it is the most important project in Minjiang river development. The dam form is a rockfill dam, the height of the dam is 162m. The Ningjiangbao landslide is the most important landslide since it is near (1400~1600)×10^5 m^3 and only 9km to the selected dam foundation. From the discussion above, the landslide is a high speed and long distance landslide, so, it is obvious that the surge and siltation will attack the rockfill dam.

(1) Siltation of the reservoir. Most of the sliding mass will siltate to the reservoir, according to the valley shape of Mingjiang river and sliding mass property, the storage capacity of the Shilipu reservoir will decrease (1400~1600)×10^5 m^3.

(2) Surge height prediction. When calculating surge height of landslide moving on ground surface of the reservoir, it is necessary to separate the sliding mass into several blocks, then, the surge height of every block can be calculated based on the law of momentum. For every block of landslide, suppose the weight of one block is m_i, the velocity of one block move into the reservoir is v_i, the time of the block moving under water is t_i, the block height of landslide is H_i, the width of the block is B_i, the average pressure on section face is p_i, then, the formula (5) can be obtained based on the law of momentum (CHEN X D, 1984; WAN et al.2003).

$$m_i v_i = p_i H_i B_i t_i \tag{5}$$

So, the η_i can be deduced by the following equation.

$$\eta_i = \frac{m_i v_i}{H_i \bullet B_i \bullet t_i \bullet \rho \bullet g} \tag{6}$$

Where η_i is the distance from the block center to the highest of the surge, if the water level before the sliding is h_0, the initial surge height of one block is determined by the equation (7).

$$\eta = \eta_i + H_i / 2 - h_0 \tag{7}$$

For Ningjiangbao landslide, on considering the height and width, 20 blocks were separated, the calculating result is shown in figure 2. From figure 2, the highest surge is more than 30m.

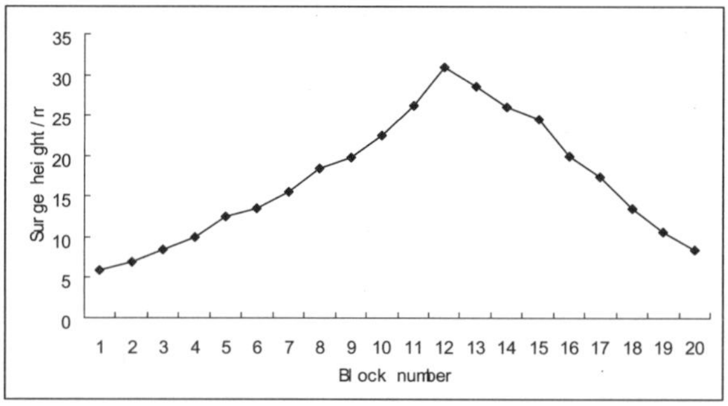

Fig.2 Surge height distribution for every blocks of Ningjiangbao landslide

CONCLUSION

The Ningjiangbao landslide will be caused by the water level of the Shilipu dam project on the slide body. In this reason, the stability analysis based on 3D improved Sarma's method and possible destroy results of the slide is evaluated in this paper. The analysis result shows that the Ningjiangbao landslide will be sliding when the reservoir water level increase to 1861m, the risk to the Shilipu dam project can be summarized as siltation of the reservoir and surge attacking to the rockfill dam.

ACKNOWLEDGEMENTS

This work was supported by the foundation of Hohai University "211" project (Xk.2.4.3).

REFERENCES

CHEN X D. (1984). "Experiential method and program design of surge triggered by landslide in reservoir." *Wuhan Science and Research School in Zhongnan Institute of Reconnaissance and Design of the Ministry of Water Conservancy and Electric Power,* 1 - 18.

Chen, Z. Wang, X, Haberfield, C, Yin, J, Wang, Y. A. (2001). "Three-dimensional slope stability analysis method using the upper bound theorem, Part I: theory and methods, *International Journal of Rock Mechanics and Mining Sciences*, Vol. 38, 369-378.

Wang Y. and Yin K. L. (2003). "Analysis of Movement Process of Landslide in Reservoir and Calculation of Its Initial Surge Height." *Earth Science —Journal of China University of Geosciences*, Vol .28, No.5.

TSUNAMIS

International Conference on Geotechnical Engineering for Disaster Mitigation & Rehabilitation
Chu, Phoon & Yong (eds) © 2005 World Scientific Publishing Company ISBN 981-256-469-1

FAILURE INVESTIGATION OF TSUNAMI DAMAGED STRUCTURES IN SRI LANKA

P.B.R DISSANAYAKE
Department of Civil Engineering, University of Peradeniya
Peradeniya, 20400, Sri Lanka.

The earthquake of the Sumatran coast on 26th December 2004 which recorded in 9.0 on Richter scale generated a tsunami that affected many countries around the Indian Ocean including Sri Lanka. The coastline from Colombo through Galle, Matara, Hambanthota, Ampara, Trincomalee, Mulathivu and Jaffna was affected by a combination of direct diffracted and reflected waves. Wave heights were measured to have reached up to 11m from the mean sea level. Depending on the wave height various types of structures behaved differently. The objectives f this paper is to discuss about different failure modes affected on structures located on the coast. At relatively low wave heights, it was mainly boundary walls and causeways that were affected. As wave height increased, single storey structures and two storey structures were significantly damaged. Where even higher energy waves were accompanied by significant scouring, most of the two storey structures in tsunami prone areas failed through the undermining of foundation. The impact of the tsunami on transport infrastructure was also very much significant.

INTRODUCTION

A tsunami is a series of long waves generated by a sudden displacement of a large volume of water. Tsunami are caused by submarine earthquakes, submarine volcanic eruptions, underwater landslides or slumps of large volumes of earth, meteor, impacts and even onshore slope failure that fall into the ocean or a bay.

The earthquake off the Sumatran coast on 26th December 2004 measuring 9.0 on Richter scale generated a tsunami that affected many countries around the Indian Ocean, including Sri Lanka. The coastline from Colombo to Jaffna peninsular in eastern side of the country was affected by a combination of direct, diffracted and reflected waves. However the direct waves caused sever damages for most of the structures. In Sri Lankan context wave heights were estimated to have reached up to 11m from mean sea level in Hambanthota (south east coast). However, the maximum height of water experienced by structures would have been some what lower than 11m. In most cases water level that caused significant damage to structures ranged firm around 2 to 3m from the plinth level.

The prime objective is to identify more resistant structures against tsunami attack. Depending on the wave height (hence loading) and speed various types of structures were affected. At relatively low wave heights. It was mainly boundary walls and causeways that were affected. As wave heights increased single storey structures were significantly damaged where even higher energy waves were accompanied by significant scouring, two or more storey structures failed though the undermining of foundation. This was also caused due to the eddy formation of water flow. The behavior of water tanks and well were different from other kind of structures. The impact on transport infrastructure was very much higher, including the world largest railway accident, which caused 1500 deaths to the erosion of roadways in small scale.

BOUNDARY WALL

In the tsunami prone areas most of the boundary walls were collapsed due to the tsunami waves. Those collapsed boundary walls were constructed of 112.5 mm thickness bricks and rarely in 225 mm thickness. Most of these were not designed for lateral loads due to that at the time of wave hit the walls. Walls were completely collapsed without any resistance. In some cases, walls of up to 2m height had been constructed in 100mm thick masonry with hardly any lateral or vertical bracings. Most of these walls collapsed by overturning mode of failure during the tsunami. Intermediate horizontal bracing and vertical buttress could have prevented the failure of even some of boundary walls.

This kind of failure gives following guidelines fir future construction works.

- The wall should be constructed with vertical and lateral bracings at regular intervals.
- The wall thickness should not be less than the standard brick length (225mm).
- The foundation depth should be enough to avoid the overturning failure due to scouring.

WATER TANKS

Many middle class dwelling units have water tanks or so called overhead tank located above the chimneys. During the tsunami, these structures behaved differently due to the weight and the shape. In many cases, water tanks behaved in such a way that the full house had collapsed but the part of the house near the water tank was not intact (undamaged).

The water tanks which independently supported by four columns were found to perform very poorly. Mostly of this kind of water of water tanks collapsed due to undermining or scouring of foundation. (Due to scouring or undermining the water tanks collapse as a result of overturning mode of failure)

SINGLE STOREY STRUCTURES

Most of the single are made out of cement, sand, bricks and timber generally on random rubble masonry foundations. In Sri Lankan context, little or no reinforced concrete is used in the construction of singe storey structures. As a result of this, there is no proper bond between sub structure and supper structure. In many cases these kinds of structures were completely pushed off their foundation in sliding mode of failure due to above reason.

In damaged single storey structures, most of the outer walls as well as inner walls had only 112.5mm thickness. The thickness of walls is unable to withstand any kind of wave forces during tsunami. Due to inadequate thickness almost all walls of damaged single storey structures were collapsed.

The remaining single storey structures in tsunami prone areas had some special features like cross walls bracing, reinforced concrete at least in corners of the structure. The single storey structures with cross walls bracing the sea facing wall performed better than those without such walls, especially where wave heights were not high enough to cause complete sliding failure. On the other hand, very robust small structures suffered overturning mode of failure due to robustness especially if their foundations were too shallow.

MULTI STOREY STRUCTURES

Multi storey structures consist of reinforced columns and slab basically. So compared to single structures, these multi storey structures performed much better. Due to greater weight and frame structure, the failure modes like sliding and overturning were minimized. Furthermore, a large part of that weight was not submerged and hence made a greater contribution to stability against lateral load.

Most of the multi storey structures have infill walls perpendicular to the advancing wave direction had been punched out especially if they were 112.5mm thickness. However, infill walls parallel to the advancing wave direction remained intact with virtually no cracking, bracing the frames against the horizontal waves loading.

The more significant problem for multi storey structure was scouring of sandy soil at the corners and other footings of building. This scouring occurs probably as a result of eddy formation as the wave advances and recedes across the building. Due to the heavily scouring some of multi storey structures were completely collapsed.

TRANSPORT INFRASTRUCTURES

This includes highways, causeways, bridges, culverts and rail tracks. Most of the transport infrastructures were damaged due to tsunami. The highways were damaged due to embankment had been washed away and heavy scouring near the retaining walls along the highways. The causeways are shallow structure. When tsunami waves hit, most of those were submerged and heavy scouring has resulted due to shear impact of the waves. In reality causeways do not posses adequate lateral resistance to the wave impact. Most of places, scouring and partial sliding failure had occurred.

Concerning the bridges, in some cases bridge decks had been pushed off their bearings due to wave loads. On the other hand scouring near the abutment was very much significant. The rail tracks and sleepers were highly affected by the tsunami in Sri Lanka.

In some cases the rail tracks and sleepers had been lifted off their bedding of coarse aggregate and in others the entire railway embankment had been washed away.

Regarding the culverts waves had attacked the backfill adjustment to the culverts and causing damage to the roadway.

TELECOMMUNICATION TOWERS

The towers situated in tsunami prone areas were affected by the impact created impulsive forces to tower and those towers were failed and collapsed. Due to inadequate depth of foundation telecommunication towers had collapsed during tsunami.

CONCLUSIONS

Due to the tsunami waves building structures can be failed by either sliding mode of failure or overturning mode of failure. These two modes of failures are the common threats to all single storey structures. When designing building in future, the more attention should be paid for overcoming these problems for single storey structures.

Apart from these two failure modes for multi storey structures, the scouring or undermining near the foundation was a very critical issue. So in designing multi storey structures for coastal zone the scouring effect also should be considered to avoid failure of structures due to scouring or undermining.

Due to inadequate thickness of walls and lateral or vertical bracings the boundary walls failed due to tsunami waves. The designing of these kinds of things in future failure modes like overturning, sliding and scouring have to be considered.

REFERENCE

Designing for tsunami, National Tsunami Hazard Mitigation Program, (2001 March), Washinton D.C., www.fema.gov/pdf/hazads/nhp_fema55.pdf

Guidelines for Reconstruction of Houses Affected by tsunami in Tamil Nadu (2005), India, www.tn.gov.in/tsunami

International Conference on Geotechnical Engineering for Disaster Mitigation & Rehabilitation
Chu, Phoon & Yong (eds) © 2005 World Scientific Publishing Company ISBN 981-256-469-1

PUBLIC OPINION ON VARIOUS ASPECTS OF 2004 TSUNAMI DISASTER

AGHJ EDIRISINGHE
Faculty of Engineering, University of Peradeniya,Peradeniya 20400, Sri Lanka

R. YATABE, NP BHANDARI
Faculty of Engineering, Ehime University, Ehime 790-8577,Japan

Frequency of natural disasters of significant magnitude is becoming high in all over the world. For Sri Lankans, natural disasters are not uncommon as nearly 30 valuable lives per year are lost due to natural disasters. Landslides, floods, cyclones are common types of natural disasters experience by Sri Lankans until the recent past. Though earthquakes of smaller magnitude were reported few times in Sri Lanka, tsunami was a new experience to many Sri Lankans until 26.12.2004.

Tsunami occured due to an earthquake of magnitude 9.0 richter scale triggered closer to Sumathra took nearly 270,000 valuable lives in few Asian countries. Contribution from Sri Lanka was more than 40,000. As it is a new experience and the engineering applications proved the inability to prevent damages fully, an attempt was made in the present paper to investigate how Sri Lankans respond to a natural disaster based on a questionnaire. Results of feed back from nearly 600 eye witnesses or victims of 2004 tsunami disaster is reported in this paper.

INTRODUCTION

Natural disasters are common in developed as well as developing countries of the world. Developed world has better warning systems, evacuation plans and recovery plans compared to developing world. But some recent natural disasters occurred in developed countries also proved the foolproofness of some of their systems. Therefore hard engineering is always not adequate to provide a perfect solution to natural disasters.

Main aspect of developed countries leading developing countries is level of public awareness and preparedness for natural disasters. As the public is well aware about magnitude of natural disaster, their risk taking level is relatively low and unwanted casualties after natural disasters are minimal. Though the direct deaths due to natural disasters are less preventable, deaths due to unwanted risk taking can be minimized through proper public awareness. This paper deals with a public opinion survey conducted in tsunami affected areas in Sri Lanka.

ASIAN TSUNAMI 2004

Due to an earthquake triggered closer to Sumatra with a magnitude of 9.0 in the richter scale resulted mega tsunami in Indian ocean according to the available records and it is nearly after 100 years of the previous mega tsunami recorded in the world history.

Sri Lanka was one of the main victims in the disaster. More than 40,000 lives were taken by this tsunami according to the available records. Many more left homeless immediately after the incident and more than 2/3 of coastal line stranded due to damages to road system including heavy damages to many bridges, telephone system, electricity and water supply. Rescue operation to safeguard those who were survived was another important task and providing basic facilities like water, food and medicine even was a big problem.

Figure 1 Some damages due to tsunami

Figure 1 shows some of the damages photographed within first week and some others within first month of the occurrence.

One reason for heavy losses was design standards of most of the structures and lack of preparedness of the general public and responsible authorities.

Human behaviour in a disastrous occasion like this is also very important factor to study.

QUESTIONNAIRE SURVEY CONDUCTED

A questionnaire survey was conducted covering six districts affected in tsunami 2004. Total number of questionnaires distributed in Batticaloa, Ampara, Hambantota, Matara, Galle and Kalutara districts were 602. Some affected districts could not be covered due to various reasons including free accessibility and security.

A group of students from the Faculty of Engineering, University of Peradeniya assisted in data collection from field. In average, they spend one day per district visiting various internally displaced camps and other possible places to interview eye witnesses and victims of the incident. Details were collected filling a standard questionnaire and allowing respondents to express their views and experience freely to record it using a digital voice recorder. Results of details collected from questionnaire are presented in this paper while the other details are going to be published in a form of a book.

FINDINGS OF QUESTIONNAIRE

Some questions were asked to check where the respondents were at the time of tsunami, and whether they saw waves coming to land. From those who saw the waves, their candid feeling was also asked. Figures 2 and 3 show the results of responses.

According to Fig. 3, 16% of respondents wanted to stay back and observe it and another 10% wanted to call their friends to experience this uncommon phenomenon.

This behavior of Sri Lankans may be due to risk taking mentality of south east Asians and/or lack of awareness about danger of natural disasters in this magnitude. Anyway, one should not neglect the fact that, 2004 tsunami is the first mega tsunami reported in Indian ocean and the first one in the last century. That also contributes significantly for taking this matter lightly and resulting heavy casualties.

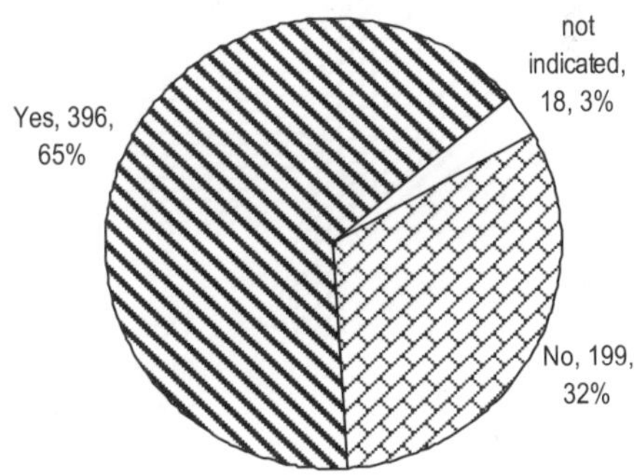

Figure 2 Did you see the wave Coming to land ?

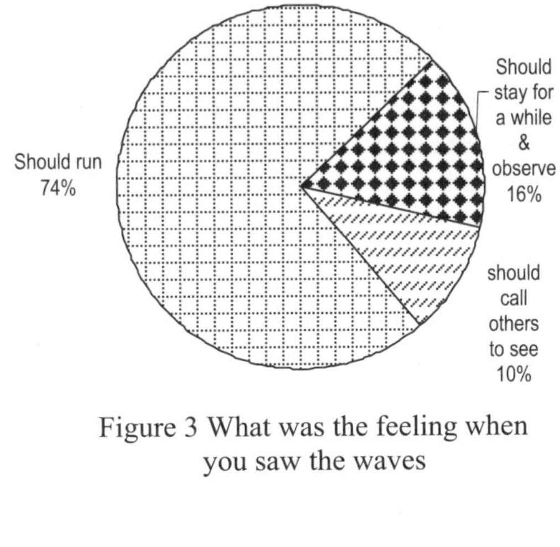

Figure 3 What was the feeling when you saw the waves

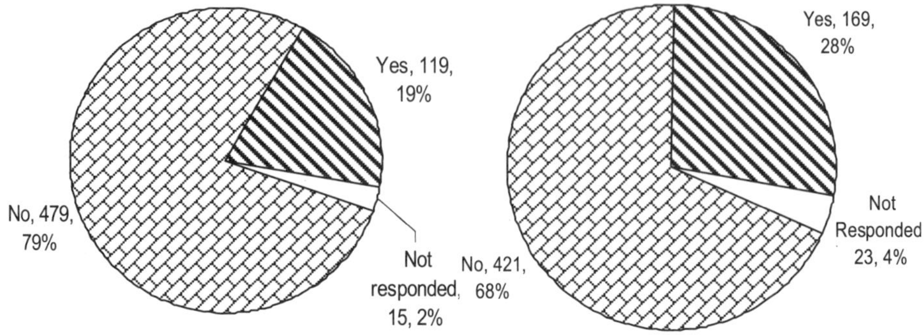

a) Relatives b) Friends

Figure 4 Number of relatives and friends of respondents went to beach to see the tsunami

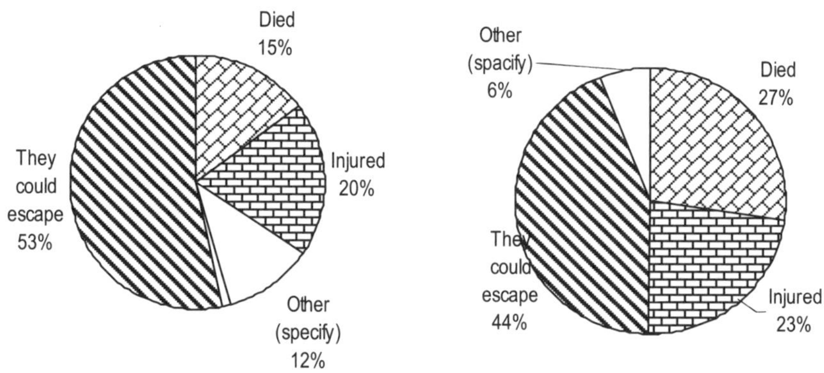

a) Relatives b) Friends

Figure 5 What happened to your relatives and friends went to see tsunami

Later in the questionnaire, whether the relatives and friends of respondents went to beach to see the tsunami was asked. The responses are given in Fig. 4, It is clear that a large number of people seems to have gone to experience this disaster either without knowing the magnitude or with the feeling of helping others. Figures 5 gives the percentages of died and injured after going to beach to see the tsunami.

As indicated above, one of the reasons for running towards a disaster is unawareness of magnitude of the disaster. In the questionnaire, whether there was any kind of public awareness program in the area was asked. As shown in Fig. 6, 74% answered that there was not a public awareness program about any natural disaster. As indicated above, this tsunami was the first in the Indian ocean and the only one in the last century. With such a frequency, one might question about the necessity of a public awareness on rare incidents like tsunami, earthquake etc.. But according to the results, it is clear that, a considerable number of deaths and injuries are reported due to unwanted risk, lack of planning in evacuation, lack of awareness on facilities available in each area. Therefore it is important to keep public aware about the possible paths of evacuation in different kinds of disasters.

Then, the next question was, if anybody has told about a tsunami in advance, would you have been believed it or not. Figure 7 is the response for that. Now it is important to know whether they feel a public awareness program is important. Will the public listen to instructions given by an authority is another important factor to consider. Figures 8 and 9 give the results of responses of the survey.

Due to this tsunami, in addition to more than 40,000 deaths, many homes, infrastructure facilities were destroyed. Even those who did not have any direct impact due to tsunami had to suffer a lot due to lack of facilities like electricity, water, transport, food, etc.. It is a known fact that there were many people affected due to lack of supply of food items and other basic facilities due to disturbances to road and other transport systems. During the questionnaire survey, the respondent's preference of the re-construction order was tested. As shown in Fig. 10, 74% wanted to have their houses done as the priority one. Priority 2 for many respondents was roads and bridges in the area.

Asian culture is always to help the neighbours. In a disaster, they never run away alone. Even if their life is in danger, they normally try to help (assist) their neighbours also. In the questionnaire, whether the respondents of this questionnaire are willing to help their neighbours were asked. 97% respondents said that they will help their neighbours at a disaster.

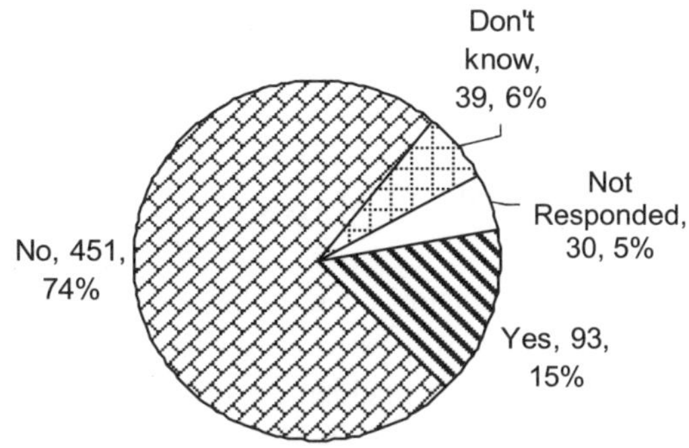

Figure 6 Was there a public awareness program in your area?

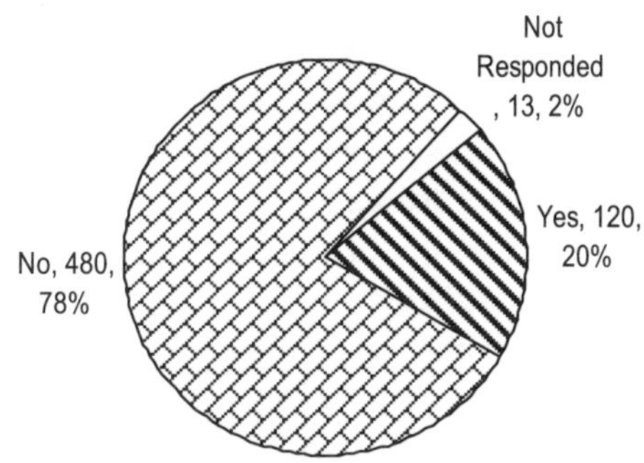

Figure 7 If anybody told you in advance about this tsunami, would you have been believed?

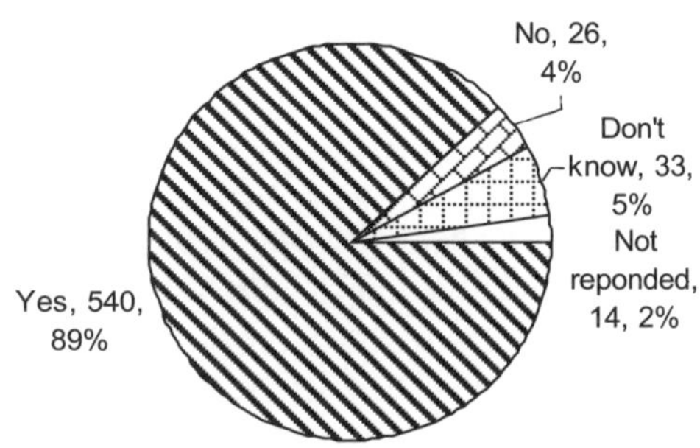

Figure 8 Do you think that public awareness programs are important?

According to that, if the general public kept educated about proper rescue methods, results will be more effective.

CONCLUDING REMARKS

2004 mega tsunami occurred as a result of the earth quake triggered closer to Sumatra was the first such incident in Indian ocean and the first such experience for Sri Lanakns. According to the available records in the history, this is the worst disaster in Sri Lanka.

This paper highlighted results of a public survey conducted based on an questionnaire. Results of questionnaire shows the level of risk taking in an incident like this and the outcome is also clear in this study. Lack of public awareness is one main reason for this behaviour. Therefore improving public awareness will be beneficial as a whole.

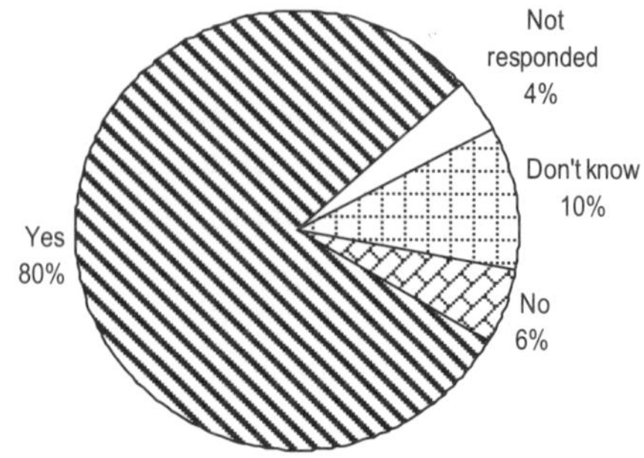

Fig. 9 Do you think the public will follow instructions given to them in an emergency

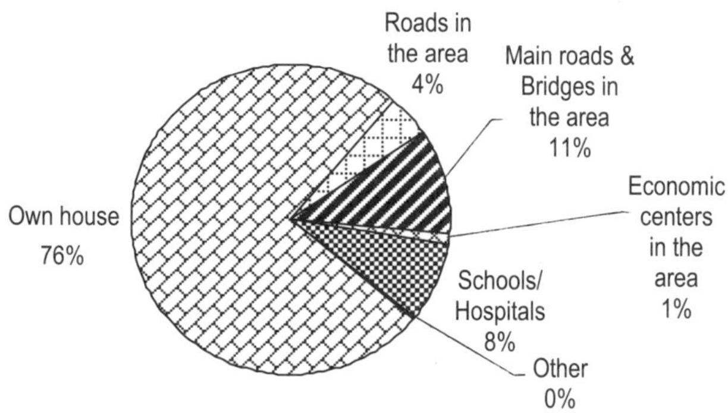

Fig. 10 Order of preference in reconstruction process

REFERENCES

AGHJ Edirisinghe, R.Yatabc (2003): "Responses of victims of 2003, may natural disasters in Sri Lanka", Proc. International seminar on disaster mitigation in Nepal, Katmandu, Nov. 2003, pp. 80-84

R Yatabe, N. Bhandary, H. K. Shrestha (2003): "Public awareness enhancement and effective networking –Keys to landslide disaster mitigation in Nepal", Proc. International seminar on disaster mitigation in Nepal, Katmandu, Nov. 2003, pp. 85-90

International Conference on Geotechnical Engineering for Disaster Mitigation & Rehabilitation
Chu, Phoon & Yong (eds) © 2005 World Scientific Publishing Company ISBN 981-256-469-1

DECEMBER 26, 2004 INDIAN OCEAN TSUNAMI: VULNERABILITY FUNCTIONS FOR LOSS ESTIMATION IN SRI LANKA

DR NAVIN PEIRIS
Arup, 13 Fitzroy Street, London, W1T 4BQ,UK

ANTONIOS POMONIS
Risk Management Solutions Inc., Athens Greece

On December 26, 2004 a great earthquake of magnitude $9.2M_w$ occurred in the Sunda Trench subduction zone to the west of Sumatra Island. The sudden violent vertical displacement of the main thrust fault resulted in tsunami waves propagating rapidly across the Indian Ocean causing destructive damage to many of the exposed countries. Sri Lanka about 2000km away from the epicenter was ranked the second worst affected country with heavy casualties and economic damage. This paper presents vulnerability functions developed for masonry structures commonly found as residential buildings in the Sri Lankan coastal areas and vulnerability functions for casualty estimation. These functions could be used to develop a loss estimation methodology for tsunami risk similar to HAZUS developed for seismic risk. The vulnerability functions were found to be dependent on the tsunami height, flow velocity and the distance from the shoreline.

INTRODUCTION

The December 26, 2005 northern Sumatra Earthquake and the tsunami that followed resulted in 31,000 confirmed fatalities (CNO, 2005) and a direct economic loss close to US$1.0billion (ADB, 2005). The direct economic loss measured in terms of asset loss was largely in the housing sector at US$306-US$341million, which can be attributed to the location of residential buildings close to the coastal routes. For instance 54% of the damaged housing units are located within 100m from the shoreline in the densely populated southwestern coast (Department of Census and Statistics, Sri Lanka, 2005). The tsunami therefore highlighted the exposure and hence the vulnerability of the coastal population and their residential properties to tsunami hazard.

A study was therefore initiated by the authors to develop a loss estimation methodology for the tsunami risk similar to HAZUS (FEMA, 2000) for seismic risk. An essential component of the loss estimation methodology is the vulnerability of the building stock in the study area to the hazard concerned. This paper discusses the vulnerability functions developed for residential buildings, which are largely unreinforced masonry structures. The data for the study came from the published damage statistics by the Department of Census and Statistics, Sri Lanka (2005) and the observations during the Earthquake Engineering Field Investigation Team (EEFIT) field mission (EEFIT, 2005a, b) and independent surveys conducted by Pillai (2005) in Sri Lanka. Additionally, the relationship between casualty figures and damage to residential buildings was used to deduce a vulnerability function for casualty estimation.

POST-TSUNAMI FIELD OBSERVATIONS

The EEFIT mission to the southwestern part of Sri Lanka from Colombo to Matara revealed that the predominant building type exposed to the tsunami was masonry structures largely used for residential purposes. Figure 1 shows the typical damage observed in these masonry structures.

Figure 1. Damage to masonry structures in southwest of Sri Lanka; (a) structural collapse and heavy damage beyond repair, (b) heavy damage due to collapse into scour hole.

The observed damage ranged from total collapse of masonry walls and heavy damage (non-repairable) to moderate (repairable) and slight damage. The principal damage mode appeared to be the out-of-plane failure of load bearing structural walls caused by the fluid pressure from the tsunami flow. Scouring was also responsible for structural damage due to foundation collapses or subsidence in areas where the surficial soils are largely sand with little clay content. Both poor quality (semi-fired clay blocks) and good quality (solid brick or cement block) masonry structures were found in the surveyed areas. The level of damage varied along the coastline but generally diminished with distance from the coast. The survey carried out by Pillai (2005) in the eastern coastal areas from Trincomalee to Kalmunai found similar damage to masonry residential structures as observed by EEFIT in Southern and Western provinces. Structural collapse and heavy damage in the eastern coastal areas occurred over larger distances due to high tsunami water levels and large inundation distances (Choi *et al.*, 2005 and Wijetunge, 2005).

VULNERABILITY FUNCTIONS FOR MASONRY STRUCTURES

Vulnerability for structural damage of a category of buildings is characterized by a set of fragility curves each of which is expressed as the conditional probability of reaching or exceeding a particular damage state (ds) given a certain value of the demand parameter (d). The conditional probability, P(DS≥ds | D=d) is expressed by a cumulative probability distribution. The "direct" approach was followed in developing the vulnerability functions where damage statistics were correlated with the field observations of a demand parameter. The damage statistics were those published by the Department of Census and Statistics, Sri Lanka (2005) at the Divisional Secretariat (DS) division level. According to FEMA55, the forces acting on a building due to a tsunami (wave impact, hydrostatic, hydrodynamic, buoyancy, debris impact and scouring) are functions of tsunami water height and flow velocity. Now, the velocity of a tsunami upon arrival at the shoreline and its subsequent profile onshore is difficult to measure in the field. However silt traces on buildings, debris heights etc. provide ample evidence of tsunami water height at the shoreline and its variation with distance from the coast. Hence the tsunami water height was used as the demand parameter in this study and data from EEFIT (2005a), Choi *et al.*, (2005) and KUDPRI (2005) were used.

Figure 2 shows the percentage of housing damage (damage in each damage state over total number of housing units before tsunami in the affected census blocks) as a function of tsunami water height, H at the coast using damage data that represents 300m (or greater) distance from the coast. The percentage of housing represents the cumulative conditional probability, P_i where i = 1, 2, 3 corresponding to damage states partial damage with re-use, partial damage without re-use and complete damage as defined by the Department of Census and Statistics, Sri Lanka (2005).

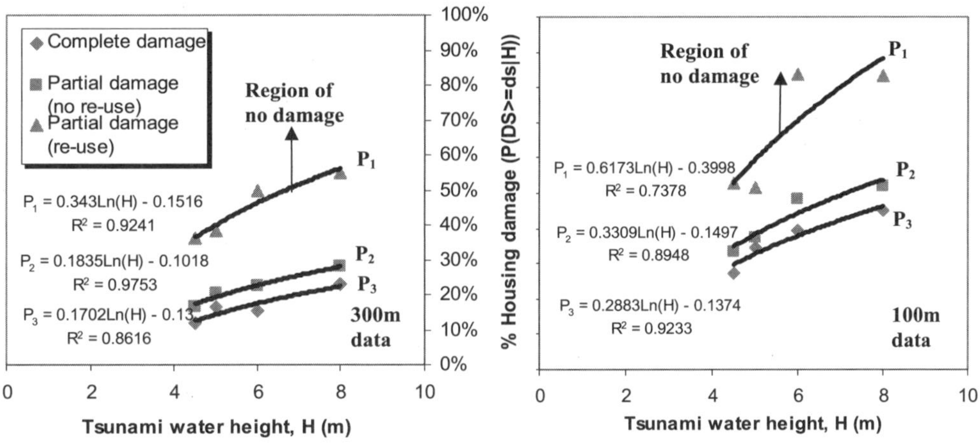

Figure 2. Vulnerability functions for masonry structures over 300m (or greater) and over the first 100m from the coast in southwest of Sri Lanka.

The trend lines show a very good correlation between P_i and H (correlation coefficient, R^2 greater than 0.8) hence the log functions for each damage state represent the observed fragility curves for damage to masonry structures over a distance of 300m from the coast. Figure 2 also shows the fragility curves for damage data over a distance of 100m from the coast. The curves indicate higher probability of damage compared to data over 300m. Hence the observed vulnerability functions for masonry structures are a function of the distance of the data set from the coast.

From a component fragility approach, the vulnerability functions of a particular structure type should not vary with location relative to the coast, as it is only a function of the demand parameter water height or velocity. Figure 3 shows the relationship between housing damage ratio (total damage at a given distance over total damage over 300m or greater) and distance from the coast. The damage ratio reduces with distance, which is consistent with the field observations. The trends in the vulnerability functions and the damage ratio suggest that the tsunami water height (and velocity) reduced with distance from the coast, which can be thought of as a tsunami exposure function for water height given a value H at the coast. This trend is expected since the strength of a tsunami dissipates as it advances onshore due to damage it causes to structures, obstructions, vegetation and topographical variations.

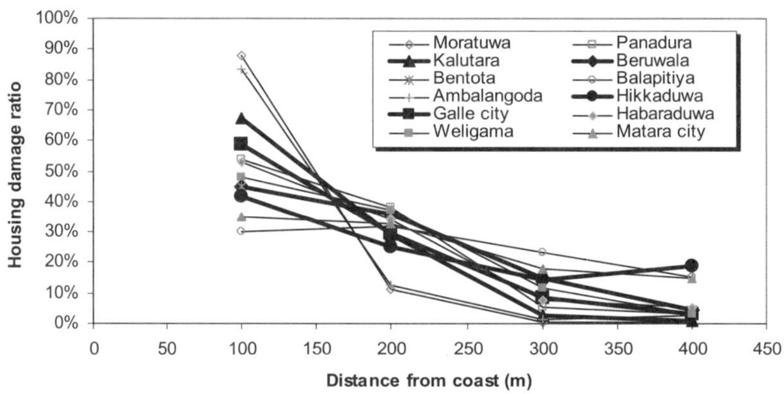

Figure 3. Relationship between housing damage ratio and distance from coast for southwest of Sri Lanka.

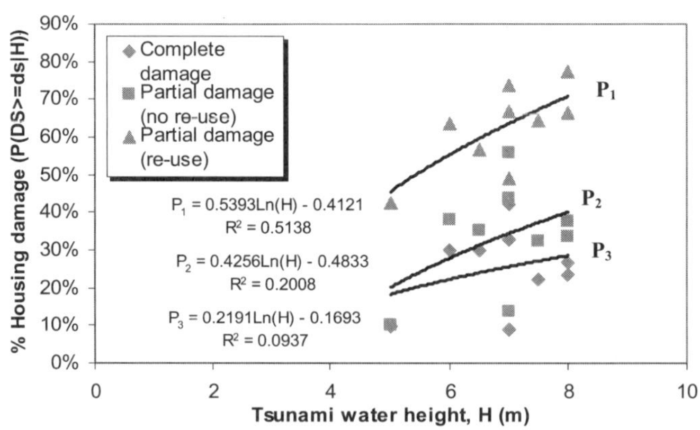

Figure 4. Vulnerability functions for masonry structures over 400m (or greater) from the coast in east of Sri Lanka.

Figure 4 shows the probability of damage in each damage state for the DS division damage data in the eastern coastal areas. The tsunami water height measured in these areas varied from 6m to 8m. The scatter in the data shows the difficulty in deriving suitable fragility curves based on observations in the eastern areas (R^2 less than 0.8). This may be attributed to the data used being preliminary estimates and sampling errors possibly due to large distances of damage in the eastern areas.

ESTIMATION OF CASUALTY VULNERABILITY

EEFIT (2005b) deduced a good correlation between total casualties (sum of deaths, missing and injuries) and total damage as shown in Figure 5. The data for this correlation comes from district level (higher than DS division) damage and casualty figures published by the Department of Census and Statistics, Sri Lanka. Estimates of total damage from vulnerability function P_1 could be used to estimate the casualties using the relationship given in Figure 5. However further studies are necessary to deduce the deaths, missing and injuries from the casualty estimates.

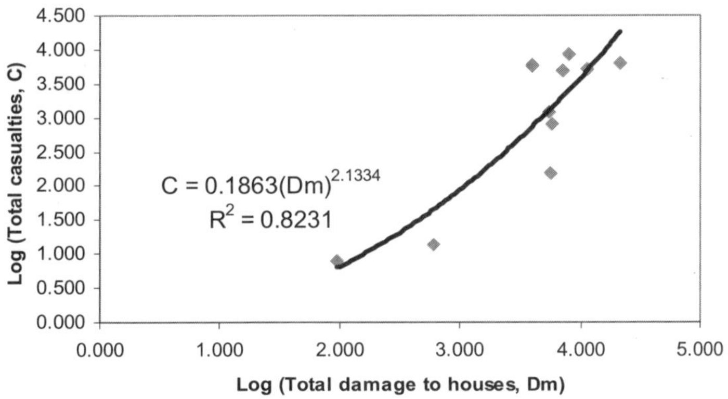

Figure 5. Relationship between casualties and number of damaged houses in Sri Lanka.

CONCLUSIONS

Vulnerability functions were developed for masonry residential structures in the coastal areas of Sri Lanka based on the published damage statistics and field mission observations. The vulnerability functions were found to be a function of the distance of the data set from the coast. The total damage ratio shows a reduction with distance from the coast. Hence a tsunami exposure function could be postulated where the tsunami water height and velocity reduces with distance from the coast. This exposure function together with the vulnerability functions could be used to estimate damage to masonry houses in the coastal areas of Sri Lanka from a future tsunami and hence the direct economic loss similar to HAZUS for seismic risk. Further data are necessary to refine the vulnerability functions presented, particularly for the eastern areas of Sri Lanka where there was a considerable scatter in the vulnerability data. Estimation of casualties could be made from total damage estimates using a relationship with casualty figures. However further data and studies are necessary to deduce estimates of deaths, missing and injuries at a local level.

ACKNOWLEDGMENTS

The lead author wishes to thank members of the EEFIT team that visited Sri Lanka in January 2005, Antonios Pomonis (team leader), Tiziana Rosetto (University College, London), Sean Wilkinson (Newcastle University, UK), Navin Peiris (Arup, London), Dominico Del Re (Buro Happold, London) and Stuart Gallocher (Halcrow, UK), and Anton Pillai and his colleagues who have conducted the independent survey in the eastern coastal areas of Sri Lanka.

REFERENCES

ADB (2005). "Preliminary Damage and Needs Assessment" 2005 Post-Tsunami Recovery Program, prepared by Asian Development Bank, Japan Bank for International Cooperation and World Bank, January 10-28, 2005.

Choi, B. H., Siripong, A., Sundar, V., Wijetunge, J. and Diposaptono, S. (2005). "When the Sea Strikes Back: December 26, 2004 Earthquake Tsunami of Indian Ocean – Post Runup Survey". Workshop on Indian Ocean Tsunami, 13th Pacific Ocean and Asian Marginal Seas (PAMS)/JECSS Meeting, Bali, Indonesia.

CNO (2005). Center for National Operations, Sri Lanka (www.cno-srilanka.org).

Department of Census and Statistics, Sri Lanka (2005). Reports on Census of Buildings and People Affected by the Tsunami - 2004 (www.statistics.gov.lk).

EEFIT (2005a). "The 26 December 2004 Indian Ocean Tsunami: A Preliminary Reconnaisance Report (www.eefit.org.uk).

EEFIT (2005b). "The Indian Ocean Tsunami 26 December 2004", final report prepared by EEFIT team, reviewed and approved for publication.

FEMA (2000). "Earthquake Loss Estimation Methodology, HAZUS 99 Service Release 2 (SR2) Technical Manual" *Federal Emergency Management Agency*, Washington DC, USA.

FEMA55. *"Coastal Construction Manual"*, Third Edition, Federal Emergency Management Agency

Pillai, A (2005). Private communication following a field survey conducted in the Eastern province of Sri Lanka in February 2005.

Wijetunge, J. (2005). *"How far inland"*, Article in "The Sunday Times", March 20, 2005 by Dr J. Wijetunge, Faculty of Engineering, University of Peradeniya, Sri Lanka.

International Conference on Geotechnical Engineering for Disaster Mitigation & Rehabilitation
Chu, Phoon & Yong (eds) © 2005 World Scientific Publishing Company ISBN 981-256-469-1

DAMAGES OF INFRASTRUCTURES IN THE DISASTER OF 26 DEC 2004 EARTHQUAKE AND TSUNAMI IN ACEH-INDONESIA

PAULUS P. RAHARDJO
Departement of Civil Engineering, Parahyangan Catholic University
Bandung, Indonesia

The Sumatera earthquake and tsunami disaster of Dec 26, 2004 was tremendous. The destruction and death were over whelming, causing all nations to be awakened by the fact that this is the largest tsunami disaster in the century. This paper describes the seismotectonic of Sumatera and some aspects on the damages from geotechnical concern.

INTRODUCTION

Indonesia is located in the mid of tectonic plate collision of Indo-Australia, Eurasia and Pacific Plates where subduction zone was formed and generates active volcanoes and active faults resulting earthquakes and volcanic eruptions. As the results, Indonesia is prone to geologic hazard which are quite disastrous and destructive.

In the 26 Dec 2004 Sumatera earthquake, more than 280,000 people were swept to death, many infrastructures and buildings were flattened or washed out to sea. The destruction was the most severe in Indonesia, mainly in the city of Banda Aceh and Meulaboh. This paper is intended to describe the seismotectonics of Sumatera which caused the earthquake and to discuss the geotechnical aspects on the damages.

SEISMOTECTONICS AND SEISMICITY OF SUMATERA

The west coast of Sumatera and Aceh is geologically instable. One of the key to understand the crustal deformation of this area is by observation of the mechanism of Sumatera fault system, which stretches out from Aceh to Sunda strait. Sumatera island is part of the Sunda Arc (Hamiltaon, 1979) and the seismicity of Sumatera is characterized as Western Sunda Arc (Puspito and Shimazaki, 1995). The tectonic pattern of Sumatera is controlled by two activities elemets, i.e. (1) subduction zone in the Sumatera-Java trench and (2) shear zone in the Sumatera-Java trench and (2) shear zone along Sumatera island.

As shown in Fig. 1, Sumatera is like other islands of Indonesia, representing and island arc where the formation was related to the activity of subduction zone. In Sumatera, the subduction is oblique (not perpendicular) with direction of approximately N 20° E, and the impact of this movement cause faulting of the west side of Sumatera island (Fitch, 1972, Beck, !983; Huchon and Le Pichon, 1984). This fault is almost parallel to Sumatera island and according to Tjia (1977) this fault consist of a few segments.

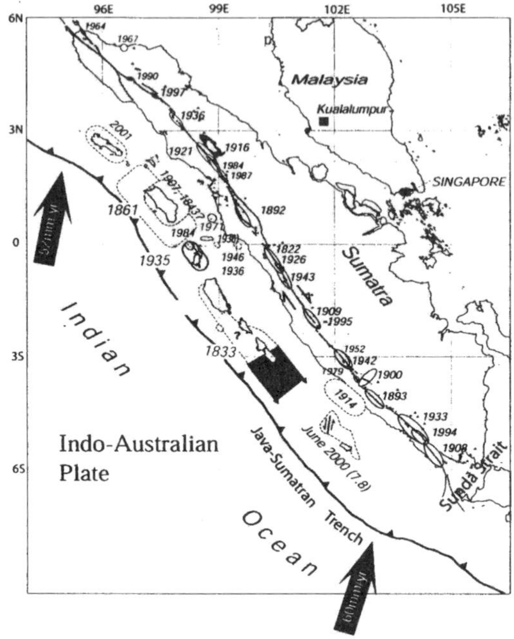

Figure 1. Neotectonic and Seismicity of Sumatera (Natawidjaja & Sieh, 1994)

The India plate is part of the huge Indo-Australian Plate which underlies the Indian Ocean and the Bay of Bengal and is drifting north-east at a rate of about 60 mm per year. The India plates meets the Burma Plate (part of the Eurasian plate) at the Sunda Trench off Sumatera and has been slipping beneath the Burma Plate for million of years. The existence of small islands in the west of Sumatera has close relationship with the Sumatera fault. Mentawai islands and some other islands including Batu, Siberut, Nias and Simeuleu were formed through this mechanism and the formation take place in thousands or even millions of years. Due to collisions between ocean plates and the continental plate, these islands are moving up by 2-3 cm/year. At plate boundaries, the thin oceanic plates slip under continental plates known as subduction. The epicenter of the earthquakes in this area, mostly occur just below the islands such that occurred in Nias last March 2005.

As understood from the seismotectonic of Sumatera, the seismicity of Sumatera is controlled by the subduction zone. Figure 2 shows the epicenters of earthquake magnitude M > 5.0 since 1975. The distribution of earthquakes in Sumatera is dominated by shallow and medium depth earthquake (depth = 0 – 300 km). The depth of earthquakes foci in the subduction zone is maximum 100 km in the Andaman islands to about 300 km south of Sumatera, this means that the subduction period in the north is more recent compared to the south. The earthquakes along the shear zones of Sumatera are generally less in magnitude, however the ground motion are capable of damaging the human settlement because the foci are shallow (<30 km) such as Liwa (1994) and Kerinci (1995) earthquakes, both with M = 7.0

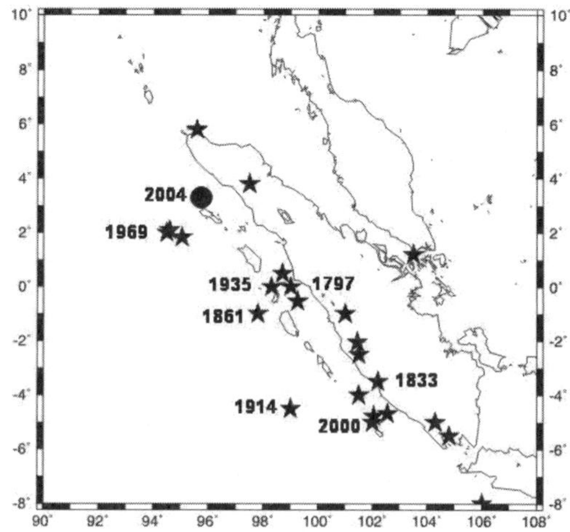

Figure 2. Sumatera Earthquakes with M > 7

THE DISASTER OF 26 DEC 2004 EARTHQUAKE AND TSUNAMI

At 00.58:53 GMT and 07.58:53 local time, the slow build up of pressure caused by the continuous grinding of the two plates resulted in the giant earthquake. About 1200 km of the fault line slipped 15 m along the subduction zone. The huge rupture is known as a "megathrust" and produced an earthquake that lasted for more than four minutes and measured 9.0 on the Richter Scale, making it one of the most violent record. In addition to the sideways movement between the plates, the seabed is estimated to have risen by several meters as a result of the quake, thereby displacing millions of tones of water above it (Tibballs, 2005).

Within just 15 minutes after the earthquake, the big wave struck Aceh Province, 248 km north-east of epicenter, 15 minutes later, the wave hit Andaman and Nicobar Island and then Malaysia. About one and a half hour, Thailand was swept and then after two hours it attacked Srilanka and the south east coast of India. Next Maldives and incredibly, seven hours later the tsunami arrived in East Africa, 4500 km from epicenter with sufficient destructive energy.

Based on the extent of main shocks and after shock, the rupture area can be determined, and this has very close relationship with the magnitude of the earthquake. Aceh earthquake of 26 Dec 2004 is one of the largest in the last 50 years after Chilie earthquake 1960 (Mw = 9.5, Mo = 20x1029 dyne cm) and Alaska earthquake of 1964

(Mw = 9.2, Mo = 8 x1029 dyne cm). Thc three earthquakes as mentioned above caused damaging tsunami (Fig. 3).

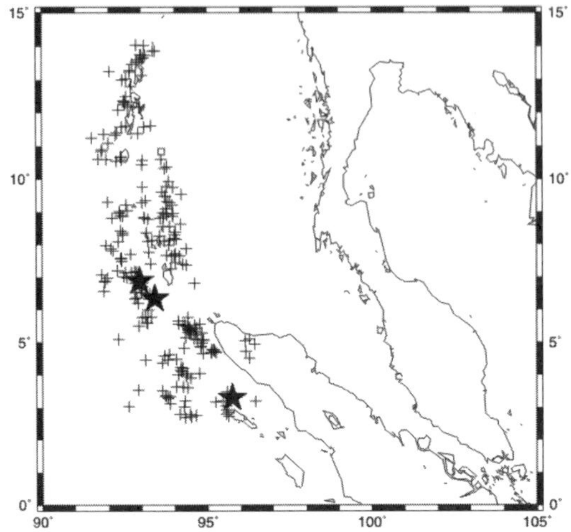

Figure 3. Main shocks and After-shocks depicting the rupture area

DAMAGES CAUSED BY THE TSUNAMIS IN ACEH

The damages by earthquake and tsunami in Nangroe Aceh Darussalam were due to ground vibration and wave energy of tsunami which intensely ruin infrastructures and buildings in the city. Since the earthquake epicenter is a relatively far from the city, the tsunami effect is more significant. The area affected by wave run up was estimated 10,000 km2 as shown on Figure 4. The damages were more in Banda Aceh and Meulaboh because the elevation of the cities are mostly less than 20 m from sea level. The appearance of the affected area were drastically changed due to destruction and washing out after tsunami. The green appearance changed into brown (soil) or black (mud). The earthquake has caused disconnection and mal-function of communication facilities and electricity. Other life lines such as roads and bridges, airports and harbors were practically washed out or filled with black muddy material. At the airport the runway were flooded by post tsunami event and in the harbor, all of the sea traffic signs were gone. The jettys were full of debris and water channel can not be identified.

The damages after tsunami will continue due to sea water flooding that damaged also city drainage system, causing innundation in the city. Water innundated low areas and the soils were saturated. When the rains come, this area can no longer be able to absorb water and flooding could easily occur. Due to the damages of buildings and infrastructures, these facilities need to be evaluated for future use and properness. This will take time before they can be reutilised. The roads damages in NAD may be more than 280 km, the worst being Aceh-Meulaboh (223.5 km) where 10 bridges collapsed.

The damages in Western area of NAD are the most serious. It is interesting to view the damages in terms of geotechnical aspects. Most western and northern part of Banda

Aceh and North Sumatera mainly consist of loose fine sand sediments. The geotechnical concern on this type of material is the problem of liquecfaction and wash out material. As shown in the Figure 5, the collapse of building and bridges are due to the tsunami energy but liquefaction may occur prior to the arrival of the wave.

Figure 4. Tsunami Affected Area in Aceh and North Sumatera

Banda Aceh is located in sedimentary area with V shape. The actual run up could reach more than 10 km although most of the damages are in the area of about 3 km from the coast line. The wave entered Banda Aceh through low land and two channels, one is from the east of the city. The drainage canal provided waterways for incoming water so that tsunami flooding could reach far into the city.

In fact the channel that was constructed in 1981, was meant for flood control at the city of Banda Aceh due to surplus of water from Krueng Aceh Lama. This channel brought the water into the heart of the city. Although the grand mosque was located 3 km away from the shoreline, when the water came in secondary channel, water turbulance were created causing even more damages. The damages of Banda Aceh were tragic as shown from sattelite photograph.

Figure 5. Photograph, the damages in the city of Banda Aceh

SUMMARY

- Knowledge on geology and seismotectonics of Sumatera is very useful to understand the mechanism of earthquake and tsunami in Aceh.
- Coastal morphology and the soil condition may have significant effect on the damages by tsunami. Since the soil condition in Aceh is predominantly loose sandy sediments, the material may have been liquefied and washed out easily upon wave return to the sea and hence light buildings or infrastructures can not survive.

REFERENCE

Eddy Gafar, Igna Hadi, S., Heri Harjono, Praptisih, S. Indarto dan Kardi Suharyono, "Identifikasi Sasar Aktif Segmen Semangko di Lampung Selatan, Sumatera", Laporan Penelitian No. 05/PPPG/90, Pusat Penelitian dan Pengembangan Geoteknologi, Lembaga Ilmu Pengetahuan Indonesia, 1990.

Fitch T.J., "Plate convergence, transcurrent faults and internal deformation adjacent to Southeast Asia and Western Pacific", Journal Geophysics no. 77, pp.4432-4460, 1972.

Hedervari, P., Papp, Z., "Seismicity Maps of The Indonesian Region", Contribution No. 04/08/IRGTCP, Elsevier Scientific Publishing Company, 1980.

Kiekhefer, R.M., "Geophysical Studies of the Oblique Subduction Zone in Sumatera," Ph.D. thesis, University of California San Diego, p. 119, 1980.

Mc. Caffrey, R., "Slip Vectors and Stretching of the Sumatran Forearc," Geology no 19, pp. 881-884, 1991.

Natawijaya, D. H., Sieh, K. E., "Neotektonik Sistem Sesar Sumatera," Prosiding Hasil hasil Penelitian Puslitbang Geoteknologi, LIPI, 1994.

Puspito, N.T. "Gempa Bumi Aceh dan Kegempaan di Sumatera". Diskusi Mitigasi Pasca Bencana Alam Gempa Bumi & Tsunami Aceh – Bandung, ISBN 979-99212-0-1, 18 Januari 2005.

Rahardjo, P.P. "Dampak Kerusakan akibat Gempa Bumi & Tsunami di Nanggroe Aceh Darussalam". Diskusi Mitigasi Pasca Bencana Alam Gempa Bumi & Tsunami Aceh – Bandung, ISBN 979-99212-0-1, 18 Januari 2005.

Sampurno and Rahardjo, P.P. "Pokok-Pokok Bahasan Gempa Bumi & Tsunami". Diskusi Mitigasi Pasca Bencana Alam Gempa Bumi & Tsunami Aceh – Bandung, ISBN 979-99212-0-1, 18 Januari 2005.

Siech, K. and D. Natawidjaja, "Neotectonic of the Sumatran Fault, Indonesia", Journal of Geophysical Research 105 (B12), 2000.

Zen, M. T., Dahrin, D., Diament, M., Harjono, H., Karta, K., Deplus, C., Gerard, M., Lassal, O., Malod, J. A., and Martin, A., "Mentawai 90-cruise result : The Sumatra Oblique Subduction and Strike Slip Fault Zones," dalam H. Prasetyo (ed.), Geodynamic Prcesses in the Forearc Silver Plate and General Topics, HAGI, 1991.

International Conference on Geotechnical Engineering for Disaster Mitigation & Rehabilitation
Chu, Phoon & Yong (eds) © 2005 World Scientific Publishing Company ISBN 981-256-469-1

INITIAL CONDITIONS ON NUMERICAL SIMULATIONS OF TSUNAMI PROPAGATION

ROSALES-SIERRA VICTOR
Maritime Research Centre, Nanyang Technological University, 50 Nanyang Av.(N1-B4b Rm03),639798,Singapore. victorrosales@ntu.edu.sg

In this research, we seek to systematically study into the effects of different kinds of initial conditions on numerical simulations of tsunami propagation. Nowadays, powerful Navier Stokes codes with non hydrostatic and non linear terms are available to study the tsunami propagation over complex bathymetries. The study case will be the India Ocean, body leading to Banda Aceh, Andaman Sea and Malacca Straits for its location as a corridor for tsunami propagation with well determined limits. A model based on finite elements is used in this study. The verification of the code is established using available run-up data from the last 2004 Indonesian tsunami. A discussion on the results is done and some preliminary conclusion is then attempted on the relative merits of the different kinds of initial condition in the light of the 2004 Indonesian tsunami data.

INTRODUCTION

Initial conditions are obtained from 3 different analytical models on tsunami wave generation. In this study the input of these conditions are tested in a 3D Navier Stokes code. The convergence of the code is verified by the error reduction in refined grids, while the validation of the simulations is considered by comparing the results with the visual estimates of the run-up in Banda Aceh in the recent tsunami event of Sumatra at 26 December 2004.

The topobathymetry needed to set up the model for the 3D Navier Stokes code is taken from the world data of NOAA, N. G. D. C. N., 2005. Local available charts are used to complement this information. A grid of unstructured elements is then designed to take into account the details of the zones where the comparison takes place: west coast of Banda Aceh.

INITIAL CONDITIONS

Three methods for initial conditions are investigated. The wave profiles are shown in Figure 1. The first initial condition is the solution for Cnoidal waves, Laitone, 1960. It is a wave of translation relative to the water mass and its surface lies entirely above the still water level. Its celerity depends on the water depth and on the wave height. The expressions for the free surface and celerity are:

$$\eta_{cnoidal} = H\left[\sec h\sqrt{\frac{3H}{4d^3}}(x)\right]^2 \tag{1}$$

$$c = \sqrt{g(d+H)} \tag{2}$$

H is the wave height, g is the gravity acceleration and d is the depth.

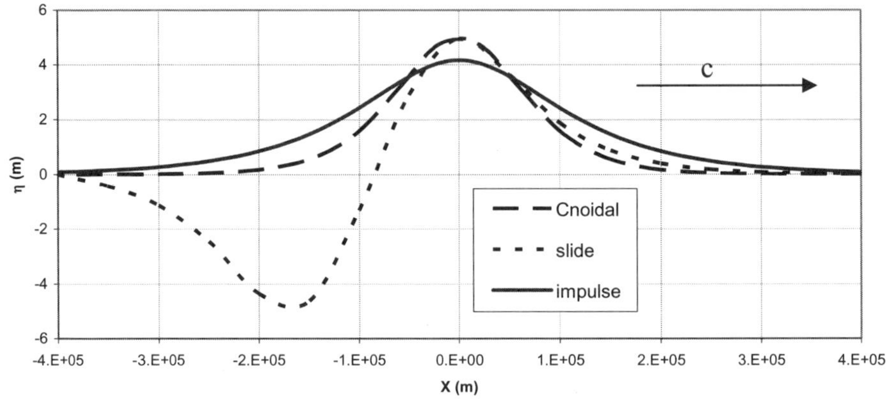

Figure 1. Free surface Initial conditions above the epicenter.

The second initial condition is the free surface profile generated by a submarine gravity mass flow (land slide). The analytical 2D solution is found with a simplified rectangular slide at a prescribed sub-critical velocity. Two identical waves are generated at the front and the rear of the slide. The characteristics of the wave depend on the slide geometry and initial acceleration; Haugen, K. B., Lovholt, F., and Harbitz, C. B., 2005. The expression for the free surface is:

$$\eta_{slide}(x,t) = \frac{H_{heav}(t-\tau(x_0))\xi(x_0)dx_0}{4d} \times \left[\sec h\left(\frac{\pi(x-x_0)+c_0(t-\tau(x_0))}{2d}\right) + \sec h\left(\frac{\pi(x-x_0)-c_0(t-\tau(x_0))}{2d}\right)\right] \tag{3}$$

H_{heav} is the Heaviside step function to limit the problem to one disturbance disturbance, t is the absolute time, $\tau(x_0)$ is the initial time of slide movement, the initial celerity is $c_0=(gd)^{1/2}$. $\xi(x_0)$ is the wide of the rectangular slide which is positive in the front and negative in the rear. The final surface elevation is the addition of both front and rear components.

The third initial condition is a wave generated by a curve bottom impulse explained by Tyvand, P. A. and Haugen, K. B., 2005. Even if Green functions for the initial flow for various bottom topographies in two and three dimensions have been developed in practice it's more useful to use asymptotic Green functions that cover arbitrary bottom geometries with moderate slope and curvature that is the case of this application. The formulation is valid for a time scale of deformation smaller than the local gravitational time, and it is relevant for rapid bottom deformations due to earthquakes. The expression for the free surface is:

$$\eta_{impulse} = \frac{q\sin(\alpha)}{2\alpha\left(x\sin(\alpha)+d\cos(\alpha)\right)}\operatorname{sech}\left(\frac{\pi}{2\alpha}\log\left(\frac{x}{d}\sin(\alpha)+\cos(\alpha)\right)\right) \tag{4}$$

α is the bottom slope, q is the volume displaced by the impulse in the bottom.

Reproducing a tsunami wave in the code

The tsunami wave is produced using equations 1, 2, 3 and 4 as initial free surface at instant zero. Only for the case of the wave generated by the land slide the initial time was different from zero as the slide needs to complete its movement.

NUMERICAL SIMULATIONS OF WAVE PROPAGATION

Numerical simulations have been carried on with the TELEMAC-3D system developed by Janin, J. M., Lepeintre, F., and Pechon, P., 1992. It's a three-dimensional computation code that describes the 3D velocity field (u, v, w) and water depth h (or free surface S measured from the bed) at each time step. The code solves the three-dimensional hydrodynamic equations under the following assumptions:

- Navier-Stokes 3D equations with free surface changing in time,
- Negligible density variation in the mass conservation equation,
- Hydrostatic (general) and non hydrostatic pressure assumed,
- Boussinesq approximation for momentum.

The system comprises many modules which could allow for external code-interfacing. Because of the structure, this then allows for the introduction of different initial boundary conditions into its modeling and simulation.

Numerical domain

The domain is composed by an irregular grid of 4272 nodes. Refinement with smaller elements is localized in the region around Banda Ache (North West of Sumatra) to simulate more accurately the run up of the tsunami. The topobathymetry was obtained from the NOAA, N. G. D. C. N., 2005. Detailed information was added from local charts. The wave propagation over the domain is shown in Figure 2. Three points for comparison were extracted as shown in the enclosed area. These points are located in the west coast of Banda Aceh between the latitudes N5°24' and N5°30'.

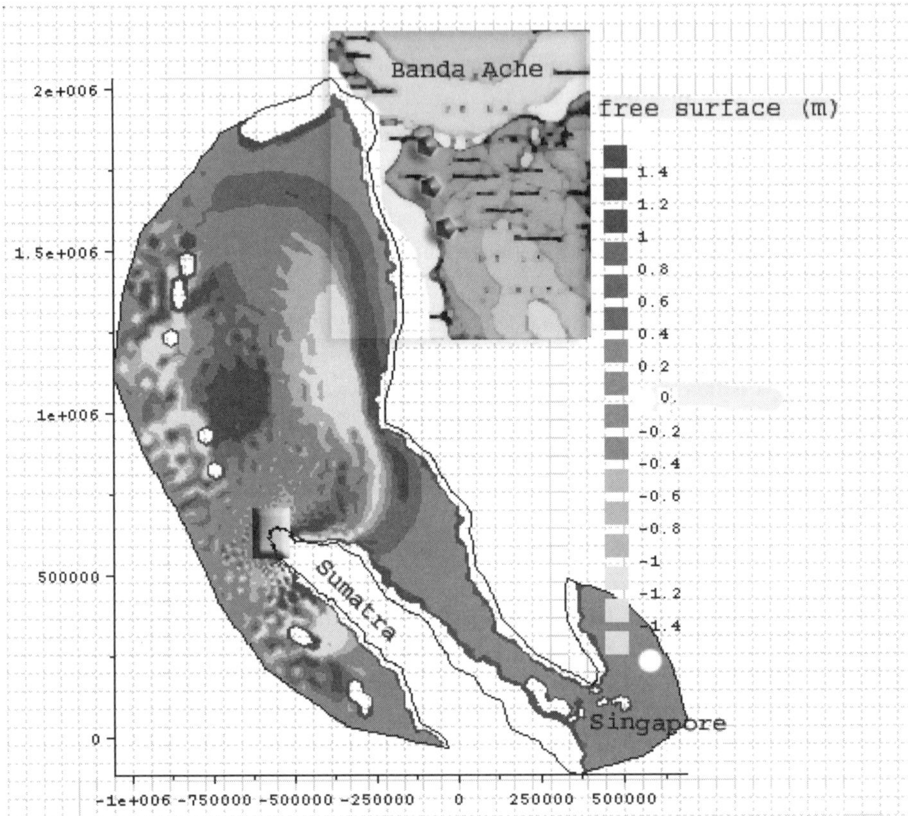

Figure 2. Domain model and Tsunami propagation.

Convergence study

A convergence study was set up to analyze the accuracy of the numerical solution. Three grids were used to compute the numerical solutions and compare their solutions to find numerical dispersion. An error can be found with the difference of the maximum run up between two grids. The results were all satisfactory due to the fact that the error decreases within the refinement. Because larger errors were found in the shallow regions, further refinement is then needed to increase the accuracy of the simulations in shallow zones. In this study a lack of detailed topobathymetry data impede additional refinement. Table 1 shows the average error in the solution of the three grids. The reduction of the error verifies the convergence of the numerical solution as established by ROSALES, V., GARAPON, A., and LEVACHER, D., 2002

Table 1. Characteristics and error of the grids used for the convergence analysis.

Grid	Nodes	Δx min (m)	Δt (s)	Error %
Coarse	2855	4240	40	15.2
Medium	4274	3000	60	4.5
Fine	6932	2120	80	-

RESULTS

The validation of the model was made by comparing the simulated run up results with visual measurements of the maximum water level after the tsunami in Banda Aceh by EERI, 2005. Figure 3 shows the maximum run up of the tsunami in three points selected of the west coast of Banda Aceh (see Figure 2). The triangle marks represent the visual measurements. The measurements were made some days after the tsunami by visual inspections of water marks in buildings and trees.

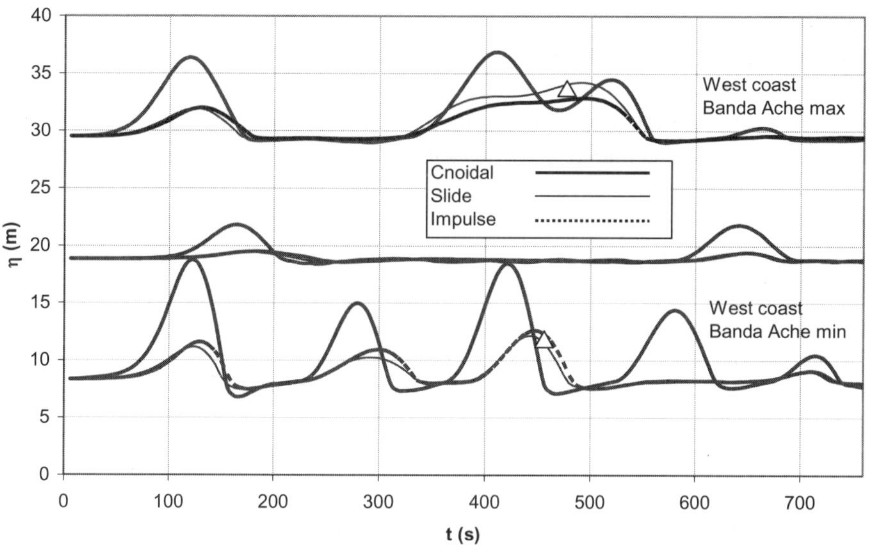

Figure 3. Comparison of tsunami run up in three locations on the west coast of Banda Ache. Maximum Run up and visual measurements are plotted.

From the comparison it's observed that the cnoidal solution over predicts the maximum run up while the land slide and impulse initial conditions are closer to the visual measurements. One remark on the land slide approximation is that its celerity of the wave computed is the same as that of the cnoidal wave. This assumption can be questionable as the tsunami was not originated by a land slide. For this reason, it is more opportune to consider the solution of the bottom impulse as more acceptable here and also because its theoretical basis as given by Tyvand, P. A. and K. B. Haugen, 2005 is closer to the physics of the tsunami generated by an earthquake.

CONCLUSIONS

The analytical solution for a tsunami due to impulsive deflection given by Tyvand, P. A. and Haugen, K. B., 2005 appears to be more suitable as input for initial condition for tsunami wave propagation in a 3D numerical model. The convergence analysis reveals that Telemac3D is able to simulate this kind of propagation in shallow waters for wave run-up prediction.

More accurate information of the topobathymetry is still required to increase accuracy of run-up prediction in shallow zones. The next phase of the study will be to reproduce tsunami wave by a direct impulse as a form of source input with volume assumed equal to the bottom displacement.

REFERENCES

EERI, 2005: Distribution of the Tsunami Heights of the 2004 Sumatra Tsunami in Banda Aceh measured by the Tsunami Survey Team.

Haugen, K. B., F. Lovholt, and C. B. Harbitz, 2005: Fundamental mechanisms for tsunami generation by submarine mass flows in idealised geometries. *Marine and Petroleum Geology*, **22,** 209-217.

Janin, J. M., F. Lepeintre, and P. Pechon, 1992: TELEMAC 3D : A finite element code to solve 3D free surface flow problems. *Computer modelling of seas and coastal regions*, Southampton, (UK).

Laitone, 1960: The second approximation to cnoidal and solitary waves. *Journal of fluid mechanics*, **9,** 430-444.

NOAA, N. G. D. C. N., cited 2005: Multibeam Bathymetry. [Available online from http://www.ngdc.noaa.gov/mgg/bathymetry/multibeam.html.]

ROSALES, V., A. GARAPON, and D. LEVACHER, 2002: Simulation numérique d'un écoulement stratifié turbulent. *VII journées nationaux génie civil génie*, Biarritz; France.

Tyvand, P. A. and K. B. Haugen, 2005: Initial free-surface elevation due to impulsive deflection of a two-dimensional bottom with small slope and small curvature. *Fluid Dynamics Research*, **36,** 1-8.

ACKNOWLEDGEMENTS

Important contributions to improve the contents of this study were made by Dr. Lim Bin Teck and Dr. Tan Soon Keat from MRC, Singapore. Fundamental guidance about the set up of initial conditions was offered by Dr. Jean-Michel Hervouet and Mr. Thierry Boulet from EDF, France. To all of them my sincere gratitude.

International Conference on Geotechnical Engineering for Disaster Mitigation & Rehabilitation
Chu, Phoon & Yong (eds) © 2005 World Scientific Publishing Company ISBN 981-256-469-1

ANALYSIS OF SHORE LINE PROTECTION WORKS IN TSUNAMI HIT AREAS OF EAST COAST OF INDIA

G. L. SIVAKUMAR BABU
Dept. of civil Engg., IISc, Bangalore, 560012, India

AMIT SRIVASTAVA
Dept. of civil Engg., IISc, Bangalore, 560012, India

Evaluation and design of shore protection works in the case of tsunamis assumes considerable importance in view of the impact it had in the recent tsunami of 26th December 2004 in India and other countries in Asia. The fact that there are no proper guidelines have made in the matters worse and resulted in the magnitude of damage that occurred. Survey of the damages indicated that the scour as a result of high velocities is one of the prime reasons for damages in the case of simple structures. It is revealed that sea walls in some cases have been helpful to minimize the damages. The objective of this paper is to suggest that design of shore line protection systems using expected wave heights that get generated and use of flexible systems such as geocells is likely to give a better protection. The protection systems can be designed to withstand the wave forces that corresponding to different probabilities of incidence. A design approach of geocells protection system is suggested and illustrated with reference to the data of wave heights in the east coast of India.

INTRODUCTION

The terminal effects of tsunami are observed in terms of damage to coastal areas including structural damages, flooding, scour, and deposition of both sediments and floating debris. Due to lack of awareness and the proper implementation of protective measures, the recent tsunami on 26th December 2004 caused enormous damage to the economic and social life of the people in India and other countries of Indian Ocean [BAPPENAS, (2005)]. It has also prompted various researchers and practitioners in the world to provide suitable protective measures and design guidelines for the protection of seashore area from such events in future. In the present study, the authors have proposed the use of flexible structure like reinforced earth retaining structure using geocells that will be able to meet all the design requirements in terms of its stability and strength to resist the static and the dynamic loads as well as wave forces from the sea water.

DESIGN REQUIREMENTS

Unlike wind generated waves [figure (1)], tsunamis are generated due to underwater displacement of tectonic plates, volcanic eruption or landslide and travel unnoticed across the open sea with amplitude of only few centimeters and a velocity as high as 800 km/hr. However, as they approach sea shore, they gradually build up and in areas where the topography of the seabed is especially suited they will reach heights of 10+ meters with a speed of 10 to 80 km/hr.

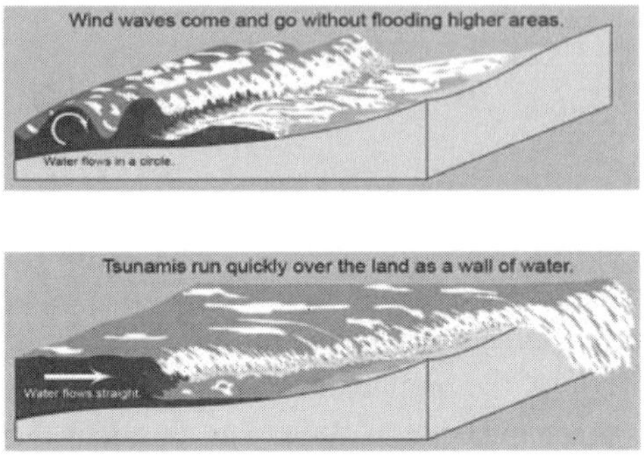

Figure 1. Difference between wind generated wave and tsunami

As tsunami propagates shoreward, large turbulent bores rush inland and cause significant damage to the coastal area. In general, the problems related to tsunami waves are studied in terms of its development, shape and propagation characteristics, run up at the shore [Whitman (1958), Keller *et al.* (1960), Yeh *et al.* (1989)] and its interaction with the structures that are constructed near the shore or that are built as barrier [Ramsden and Raichlen (1990), Ramsden (1996)]. In the present study, only the last part i.e., the tsunami-structure interaction is studied in terms of estimation of hydrodynamic forces on the structure built as a barrier and the design and safety of proposed geocells structure is studied considering the effect of dead load, live load, earthquake load in addition to the hydrodynamic effect of the water wave generated due to tsunami.

CHARACTERISTICS OF TSUNAMI (WAVE VELOCITY, WAVE LENGTH AND WAVE PERIOD)

The tsunami waves are treated as long wave (shallow water waves) and the phase velocity (c) and group velocity (c_g) are dependent on depth of sea floor (h) which is obtained using the Eq. (1).

$$c = c_g \approx \sqrt{gh} \tag{1}$$

The estimation of design wave height and wave period need the statistical analysis of the past records. The significant wave height (H) is defined as the average height of the highest one-third of the waves in a wave train. In the similar way, H_{10} and H_1 can be defined as the average of the 10 % and 1% of the highest waves in a wave train respectively. Assuming a Rayleigh distribution for the wave heights in a wave train, the values of H_{10} and H_1 will be approximately 1.27 H and 1.67 H.

The different wave periods are defined as the period of the peak energy density of the spectrum (T_p), average wave period (T_z) and significant wave period (T_s). In the design, an appropriate choice of design wave height (H) and wave period (T) depends on the designer choice or the recommendations given by the codes or design manuals. For the known value of wave period (T) and depth of sea floor below the SWL (h), the estimation of wave length (λ) can be made using the Eq. (2) given by Eckart (1952).

$$\lambda = \frac{gT^2}{2\pi}\sqrt{\tanh\left(\frac{4\pi^2 h}{T^2 g}\right)} \tag{2}$$

DESIGN OF VERTICAL BARRIER

The Height (H_s) of the structure

The total height of the structure above the still sea water level, as shown in Figure (2), is the sum of design wave height (H) plus extra allowance given for the wave run up, anticipated settlement of the structure and additional freeboard to avoid overtopping.

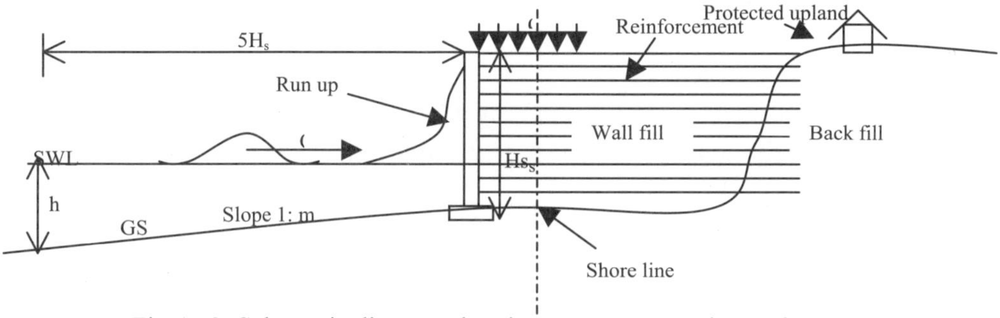

Figure 2. Schematic diagram showing wave propagation and run up

The height of wave run up (R) on the vertical wall can be estimated using Eq. (3).

$$\frac{R\,g}{c^2} = 1.467\,(H/h)^{-0.0504} \tag{3}$$

Estimation of wave force

In the literature analytical, experimental and numerical solutions (using finite difference codes) are available for estimating the hydrodynamic force on the wall due to tsunami waves (Cross 1967, Wiegel 1970, Ramsden and Raichlen 1990, Ramsden 1996, Hamzah 2000). Ramsden (1996) gave the empirical relationships for finding the hydrodynamic Force (F) and moment (M) on the vertical wall type barrier.

$$\frac{F}{F_1} = 1.325 + 0.347\left(\frac{H}{h}\right) + \frac{1}{58.5}\left(\frac{H}{h}\right)^2 + \frac{1}{7160}\left(\frac{H}{h}\right)^3 \tag{4}$$

$$F_1 = \frac{1}{2}\gamma\, b(2H + h_w)^2 \tag{5}$$

$$\frac{M}{M_t} = 1.923 + 0.454\left(\frac{H}{h}\right) + \frac{1}{8.21}\left(\frac{H}{h}\right)^2 + \frac{1}{808}\left(\frac{H}{h}\right)^3 \tag{6}$$

$$M_t = \frac{1}{6}\gamma\, b(2H + h_w)^3 \tag{7}$$

The values of F_1 and M_1 are calculated assuming hydrostatic distribution for value of the wave run up height (R) equal to two times the wave height (H). The table (1) shows the calculation for the wave length (λ), run up (R) and forces.

Table 1. Estimation of wave run up and forces on the wall

Depth of sea water at the wall (h_w)	**1.50**	m
Wave height (H)	**3.50**	m
Time period (T)	**1000.00**	sec
Celerity (C)	3.84	m/sec (13.8 km/hr)
Wave length (λ)	3836.01	wave length of tsunami = 3.8 km
Run up on the wall (R)	2.12	m
Free board	2.00	m
Total height of the structure (H_s)	9.12	**height of structure 9.15 m**
unit weight of water (γ_w)	9.79	kN/m^3
Unit weight of the reinforced soil (γ_s)	22.00	kN/m^3
slope (1: m)	20.00	
Estimation of hydrostatic force (F)		
Estimation of hydrostatic force (F_1)	**353.56**	kN
H/h	1.47	
F/F_1	1.87	
Hydrodynamic force on the structure	**662.54**	kN
Estimation of Moment (M) on the wall		
Estimation of Moment M_t	**1001.74**	KN-m
M/M_t (using equation 11)	2.86	
Moment M	**2865.52**	KN-m

For the design of vertical reinforced soil wall under static and dynamic conditions, the procedure given in IS: 1893-1984 is followed. As shown in table (2), the external stability of the wall is checked against sliding, overturning and bearing failure both under static and dynamic conditions. Based on the analysis, the proposed base width is 15.0 m. The internal stability of the wall is checked against tension failure and pull out failure. Table (3) and (4) shows the calculations for tension failure under static and dynamic conditions respectively. From the pull out failure criterion, the maximum tension (T_i) in the reinforcement is obtained as 20.5 kN and 24.52 kN under static and dynamic conditions respectively. The results of the analysis shows that the wall is safe and stable (internally and externally) for the forces coming on it from the backfill and surcharge load.

TOE PROTECTION

The toe of the structure should be protected from the scouring and undercutting due to passage of waves in order to make the structure stable. Figure (3) shows the arrangement for the toe protection. The total weight of the toe stone can be calculated using the following Eq. (16) given by Brebner and Donnelly (1962).

$$W = \frac{\gamma_t H^3}{N_s^3 (S_t - 1)^3} \tag{8}$$

Table 2. External stability analysis of the wall uder static and dynamic loading

Stability Analysis of the reinforced soil wall
Soil properties for wall fill (c_1=0,πN=30, π_{t1}=18 kN/m3, π_1=20,πN$_{s1}$ = 22 kN/m3 and π_{b1}=12 kN/m3)
Soil properties for wall fill (c_2=0,πN$_2$=27, π_{t2}=17 kN/m3, π_2=18, π_{s2} = 20 kN/m3 and π_{b2}=10 kN/m3)
Surchagre load on the wall = 10 kN/m2
External stability Analysis
The coefficient of active earth pressure (Ka) in static case above the water table = 0.326
The coefficient of active earth pressure (K'a) in static case below the water table = 0.349
The total static earth pressure (PT) = 438.73 kN/ m width of wall
The total Moment of static pressure about the base (MT) = 3905.86 kN-m
Dynamic Increment
The horizontal (πh) and vertical seismic coefficient (πv)= 0.08 and 0.04 respectively
The coefficient of active earth pressure for the soil above the water level in dynamic case = 0.404
The value of lateral dynamic increment in active case (Paπi)= 96.96 kN/m width of the wall
The value of additional dynamic increment due to uniform surcharge (Paqi) = 13.94 kN/ m width of the wall
The moment of dynamic increment about the base of the wall (Maπi) in active case = 628.62 kN-m
The moment of dynamic increment about the base of the wall (Maqi) in active case due to surcharge q=115.05 kN-m
Assuming the base width of the wall = L
The weight of the wall (Ww) = 235.8 L
The effective weight of the wall (W'w) = 190.8 L
The moment of weight of wall about toe (Mw) = 117.9 L^2
The surcharge force on the wall (Q) = 10 L
The moment of surcharge force about toe (Mq) = 5 L^2
Seismic force on the weight of the wall
In horizontal direction = 18.86 L
In vertical direction = 9.43 L
The moment of seismic forces acting on the weight of the wall (Msw) = 292.06 L $\pm$ 4.716 L^2
Seismic forces acting on the wall due to surcharge load
Horizontal seismic force = 0.8 L
Vertical seismic force = 0.4 L
The moment of seismic forces due to the surcharge acting on the wall about toe (Msq) = 9.68L $\pm$ 0.2 L^{π}
The Factor of safety against sliding = 2.0
Static condition: The length L is obtained as 13.44 m
Dynamic condition: The length L is obtained as 14.87 m
Factor of safety against overturning = 2.0
Static condition: The length L is obtained as 9.15 m
Dynamic condition: The length L is obtained as 11.8 m
From the above analysis, the base width of the wall (L) is provided as 15 m
Check for tilting/bearing failure
Static condition: qmax = 241.00 kpa and qmin = 160.45 kPa
Dynamic condition: qmax = 357.03 kPa and qmin = 340.96 kPa

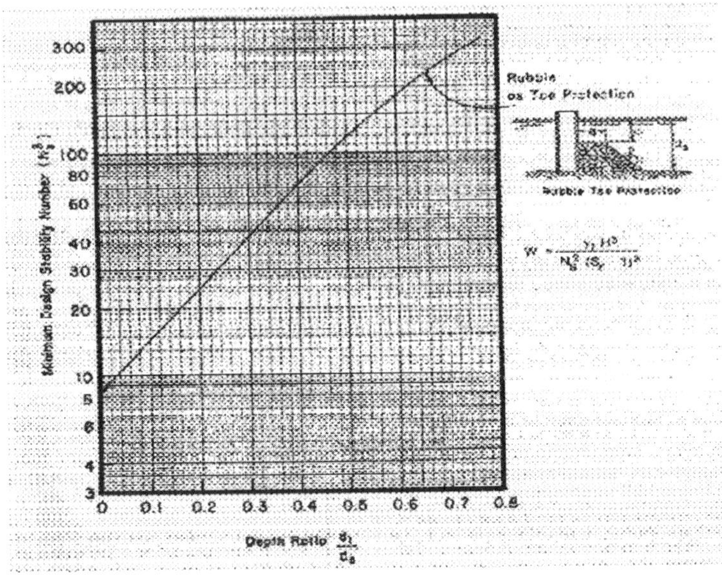

Figure 3. Value of Ns, toe protection design for vertical walls (Brebner and Donnelly 1962

Table 3. Checking the internal stability of the wall against tension failure (Static condition)

Tension Failure (Static condition)								
h_i	M_1	σ_{vi}	Ti	$M_{a\gamma}$	M_{aq}	M_2	σ_{vi}	T_i
0.50	9.88	19.09	2.86					
1.00	39.96	28.38	4.33					
1.50	90.96	37.87	5.87					
2.00	163.55	47.57	7.49					
2.50	258.44	57.48	9.20					
3.00	376.31	67.61	10.99					
3.50	517.86	77.97	12.87					
4.00	683.78	88.56	14.83					
4.50	874.76	99.40	16.90					
5.00	1091.50	110.48	19.06					
5.50	1334.69	121.81	21.32					
6.00	1605.02	133.41	23.68					
6.50	1903.19	145.27	26.15					
7.00	2229.89	157.41	28.73					
7.50	2585.81	169.82	31.42					
8.00				473.81	17.99	491.80	158.72	29.76
8.50				571.85	31.76	603.61	168.79	32.07
9.00				684.72	46.41	731.13	179.02	34.46
9.50				813.30	61.92	875.22	189.40	36.93
10.00				958.45	78.31	1036.77	199.95	39.49
10.50				1121.06	95.58	1216.64	210.68	42.14
11.00				1301.99	113.71	1415.70	221.59	44.87
11.50				1502.11	132.72	1634.83	232.69	47.70
12.00				1722.30	152.60	1874.90	244.00	50.63

Table 4. Checking the internal stability of the wall against tension failure (Dynamic condition)

Tension Failure (seismic condition)															
h_i	M_3	M_4	M_{Tni}	σ_{vi+}	σ_{vi-}	T_{i+}	T_i -	h_i-(H-H_w)	M_5	M_6	M_{Tni}	σ_{vi+}	σ_{vi-}	T_{i+}	T_i -
0.50	0.08	0.20	24.65	20.76	17.72	3.11	2.66								
1.00	1.33	0.78	80.07	31.01	26.53	4.73	4.05								
1.50	6.71	1.76	169.93	41.59	35.67	6.45	5.53								
2.00	21.22	3.12	299.89	52.56	45.20	8.28	7.12								
2.50	51.80	4.88	477.61	63.99	55.19	10.24	8.83								
3.00	107.41	7.02	712.73	75.96	65.72	12.34	10.68								
3.50	198.98	9.56	1016.89	88.60	76.92	14.62	12.69								
4.00	339.46	12.48	1403.71	102.04	88.92	17.09	14.89								
4.50	543.74	15.80	1888.80	116.41	101.85	19.79	17.31								
5.00	828.75	19.50	2489.75	131.90	115.90	22.75	19.99								
5.50	1213.37	23.60	3226.16	148.69	131.25	26.02	22.97								
6.00	1718.50	28.08	4119.60	166.99	148.11	29.64	26.29								
6.50	2366.99	32.96	5193.64	187.02	166.70	33.66	30.01								
7.00	3183.73	38.22	6473.84	209.03	187.27	38.15	34.18								
7.50	4195.55	43.88	7987.73	233.28	210.08	43.16	38.87								
8.00								0.40	171.97	49.92	714.68	152.81	152.11	28.65	28.52
8.50								0.90	229.68	56.35	890.64	165.94	164.36	31.53	31.23
9.00								1.40	342.13	63.15	1137.41	179.75	177.29	34.60	34.13
9.50								1.90	540.86	70.33	1487.41	194.55	191.21	37.94	37.29
10.00								2.40	859.79	77.88	1975.43	210.68	206.45	41.61	40.77
10.50								2.90	1335.17	85.79	2638.59	228.48	223.38	45.70	44.68
11.00								3.40	2005.60	94.06	3516.35	248.35	242.36	50.29	49.08
11.50								3.90	2911.97	102.69	4650.49	270.68	263.81	55.49	54.08
12.00								4.40	4097.49	111.68	6085.08	295.89	288.14	61.40	59.79

Where Ns is the minimum design stability number obtained from the curve provided in figure (5). γ_t and S_t are the unit weight and specific gravity of the stone material. H is the design wave height taken between H_1 and H_{10}. Using friction circle method, the passive resistance of the backfill material is obtained as 1354.21 kN which is approximately 2.0 times more than the expected hydrodynamic force due to tsunami waves.

CONCLUDING REMARKS

In the present study, a vertical reinforced retaining wall using geocells is proposed for protecting the sea shore area in the event like tsunami. In the analysis and design, the discussions are restricted to geotechnical and hydrological aspects only. The procedures are given to find the hydrodynamic force expected to act on the structure and to check the external stability of the structure from different modes of failure. The advantage of using reinforced earth retaining wall can be said in terms of the economy and the flexibility of the structure. Being flexible, it has the capability of absorbing some part of kinetic energy of the water waves coming on it but the quantification of the same could be an active area of research in this field. In addition, the use of numerical tools and model testing will give extra inputs in studying the behavior of structure. In addition to these engineering barriers, development of effective tsunami warning system and general awareness program conducted by the local government will definitely help in saving the life of people living in the coastal region.

REFERENCES

BAPPENAS (2005). Indonesia: Preliminary damage and loss assessment, the December 26 2004 natural disaster. A technical report prepared by BAPPENAS and the international donor community.

Brebner, A., and Donnelly, P. (1962). "Laboratory study of rubble foundations for vertical breakwaters," Engineer Report No. 23, Queen's University at Kingston, Ontario.

Cross, R. H. (1967). "Tsunami surge forces", Journal of waterways and harbor division ASCE, vol. 93, No. 4, 201-231.

Cumberbatch, E. (1960). "The impact of a water wedge on a wall", Journal of fluid mechanics, Vol. 7, No. 3, 353-373.

Eskart, C. (1952). "The propagation of gravity waves from deep to shallow water", Natl. bur. Standards, Circular 521, Washington, DC, 165-173.

Hamzah, M.A., Mase, H., and Takayama, T. (2000). "Simulation and experiment of hydrodynamic pressure on a tsunami barrier", Coast. Engrg., 1501-1507.

Hoyt, J. W. and Sellin, R. H. (1989). "Hydraulic jumps as a 'mixing layers'", Journal of hydraulic engineering division, ASCE, Vol. 115, No. 12, 1607-1614.

IS 1893 : 1984. "Criteria for earthquake resistant design of structures", Bureau of Indian standards.

Kamel, A.M. (1970). "Laboratory study for design of tsunami barrier", J. Waterway. Horbor and Coastal Engrg. Division, Vol. 96, No. WW4, 767-779.

Keller, H. B., Levine, D. A. and Whitham, G. B. (1960). "Motion of a bore over a sloping beach",. Journal of fluid mechanics, Vol. 7, No. 2, 302-316.

Ramsden, J. D. (1996). "Forces on a vertical wall due to long waves, bores, and dry-bed surges", Journal of waterway, port, coastal, and ocean engineering, ASCE, Vol. 122, No. 3, 134-141

Ramsden, J. D. and Raichlen, F. (1990). "Forces on vertical wall caused by incident bores". Journal of waterway, port, coastal, and ocean engineering, ASCE, Vol. 116, No. 5, 592-613

Shen, M.C. and Meyer, R.E. (1963). "Climb of a bore on a beach Part 3: Run-up". Journal of fluid mechanics, Vol. 16, No. 1, 113-125.

Titov, V. V. and Synolakis, C. E. (1995). "Modeling of breaking and nonbreaking long-wave evolution and runup using VTCS-2", Journal of waterway, port, coastal, and ocean engineering, ASCE, Vol .121, No.6, 308-316.

Whitham, G.B. (1958). "On the propagation of shock waves through regions of non-uniform area of flow", Journal of fluid mechanics, Vol. 4, No. 4, 337-360.

Yeh, H.H., Ghazali, A. and Marton, I. (1989). "Experimental study of bore runup". Journal of fluid mechanics, Vol. 206, 563-578.

MITIGATION & REHABILITATION

International Conference on Geotechnical Engineering for Disaster Mitigation & Rehabilitation
Chu, Phoon & Yong (eds) © 2005 World Scientific Publishing Company ISBN 981-256-469-1

STONE COLUMNS FOR SEISMIC LIQUEFACTION MITIGATION

K. ADALIER
Department of Civil and Environmental Engineering, Florida State University, Panama City, FL, USA

A. ELGAMAL
Department of Structural Engineering, University of California, San Diego, CA, USA

Soil liquefaction has been a major source of damage during many past earthquakes. The risk of liquefaction and associated ground deformation can be reduced by various ground-improvement methods including the stone column (gravel drains) technique. Currently, there is a great need for better understanding of stone column liquefaction hazard mitigation mechanisms, particularly in silty soils. In order to address this issue, four dynamic centrifuge model experiments were conducted. Response of saturated silt strata with and without stone columns was analyzed under base dynamic excitation conditions. The underlying mechanism and effectiveness of the stone columns are discussed based on the recorded responses. The test results demonstrated that stone columns can be an effective technique in the remediation of liquefaction induced settlement of non-plastic silty deposits particularly under shallow foundations, or at depths of about 5 m in the free field.

INTRODUCTION

The risk of seismically induced liquefaction and associated ground deformation can be reduced by various ground improvement methods including the stone column technique. A comprehensive literature review by the authors on stone columns (Adalier and Elgamal, 2004) revealed that there is a great need for better understanding of stone column liquefaction hazard mitigation mechanisms, particularly when constructed in silty soils. The possible benefits of stone columns include densification of surrounding non-cohesive soil, dissipation of excess pore water pressure (EPWP) and re-distribution of earthquake-induced or pre-existing stress (due to introduction of the stiffer columns).

When dealing with non-plastic silty soils, only the third benefit can be primarily expected to mitigate liquefaction. Even with the vibro-flotation installation method, densification due to vibrations in silts can be impractical to achieve. In addition, due to very low permeability of the silt, the drainage function of the stone column is practically negligible. Shear stress re-distribution concepts have been previously proposed (Baez, 1995) as means to assess stone columns as a liquefaction countermeasure in such non-plastic silty soils. In order to address this issue, a centrifuge experimental program was conducted.

This paper briefly reports the results of this experimental study which was focused on the possible stiffening benefit, rather than improved drainage and densification due to stone column installation. Saturated silt strata of 8 m and 10 m in thickness (prototype scale) were studied. Using the centrifuge at Rensselaer Polytechnic Institute, Troy, NY-USA, two benchmark model tests were performed first to document the dynamic response characteristics of a silty stratum with and without a surface foundation surcharge. Under the same shaking conditions, the responses of these models, remediated with stone columns, was studied and compared to the benchmark models. Free-field and surface foundation surcharge situations were investigated. Settlement, acceleration, and EPWP in the soil models were recorded. Based on the recorded dynamic responses the underlying mechanism and effectiveness of the stone columns are briefly discussed herein. More detailed discussions on these tests can be found in Adalier *et al.* (2003).

CENTRIFUGE PHYSICAL MODEL TESTING

The main principle in centrifuge modeling is that a 1/N scale model subjected to a gravitational acceleration of Ng (g is acceleration of gravity) experiences the same stress as the prototype. Thus, stress-strain relationships at similar points in the model and prototype will be equivalent and the behavior of the model will mimic the behavior of the prototype. Consequently, with the help of scaling laws measurements in centrifuge tests can be related directly to an equivalent full-scale prototype. Unless otherwise indicated, all dimensions reported in this paper are in prototype scale, obtained from the actual model units following basic scaling relations. This means that all linear dimensions, including measured deformations, as well as the model time were multiplied by N, and the actual model shaking acceleration and frequency divided by N.

Model 1 test explored the response of a 7.8 m thick, pure silt saturated stratum (relative density, D_r= 60%). In Model 2 (Fig. 1), a total of 45, 1.27 m diameter columns were placed (2.5 m center-to-center in a square pattern) vertically in the model laminar container, giving an area replacement ratio (A_r) of 20%. Note that the Model 1 configuration is similar to that of Model 2 except it is without stone columns.

In the Model 3, the response of a 10 m thick saturated silt stratum (D_r = 65%) with a rigid footing surcharge (rigid steel rectangular block applying an average vertical contact pressure of 144 kPa) was studied. This surcharge simulated approximately the vertical pressure transmitted by a 10-15 story reinforced concrete building. Model 4 (Fig. 2) investigated the response of the same system employed in the Model 3 but with the inclusion of 36, 1.6-m diameter, stone columns (2.55 m center-to-center spacing). This configuration provided an A_r of 30% within the instrumented zone below the footing. Models 1 and 2 were tested at a 50g gravitational acceleration field, whereas Models 3 and 4 were tested at 63g. The soil container used in the Model 1 and 2 tests was a rectangular, flexible-wall laminar box. Models 3 and 4 were constructed in a rigid-wall container. Model response was measured by a large number of miniature transducers, including horizontal accelerometers, pore pressure, and displacement transducers.

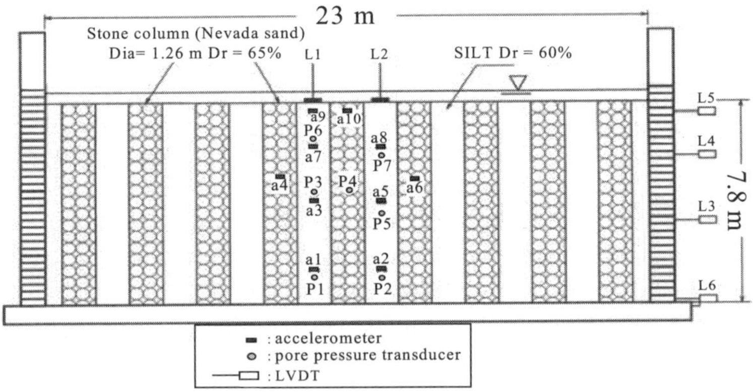

Figure 1. Cross-sectional view of Model 2 configuration (45 stone columns, $A_r = 20\%$).

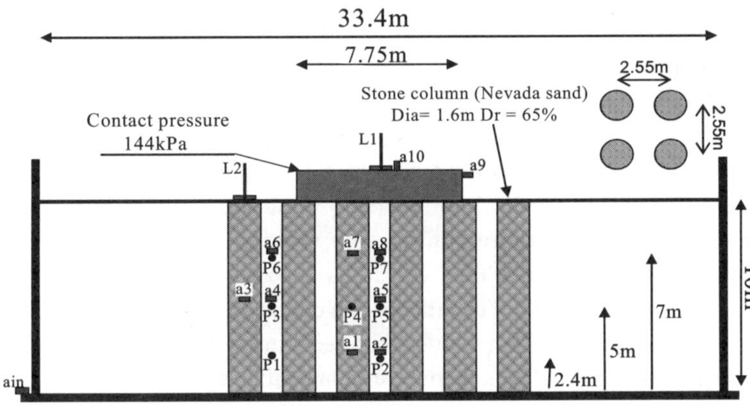

Figure 2. Cross-sectional view of Model 4 configuration (36 stone columns, $A_r = 30\%$).

A 100% silt size material "Sil-Co-Sil 120" (Walker and Stewart, 1989) was employed to construct the ground layer in all models. The material representing the stone columns was Nevada 120 sand (http://geoinfo.usc.edu/gees/velacs/). The sand columns had D_r of about 65%, although in the field it is possible to achieve D_r as high as 90% with crushed stone. Lower D_r were desired in this experimental program to have a G_r ratio of 5-6 ($G_r = G_{SC}/G_S$, where G_{SC} is stone column shear modulus, and G_S is silt shear modulus). This ratio is a critical parameter for stress concentration or stiffening effects due to introduction of a stone column system (Baez, 1995). The models were saturated with de-aired water under vacuum. Detailed description of model construction and instrumentation is provided by Adalier *et al.* (2003). In all cases, due to the very low silt permeability, the stone columns did not increase overall drainage or decrease the EPWP build-up rate during the shaking phase in any appreciable way. Therefore, for all tests any change in the behavior of the remediated ground (relative to the unremediated ground) is primarily a result of the stiffening effect of the stone columns.

TEST RESULTS

Model 1 and 2 results

Due to severe space limitation, the test results will be only briefly presented herein. Comprehensive discussions for all model tests are provided by Adalier *et al.* (2003).

Even though both models attained high EPWP, their dynamic behavior was noticeably different. The decay of accelerations (i.e., loss of strength) in Model 1 was significantly quicker. At corresponding locations, both the softening-induced initial amplification and the subsequent severe attenuation phases are significantly delayed in Model 2 relative to Model 1. This can be attributed to the reinforcing-stiffening effect of the installed stone columns. In general, Model 2 behaved in a stiffer manner. The EPWP traces measured in the top half of both models at corresponding locations, showed fairly close similarity (with Model 2 EPWP build-up being somewhat slower). However, in the bottom half, EPWP build-up in Model 2 was considerably slower than that observed in Model 1. Therefore, the difference in the rate of EPWP build-up between Model 1 and Model 2 soil was more pronounced at depth. Moreover, the entire silt stratum completely liquefied at the end of 12th base input cycle in the Model 1, whereas even at the end of shaking (i.e., 20th cycle), only the top-half of the silt stratum liquefied in Model 2. These EPWP records are consistent with the recorded accelerations, which exhibit much stronger response in the bottom half of Model 2 compared to Model 1. Even at the top half of the silt stratum, it took about two to three times more shaking cycles for the silt to show significant strength degradation in Model 2 compared to Model 1. Therefore, although liquefaction was not prevented by the installed stone columns (in the upper half of the silt stratum) under the strong base input motion applied during these tests, the composite ground had a considerably higher liquefaction resistance than the uniform silt.

Based on reinforcement concepts proposed by Baez (1995), a system of stone columns and silt stratum with parameters such as the one tested in Model 2 (G_r= 6, pre-treatment liquefaction Factor of Safety-FS_{pre}= 0.5), an A_r of 20%, would have been sufficient to prevent the occurrence of liquefaction (FS $\geq$ 1) in the silt profile. However, the above observations suggest that sufficient vertical stress or confining pressure might be required to "engage" the reinforcing effect of the stone columns as suggested by Baez (1995). The centrifuge tests of Models 1 and 2 indicate that such vertical effective stress might need to exceed about 45 kPa in order for the stone columns to provide significant stress re-distribution and mitigate the liquefaction of the loose silt. In practice, this confinement could be obtained with the weight of the structure. Models 3 and 4 test this hypothesis.

Model 3 (model ground with surcharge) results

Models 3 and 4 were subjected to three sinusoidal shaking events. The first shaking event (Shake1) simulated a moderate level of earthquake excitation (10 cycles of 0.08g). Shake2 was stronger excitation (30 cycles of 0.18g) and more clearly revealed the important response characteristics. Shake3 was essentially similar to Shake2. Accordingly, for the sake of brevity Shake2 event will be emphasized herein.

During Shake2, asymmetry in accelerations indicative of lateral deformations (Aydingun and Adalier, 2003) was observed at a6 and a3 (both under the footing edge at 3 and 5 m depth respectively) followed by a significant attenuation phase. At a8, 3 m beneath the foundation, notable gradual attenuation of accelerations was also observed. Likewise, the footing accelerations were gradually attenuated as the foundation soils became softer due to EPWP development. Compared to Shake1, a stronger (both in magnitude and spatial extent) negative EPWP build-up tendency was observed at the central foundation zones (as the magnitude and the spatial extent of horizontal normal strains in the foundation grew with stronger base input excitation). However, away from these expansive zones, significant positive EPWP was attained (i.e., P1, P2, P3, and P6).

The footing was observed to undergo large settlements of 0.47 m (Fig. 3) during Shake2. It is noted that in every shaking event, both in Model 3 and Model 4, over 90% of the foundation settlements occurred during shaking. These large settlements were partially a result of migration of underlying foundation soil towards the free field, where the ground surface was observed to have negligible net vertical deformations (compaction settlements were largely masked by the heave).

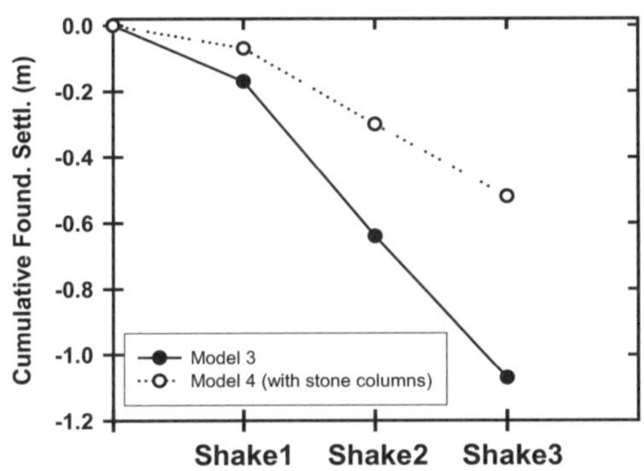

Figure 3. Accumulated foundation-footing settlements after three shaking events.

Model 4 (model ground with 36 stone columns and with surcharge)

Based on the recorded response, it may be concluded that the stone column application in the foundation layer has led to the following effects:

i) The acceleration spikes that appeared at a1, a2, and a3 in the Model 3 were not observed in Model 4. Absence of this cyclic mobility effect indicates that shear strains were smaller (Elgamal *et al.,* 1996; Aydingun and Adalier, 2003) than those in Model 3. Moreover, accelerations in the silt were slightly stronger than those measured in Model 3. Overall, the relatively high recorded accelerations (including those of the footing), clearly indicate that the stone columns largely preserved overall foundation stiffness.

ii) Overall increased foundation stiffness during shaking also reduced the outward migration of soils beneath the footing, in turn reducing the negative EPWP tendency that was strongly observed at P7, P5, and P4 of Model 3. In general, EPWP in Model 4 was somewhat slower, reached lower ultimate values, and dissipated faster than in Model 3.

iii) The seismic shaking was effectively transferred (actually with some amplification) directly from the base of the deposit to the footing (see a9 record) by the stiff composite ground (i.e., silt-stone column). Thus the composite soil block under the foundation sustained enough of its initial stiffness to transmit and amplify the base accelerations to the footing. The overall foundation shear strength also provided resistance to the heavy foundation penetration during shaking and much reduced the vertical settlements.
iv) And most importantly, foundation settlements were reduced by about 50% (Fig. 3).

CONCLUSIONS

Centrifuge model test results indicated an overall stiffer foundation material response during shaking in the models remediated by stone columns. Stone columns somewhat retarded the EPWP build-up (in the soil between columns), increased the foundation soil overall stiffness, and significantly reduced the surcharge-footing settlements. In the free-field situation, the stiffening effect provided by the stone columns was only primarily effective in reducing pore-pressures at depths below 5 m (45 kPa) approximately. In practice, this confinement level could be obtained with the weight of the structure. Near ground surface, the installed columns were only of marginal effect in reducing pore pressures. However, this issue does not substantially affect the important situations of remediation below shallow foundations, where the deformation mechanism is totally different, and the stone columns were found to reduce settlements by about 50%. Moreover, improved shear and bending stiffness (i.e., higher G_r) of columnar inclusions may further restrain the lateral outflow of EPWP-softened soils, which was observed to significantly contribute to the vertical foundation settlement.

REFERENCES

Adalier, K., Elgamal, A.., Meneses, J. and Baez, I.J. (2003). "Stone columns as liquefaction countermeasure in non-plastic silty soils", *J. Soil Dynamics and Earthquake Engineering*, Vol. 23(7), 571-584.

Adalier, K. and Elgamal, A. (2004). "Mitigation of liquefaction and associated ground deformations by stone columns", *J. Engineering Geology*, Vol. 72(3-4), 275-291.

Aydingun, O. and Adalier, K. (2003). "Numerical analysis of seismically-induced liquefaction in earth embankment foundations. Part I. Benchmark model", *Canadian Geotechnical J.*, 40(4), 753-765.

Baez, J.I. (1995). "*A design model for the reduction of soil liquefaction by vibro-stone columns*", Ph.D. Thesis, University of Southern California.

Elgamal, A., Zeghal, M., Taboada, V., Dobry, R. (1996). "Analysis of site liquefaction and lateral spreading using centrifuge testing records", *Soils and Foundations*, 36(2), 111-121.

Walker, A.J. and Stewart, H.E. (1989). "*Cyclic undrained behavior of nonplastic and low plastic silts*", Technical Report, NCEER-89-0035, 220pp.

International Conference on Geotechnical Engineering for Disaster Mitigation & Rehabilitation
Chu, Phoon & Yong (eds) © 2005 World Scientific Publishing Company ISBN 981-256-469-1

IMPROVING SEISMIC RESPONSE OF FINE GRAINED SOILS

L. SEBASTIAN BRYSON
Department of Civil Engineering, Ohio University, 141 Stocker Center
Athens, Ohio 45701, USA

Strength softening and liquefaction of soft clayey soils are a major source of damage to civil engineering works in earthquake-prone zones. Extensive liquefaction research over the past few decades has helped determine some of the parameters influencing the liquefaction potential of finer granular soils and silty sands, and procedures to identify these liquefaction-prone deposits have been developed, combining field and laboratory tests. However, very little effort has been dedicated to the understanding and modeling of strength softening and liquefaction of soft and saturated clayey soils and in proposing solutions to minimize damage during seismic events. Evidence of strength softening and liquefaction of clayey soils has been reported during well-studied seismic events such as in the Marina District during the Loma Prieta earthquake in San Francisco, CA in 1989, the Chi Chi earthquake, Taiwan in 1999 and the Kocaeli earthquake, Turkey in 1999, among others. Devastating damage to structures founded on soft clayey deposits was reported during these and other earthquakes, thus challenging researchers to better understand the phenomena, improve available methods to predict their occurrence and to propose viable solutions to minimize their effects, if a deposit is identified to be prone to the development of either or both of the two phenomena. This paper evaluates the effectiveness of several treatment methods in mitigating seismic hazards such as liquefaction potential, strain softening, and seismic amplification. This paper also investigates the influence of treatment zone geometry on performance parameters for a given site condition and earthquake loading.

INTRODUCTION

It has long been recognized that relatively "clean" sandy soils, with few fines, are potentially vulnerable to seismically-induced liquefaction. There has, however, been significant controversy and confusion regarding the liquefaction potential of silty soils (and silty/clayey soils). Evidence of of clayey soils has been reported during well-studied seismic events such as in the Marina District during the Loma Prieta earthquake in San Francisco, CA in 1989, the Chi Chi earthquake, Taiwan in 1999 and the Kocaeli earthquake, Turkey in 1999, among others. Devastating damage to structures founded on soft clayey deposits was reported during these and other earthquakes, thus challenging researchers to better understand the phenomena, improve available methods to predict their occurrence and to propose viable solutions to minimize their effects, if a deposit is identified to be prone to the development of either or both of the two phenomena (strength softening and liquefaction).

Although the use of Ground Improvement (GI) is growing especially for seismic mitigation, few guidelines exist for the selection and implementation of various schemes. Limited data exist on the effect of GI on liquefaction and ground motions. However, recent observations of improved ground performance have already made a great impact on understanding the benefits of this mitigation technique and have provided necessary data to expand the state-of-the-practice into wider areas, such as performance-based design.

SEISMIC IMPROVEMENT METHODS FOR FINE GRAINED SOILS

There are many methods and variations on methods currently used for seismic ground improvement. Comprehensive treatments of many of the improvement methods are available in a number of references (e.g. Mitchell et al., 1995; Hayden and Baez, 1994; Xanthakos et al., 1994; Hausmann, 1990). The most commonly used methods for treatment of fine grained soils are presented in Table 1.

Table 1. Seismic Ground Improvement Methods for Fine Grained Soils

Method	**Principle**
<u>Densification:</u> Stone Columns Sand Columns Sand Compaction Piles Aggregate Piers	Hole jetted or drilled into soil and backfilled with densely compacted gravel or sand. Columns provide vertical support and shear resistance in the horizontal direction. Also, provides additional drainage pathway.
<u>Pore Pressure Relief</u>: Gravel Drains Sand Drains Wick Drains	Relief of excess pore water pressure by installing material that is more permeable than surrounding soil.
<u>Stabilization:</u> Deep Soil Mixing	Lime, cement, or grout introduced through rotating auger or special in-place mixer. Increased stiffness and shear resistance in treated ground
<u>Injection and Grouting:</u> Jet Grouting	High-speed jets at depth then excavate, inject, and mix grout slurry with soil to form columns or panels. Increased strength and stiffness of treated ground, but no drainage allowed.

CENTRIFUGE MODELING OF SEISMIC GROUND IMPROVEMENT

Applicability of Seismic Ground Improvement for Fine Grained Soils

Adalier et al. (2003) studied the liquefaction mitigation effectiveness of stone columns in nonplastic silty soils through a series of highly instrumented geotechnical centrifuge model tests. The study was focused on the possible stiffening benefit, rather than improved drainage and densification due to stone column installation. Settlement, acceleration, and excess pore water pressure data indicated an overall stiffer material response during shaking in the models improved with stone columns. Stone columns also considerably retarded excess pore pressure build-up in the soil between columns, increased the soil overall stiffness, and significantly reduced the footing settlements. Footing settlement recorded for the treated model tests were approximately 50 percent of those recorded for the untreated tests.

Adalier et al. (2003) concluded that a major gain from stone columns comes from their ability to: (i) densify surrounding soils during construction, (ii) restrict shear deformation and offer containment of the "encapsulated' soils, and (iii) provide stiffening matrix effects (i.e. reducing the stresses in adjacent soil). However, they also concluded that while theoretical analyses and model testing results indicate that stone column methods can be efficient in mitigating liquefaction, well-documented case histories of successful utilization under seismic conditions were rather limited. They state that there remains a great need for well-documented data sets of field performance scenarios. Also, there remains a need for research-based guidelines for design of stone columns in silty soil deposits.

Influence of Improvement Zone Geometry

Hausler (2002) performed a series of dynamic centrifuge tests using the centrifuges located at the Public Works Research Institute (PWRI) in Japan and at the University of California at Davis. The purpose of the research was to explore the influence of the depth and lateral extent of the remediation zone on the acceleration, settlement, and tilt of a rigid structure on an embedded mat foundation during seismic loading.

Hausler's data shows that only minor reductions in seismic-induced settlement were (approximately 10 percent) observed for improved thicknesses between 0 and 30 percent of the total liquefiable thickness. However, for improved thicknesses between 30 percent and 70 percent of the total liquefiable thickness, a 50 percent reduction in seismic-induced settlement was observed. Beyond 70 percent, there was apparently no improvement.

SEISMIC GROUND IMPROVEMENT CASE HISTORY - CARREFOUR SHOPPING CENTER (KOCAELI 1999)

Martin and Olgun (2003) presented a case history demonstrating the effectiveness of ground improvement for liquefaction mitigation. Settlement and pore pressures were recorded at the Carrefour Shopping Center located in the city of Izmit, Turkey during the 17 August 1999 Kocaeli earthquake (Mw = 7.4). Peak ground accelerations in this area were approximately 0.2g. The liquefaction susceptibility assessment of the Carrefour soils based on the criteria recommended by Seed et al. (2003) is shown in Figure 1. The figure shows that the silty clay and the clay-silt-sand layers were susceptible to liquefaction.

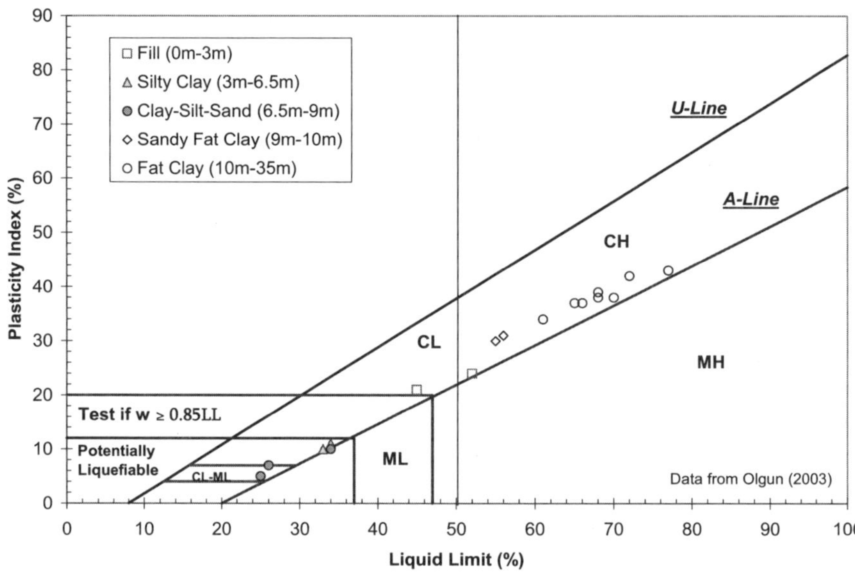

Figure 1. Liquefaction susceptibility assessment of the soils at the Carrefour site.

Foundations and Ground Improvement

The soils were improved to increase bearing support for shallow foundations and reduce liquefaction potential of the sand layers. Surcharge fills with wick drains were used to improve the soft clays, and jet-grout columns were used to provide increased bearing support in the clays and prevent liquefaction of the loose sands. Figure 2 show the penetration of the jet-grout columns at the site. Jet-grout column spacings and diameters were selected mainly on the basis of footing spacing, footing loads, floor slab loads, and judgment. No rigorous analyses were performed to modify the spacings for liquefaction performance criteria. A primary and secondary grid of columns was installed in a rectangular pattern to provide blanket treatment. The columns in the primary grid were 0.6 m in diameter with a center-to-center spacing of 4 m. These columns extended from the ground surface to a depth of 9.0 m. The secondary grid consisted of shorter, 2.5 m-long grouted columns that were installed in between the primary columns to further increase the liquefaction resistance of the sand stratum, extending from a depth of 6.5 m to 9.0 m. In addition to the primary and secondary grids, 0.6 m-diameter columns were also installed at each spread footing location of the supermarket building.

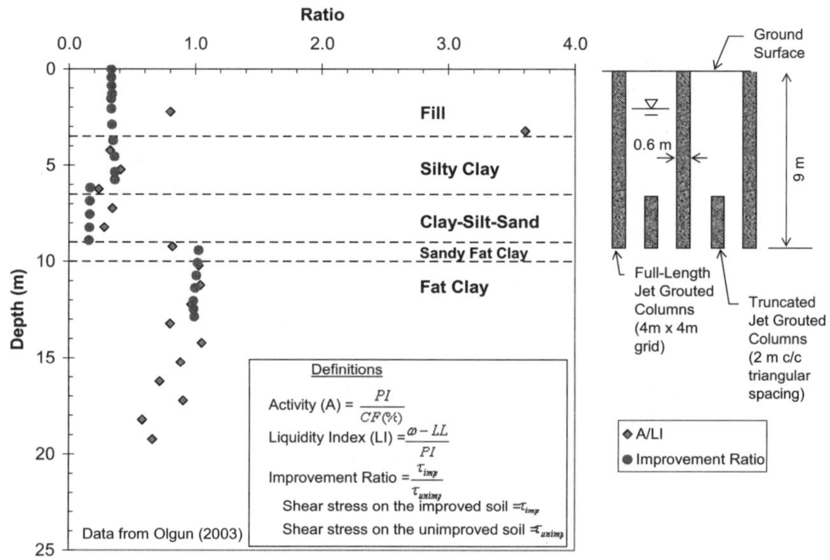

Figure 2. Typical ground improvements at the Carrefour site.

Figure 2 shows an attempt to quantify the effectiveness of the jet-grouting during the design phase. The figure shows that the degree of improvement can be characterized by the improvement ratio, which is the ratio of the shear stress on the improved soil to the shear stress on the unimproved soil. The elastic shear properties of the improved soil were approximated using the equivalent homogenization approach presented by Olgun (2003). The approach is based on the replacement ratio, which is the ratio of the sum of the jet-grout column areas to the area of the soil mass. The data shown in the figure assumed a peak ground acceleration (pga) =0.15. It is noted in the figure that the activity normalized by the liquidity index tends to bound the estimated improvements effects. The may suggest that this normalization may can be used to estimate the amount of improvement needed for a particular soil layer.

Seismic Performance

Martin and Olgun. (2003) reported that visual field inspections following the earthquake indicated that no structural damage occurred and no noticeable settlements or ground damages were observed anywhere at the site except in the untreated portions. The untreated areas experienced estimated settlements of 70 mm to 100 mm. Thus, it appears that the surcharge fills and the wick drains alone did very little to increase liquefaction resistance and reduce cyclic shear strains. Field reconnaissance of the site also revealed that structures in an apartment complex located across the street from the shopping center typically experienced settlements of 50 mm to 100 mm. Thus, Martin and his co-workers concluded that jet-grout columns and stone columns were generally effective in mitigating seismic damages, especially liquefaction-related damages. According to

Martin and Olgun (2003), the stone columns and jet-grout columns most likely decreased shear strains (and thus pore pressure development) in the soils. Also, the jet-grout columns appeared to have greatly reduced post-earthquake reconsolidation settlements in the sands.

CONCLUSIONS

Data from geotechnical centrifuge testing and from observations of sites of unimproved ground located adjacent to improved sites have been presented in an effort to gauge the effectiveness of ground improvement on seismic mitigation. Specifically, conclusions are as follows:

1. There were a number of sites within the Kocaeli region affected by the earthquake that contained liquefiable sands and medium to soft clays that were improved for seismic mitigation purposes. The soils at these improved sites performed better compared to nearby/adjacent unimproved soils.
2. Jet-grout columns and stone columns were effective in mitigating seismic damages, especially liquefaction-related damages.
3. The presence of wick drains at the Carrefour site appeared to have had negligible effects on preventing liquefaction-induced settlement.
4. Geotechnical centrifuge testing showed that ground improvement significantly minimized liquefaction-induced settlement of structures. Specifically, excess pore water pressure generation is minimized by ground improvement, with no liquefaction in the improved zones. Excess pore water pressure is also inhibited in the near surface by the presence of the structures.

REFERENCES

Adalier, K., Elgamal, A., Meneses, J., Baez, I.J., (2003). "Stone columns as liquefaction counter-measure in non-plastic silty soils." *Journal of Soil Dynamics and Earthquake Engineering,* 23 (7), 571– 584.

Hausler, E. A. (2002). "Influence of Ground Improvement on Settlement and Liquefaction: A Study Based on Field Case History Evidence and Dynamic Geotechnical Centrifuge Tests." *Ph.D. Dissertation*, University of California, Berkeley.

Hausmann, M. R. (1990). *Engineering Principles of Ground Modification*, McGraw-Hill Publishing Co., New York, 632 pp.

Hayden, R.F. and Baez, J.I., (1994). "State of Practice for Liquefaction Mitigation in North America," *Proc., 4th U.S.-Japan Workshop on Soil Liquefaction: Remedial Treatment of Potentially Liquefiable Soils*, Public Works Research Institute, Japan: 27-48.

Martin, J.R. and Olgun, C.G. (2003). "Unexpected 'Liquefaction-type' Behavior of Fine-grained Soils During the 1999 Kocaeli EQ", *U.S.-Taiwan Workshop on Soil*

Liquefication, National Chiao Tung University, Hsin-Chu, Taiwan, 3-5 November, 2003.

Mitchell, J.K., Baxter, C.D.P., and T.C. Munson. (1995). "Performance of Improved Ground During Earthquakes", Proc., *Soil Improvements for Earthquake Hazard Mitigation*, Geotechnical Special Publication No. 49, ASCE, pp. 1-36.

Mitchell, J.K., Cooke, H.G., and J.A. Schaeffer. (1998). "Design Considerations in Ground Improvement for Seismic Risk Mitigation", *Geotechnical Earthquake Engineering and Soil Dynamics III*, Geotechnical Special Publication No. 75, ASCE, August, pp. 580-613.

Olgun, C. G., (2003). "Performance of Improved Ground and Reinforced Soil Structures during Earthquakes – Case Studies and Numerical Analyses," *Ph.D. Dissertation*, Department of Civil and Environmental Engineering, Virginia Polytechnic Institute and State University, Blacksburg, VA, December.

Xanthakos, P. P., Abramson, L. W. and Bruce, D. A. (1994). *Ground Control and Improvem*ent, John Wiley & Sons, Inc., New York, 910 pp.

International Conference on Geotechnical Engineering for Disaster Mitigation & Rehabilitation
Chu, Phoon & Yong (eds) © 2005 World Scientific Publishing Company ISBN 981-256-469-1

SOIL IMPROVEMENT USING A COMBINED VACUUM AND FILL SURCHARGE PRELOADING METHOD

JIAN CHU
School of Civil and Environmental Engineering, Nanyang Technological University, Singapore 639798

SHUWANG YAN
Geotechnical Research Institute, Tianjin University, Tianjin, 300072, China

A vacuum preloading technique and its implementation in soil improvement projects are introduced in this paper. A case study using the combined vacuum and fill surcharge preloading method is discussed to show how vacuum load can be combined with fill surcharge to consolidate soil when a surcharge higher than 80 kPa is required. Settlements and pore pressures data measured during the preloading are analyzed and used to indicate that the vacuum preloading method and the combined vacuum and fill preloading scheme adopted are effectiveness.

INTRODUCTION

One of the commonly used soil improvement methods for soft clay is vacuum preloading. This method has been successfully used in a number of countries for land reclamation and soil improvement work (Holtz 1975; Chen and Bao 1983; Bergado et al. 1998; Chu et al. 2000; Yan and Chu 2005). Sand drains and recently prefabricated vertical drains (PVDs) have often been used to distribute vacuum load and discharge pore water. A vacuum load of 80 kPa or above can be applied and maintained as long as it is required. When a higher surcharge load is required, a combined vacuum and fill surcharge can be applied. Compared with the fill surcharge method for an equivalent load, the vacuum preloading method is cheaper and faster (Chu et al. 2000). The vacuum preloading method has also been incorporated in the land reclamation process when clay slurry dredged from seabed is used as fill material for land reclamation. As the clay slurry fill is too soft for fill surcharge to be applied, the vacuum preloading method is ideally used for the consolidation of the clay slurry. Thousands of hectares of land have been reclaimed in Tianjin, China, using this method. When the reclaimed land is subsequently used for industrial or infrastructure developments, the vacuum preloading method is used again to improve the foundation soil that consists of a layer of consolidated slurry fill and the underlying seabed marine clay (Chu et al. 2000; Yan and Chu 2003).

The principles and mechanism of vacuum preloading have been well explained in the literature, e.g., Kjellman (1952), Holtz (1975) and Chu et al. (2000). For illustration, the pore water pressure and effective stress change processes in both fill and vacuum preloading cases can be explained using a spring piston analogy model shown in Fig. 1.

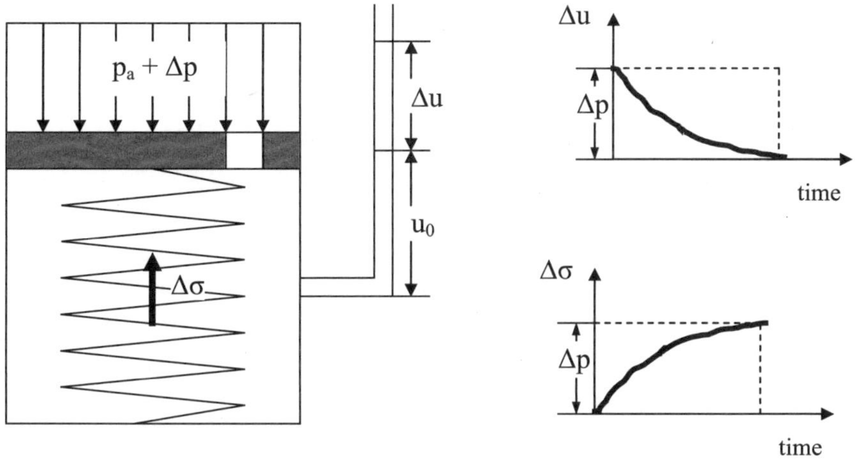

$$u_0 = p_a$$
$$\Delta\sigma = p_a + \Delta p - (u_0 + \Delta u) = \Delta p - \Delta u$$

(a)

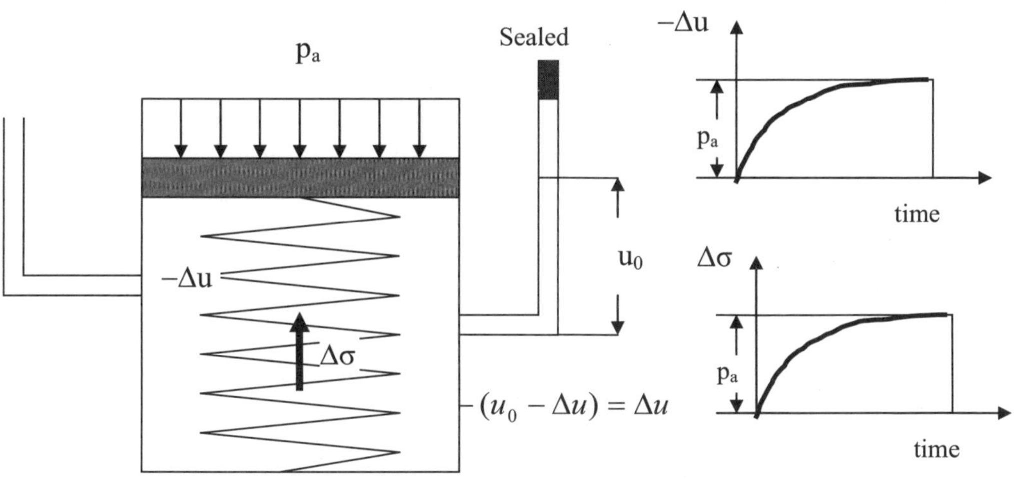

(b)

Figure 1 Spring analogy models for consolidation process (a) under fill surcharge; (b) under vacuum load

The consolidation process of soil under a surcharge load has been well understood and can be illustrated using the model shown in Fig. 1(a). For the convenience of explanation, the pressures in Fig. 1 are given in absolute values and p_a is the atmospheric pressure. As shown in Fig. 1(a), the instance when a surcharge load, Δp, is applied, it is the excess pore water pressure that takes the load. Therefore, for saturated soil, the initial excess pore water pressure, Δu_0, is the same as the surcharge Δp. Gradually, the excess pore water pressure dissipates and the load is transferred from water to the spring (i.e., the soil skeleton) in the model shown in Fig. 1(a). The amount of effective stress increment equals to the amount of pore water pressure dissipation, $\Delta p - \Delta u$ (Fig. 1(a)). At the end of consolidation, $\Delta u = 0$ and the total gain in the effective stress is the same as the surcharge, Δp (Fig. 1(a)). It should be noted that the above process is not affected by the atmospheric pressure, p_a.

The mechanism of vacuum preloading can also be illustrated in the same way using the model shown in Fig. 1(b). When a vacuum load is applied to the system shown in Fig. 1(b), the pore water pressure in the soil reduces. As the total stress applied does not change, the effective stress in the soil increases. The instance when the vacuum load, $-\Delta u$, is applied, the pore water pressure in the soil is still p_a. Gradually the pore pressure is reducing and the spring starts to be compressed, that is, the soil skeleton starts to gain effective stress. The amount of the effective stress increment equals to the amount of pore water pressure reduction, Δu, which will not exceed the atmospheric pressure, p_a, which is normally 80 kPa in practice.

VACUUM PRELOADING TECHNIQUE

The schematic arrangement of the vacuum preloading method adopted is shown in Fig. 2. The construction procedure can be described as follows. 1). PVDs are installed at a closing spacing of normally 1.0 m in square grid through the entire depth of the soil that needs to be treated. If the ground is too soft, a fill of 1 to 2 m can be applied using sand or other competent fill materials to form a working platform for the PVD installation rigs. A layer of geogrid can be used in lieu of fill. Alternatively, a portable, lightweight drain installation rig may be used. 2). After the installation of PVDs, main corrugated flexible pipes of 100 mm diameter are laid horizontally to link the PVDs to the main vacuum pressure line. The pipes were perforated and wrapped with a permeable fabric textile to act as a filter layer. 3). A sand blanket of 0 .3 m on top is then placed to cover the pipes. 4). Three layers of thin PVC membrane are laid to seal each section. The membranes are anchored in a trench at the four sides and backfilled with clay to form a small dike of 0.3 to 0.5 m to contain the water that will drain out the PVDs. To cover the membrane with a layer of water can slow down the deterioration of the membrane caused by direct sun exposure. More importantly, when the membrane is immersed in water, leakage can be identified easily by the formation of air bubbles. 5) Vacuum pressure is then applied using a patented vacuum pumping system. The vacuum pressure will be applied continuously for several months until the required soil improvement objectives are achieved. A vacuum pressure of 80 kPa or above can be achieved and maintained easily using this system. The total surcharge applied will then be the vacuum pressure plus the fill or sand blanket used. When a higher surcharge is required, fill surcharge can be applied in stages after some consolidation of the soil under the vacuum load.

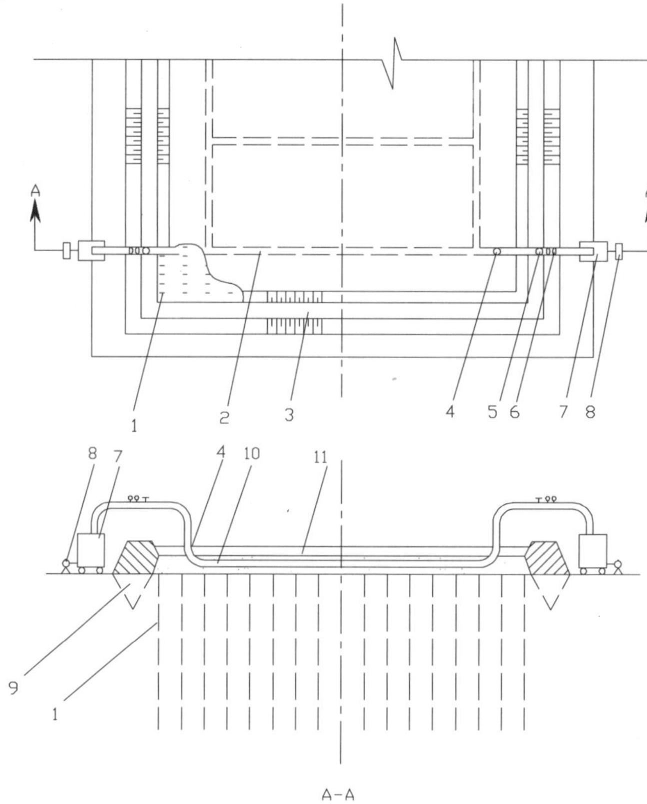

1, drains; 2, pipe with filter; 3, revetment; 4, water outlet; 5, valve; 6, vacuum gauge; 7, jet pump; 8, centrifugal pump; 9, trench; 10, horizontal piping; 11, sealing membrane.

Figure 2 Schematic arrangement of the vacuum preloading method.

Fielding monitoring of vacuum pressure applied to the drain, settlements and pore water pressure changes are an essential part of a vacuum preloading project, as whether the method is effective or implemented properly or not can only be evaluated using the field monitored data. The field monitoring scheme and the methods of evaluation will be discussed using the following case study.

CASE STUDY

The described vacuum preloading method has been used in many projects in China and elsewhere (Shang et al. 1997; Chu et al. 2000; Yan and Chu 2003). The case presented by Yan and Chu (2005) is used as an example to illustrate the use of the vacuum preloading technique in soil improvement works.

A storage yard of 7433 m^2 was to be constructed at the Tianjin Port, China. The storage yard was located on a 16 m thick soft clay layer. The top 3 to 4 m of the clay layer was reclaimed recently using clay slurry dredged from seabed. The rest 16 to 19 m was original seabed clay. The soil in both layers was soft and was still undergoing consolidation. This soft clay layer needed to be improved before the site could be used as a storage yard.

Preloading using fill surcharge alone was not feasible as it was difficult to place a fill embankment several meters high on soft clay. The vacuum preloading method could be used. However, the nominal vacuum load of 80 kPa was not sufficient for this project. Therefore, a combined vacuum and fill surcharge preloading method was adopted. Fill surcharge of a height ranging from 2.53 to 3.50 m was applied in addition to the vacuum load and a 0.3 m of sand blanket. The fill was applied in stages partially for stability consideration and partially due to practical constraints in transporting fill. The fill used was a silty clay with an average unit weight of 17.1 kN/m^3.

The layout of the storage yard is shown in Fig. 3. It was a L-shape with a total area of 7433 m^2. For the convenience of construction, the site was divided into three sections, I, II and III, as shown in Fig. 3. The soil conditions in the 3 sections were very similar. The idealized soil profile together with the typical liquid limit (LL), plastic limit (PL), water content (w/c), and the field vane shear strength profiles as measured for Section II are shown in Fig. 4. The water content of the soil was higher than or as high as the liquid limit at most locations in the soft clays. The field vane shear strength of the soil was generally between 20 to 40 kPa.

The soil improvement work was carried out by following the method described in the preceding section. The loading sequence and the ground settlements induced by the vacuum and surcharge loads for Section II are shown in Fig. 5. The vacuum load was

Figure 3 Project site layout and plan view of instrumentation

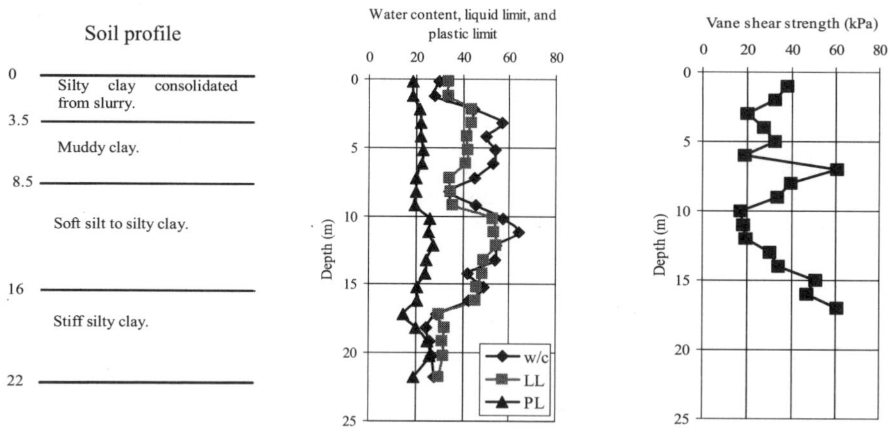

Figure 4 Basic soil properties at Section II

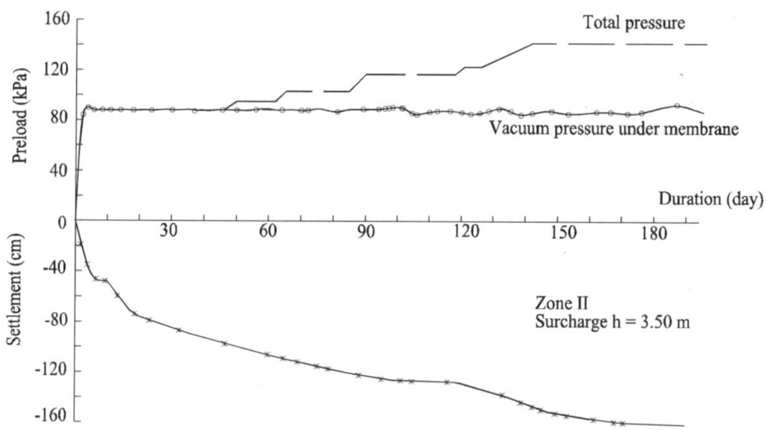

Figure 5 Loading sequence and ground settlement measured at Section II

Instruments including surface settlement plates, multi-level settlement gauges, inclinometers and standpipes were used. The settlements monitored by the settlement gauges installed at different depths during vacuum and surcharge loadings are plotted versus duration for Section II in Fig. 6. The reductions in the pore water pressures at different depths are also measured by the piezometers. Based on the pore water pressure monitoring data, the pore water pressure distributions with depth at the initial stage, 30 and 60 days and the final stage are plotted in Fig. 7 for Section II. The hydraulic pore water pressure line and the suction line are also plotted in Fig. 7. Before the application of vacuum and surcharge loads, the initial pore water pressures, $u_0(z)$, were greater than the hydrostatic pore water pressure, indicating that the subsoil was still under consolidation. These initial excess pore water pressures were mainly the remaining pore water pressures generated during land reclamation and the pore water pressure induced by the placement of the sand blanket. The total fill surcharge was about 60 kPa for Section II. The initial pore water pressure distribution after the application of the fill surcharge is shown as $u_0(z) + \Delta\sigma$ in Fig. 7. The suction line for a suction of –80 kPa is also plotted in Fig. 7 as the line u_s. The pore water pressure distributions at 30, 60 days and the end of preloading ($u_f(z)$) are also shown in Fig. 7. These curves show the changes of the pore water pressure profiles with time. The area bound by the final pore water pressure curve, $u_f(z)$, and the suction line, u_s, represents the remaining excess pore water pressures that have not dissipated.

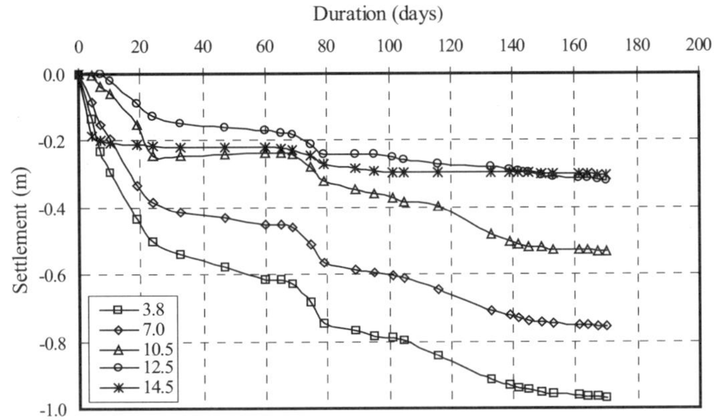

Figure 6 Settlement measured at different depths against duration at Section II.

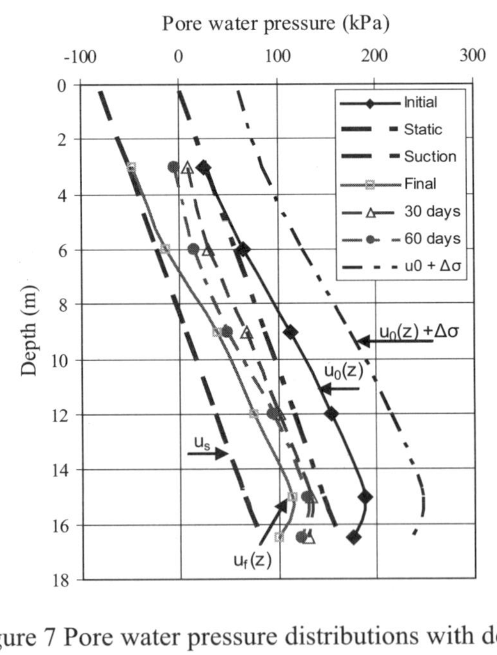

Figure 7 Pore water pressure distributions with depth

CONCLUSIONS

A vacuum preloading technique and its application in soil improvement are introduced in this paper. A case study of using vacuum preloading combined with fill surcharge is presented. The study shows that the vacuum distribution system comprising PVDs at a square grid of 1.0 m together with horizontal 100 mm diameter corrugated flexible pipes was effective in distributing the vacuum pressure and collecting drained water. Vacuum load can be combined with fill surcharge to provide a combined surcharge much higher than the vacuum load alone. In the case presented, the ground settled for more than 1.6 m. The pore water pressure reduced substantially. The case study has proven that the vacuum preloading technique or a combined vacuum and fill surcharge scheme is effective for soil improvement works.

REFERENCES

Asaoka, A. (1978). "Observational procedure of settlement prediction." *Soils and Foundations*, Vol. 18, No. 4, 87-101.

Bergado, D. T., Chai, J. C., Miura, N., and Balasubramaniam, A. S. (1998). "PVD improvement of soft Bangkok clay with combined vacuum and reduced sand embankment preloading." *J. Geot. Eng., Southeast Asian Geot. Soc.*, Vol. 29, No. 1, 95-122.

Chen, H. and Bao, X.C. (1983). "Analysis of soil consolidation stress under the action of negative pressure." *Proceedings 8th European Conference on Soil Mechanics and Foundation Engineering*, Helsinki, Vol. 2, 591-596.

Chu, J. Yan, S.W., and Yang, H. (2000). "Soil improvement by the vacuum preloading method for an oil storage station." *Geotechnique*, Vol. 50, No. 5, 625-632.

Chu, J. and Yan, S.W. (2005). "Estimation of Degree of Consolidation for Vacuum Preloading Projects." *International Journal of Geomechanics, ASCE*, Vol. 5, No. 2, 158-165.

Holtz, R. D. (1975). "Preloading by vacuum: current prospects." *Transportation Research Record*, No. 548, 26-79.

Kjellman, W. (1952). "Consolidation of clayey soils by atmospheric pressure." *Proc. of a Conf. on Soil Stabilization*, Massachusetts Institute of Technology, Boston, 258-263.

Shang, J.Q., Tang, M. and Miao, Z. (1998). "Vacuum preloading consolidation of reclaimed land: a case study." *Canadian Geotechnical Journal*, Vol. 35, No. 5, 740-749.

Yan, S.W. and Chu, J. (2003). "Soil improvement for a road using the vacuum preloading method." *Ground Improvement*, Vol. 8, No. 2, 35-44.

Yan, S.W. and Chu, J. (2005). "Soil improvement for a storage yard using the combined vacuum and fill preloading method." *Canadian Geotechnical. Journal*, Vol. 42, No. 4, 1094-1104.

International Conference on Geotechnical Engineering for Disaster Mitigation & Rehabilitation
Chu, Phoon & Yong (eds) © 2005 World Scientific Publishing Company ISBN 981-256-469-1

SETTLEMENT RESPONSE OF EMBANKMENT ON MULTI LAYER GEOSYNTHETIC-REINFORCED RECLAIMED GROUND

KOUSIK DEB
Research Scholar, Department of Civil Engineering, Indian Institute of Technology, Kanpur, Kanpur 208016, India, Email: kdeb@iitk.ac.in

SARVESH CHANDRA
Professor, Department of Civil Engineering, Indian Institute of Technology, Kanpur, Kanpur 208016, India, Telephone: +91-512-259-7667, Telefax: +91-512-259-7395, Email: sarv@iitk.ac.in

P. K. BASUDHAR
Professor, Department of Civil Engineering, Indian Institute of Technology, Kanpur, Kanpur 208016, India, Telephone: +91-512-259-7029, Telefax: +91-512-259-7395, Email: pkbd@iitk.ac.in

The paper presents a study on the settlement response of embankment on multi-layered geosynthetic reinforced granular fill over reclaimed soil. The granular fill and the soil have been idealized by the Pasternak shear layer and a layer of Winkler springs, respectively. Stretched rough elastic membranes represent the geosynthetic reinforcement layers. Plane strain conditions are considered for the loading and reinforced foundation soil system. An iterative finite difference scheme is applied for obtaining the solution and results are presented in nondimensional form. Significant reduction in the settlement is observed when the number of reinforcement layer is increased.

INTRODUCTION

Coastal lowlands are very soft and highly compressible soils and get inundated by the fluctuating water levels. Reclamation of these lowlands requires raising the ground to a level and makes it safe from the effects of tides and inundation. Reclamation and soil improvement works in these areas are very challenging. To develop these grounds, they have to be reclaimed by placing suitable material over soft ground to improve bearing capacity and to reduce settlements of the structures constructed on them. The embankments are required for highways and railways and are one of the most critical structures to be constructed. The common form of reclamation is to spread and compact granular fill over the soft sub-grade soil. However, this conventional reclamation approach requires huge amount of fill, as fill tends to displace and sinks non-uniformly into the soft soil. To reduce the spreading of the granular material, geosynthetic reinforcement at the base of the granular fill can be used to make the process of reclamation more effective. The use of geosynthetic reinforcements reduces the settlement of the reclaimed soft ground under loading, distributes the loads uniformly and prevents non-uniform sinking of the granular material into the ground. The behavior of the reclaimed ground can be further improved by providing single or multi layer of reinforcements within the granular fill. In recent years, many researchers have studied load-settlement behavior of such reinforced foundation beds based on lumped parameter modeling (Madhav and Poorooshasb, 1988; Ghosh and Madhav, 1994; Shukla and Chandra, 1995; Yin, 1997; Ramu, 2001; Maheshwari et al., 2004).

However, most of the models reported in the literature are developed for single layer reinforced systems. In this paper, the settlement response of embankment on multi-layered geosynthetic reinforced granular fill over reclaimed soft soil has been studied. Parametric studies have been carried out to find the load-settlement behavior of the multi layer geosynthetic-reinforced soil as well as that of the unreinforced soil.

STATEMENT OF THE PROBLEM

A multi layer geosynthetic-reinforced granular fill on reclaimed soft foundation soil is shown in Figure 1. The proposed foundation model as shown in Figure 2 may idealize the behavior of such a system. The stress-strain behavior of the granular fill and the soft soil is considered as linear. Three geosynthetic layers are considered in the model and they divide the shear layer into four equal parts. The shear modulus of the granular fill is taken as constant throughout the study. It is assumed that the geosynthetic reinforcement layers are rough enough to prevent slippage at the interface with soil.

An embankment loading is applied over a width of *2B* on the multi layer geosynthetic-reinforced granular fill of width *2L* over soft soil as shown in Figure 2. The intensity of loading depends on the height of the embankment and density of the embankment soil. Higher height will be more loading intensity. Additional surcharge loading may also act on the embankment as traffic load. The analysis of the problem so defined is carried out by extending the single layer reinforcement model reported by Shukla and Chandra (1995) for multi-layered reinforced system.

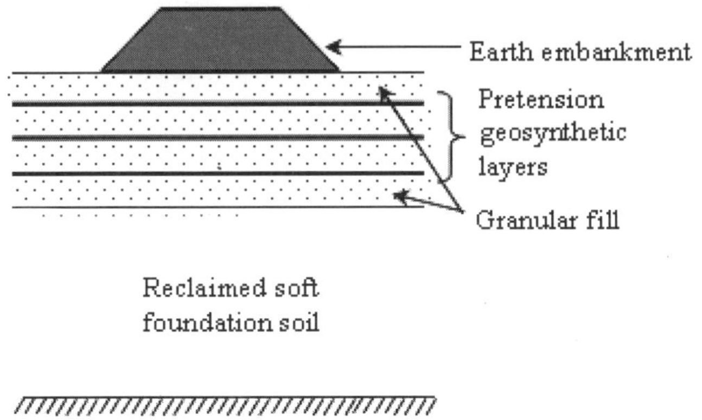

Figure 1. Multi layer geosynthetic-reinforced granular fill-soft soil system

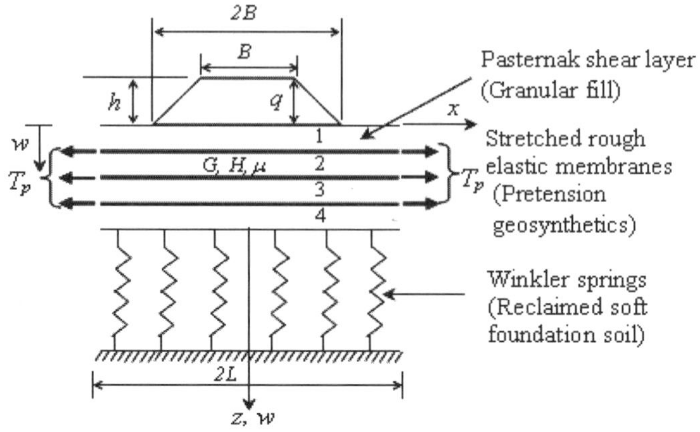

Figure 2. Proposed foundation model

Following the procedures described by Shukla and Chandra (1995) and using the nondimensional parameters as: $X=x/B$; $W=w/B$; $G^*=GH/k_sB^2$; $T_j^*=T_j/k_sB^2$; $T_p^*=T_p/k_sB^2$ and $q^*=q/k_sB$, one gets the following governing equations for multi-layered reinforced system,

$$q^* = A^3W - \left[G^*(1+A+A^2+A^3) + B(T_p^*+T_1^*)\cos\theta + AB(T_p^*+T_2^*)\cos\theta + A^2B(T_p^*+T_3^*)\cos\theta\right]\frac{d^2W}{dX^2} \tag{1}$$

$$\frac{dT_1^*}{dX} = -C\left(q^* + G^*\frac{d^2W}{dX^2}\right) - D\left[A^2W - \left\{G^* + AG^* + A^2G^* + B(T_p^*+T_2^*)\cos\theta + AB(T_p^*+T_3^*)\cos\theta\right\}\frac{d^2W}{dX^2}\right] \tag{2}$$

$$\frac{dT_2^*}{dX} = -C\left[\frac{1}{A}\left\{q^* + \left\{G^* + B(T_p^*+T_1^*)\cos\theta\right\}\frac{d^2W}{dX^2}\right\} + G^*\frac{d^2W}{dX^2}\right] - D\left[AW - \left\{G^* + AG^* + B(T_p^*+T_3^*)\cos\theta\right\}\frac{d^2W}{dX^2}\right] \tag{3}$$

$$\frac{dT_3^*}{dX} = -C\left[\frac{1}{A}\left\{\frac{1}{A}\left\{q^* + G^* + B(T_P^*+T_1^*)\cos\theta\frac{d^2W}{dX^2}\right\} + \left\{G^* + B(T_P^*+T_2^*)\cos\theta\right\}\frac{d^2W}{dX^2}\right\} + G^*\frac{d^2W}{dX^2}\right] - D\left(W - G^*\frac{d^2W}{dX^2}\right) \tag{4}$$

where $A = \dfrac{1+K_0\tan^2\theta - (1-K_0)\mu\tan\theta}{1+K_0\tan^2\theta + (1-K_0)\mu\tan\theta}$; $B = \dfrac{1}{1+K_0\tan^2\theta + (1-K_0)\mu\tan\theta}$

$$C = \mu\cos\theta(1 + K_0\tan^2\theta) - (1 - K_0)\sin\theta\,;\quad D = \mu\cos\theta(1 + K_0\tan^2\theta) + (1 - K_0)\sin\theta$$

and H and G are the thickness and the shear modulus of the granular layers, respectively; w is the vertical displacement; q is the loading intensity; T_1, T_2 and T_3 are the tension mobilized in the top, middle and bottom geosynthetic layer, respectively; T_p is the pretension force applied to the geosynthetic layers; μ is the interface friction at the top and bottom of the geosynthetic layers; θ is the slope of the membrane and $\tan\theta = -dw/dx$; K_0 is the coefficient of lateral earth pressure at rest is assumed to be equal to $1 - \sin\phi$ (Brooker and Ireland, 1965); and k_s is the modulus of the subgrade reaction (spring constant) for the Winkler spring layer as shown in Figure 2.

METHOD OF SOLUTION

Finite difference formulation

Finite difference method has been employed to solve the differential equations [Equations (1)-(4)]. In these equations, the derivative d^2W/dX^2 has been expressed by central difference scheme while dT_j^*/dX have been expressed by forward difference scheme. The length L/B may be divided into n number of the same increment length with (n+1) number of node points (i = 1, 2, 3, 4.........., n).

Boundary conditions and loading

Due to symmetry of the system only half portion is taken into consideration. At $X = 0$, due to symmetry, the slope, dW/dX, will be zero and at $X = L/B$, the geosynthetic layers are free, so that the mobilized tension in different geosynthetic layers are zero, i.e. $T_j^* = 0$. The shear stress acting on the shear layer at the edge (at $X = L/B$) will also be zero since there is no confinement, i.e. $dW/dX = 0$. The loading conditions that are considered are given as: $q_i^*(X) = q^*$ for $|X| \le 0.5$; $q_i^*(X) = 2q^*(1 - |X|)$ for $0.5 < |X| \le 1.0$.and $q_i^*(X) = 0$ for $|X| > 1.0$.

RESULTS AND DISCUSSIONS

A computer program based on the formulation as described above has been developed and solutions are obtained using an iterative technique with a tolerance value of 10^{-4}. The typical values used for this study are the angle of shearing resistance for the granular fill, $\phi = 36°$; the coefficient of lateral stress, $K_0 = 0.41$; the interface friction coefficients, μ = 0.5; G^*=0.05; T_p^*=0; and $L/B = 2.0$.

Figure 3 shows the settlement profiles of the embankment for different number of reinforcement layer as well as for unreinforced soil for various loading intensities. For loading intensity 0.3, as the number of geosynthetic layer increases from zero to three, the settlement at the centre of the loaded region is decreased by 19.7%, whereas for loading intensity 0.8, this reduction is 33.1%. Thus, for higher loading intensity multiple geosynthetic reinforcements reduce the vertical settlement more effectively than for lower loading intensity. This is due to the fact that tension mobilization of geosynthetic reinforcements increases with increasing load. It is further observed that at $q^* = 0.8$, the reduction of the settlement at the centre of the loaded region is 18.2%, 26.6% and 33.1% from unreinforced to one, two, and three layer geosynthetic reinforced soil system, respectively. However, at the edge of the loaded region the difference of the settlement for different number of reinforcement layer is less that 1%. The results indicate that application of multiple geosynthetic reinforcements reduce both the total and differential settlements of the loaded region.

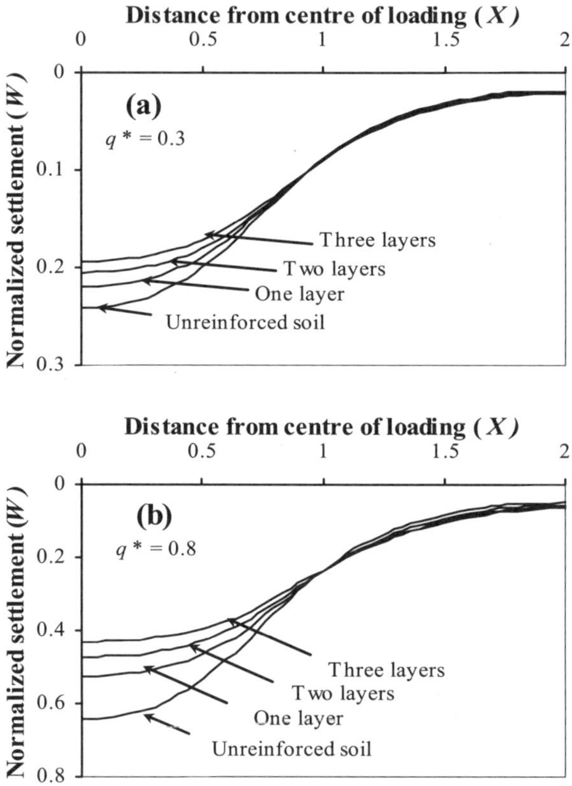

Figure 3. Settlement profiles for different layer of geosynthetic reinforcements (a) $q^* = 0.3$ (b) $q^* = 0.8$

Figure 4 shows the mobilized tension in the reinforcement layers of a three layer reinforced system for a particular loading intensity ($q^*= 0.8$). It can be observed that the top geosynthetic layer is subjected to higher tension up to $x/B \leq 0.3$, whereas beyond that region the mobilized tension in bottom geosynthetic layer is higher. It can be further seen that, beyond a distance equal to the width of the embankment reinforcements are not very effective as there is no significant amount of mobilized tension beyond that length.

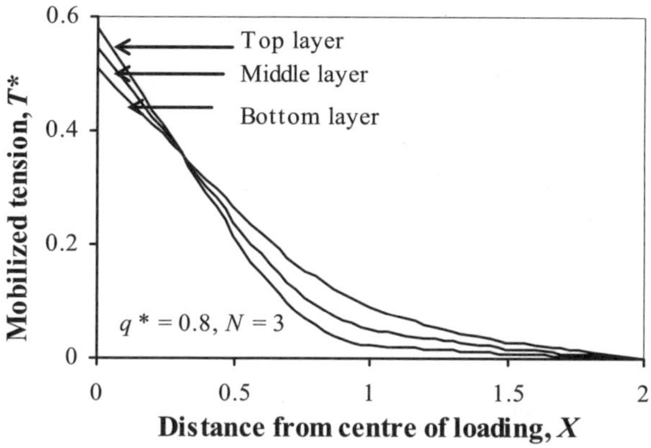

Figure 4. Mobilized tension in multi layer (three) geosynthetic reinforcements

CONCLUSIONS

Application of multiple geosynthetic reinforcements reduces the total as well as the differential settlements of the loaded region of the reclaimed soil and the use of multiple geosynthetic-reinforced systems is more effective at the higher load intensities. As the number of reinforcement layer increases the maximum settlement of the embankment decreases in a decreasing rate. The top reinforcement layer is subjected to higher mobilized tension than the lower layers and it is maximum at the center of the loaded region.

REFERENCES

Brooker, E. W. and Ireland, H. O. (1965). "Earth pressure at rest related to stress history", *Canadian Geotechnical Journal*, Vol. 2, No. 1-2, 1-15.

Ghosh, C. and Madhav, M. R. (1994). "Settlement response of a reinforced shallow earth bed", *Geotextile and Geomembranes,* Vol. 13, No. 9, 643-656.

Madhav, M. R. and Poorooshasb, H. B. (1988). "A new model for geosynthetic-reinforced soil", *Computers and Geotechnics,* Vol. 6, No. 4, 277-290.

Maheshwari, P., Basudhar, P. K. and Chandra, S. (2004). "Analysis of beams on reinforced granular beds", *Geosynthetics International,* Vol. 11, No. 6, 470-480.

Ramu, K. (2001). *"Modeling approaches for & analysis of reclamation process and response of reclaimed ground",* PhD thesis, Department of Civil Engineering, IIT Kanpur.

Shukla, S. K. and Chandra, S. (1995). "Modeling of geosynthetic-reinforced engineered granular fill on soft soil", *Geosynthetics International,* Vol. 2, No. 3, 603-617.

Yin, J. H. (1997). "Modeling geosynthetic-reinforced granular fills over soft soil", *Geosynthetics International,* Vol. 4, No. 2, 165-185.

International Conference on Geotechnical Engineering for Disaster Mitigation & Rehabilitation
Chu, Phoon & Yong (eds) © 2005 World Scientific Publishing Company ISBN 981-256-469-1

GROUND IMPROVEMENT OF SOFT MARINE CLAY LAYER BY SUPER WELL POINT METHOD

K. MIYAZAKI, T. HAGIWARA AND S. IMAMURA
Nishimatsu Construction Co., Ltd, Tokyo, JAPAN

This paper describes an in-situ trial experiment on ground improvement of soft marine clay in Singapore by using a newly developed vacuum drainage method, or "Super Well Point method". It is demonstrated that the vacuum drainage by the Super Well Point method is an effective method to in order to improve the soft clay ground.

INTRODUCTION

Nowadays, a wide variety of ground improvement techniques are used for soft clay grounds. Among them, a vacuum drainage method has gradually been used as a promising one, because it is environmentally friendly and no need for heavy machinery.

In recent years, a new vacuum drainage method ("Super Well Point method") has been developed (Association of SWP method, 2000) and broadly used for a wide range of ground conditions in order to lower the ground water level. Up to now, this method has construction performance over 90 cases as very useful method in order to lower the ground water table. However, there are few case histories for the application of the vacuum drainage method in soft clay grounds so far. Therefore, further study is needed to ensure weather the Super Well Point method can apply to the ground improvement of soft clay layers as a vacuum consolidation method by the field tests and/or the lab tests.

This paper examines an in-situ trial experiment on the ground improvement of the uniform marine clay soil in Singapore by the Super Well Point method. The trial test was performed in an attempt to dewater from Upper marine clay layer. Throughout the test, a series of in-situ monitoring, including pore water pressure, vacuum pressure and soil strength at the site, was extensively conducted. At the samc time, a series of centrifuge model tests was carried out to evaluate the effectiveness of the vacuum drainage on dewatering from soft clay layer and reinforce the results of the field test.

In this paper, the outline of the in-situ trial experiment and monitoring results are only described and discussed due to limited space.

OUTLINE OF SUPER WELL POINT METHOD

The Super Well Point (hereafter referred to as SWP) method is a newly developed dewatering method, which supersedes the conventional dewatering methods, such as the Well Point method, the Deep Well method and the Vacuum Deep Well method. By utilizing a special strainer section made of double tubes (a separation screen), forced dewatering, keeping a high vacuum pressure in the well can be achieved. The separation screen is able to suck only water, preventing air inflow. This brought to much higher efficiency of the dewatering, compared with the existing drainage methods. Depending on the ground conditions, the dewatering volume of the SWP method is 1.2 to 10 times larger than those of the other methods. This method also can lower the groundwater table over a wider area in a shorter period of time. Further details on the SWP method are described elsewhere (Association of the SWP method, 2000).

IN-SITU TRAIL TEST BY SWP METHOD

Objectives of the in-situ trail test

The main objectives in relation to the in-situ trail test are as follows;

(1) Materials and machinery, which require to make the SWP system, are procured in Singapore.

(2) During the SWP operation, a series of the in-situ monitoring, including water flow from the well, vacuum pressure and pore water pressure, is conducted.

(3) The quantitative evaluation of the ground improvement in the test ground is performed by the site investigations and laboratory tests.

Summary of In-situ trial investigation

Figure 1 shows the soil profile, soil properties, and section through the instrumented well at the test site. The properties of Upper marine clay are summarized in Table 1. As shown in Figure 1, the test site was at a ventilation shaft in the station and surrounded by diaphragm wall and sheet-piles. The test ground right below the tip of the well had been improved by the jet grouting piles (JGP). The test area is 13.55 m wide, 18.00 long and 12.00 m deep. In this trial, dewatering from Upper marine clay layer was planned and conducted. The well of the SWP which has a special strainer (0.4 in diameter and 7m in depth) was installed.

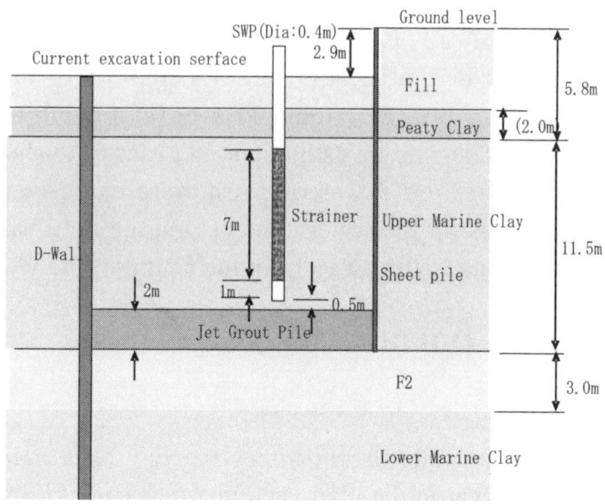

Figure 1. Soil profiles and cross section of the test site.

Table 1. Properties of Singapore Upper marine clay.

Parameters	Values
Specific gravity	2.61
Natural water content	60—70 %
Plastic limit	25—30 %
Liquid limit	70—80 %
Unit weight	16 kN/m^3
Compression index	0.75
Undraind shear strength	Su=5+1.5z z: depth below the ground level
Coefficient of permeability	10^{-6} —10^{-8} cm/sec

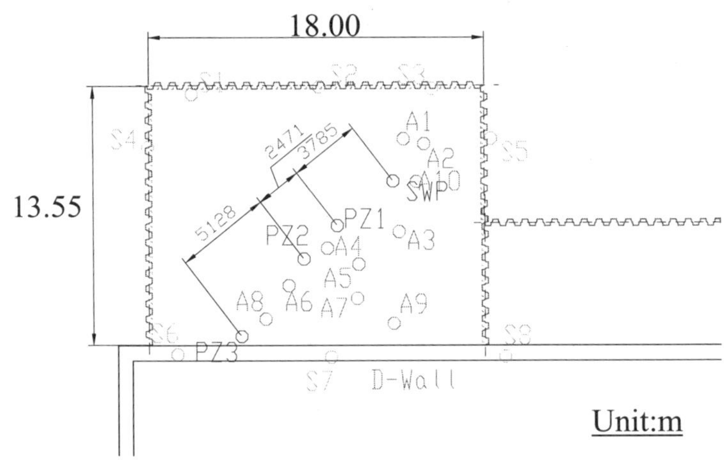

Figure 2. Plan view of the test site and layout of monitoring instruments.

Instrumentation and Monitoring program

The layout of the monitoring instruments is shown in Figure 2. The instruments consisted of three piezometers(PZ1, PZ2 and PZ3) and 10 settlement markers(A1—A10). Three piezometers were located at 3.8m (at a depth of 8 m from the original GL(PZ1)),6.3 m (at a depth of 8 m(PZ2)) and 11.4 m (at a depth of 10 m(PZ3)) away from the centre of the well. The site investigations in order to evaluate the soil strength of the Upper marine clay after the dewatering test were extensively conducted. They consisted of the measurements of the water content and the field vane shear tests. They were located around 2 m, 5m and 10m away from the centre of the well (see Figure 2.).

MONITORING RESULTS AND DISCUSSIONS

Water flow from the well

Drainage work was carried out one or two times per day by a manual operation. The SWP system was operated, keeping the high vacuum pressures of 75 to 85 kPa all day.

The cumulative water flow from the well of the SWP is shown in Figure 3. In this

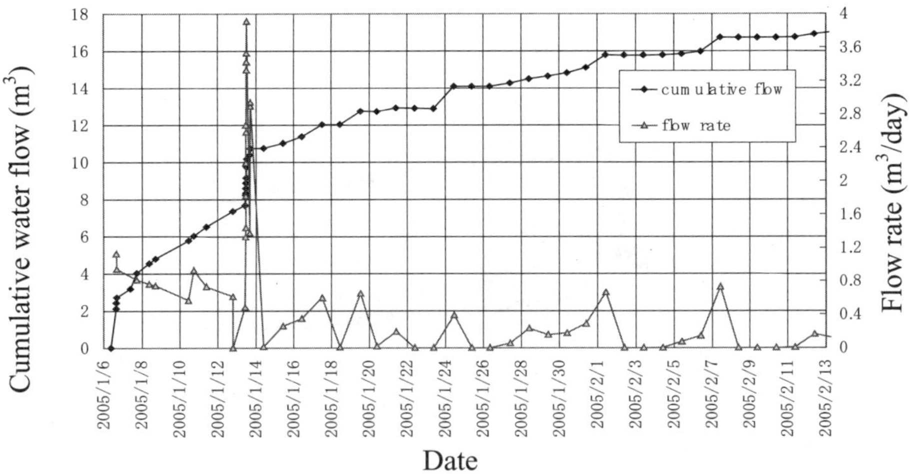

Figure 3. Water flow from the well during the SWP operation.

figure, the outflow rate of water, or water volume per day, is also shown. The water flow volume is relatively large at an early stage of the operation, keeping the flow rate of 0.6 to 0.8 m^3/day. This is because water due to the water wash of the well is also sucked. After the middle of January, the water flow rate is almost constant, keeping the values of 0.2 to 0.3 m^3/day, although the flow rate shows some scatter.

Pore water pressure behaviour

The monitoring results of the pore water pressure in marine clay ground are shown in Figure 4. After the equilibrium was reached on January 3, the SWP operation was started. First of all, the washing work of the well of the SWP was carried out. At this time, sharp increase in the pore water pressure was monitored. After the water wash, the pore water pressure continued to decrease at a constant ratio of about 0.2 kPa/day. The decreasing of the pore water pressure from the middle of January to the middle of March (approximately two months) varies from 10 kPa to 15 kPa.

Results of field vane shear tests

Figure 5 shows the results of the field vane shear tests. As can be clearly seen, the values of the shear strength in the ground after the drainage increase, compared with those before the test, particularly at the depths of 4 m to 8 m in close proximity to the well. The increase ratio of the shear strength at 2m, 5m and 10 m away from the well is as follows;
Distance 2m: Ratio 2.0 (Before operation 18kN/m^2 and after operation 36kN/m^2)
Distance 5m :Ratio 1.3 (Before operation 18kN/m^2 and after operation 24kN/m^2)
Distance10m:Ratio1.0 (Before operation 18kN/m^2 and after operation 18kN/m^2)
According to the above results, the nearer the well is, the more the values of the shear strength after the SWP operation increase. Therefore, from these results, it is concluded that the vacuum drainage by the SWP method is effective for the improvement of the uniform marine clay.

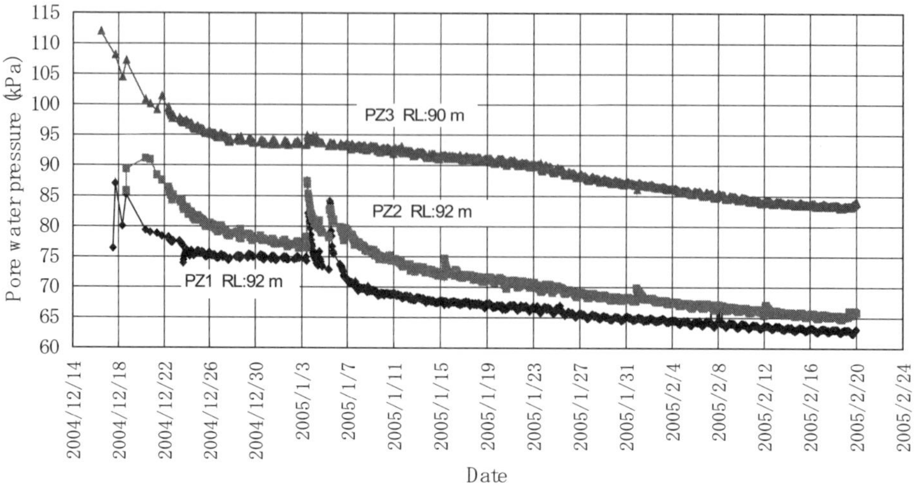

Figure 4. Pore water pressure behaviour during the SWP operation.

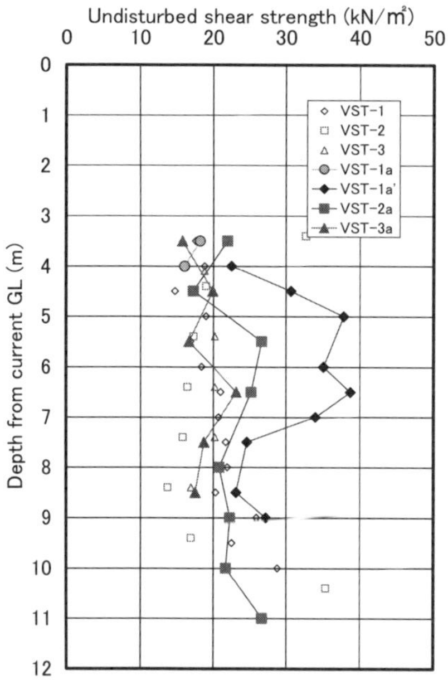

Figure 5. The results of the field vane shear tests.

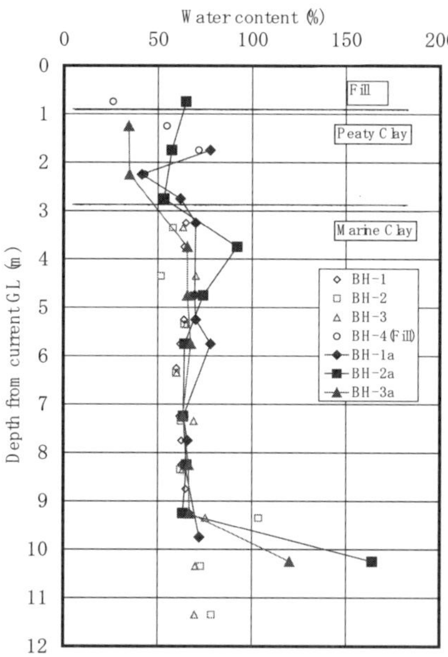

Figure 6. Comparison of water contents before and after the SWP operation.

Change of water content

The change of the water content in marine clay due to the operation of the SWP system is shown in Figure 6. The results indicate that the water contents of marine clay after operation of the SWP are almost the same as those before the test. There are two reasons for this. One is due to scatter of measurements of the water content. As illustrated in Figure 6, the measured values of the water content before the test also show large scatter. In this case history, the change of the water content after the SWP operation may exist within the scatter. The other is due to short period of the SWP operation, just two-month.

Applicability of the SWP method for soft clay ground

Because of the first trial test of the SWP at abroad, there were some miner troubles at the first stage. And, the two-month operation time of the SWP method might be a bit shorter for Singapore marine clay, compared with the time conducted by the usual methods.

However, in spite of these limited conditions, the in-situ experiment on the soil improvement of the marine clay soil by the SWP method was fairly well conducted in Singapore, indicating that the effectiveness of the SWP system for dewatering from soft clay. In addition, the effectiveness of vacuum drainage for soft clay soil also verified from the results of the lab-test using the centrifuge, although they could not be described in this paper.

In order to promote the ground improvement faster by the SWP method, some auxiliary measures may be needed such as applying atmospheric pressure to soil skeleton and putting some surcharge pressures on the ground surface. Further research and development, including more in-situ trial tests, are urgently needed to put the SWP method to practical use for the ground improvement of a wide variety of soft clays.

CONCLUSIONS

Based on the in-situ monitoring results, the following summary and conclusions are made.

1. Sucking water from Singapore Upper marine clay that has very low permeability could be satisfactory carried out at the extraction flow rate of 0.2 to 0.3 m^3/day during nearly two-month SWP operation.
2. After about two-month continuous vacuum operation by the SWP system, the shear strength of the marine clay sufficiently increased, showing a trend that the nearer the well, the more the shear strength increases.
3. Clear decrease trend of the water contents of the clay ground in the site after the SWP trial test was not confirmed due to the measured value's scatter. However, it is assumed that some decrease of water content occurred, because the increase of the shear strength in the ground was confirmed by the site investigations.

REFERENCES

Association of the Super Well point method (2000). "Super Well Point Method technical report" (in Japanese).

International Conference on Geotechnical Engineering for Disaster Mitigation & Rehabilitation
Chu, Phoon & Yong (eds) © 2005 World Scientific Publishing Company ISBN 981-256-469-1

GEOSYNTHETICS IN MITIGATING SEISMIC-INDUCED WASTE CONTAINMENT COVER INSTABILITY

VINCENT O. OGUNRO
Department of Civil Engineering, and Global Institute for Energy and Environmental Systems (GIEES), University of North Carolina at Charlotte, 9201University City Blvd, North Carolina 28223, USA

MUTIU G. AYOOLA
Department of Civil Engineering, and GIEES, University of North Carolina at Charlotte, North Carolina 28223, USA

HILARY .I INYANG
Director, GIEES, University of North Carolina at Charlotte, North Carolina 28223, USA

GUSTAVO B. MENEZES
Department of Civil Engineering, and GIEES, University of North Carolina at Charlotte, North Carolina 28223, USA

Geosynthetics are now routinely employed to improve static stability of many geotechnical and geoenvironmental structures. Lack of available data on the impact of seismic actions on the interface characteristics of geosythetics may have resulted in the limited studies performed on the long-term impact of geosynthetics on the seismic-induced instability of waste containment cover. Water mounding resulting from earthquake-induced warping of covers of waste containment systems may result in time-dependent reduction of post-earthquake factor of safety against instability (FS). The reduction rate of FS is dependent on whether slope stability is achieved with or without geosynthetic reinforcement. For a selected case study, FS of a cover (with slope of 15°) was found to fall abruptly from 1.93 to 1.1089 immediately after the occurrence of seismic event, but decreased gradually to 0.98 over 50 years under mounding effect of infiltrated water. Analysis of a similar system (but with a slope of 18°) with geosynthetic reinforcement of tensile strength of 37.5 kN/m showed that FS decreased abruptly from 1.93 to 1.212 immediately after impact by seismic of the same loading characteristics. However, 50 years after earthquake occurrence, the FS gradually decreased to 1.097. The analyses show that in addition to increase in waste storage capacity, geosynthetic-reinforced cover slopes tend to be more stable than unreinforced system immediately and long-term after earthquake occurrence. The percentage improvement in cover stability (P_R) defined herein as the ratio of the long-term factor of safety of geosynthetic-reinforced cover slope to that of the unreinforced structure may be as high as 150% for the range of seismic events covered in hazard mitigation programs.

INTRODUCTION

Waste containment systems are used to manage municipal and industrial solid wastes, and semi-liquid wastes generated during industrial activities such as mining, agriculture, petroleum production etc. Containment systems are in various forms and are generally multi-component in configuration. Cover system is one of the vital components of all waste management systems, except conventional impoundments and waste piles. However, in some instances where the configuration is favorable, non-active waste piles are capped to minimize the production of acid mine drainage, which results when the waste minerals are exposed to water and air.

A granular drainage layer and one or more geosynthetic interfaces are usually present in covers to provide liquid drainage, and containment and/or reinforcement functions, respectively. Dynamic and static stability of covers remain a major challenge in the design of containment systems. Extensive work by De and Zimmie (1998), Yegian and Lahlaf (1992), Ground and Beech (1989), Koerner and Hwu (1991) and Soong and Koerner (1996) have been carried out to investigate cover stability issues. Configuration parameters that are of interest in the stability aspect of cover design include the slope and length, density and thickness of cover soil, presence of seepage forces, static and dynamic interfacial angle etc. Depending on the magnitude of these parameters, stability may be achieved with or without the use of reinforcing geosynthetic.

METHODOLOGIES FOR COVER STABILITY ANALYSES

Prior to 1993, seismic effects are seldom considered in the design of waste containment systems. Consequently, the number of earthquake-induced catastrophic failure of the system and the corresponding environmental impact was significant. Presently, the U.S. Environmental Protection Agency (U.S. EPA) requires that all waste containment systems to be located in regions classified as seismically active be designed to withstand seismic-induced stresses, in addition to geostatic and hydrostatic stresses. As stated in the introductory section, numerous studies have been carried out on both static and dynamic stability of cover; a comprehensive review of these studies is presented in Ayoola et al. (2005). In most of the cover stability analyses techniques described in literature, the potential for liquid mounding in malfunctioning drainage layer, and the effect of resulting additional hydrostatic stresses on the stability of cover are not addressed.

In the case of cover system, seismic-induced warping of liquid collection and removal pipes and the entire drainage layer leads to liquid mounding in the drainage layer. Apparently, the magnitude of liquid mounded and the resulting hydrostatic stresses will increase with time with corresponding decrease in stability of the cover system. The authors in separate study (Inyang et al. unpublished manuscript, 2005) developed Eq. (1) for quantifying the magnitude of liquid head (h_w) in seismic-disrupted drainage layer of covers. The corresponding effect of h_w on the cover slope stability (assessed through the time- dependent factor of safety against instability, $FS_{(t)}$ is investigated in yet another study (Ayoola et al. unpublished manuscript, 2005) and is quantitatively given by Eq. (2)

$$h_w = \frac{L \times k}{A'\ 2l}\left[a \times \left(\frac{T_n}{2\pi}\right)^2 \cos\left(\frac{\pi}{2D}Z\right)\right]^2 \times t \times \cos^3\alpha \tag{1}$$

$$FS = \frac{P + C_A + T_A + k_v W_A \sin\alpha + U_h \cos\alpha}{W_A(k_h \cos\alpha + \sin\alpha)} \tag{2}$$

Where

$$N_A = W_A[(1 - k_v)\cos\alpha - k_h \sin\alpha] - U_{NA} + U_h \sin\alpha \tag{3}$$

$$W_A = 0.5 \times h \times [\gamma_{sat}(2h_w - 2h_f - h\sec\alpha) + \gamma(2H - 2h_w - h\sec\alpha)]\ \operatorname{cosec}\alpha \tag{4}$$

$$U_{NA} = \gamma_w \times h \times (h_w - h_f)\operatorname{cosec}\alpha \tag{5}$$

$$U_H = 0.5 \times \gamma_w \times h^2 \tag{6}$$

$$C_a = c_a L_a = c_a(H - h_f - h\sec\alpha)\operatorname{cosec}\alpha \tag{7}$$

In Eq. (1) through (7) A' is the cross-sectional area of the waste containment system (L^2); h_w is the average water head developed in the seismic-disrupted drainage layer (L); k is the permeability of the granular drainage layer (LT^{-1}); α is the cover slope (°); a is seismic-induced ground acceleration (LT^{-2}); T is the ground vibration period (T); L is the length of the landfill (L); t is time (T); and l is half the liquid collection pipe spacing (L). Other parameters and notations are defined in the appendix.

Typically, in scenarios where waste storage capacity requirement is high and cover slope is very steep, stability is achieved by incorporating geosynthetics to provide reinforcing functions. In the following section, by using two (2) cover systems originally designed to the same level of factor of safety, it is shown that the geosynthetic-reinforced cover slope exhibit better stability, in addition to higher waste storage capacity, during and long-term after earthquake occurrence. It is assumed that owing the damping effects of a waste system, no rupture of the geosynthetic material occurs.

CASE STUDY

Consider the two (2) cover systems with the following configuration parameters summarized in Table 1, subjected to the same level of seismic-induced stresses, but originally designed to the same level of factor of safety. By using Eq. (1) through (7), the time-dependent magnitude of factor of safety for the two scenarios immediately and long-term after earthquake occurrence is computed and presented in Figure 1. As evident in

Figure 1, the FS of cover type A (geosynthetic-reinforced) reduced from 1.93 to 1.212 immediately after earthquake occurrence and gradually to 1.097 (stable, FS slightly greater than 1.0) over 50 years. On the other hand, in the case of cover type B, (unreinforced), the FS reduced from 1.93 to 1.1089 immediately after earthquake occurrence and steadily to 0.98 after 50 years (failure, FS lesser than 1.0).

Table 1: Range of configuration and seismic parameters magnitude for case study cover slope

cover and seismic parameter	definition	unit	magnitude	
			geosynthetic-reinforced (type A)	unreinforced (type B)
α	Slope angle	(°)	18	15
T_{req}	tensile strength of geosynthetic	$MLT^{-2}L^{-1}$	37.5 kN/m	0.0
A'	area	L^2	300m by 200m	
γ	cover soil density	$ML^{-1}T^{-2}$	18 kN/m^2	
h	thickness of cover soil	L	0.65m	
k	permeability of drainage layer	LT^{-1}	0.0875 cm/s	
c	cohesion of cover	$ML^{-1}T^{-2}$	0.0	
l	drainage pipe spacing	L	20m	
a	seismic-induced ground acceleration	LT^{-2}	0.15g	

CONCLUSION

The percentage improvement in cover stability (P_R) defined herein as the ratio of long-term factor of safety of geosynthetic-reinforced cover slope to that of the unreinforced structure may be as high as 150% for the range of seismic events covered in hazard mitigation programs. It is therefore recommended that geosynthetic reinforcement should be used in covers of waste containment systems located in seismically active zones.

ACKNOWLEDGEMENT

Resources for this analysis were provided by the U.S. National Science Foundation (NSF) through a grant awarded to the authors at the Global Institute for Energy and Environmental Systems (GIEES) and the Department of Civil Engineering of the University of North Carolina, Charlotte, NC. The authors appreciate the opportunity provided by NSF for engagement in this research.

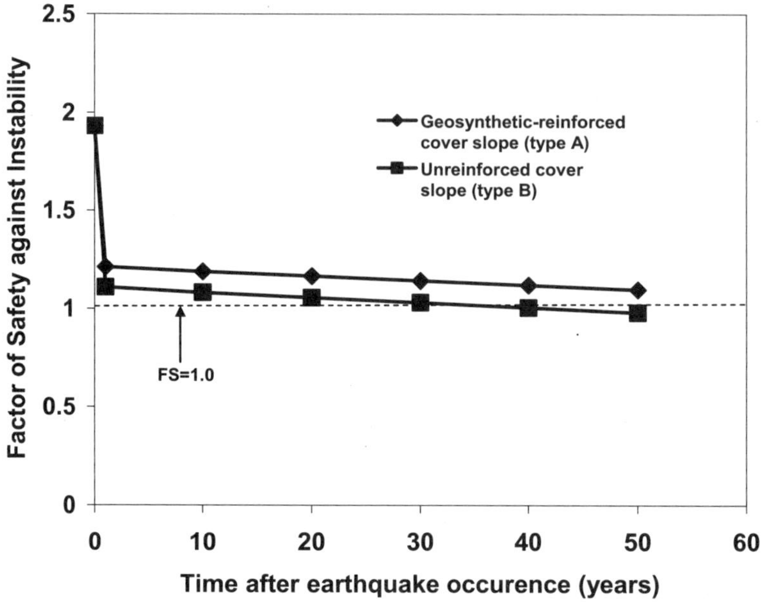

Figure 1: Long-term stability of geosynthetic-reinforced and unreinforced cover slopes with identical configuration parameter and subjected to same level of seismic stresses.

REFERENCES

Ayoola, M.G., Inyang, H.I., and Ogunro, V.O. (2005).”Analyses of Seismic Damage to Interfaces in Waste Containment Systems: A review”, ASCE Practice Periodical of Hazardous, Toxic, and Radioactive Waste Management, special edition on Innovative Barrier Systems for Waste Containment, , Vol. 9, No 4 , Pp. 292-304, October, 2005.

De, A. and Zimmie, T.F. (1998).” Estimation of Dynamic Interfacial Properties of Geosynthetics” Geosynthetic International, Vol. 5, Nos 1-2, pp 17-39.

Koerner, R.M. and Hwu, B-L (1991).”Stability and Tension Consideration Regarding Cover Soils on Geomembrane Lined Slopes”, Geotextile and Geomemebranes, Vol. 10, Pp. 335-355.

Soong, T-Y. and Koerner, R.M. (1996).”Seepage Induced Instability”, Geotextile and Geomembranes, Vol. 14, Pp. 425-445.

Yegian, M.K. and Lahlaf, A.M. (1992).”Dynamic Interface Shear Strength Properties of Geomembranes and Geotextiles.” ASCE Journal of Geotechnical Engineering, 118(5): 760-779.

APPENDIX

h = thickness of the soil layer perpendicular to the slope (L)
γ_{sat} = saturated unit weight of the cover soil ($MLT^{-2}L^{-3}$
α = cover slope (degree)
θ = angle of failure plane to the horizontal (degree)
v_f = vertical distance from the toe to failure location (L)
γ = dry unit weight of the cover soil ($MLT^{-2}L^{-3}$)
H = vertical height of the cover measured from the toe (L)
U_h = resultant of the pore pressures acting on the interwedge surface ($MLT^{-2}L^{-1}$)
U_{NP} = resultant of the pore pressure acting perpendicular to the passive wedge slope ($MLT^{-2}L^{-1}$)
U_{NA} = resultant of the normal component of pore pressures acting on the active edge ($MLT^{-2}L^{-1}$)
γ_w = unit weight of water ($MLT^{-2}L^{-3}$);
$k_v = \frac{a_v}{g} =$ vertical component of seismic coefficient (-);
$k_h = \frac{a_h}{g} =$ horizontal component of seismic coefficient (-);

International Conference on Geotechnical Engineering for Disaster Mitigation & Rehabilitation
Chu, Phoon & Yong (eds) © 2005 World Scientific Publishing Company ISBN 981-256-469-1

SETTLEMENT OF SHALLOW FOUNDATION ON GEOGRID REINFORCED SAND

MAHER OMAR
Department of Civil Engineering, University of Sharjah
Intelligent Systems and Infrastructure & Energy and Environment Research Groups
P.O. Box 27272, Sharjah, United Arab Emirates
momar@sharjah.ac.ae

The permanent settlement of a square model foundation supported by geogrid-reinforced sand and subjected to cyclic loading has been evaluated. The model foundation was initially subjected to an allowable static load and cyclic load was superimposed on it. For a given static allowable load, the ultimate permanent settlement increases with the amplitude of the cyclic load. The results indicate that the ultimate permanent settlement is markedly reduced due to geogrid reinforcement. The results of the study can be used in shallow foundation construction to reduce earthquake related settlement.

INTRODUCTION

Several laboratory model test results are presently available in the literatures which are directed towards the evaluation of the ultimate and allowable bearing capacities of square and strip foundations supported by sand reinforced with materials such as metal strips, metal bars, rope fibers, geotextiles, and geogrids. Figure 1 shows a square surface foundation measuring $B \times B$ in plan on a sand layer which is reinforced with N layers of geogrid. The dimensions of the geogrid layers are $b \times b$. The first layer of geogrid is located at a distance u from the bottom of the foundation. The distance between each consecutive layer of geogrid is h. referring to Figure 1, the distance d measured from the bottom of the foundation to the bottom layer of geogrid can be expressed as

$$d = u + (N-1)h \tag{1}$$

Past model test results Binquet & Lee (1975), Guido et al. (1986) have shown that the ratio u/B should be kept to less than about 0.67 to 1 in order to derive the beneficial effects of soil reinforcement. It has also been shown that, for given values of u/B and h/B, there are maximum values of b/B and d/B beyond which the ultimate bearing capacity remains practically constant or increases by a very small amount. Hence, for all practical purposes, these maximum values may be considered to be the critical values $(b/B)_{cr}$ and $(d/B)_{cr}$. For geogrid-reinforced sand supporting a square foundation, Guido *et al.* (1986) have reported the following values:

$(b/B)_{cr} \approx 2$ to 3 and $(d/B)_{cr} \approx 1.25$

Foundations on some occasions support machines which impart cyclic load to the foundation. This will, in turn, induce some permanent settlement. A review of the existing literature shows that, at the present time, practically no studies have been conducted to evaluate the permanent settlement of shallow foundations supported by geogrid-reinforced sand. This paper presents the results of some laboratory model tests related to the permanent settlement of a square surface foundation supported by geogrid-reinforced sand and subjected to cyclic loading superimposed on a static load.

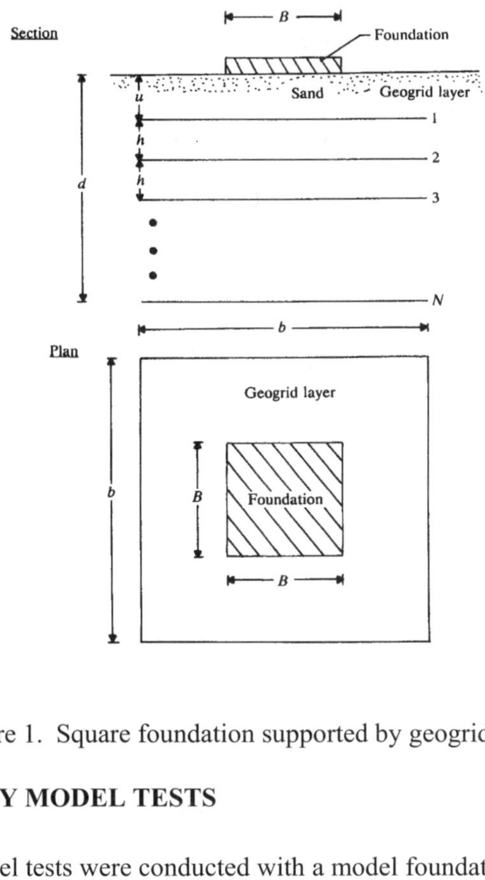

Figure 1. Square foundation supported by geogrid-reinforced sand.

LABORATORY MODEL TESTS

Laboratory model tests were conducted with a model foundation made from an aluminum plate with dimensions (B x B) of 76.2 mm x 76.2 mm. A fine, round silica sand was used. The sand used for the tests had 100% passing a 0.85-mm sieve, 26% passing a 0.425-mm sieve and 0% passing a 0.25-mm sieve. All model tests were conducted at an average relative density of compaction (D_r) of 70% The average physical properties of the sand during the laboratory tests are given in Table 1. A biaxial geogrid was used for reinforcement, and its physical properties are given in Table 2. Laboratory model tests were conducted in a box Measuring 760mm x 760mm x 760mm.

Table 1. Average Physical Properties of the Sand During the Model Tests

Parameter		Quantity
Maximum dry unit weight (kN/m^3)	according to ASTM D-4253	18.94
Minimum dry unit weight (kN/m^3)	according to ASTM D-4253	14.07
Dry unit weight (kN/m^3)		17.14
Relative density of compaction (%)		70
Angle of friction (degrees)	from direct shear test	40.3

Table 2. Physical Properties of the Geogrid

Parameter	Description / Quantity
Structure	Punctured sheet drawn
Polymer	PP/HDPE copolymer
Junction method	Unitized
Aperture size (MD/XMD)	25.4mm / 33.02 mm
Nominal rib thickness	0.762 mm
Nominal junction thickness	2.286 mm

The model foundation was given a rough base by cementing a thin layer of sand onto its base with epoxy glue. In conducting the tests, sand was poured into the box in 25.4-mm thick layers using a raining technique. The accuracy of sand placement and consistency of placement density were checked during raining by placing small cans with known volumes at different locations in the box. Geogrid layers were placed in the sand at desired values of u/B and h/B (h = center-to-center spacing of the geogrid layers as shown in Fig. 1). After completion of the sand placement, the model foundation was placed on the surface of the sand to start the model test. For all model tests under this program, $u/B = h/B$ was kept at 1/3. Based on static bearing capacity tests conducted under this program, it was determined that

$(b/B)_{cr} \approx 4$ & $(d/B)_{cr} \approx 1.32$
Ultimate bearing capacity on unreinforced sand, $q_u = 65$ kN/m^2
Ultimate bearing capacity on geogrid-reinforced sand, q_{uR}
[with $u/B = h/B = 1/3$; N = 4, $b/B = (b/B)_{cr} = 4$;
and $d/B = (d/B)_{cr} = 1.32$] = 175 kN/m^2

The details of the preceding test results are provided elsewhere (Omar *et al.*, 1992) and will not be repeated here. For conducting the cyclic load tests in the presence of geogrid reinforcement, the following constant parameters were adopted:
$u/B = h/B = 0.333$; $b/B \approx (b/B)_{cr} = 4$; $d/B \approx (d/B)_{cr} = 1.32$; and average relative density of sand, Dr = 70 %. A universal testing machine was used for the application of load on the model foundation. The magnitude of the load and the foundation settlement were recorded by a data acquisition system. The following are the steps for conducting a test.
Step 1: Application of a static load, q_s, per unit area on the foundation. The static load per unit area can be expressed as

$$q_s = \frac{q_{uR}}{FS} \tag{2}$$

Where q_{uR} is the ultimate static bearing capacity on reinforced sand, and FS is a factor of safety.
Step 2: Application of a cyclic load of intensity q_d superimposed on the static load. The period T of the cyclic load for all tests was 1.0s. Thus the variation of the intensity of total load (i.e. $q_{total} = q_s + q_d$) on the model foundation was of the nature as shown in Figure 2(a). Figure 2(b) shows the expected nature of the variation of foundation settlement with time. A total of seven cyclic loading tests were conducted in the laboratory. The details of the tests are given in Table 3.

TEST RESULTS AND DESCUSSION

As shown in Fig. 2(b), the settlement due to cyclic load only is given by the notation s_d. This is the permanent settlement which occurs after the settlement due to static loading (q_s) is complete. Figure 3 shows the variation of s_d/B with the number of load cycle (n) application (log scale) for test nos 1-4. These are tests in which the critical geogrid reinforcement was in place. As can be seen from these plots, for any given value of q_d/q_{uR}, the magnitude of s_d/B increases with n up to a point marked b (Figure 3), beyond which the slope of $\Delta(s_d/B)/\Delta(\log n)$ is zero or very small. In this study, the permanent settlement at b is defined as the ultimate settlement, s_{du}, due to cyclic loading and the corresponding n as n_{cr}. Based on the results shown in Figure 3, the following general conclusions can be drawn.

(1) For a given value of q_s and the number of load cycle application, the magnitude of s_d/B increases with the increase of q_d.
(2) The magnitude of n_{cr} for all tests is approximately the same, varying between 1.75×10^5 and 2.5×10^5 cycles.

Figure 4 shows the variation of s_d/B with n for test nos 5-7. These were tests in sand without reinforcement. However, the magnitude of q_s was kept the same as that in test nos 1-4. the magnitudes of q_d in test nos 5-7 were, respectively, the same as those in test nos 1, 2 and 3. This was done primarily to compare the magnitudes of s_{du}/B with and without geogrid reinforcement under the same values of q_s and q_d. It can be seen from Figure 4 that the general nature of the variation of s_d/B with n for any given value of q_d/q_{uR} is similar to that observed in Figure 3. The magnitude of n_{cr} varied between 1.5 x 10^5 and 1.75 x 10^5 cycles. The magnitudes of s_{du} /B for all tests shown in Figures 3 and 4 have been plotted in Figure 5 against their corresponding values of q_d/q_{uR} for the tests conducted with and without geogrid reinforcements. Based on this, the following general conclusions can be drawn:

(1) The value of s_{du} is substantially larger in unreinforced soil as compared to that in reinforced soil.
(2) The ratio of $s_{du\ (unreinforced)}$ / $s_{du\ (reinforced)}$ decreases with increase in q_d.

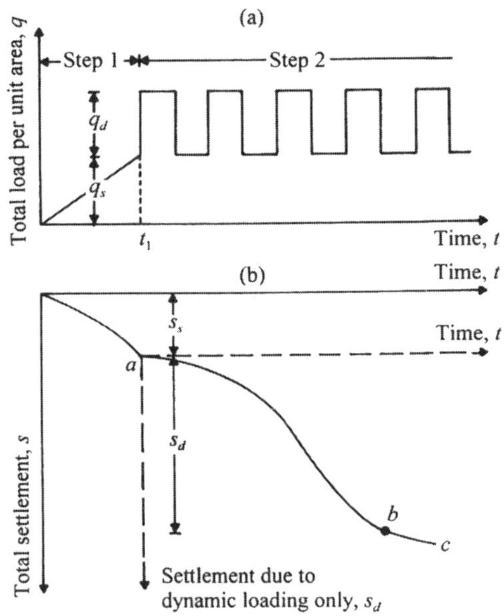

Figure 2. Variation of load intensity and foundation settlement with time

Table 3. Details of Cyclic Load Tests

Test no.	Presence of reinforcement	Reinforcement details	q_s (kN/m^2)	q_{uR}/q_s	q_d (kN/m^2)	q_d/q_{uR}
1	Yes	N = 4	58.3	3	7.63	4.36
2	Yes	u/B = 0.33	58.3	3	18.67	10.67
3	Yes	h/B = 0.33	58.3	3	25.36	14.49
4	Yes	$b/B = 4 = (b/B)_{cr}$ $d/B = 4 = (d/B)_{cr}$	58.3	3	39.1	22.34
5	No	-	58.3	3	7.63	4.36
6	No	-	58.3	3	18.67	10.67
7	No	-	58.3	3	25.36	14.49

CONCLUSIONS

A number of laboratory model tests were conducted on a square surface foundation supported by sand with and without critical geogrid reinforcement. The foundation was subjected to an initial allowable static load. A cyclic load was then superimposed on the static load, and the permanent settlement of the foundation during cyclic loading was monitored. Based on the model test results, the following general conclusions can be drawn:

(1) For a given static load (qs) and number of load cycles (n) , the magnitude of sd increases with the increase of magnitude of qd.
(2) The critical number of cyclic loading (ncr) required for ultimate Permanent settlement (sdu) is approximately the same for reinforced and unreinforced soil.
(3) For given values of q_s and q_d, the magnitude of s_{du} is substantially higher in unreinforced soil as compared to reinforced soil. However, the ratio of $s_{du\ (unreinforced)} / s_{du\ (reinforced)}$ decreases with increase in q_d.

ACKNOWLEDGEMENTS

The author wish to thank the Research and Studies Center at the University of Sharjah for providing the research funding. This research was supported by seed grant from the University of Sharjah (Seed Grant # 04-01-03)

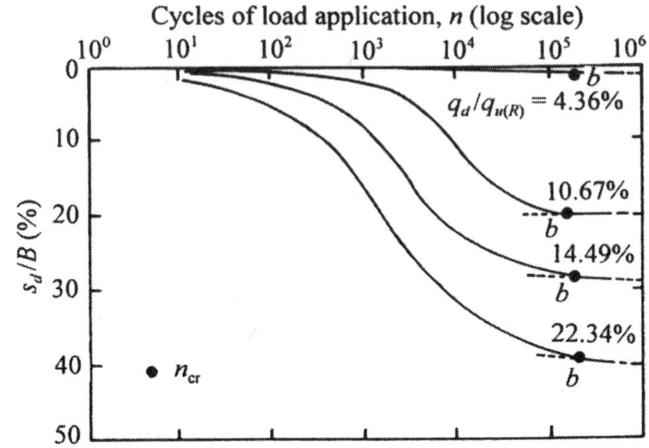

Figure 3. Variation of s_d/B with the number of load cycles (test nos 1-4)

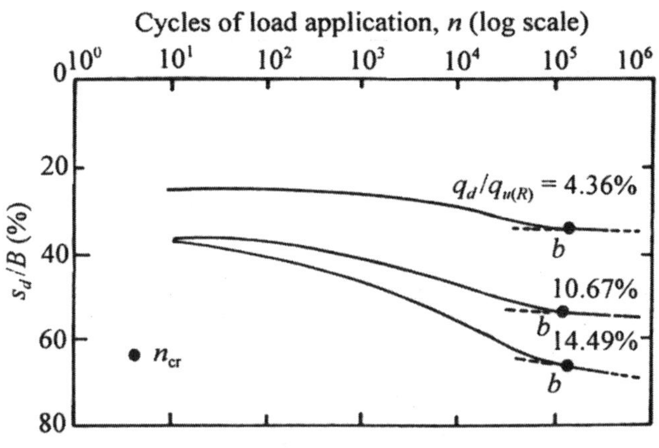

Figure 4. Variation of s_d/B with the number of load cycles (test nos 5-7)

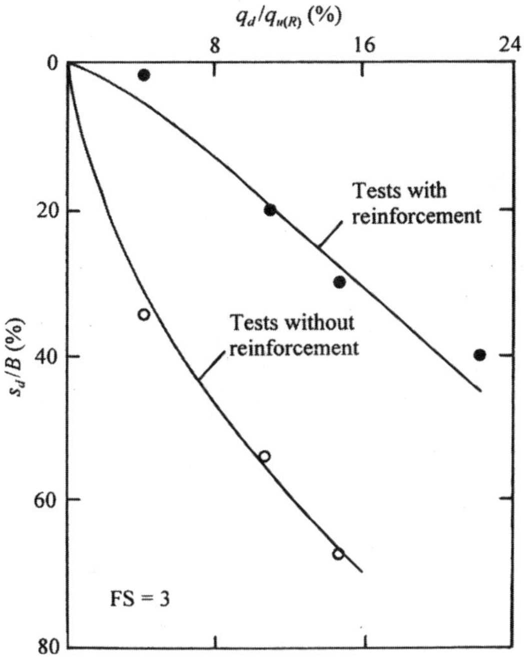

Figure 5. Variation of s_d/B with $q_d/q_{u(R)}$

REFERENCES

Binquet, J & Lee, K.L.(1975). Bearing capacity analysis of reinforced earth slabs. *J.* Geotech. Engng Div., ASCE, 101, 1257-76

Guido, V.A., Change, D.K & Sweeney, M.A, (1986). Comparison of geogrid and geotextile reinforced slabs. *Can. Geotech. J.*, 23,435-40.

Omar, M.T., Das, B.M., Puri, V.K., Yen, S. C. & Wright, M.A. (1992). A Comparison of the ultimate bearing capacity of wquare and strip foundations on geogrid-reinforced sand. In *proceedings of the 4th international Conference on Numerical Models in Geomechanics*. Balkama, Rotterdam, pp. 967-76.

International Conference on Geotechnical Engineering for Disaster Mitigation & Rehabilitation
Chu, Phoon & Yong (eds) © 2005 World Scientific Publishing Company ISBN 981-256-469-1

GEOSYNTHETICS IN THE REHABILITATION OF FAILED SLOPES

T.A. OOI
TAO Consult Sdn Bhd, 17A, Jalan Awan Hijau, Taman Overseas Union, Batu 5, Off Jalan Klang Lama, 58200 Kuala Lumpur, Malaysia

C. H. TEE
Mega Geoproducts And Services Sdn Bhd, 40-1, Jalan 2/109E, Desa Business Park, Taman Desa, Off Jalan Klang Lama, 58100 Kuala Lumpur, Malaysia

M. J. D. DOBIE
Tensar International Limited, Wisma Sejahtera, Room 407, Jalan Letjen S Parma, Kav 75, 11410 Jakarta, Indonesia

This paper discusses the use of geogrids in the rehabilitation of failed slopes in restricted sites. Failed slopes are often associated with prolonged intense rainfall and lack of slope internal drainage. The consequences of a landslide in the case of a highway or railway can be very grave. The instant reinstatement of the same is almost demanded by the general public. In one particular failed slope measuring about 36m high, sheet piles were used as a temporary measure at the slope toe to stop the movement of the sliding mass. However, with sheet piling, the flow of groundwater was cut off and pore water pressure built up. This pore water pressure built-up caused the soil strength to decrease and the sheet piles began to rotate and tilt towards the railway lines. This steady movement of the sheet piles posed an immediate danger to the safe operation of the railway and daily movements of the goods train. Geogrids were used to mitigate this continuing ground movement and prevent further slope failure. The use of geogrids in the speedy rehabilitation works has been well established. However, the methodology in the application of the right technology is still not well understood, particularly for very high slopes. This paper reports case histories of landslide occurrences and the solutions successfully applied.

INTRODUCTION

The Tsunami that struck the Indian Ocean on 26 December was a global catastrophe that caught many in the world by surprise. The most devastated area happened at Bandar Acheh where the town was completely demolished. The coastal areas of the Indian Ocean were hit by Tsunami to varying degree. The world all over started the massive relief operation by way of sending aids and donations. The whole world was shocked when faced with such unexpected nature's calamities and steps are being taken to monitor such oceanic ground movements that triggered tsunami with devastating power so as to mitigate the damages and loss of lives. Landslide is one of nature's calamities that occur on land when there is excessive amount of water in the soil slopes. The excessive amount of water in soil slope occurs usually during the incessant period of rain. This paper reports case histories of slope failures rehabilitated by using Tensar Geogrids.

MOTEL DESA SLOPE FAILURE

In early 1980s, part of the Motel Desa slope in Terengganu collapsed during the heavy monsoon rainfall period and left the kitchen of the motel hanging precariously on top of the failed slope. The failed slope was about 30m high and 25m wide. It was important to ensure that the kitchen did not collapse. The access to site was difficult and speedy rehabilitation of the slope was necessary. The technology of geogrid reinforcement of slope was new at that time in Malaysia and careful instrumentation and testing was carried out during the construction stage. In view of the difficult access and the urgency to safeguard the motel kitchen, the upper half of the failed slope was stabilized by using a combined technology of geogrid reinforced soil slope of about 5m wide x 6.7m high supported on micro-piled platform while the lower half failed slope was reconstructed using traditional method of compacted fill without any soil reinforcement but to a gentle slope of not more than 1:2. Beyond the geogrid reinforced slope at the top similar un-reinforced soil slope of not more than 1:2 was also adopted. Fig. 1 shows the general cross section of the rehabilitation scheme. The details of this landslide rehabilitation scheme were reported and discussed by Toh et al (1986) and Ooi & Tee (2004). The Motel Desa project has given designers an opportunity to make further improvement to the design of geogrid reinforced slope through in-situ monitoring of actual completed works. The rehabilitation work was completed more than 20 years ago and the slope has performed satisfactorily and remained stable.

RAILWAY LINE SLOPE FAILURE

A railway line cut slope in the east coast state of Malaysia measuring some 36m in height failed during a severe monsoon rain season. The failure threatened the safety of the railway line as the debris encroach the railway side drain and sub-ballast. Immediate action was taken to put a team of workers with plant and machinery on standby round the clock to remove any debris that come close to 4m from the centerline of the railway tracks. A speed restriction was imposed to enable the operator of the train to bring the train to an immediate stop upon sighting of any sign of further slope movement and at the same time a row of steel sheet piles was driven to refusal as an attempt to stop the flow of the debris and prevent the movement of the failed slope. Fig. 2 shows a general view of the landslide and the steel sheet piles at the slope toe near the railway line. It can be seen that the hill slope is too close for comfort to the railway line. An analysis shows that the landslide that had occurred was due to toe softening of the slope. The results of analysis show slope toe failure with factor of safety of 1.03. The provision of sheet piles driven to refusal however could not provide the embedment depth required and at the same time cut off the flow of ground water and thus caused built up of water pressure. Consequently, with accumulation of water and the loss of soil strength due to soaking of the slope toe, the sheet pile rotated and tilted forward towards the railway line. Fig. 3 shows the picture of tilted steel sheet piles and the railway line. Slope stability analysis show that the steel sheet piles are within the localized sliding circle with factor of safety of less than unity. In view of this critical situation, it was necessary to carry out urgent repair before the arrival of the next monsoon. Tensar geogrid reinforced block of 8m high was chosen as the solution to stabilize the toe area of the slope and at the same time provide support for the 36m high failed slope. Fig.4 shows the results of the slope stability analysis. In the rehabilitation of this type of high failed slope, it is very important that the correct technology with the right construction methodology is used so as to ensure successful implementation of the reconstruction of the slope. The provision of internal and external drainage system of the slope and the geogrid block remains as an important factor to be considered in design and construction of the rehabilitation works.

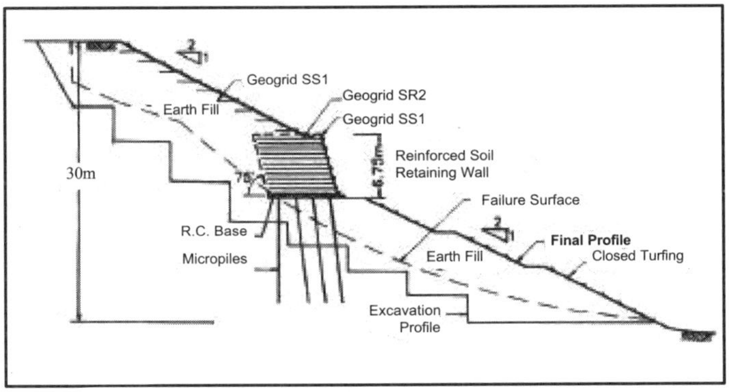

Figure 1. Model Desa slope failure rehabilitation (after Toh et al., 1986).

Figure 2. General view of landslide.

Figure 3. View of tilted sheet piles and passing train.

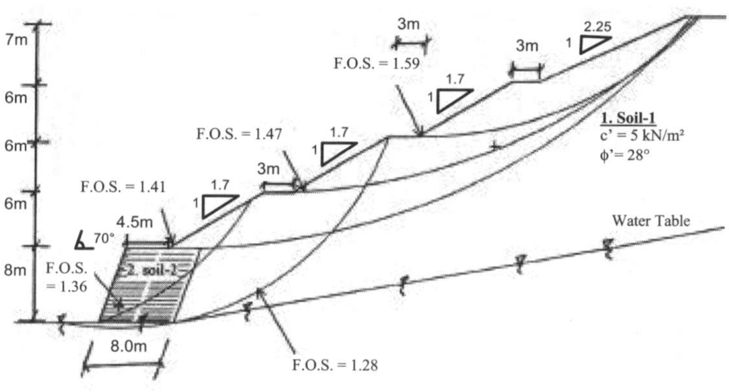

Figure 4. Result of slope stability analysis with Tensar Block.

DESIGN METHODOLOGY

The design analysis must first start with the analysis of failure mechanism based on site information and relevant soil parameters. Having established the correct failure mechanism, it is then logical to proceed to find a solution that is most economic based on the local conditions and situations. Many designers often failed to consider the effect of ground water table on the stability of slope. This can give rise to wrong solution being prescribed and resulting in distress and/or failure of the slope when the rehabilitation work is completed or during service.

CONSTRUCTION METHODOLOGY

In the rehabilitation work, the landslide areas were divided into 3 sections namely, section A, B and C as shown in Fig.5. It is important to rehabilitate the slope in sections and to ensure failure of slope does not happen during construction. The site was first drained and the top 1m of the foundation soil for the base of the geogrid block was re-compacted with two layers of geogrid reinforcement to receive the geogrid block to be constructed. Suitable drainage layer is then installed with sand columns placed at the back slope and compacted in layers with geogrid reinforcement. The geogrid block is built to a safe height before excavation of the next section is started. This second section is similarly built to a safe height before the third section is excavated. The process is repeated until all the geogrid reinforcement blocks are completed. Surface drainage system of the slope is constructed as soon as it is possible and the completed slope is covered with closed turf. Figs.6 shows the completed slope.

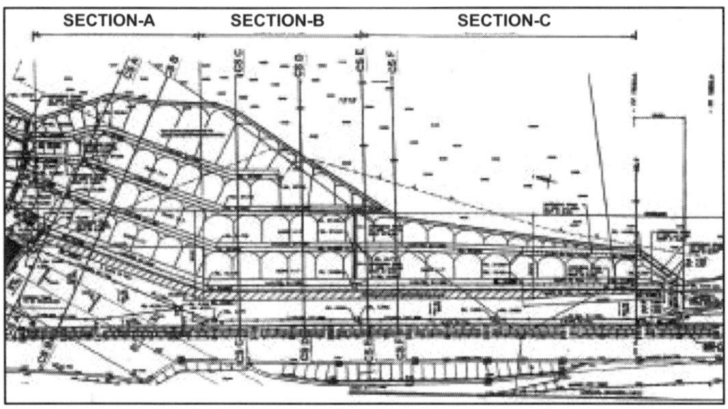

Figure 5. Construction Methodology of slope rehabilitation.

Figure 6. Closed up of completed Tensar Geogrid Slope.

CONCLUSIONS

From the case histories presented by Toh et al (1986) and Ooi & Tee (2004) as well as the railway project it can be concluded that:

1. Slope failures usually occurred during incessant raining season like that of the monsoon period.
2. Different solutions are used in different case histories even though the Geosynthetics of geogrid type were used, they were applied differently.
3. The need for foundation soil improvement was found necessary to support the geogrid reinforced block in order to reduce settlement and movement of slope during and after construction of the slope.
4. It is important to provide internal and external drainage system to ensure that groundwater table is lowered sufficiently to ensure long-term stability of slope.
5. Slope rehabilitation is often required when slope collapses threaten lives and or properties. We need to ensure the slopes are designed to withstand long-term slope stability for such cases and not wait for it to fail before something is done.
6. Combined technology is useful to provide economic solution to slope rehabilitation works.

REFERENCES

Ooi, T.A. and Tee, C.H. (2004). "Reconstruction of failed slopes using Geosynthetic Reinforcement in Malaysia", *Proceedings Malaysia Geotechnical Conference, MGC 2004*, Kuala Lumpur, P435-447.

Toh, C.T., Chee, S.K. and Ting, W.H. (1986). "Design, construction and performance of a geogrid reinforced high slope and unreinforced fill slopes", *Proceedings Joint Symposium on Geotechnical Problems*, Kuala Lumpur, P.90-111.

International Conference on Geotechnical Engineering for Disaster Mitigation & Rehabilitation
Chu, Phoon & Yong (eds) © 2005 World Scientific Publishing Company ISBN 981-256-469-1

SEISMIC LIQUEFACTION AND REHABILITATION OF LARGE EARTH DAMS IN U.S.

M. K. SHARP
U.S. Army Corps of Engineers, Engineer Research and Development Center, MS, USA

K. ADALIER
Department of Civil and Environmental Engineering, Florida State University, Panama City, FL, USA

Earthquake induced liquefaction is a major concern for embankment dam safety. The U.S. Army Corps of Engineers (USACE) has responsibility for the safe performance of over two hundred dams in high seismic hazard areas of the U.S. The USACE periodically evaluate the safety of these dams against expected seismic events in the light of new hydrologic or seismic data or changes in the state-of-the-art design or construction criteria. Those dams that are found deficient are retrofitted through appropriately designed remediation methodologies. One of the major seismic safety problems for the dams is liquefiable foundation deposits. This paper is focused on the USACE experience in dynamic stability evaluation and remediation of earth dams with liquefiable foundations. Several major earth dam remediation case histories are discussed. Furthermore, the current state-of-the-art in retrofit ground improvement for liquefiable dam soil deposits is provided.

INTRODUCTION

The USACE has responsibility for the safe performance of over two hundred dams in high seismic hazard areas of the U.S. The USACE periodically evaluate the safety of these dams against expected seismic events in the light of new hydrologic or seismic data or changes in the state-of-the-art design or construction criteria. Those dams that are found deficient are retrofitted through appropriately designed remediation methodologies. One of the major seismic safety problems for the dams is liquefiable foundation deposits. This paper is focused on the USACE experience in dynamic stability evaluation and remediation of earth dams with liquefiable foundations. Case histories on two major USACE earth dam seismic rehabilitation projects are provided. Discussions are given on the dynamic analysis of these dams and selected remediation alternatives.

USACE SEISMIC EVALUATION PROCEDURE FOR EARTH DAMS

USACE publication "Earthquake Design and Evaluation of Civil Works", designated as Engineer Regulation 1110-2-1806 (HQUSACE, 1995) provides guidance and direction for the seismic design and evaluation of all civil works projects. This document is a regulation stating that projects which retain or have the potential to retain a pool must survive and remain safe during and following the maximum credible earthquake. The policy also states that the project must be capable of remaining operational with only minor repair during and after an operating basis earthquake. It also states that re-evaluations of the project should be conducted every third periodic inspection or every 15 years, whichever comes first.

USACE publication “Dam Safety Assurance Program”, designated as Engineer Regulation 1110-2-1155 (HQUSACE, 1997) establishes the policy on seismic criteria used to make decisions on the merits of dam safety modifications related to current earthquake design criteria. The seismic safety evaluation process is specified as being (a) Seismic Safety Review (SSR), (b) Phase I Special Studies, and (c) Phase II Special Studies. The seismic safety evaluation process is designed to increase in complexity of analyses as warranted by increasing steps in the process. Therefore, the SSR would consists of simple ‘desk’ evaluations of existing data while the Phase I and Phase II studies would consists of data collection and analyses utilizing one and two dimensional methodologies and possibly progressing to three dimensional analyses. The evaluation process is intended to lead to the conclusion that the project will function in an acceptable manner during the design earthquake, or to the conclusion that modifications are required to the embankment or its foundation to ensure acceptable performance when subjected to the design earthquake.

Currently there is no USACE publication related to guidance and direction for deformation analyses of earth dams as a consequence of seismic loading. Each individual district or division of the USACE, that has responsibility for a dam, is also responsible to insure that the deformation analyses are consistent with current state of the practice. A publication related to deformation analyses will be forthcoming.

EARTH DAM FOUNDATION LIQUEFACTION REMEDIATION

The design challenge of remedial construction for safe seismic performance of embankment dams built on liquefiable foundation soils is to safely limit deformations to acceptable levels. This can be done by modifying the soil conditions to prevent liquefaction, or by letting the soil liquefy but introducing new strong elements that provide the needed resistance to deformations, or modify the geometry of the existing dam to accommodate expected deformations without catastrophic failure, or combinations of these methods. The current practice of seismic remediation of embankment dams with liquefiable soils includes the following (the engineering remedies only):

Excavate and Replace: This technique involves removal of the unsatisfactory soil and replacing it with a material more resistant to liquefaction. The major disadvantage of this method is that it is economically limited to surface soils. However, it has a major advantage over other methods in that the results are assured and are easily verifiable.

In-Situ Densification: This technique decreases the liquefaction potential by decreasing the void ratio. Soil densification is achieved by the use of vibratory probes, deep dynamic compaction, and compaction grouting. The major concern for this method is the likelihood of settlement and resultant cracking of the existing structure. Hence in many cases the densification is applied to the dam foundations after the sections of the dam was temporarily removed. The current practice is to treat the entire liquefiable soil depth. However, this may not always be technically and economically feasible. In such cases limited-depth densification treatment may be an alternative (Adalier and Elgamal, 2002; Adalier and Sharp, 2004).

Drainage: This technique provides relief of seismically induced pore water pressure (e.g., stone columns; Adalier and Elgamal, 2004). The main disadvantage is that these drains may increase seepage under static pool conditions, increasing the piping potential, and consequently reducing the static safety.

In-Situ Strengthening: This technique involves forming a composite material (soil) that is strong enough to ensure overall stability. The techniques include but are not limited to soil nailing, concrete or steel columns, deep-soil-mixing, and sheet pile walls.

Increase Freeboard: Usually adopted for marginally stable dams and/or small earthquake cases, this technique decreases the possibility of flooding due to settlement or slumping of the crest during an earthquake.

Berm/Buttress: Building upstream and/or downstream berms and buttresses will increase effective confining stresses in the dam and foundation soils; thus reducing the liquefaction potential. Furthermore, these additions will increase length of failure surface and provide a counterweight and enhance dynamic slope stability and integrity during earthquake shaking. Comprehensive discussions of the abovementioned countermeasure techniques can be found in Adalier (1996).

REMEDIATED EARTH DAM CASE HISTORIES

This section of the paper will provide a brief overview on two major USACE earth dam seismic evaluation and liquefaction retrofit projects; Mormon Island Auxiliary Dam and Sardis Dam. Analyses of the dams and the subsequent decisions to remediate are a consequence of the dam safety evaluation process as described previously. USACE personnel typically perform the analyses in-house, with contracted support as necessary.

Mormon Island Auxiliary Dam (California)

The dam was constructed (in early fifties) across a channel where gravely alluvium of about 20 m in thickness fills the channel for about 500 m of its width. These gravels were dredged for their gold content, resulting in a very loose fine materials near the base and coarser materials near the top of the channel. The dam is a zoned embankment 1,469 m long and 50 m high from core trench to crest at the maximum section. The narrow, central impervious core is a well-compacted clayey mixture founded directly on rock. Dam shells were constructed of dredged tailing gravels (slopes of 1:2).

Seismological studies predicted a design earthquake with 0.35g peak acceleration, 20 cm/s peak velocity, and 16 s bracketed duration. Seismic studies were conducted using laboratory investigations, in-situ testing, and equivalent linear dynamic stress analysis (Wahl *et al.*, 1988). These studies predicted an extensive liquefaction in the dredged gravel foundation over a 244 m long section of the dam. Analysis of post-liquefaction deformation for this project was not based on a factor of safety (FS) approach. Rather, engineers on this project expressed concerns that a FS can have different connotations depending on the dam geometry and the extent and location of liquefied zones. However, the displacements can be interpreted in light of dam-specific criteria about the allowable potential loss of freeboard or the tolerable extent of potential horizontal deformation. Sounder and more cost-effective decisions can be made based on both FS and deformation data than on FS alone. The evaluation and development of remedial measures were based on allowing a tolerable amount of deformation for the design earthquake (Ledbetter and Finn, 1993). The potential post-liquefaction deformations before and after remediation were estimated using the computer program, TARA-3FL. Deformation analysis provided guidance for the most efficient and economical remedial treatment (Ledbetter *et al.,* 1991), which included foundation treatment zones and an upstream berm (Fig. 1). Dynamic compaction was chosen as the primary method for treatment of the upstream foundation with compaction grouting as a supplementary method. The downstream foundation treatment used stone columns and a vertical gravel curtain drain for both excess pore pressure control and densification. The design allowed for earthquake induced liquefaction and strength loss in the dam and foundation material, but provided containment and control of the liquefied material such that deformations of the dam would be acceptable.

Sardis Dam (Mississippi)

Sardis Dam is a 4570 m long embankment dam with a maximum height of 36 m. The central portion of the dam, located in a river floodplain is 2590 m long. This section was constructed by hydraulic fill methods in the late 1930's and consists of a predominantly silt core surrounded by sand shells. The foundation consists of a 3 to 6 m thick zone of clay and silt (top stratum clay), which extends approximately 370 m upstream from the dam centreline. This top stratum clay is underlain by 12 m thick pervious alluvial sands and that underlain by Tertiary silts and clays. During dam construction, the top stratum clay was removed from the downstream portion to help control under-seepage.

Seismological studies predicted a design earthquake with 0.20g peak acceleration, 35-40 cm/s peak velocity, and bracketed duration of 15 s. Seismic investigations were

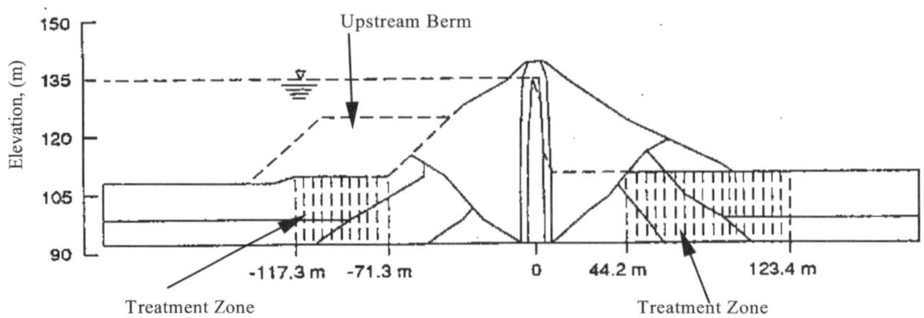

Figure 1. Mormon Island Auxiliary Dam remediation design.

conducted in two separate phases. In the first phase (Finn *et al.,* 1991) the analysis procedures included; establishing the maximum credible earthquake, determining the liquefiable soils from in-situ penetration tests, determining the residual strength of liquefied soils, performing post liquefaction slope stability analysis and estimation of deformations using rigid block analysis, and determining required remediation based on the FS. The analyses concluded that the downstream stability of the dam was adequate but that there was weak silty clay layer extending upstream with post-earthquake FS of less than one. A broad range of viable alternatives was evaluated to determine the most economical and verifiable remedial treatment, and a remedial measure of constructing a 93 m wide and 305 m long berm on the upstream slope was chosen.

During and subsequent to the design and construction of the upstream berm, studies were expanded to determine the areal extent of the weak silty clay layers beneath the remainder of the dam and to define other areas possibly requiring remediation. For this phase of the investigation, a deformation approach was adopted to clarify and supplement the slope stability FS approach. Deformations were calculated using the program TARA-3FL. The analyses indicated that large deformations of the upstream slope and dam crest could occur resulting in possible loss of the reservoir pool in areas of the foundation containing the weak silty clay. Remediation was required along an additional 823 m of the dam where the weak silty clay was discovered. The upstream berm was appropriate where it was used because the upstream extent of the weak clay layer was limited to about 116 m and does not underlie the upstream half of the berm. However, the layer extends much further upstream in other areas. Various remedial designs were evaluated and improving the shearing resistance of the weak silty clay layer was chosen. Among various methods evaluated, using driven pre-stressed concrete piles were determined to be the most feasible, economical and reliable technique to achieve this purpose. The principle of the pile reinforced zone is to provide a high shear strength plug that will transfer the load applied by the movement of the upstream dam mass from the weak silty clay through the piles into the dense substratum sands. A total of 25,940 pre-stressed concrete, 0.6 m square, heavily reinforced piles, were driven in the upstream toe area of the dam.

CONCLUSIONS

Increasingly the focus of dam engineering in the U.S. is related to seismic hazard evaluation and possible remediation methodologies. This owing in large part to the age of current dams and their lack of design for seismic stability during their construction. In spite of continual major advances, there remain serious gaps in our knowledge in the areas of earthquake hazard estimation and in the most appropriate remediation methodologies for the various dam/foundation scenarios. Although, the non-linear finite element codes have good potential for the betterment of hazard estimation and remedial designs, there still remains a significant lack of validation and verification of these routines. In the absence of well instrumented in-situ case histories, centrifuge model tests remain to offer the best sources of database for such validations.

REFERENCES

Adalier, K. (1996). "*Mitigation of earthquake induced liquefaction hazards*", Ph.D. Thesis, Dept. of Civil Engineering, Rensselaer Polytechnic Institute, Troy-NY, 659pp.

Adalier, K. and Elgamal, A. (2002). "Seismic response of adjacent saturated dense and loose sand columns," *Soil Dynamics and Earthquake Engineering*, 22(2), 115-127.

Adalier, K. and Elgamal, A. (2004). "Mitigation of liquefaction and associated ground deformations by stone columns", *J. Engineering Geology*, Vol. 72(3-4), 275-291.

Adalier, K. and Sharp, M.K. (2004). "Embankment dam on liquefiable foundation–dynamic behavior and densification remediation. *ASCE J. Geotechnical and Geoenvironmental Engineering*, 130(11), 1214-1224.

Finn, L., Ledbetter, R.H., Fleming, R., Templeton, A, Forrest, T.W. and Stacy, S.T. (1991). "Dam on liquefiable foundation–safety assessment and remediation," *Proc., 17th International Congress on Large Dams*, Vienna, 531-553.

HQUSACE (1995). "*Earthquake Design and Evaluation of Civil Works",* Engineer Regulation 1110-2-1806, Headquarters U.S. Army Corps of Engineers.

HQUSACE (1997). "*Dam Safety Assurance Program*," Engineer Regulation 1110-2-1155, Headquarters U.S. Army Corps of Engineers.

Ledbetter, R.H., Finn, L., Nickell, J.S., Wahl, R.E., Hynes, M.E. (1991). "Liquefaction induced behavior and remediation for Mormon island auxiliary dam", *International Workshop on Remedial Treatment of Liquefiable Soils*, Tsukuba, Japan, 215-233.

Ledbetter, R.H., Finn, L. (1993). "Development and evaluation of remediation strategies by deformation analysis", *ASCE Specialty Conference on Geotechnical Practice in Dam Rehabilitation*, Raleigh, North Carolina, 386-401.

Wahl, R.E., Crawforth, S.G., Hynes, M.E., Comes, G.D. and Yule, D.E. (1988). "*Seismic stability evaluation of Folsom dam and reservoir project*", U.S. Army Engineer Waterways Experiment Station, Report GL-87-14-No.8, Vicksburg, Mississippi.

International Conference on Geotechnical Engineering for Disaster Mitigation & Rehabilitation
Chu, Phoon & Yong (eds) © 2005 World Scientific Publishing Company ISBN 981-256-469-1

CONSIDERATION OF OVERSURCHARGE EFFECTS ON SOFT CLAY WITH VERTICAL DRAINS

YANG WENYI, KONG GUANGYA
Tiandi Science and Technology Co. Ltd, China

JEFFREY WANG
Tritech Consultants Pte Ltd, Singapore

The effects of over surcharge on soft clay have been assessed in a term of "equivalent time". Based on the assumption of the uniform relationship of clay in effective stress-strain-time space (Bjerrum et al., 1967), the deformation of soft clay caused by the process of applying and removing overcharge is assessed by the additional elapsed time under the load without overcharge. Based on the results of theoretical analyses and laboratory tests, a simplified method of assessment of over surcharge effects of soft clay is proposed.

GENERAL

The main considerations involved in the design of over surcharge with vertical drains are the arrangement of vertical drains and procedures of over surcharges. The procedure of over surcharges is governed by the requirements of soil strength and settlement rate after treatment. The criteria of settlement rate might be a critical issue due to the possibility of mobilizing negative skin friction when piled foundation system is selected.

A problem might be encountered in the design is to predict the consolidation state of soft clay after the removal of over surcharge. In most cases, the backfilling material is stacked on top of soft clay and sand blanket stage by stage. Once one layer of the backfilling material is loaded on the sand blanket, soft clay is subjected to the proccss of consolidation under current weight of backfilling material. The next layer is then loaded prior to the end of primary consolidation under the previous loading. This process is repeated until the over surcharge is completely loaded.

Terzaghi's consolidation theory is traditional method to calculate the degree of consolidation under multiple stages of loading. The primary consolidation of soft clay under the over surcharge may not be completed prior to the removal of the over surcharge. In this case, soft clay is still subjected to further consolidation under current reduced surcharge. As Terzaghi's consolidation theory applies to constant load, the process of applying-and-removing over surcharge normally brings difficulties for the calculation of consolidation of soft clay. The concept of "equivalent time" in effective stress-strain-time space is proposed by the author to simplify the calculation of consolidation under the process of applying-and-removing over surcharge.

CONCEPT OF EQUIVALENT TIME

The deformation properties of soft clay under various loading conditions have been widely discussed since the establishment of Terzaghi's one-dimensional consolidation theory in 1925. The soft clay was found exhibiting time–independent and time-dependent deformation. The whole process of deformation can be defined as primary and secondary stages (Fig.1). Terzaghi's theory is now recognized only to reflect the deformation at primary stage rather than secondary stage.

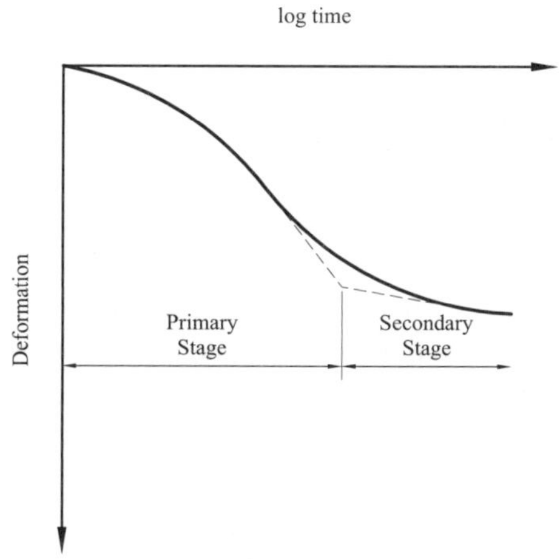

Fig. 1 Typical deformation curve of soft clay

The secondary deformation of soft clay is time-dependent and is found to be significant in the case of very soft clay. A series of studies on the secondary deformation have been carried out and a number of models proposed to describe the primary and secondary deformation of soft clay. The secondary deformation of clay under constant load is basically found to be a straight line in strain- logarithm time coordinates.

Bjerrum et al. (1967) have proposed a comprehensive model to describe the deformation properties of soft clay with assumption that void ratio of clay is function of effective stress and time. As shown in Fig. 2, the relationship of the effective stress, strain and time is represented by a set of paralleled straight lines. It is reasonably assumed that soft clay follows a uniform relationship in the effective stress-strain-time space, i.e.

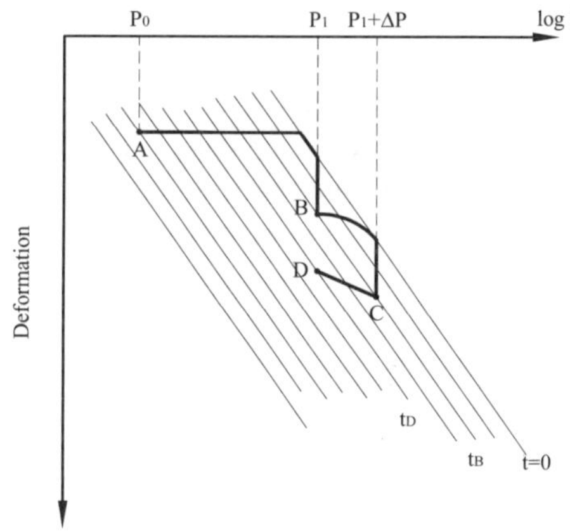

Fig. 2 Analysis of oversurcharge effects in stress-strain-time space

$$R(\sigma', \varepsilon, t) = 0 \tag{1}$$

where σ' is effective stress of clay;
ε is strain of clay;
t is elapsed time

The effect of over surcharge can be assessed based on Bjerrum's assumption. As shown in Fig. 2, the initial state of clay after experience of deformation under load P_0 to P_1, the point in the diagram representing the state of clay is moved from point A to point B after time t_B . Under this load, over surcharge ΔP is applied and then removed. The point of soil state is moved to point D instead of point B, i.e. soil is subjected to residual deformation (ε_D—ε_B).

Above process can be understood in following way: once the load is increased from P_0 to P_1, and the point of soil state in the diagram is moved from point A to point B, the deformation of clay is to be continued under the load P_1. With the additional elapsed time (t_B—t_D), the point of soil state is also moved to point D.

The above analysis of clay deformation clearly indicates that based on the assumption of uniform deformation curve of clay in effective stress-strain-time space, the residual deformation (ε_D—ε_B) due to the applying and removing over surcharge ΔP is equal to the effect of additional elapsed time (t_B—t_D) under load P_1. In other words, the effect of over surcharge can be assessed with additional elapsed time under previous loading (Fig. 3).

For this reason, it is not irrational to defined the residual deformation (ε_D—ε_B) as additional deformation for P_1 and corresponding time difference (t_B—t_D) as "equivalent time" to additional deformation for P_1 in effective stress-strain-time space. If secondary deformation is involved in the process of over surcharge, the additional deformation can be divided into two parts: i.e. "additional deformation in primary stage" and "additional deformation in secondary stage". The equivalent time could be defined as "equivalent time in primary stage" and "equivalent time in secondary stage".

If t_c is supposed to be the elapsed time of end of primary consolidation (EOP), and ε_c is the deformation of clay at t_c the equivalent time at primary stage and secondary stage can be calculated as follows:

(1) Equivalent time in primary stage

$$t = \frac{H^2}{C_V}(T_{vc} - T_{vb})$$

where

H is thickness of clayey layers;

C_v is consolidation coefficient;

T_{vc} is time factor corresponding to t_c ;

T_{vb} is time factor corresponding to t_B.

(2) Equivalent time in secondary stage

$$t = 10^{\left(\frac{\Delta_D - \Delta_C}{HC_{\alpha\varepsilon}} + \log t_C\right)}$$

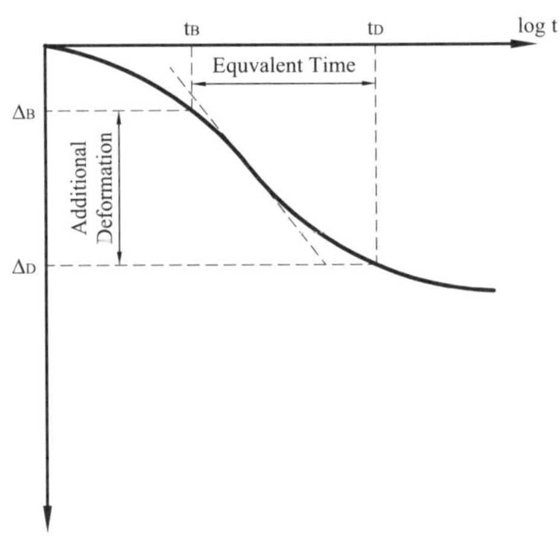

Fig. 3 Correlation between additional deformation and equivalent time

where

$\Delta_D - \Delta_C$ is additional deformation at secondary stage;

$C_{\alpha\varepsilon}$ is secondary consolidation coefficient

LABORATORY TEST RESULTS

A serious laboratory tests (total 19 number) were carried out using the standard odometers to verify above concept. The samples were selected from the same tube to minimize the discrepancy of soil properties. Two different loading procedure were adopted as shown in Fig.4. The results of tests are summarized in Table 1.

It could be seen from the results of tests that clay samples subjected to loading scheme A and B had two different types of consolidation curves (Fig.5). Coefficient of consolidation Cv was apparently influenced by the type of loading schemes, with roughly 1.8 to 1.7 times difference. Meanwhile, values of t_{90} also showed significant difference which indicates that the effect of ΔP. Under loading scheme B, the load increment ΔP caused additional deformation to the clay samples, hence equivalent to the additional lapsed time under the previous loading which reflected in the reduced of values of $t_{90.}$

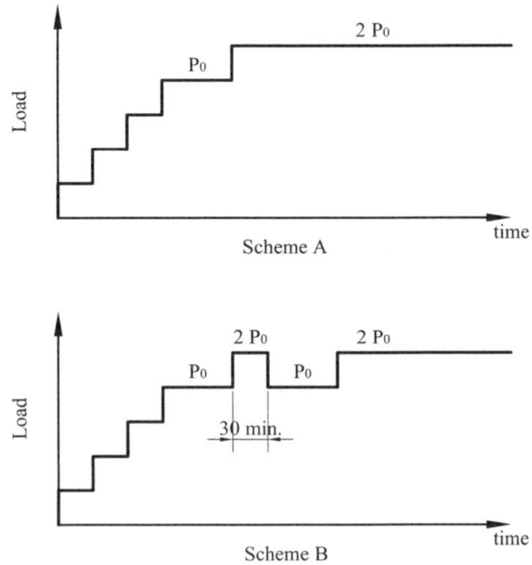

Fig. 4 Two loading schemes adopted in laboratory tests

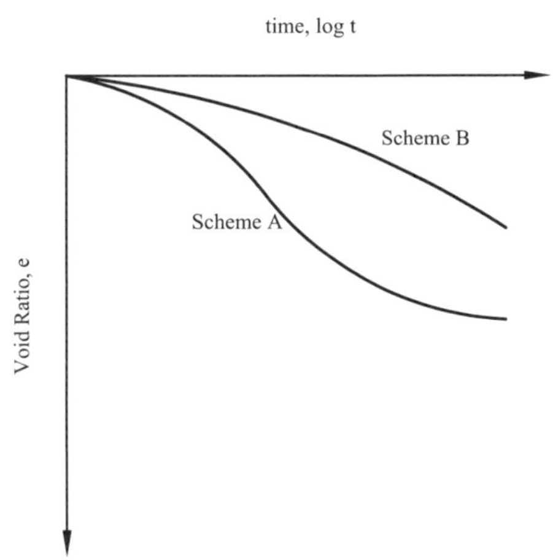

Fig. 5 Two types of deformation curves corresponding to loading schemes A and B

APPLICATION OF THE CONCEPT OF EQUIVALENT TIME

As discussed above, the effects of over surcharge can be measured as a term of equivalent time corresponding to additional deformation of clay under the over surcharge. The additional deformation can be estimated based on the magnitude of over surcharge and soil properties. The corresponding time can be calculated based on this additional deformation. The effects of over surcharge can be assessed according to calculated equivalent time:

(1) to calculate the final settlement Δf of clay layer under the load of surcharge at designed final backfilling level;

Table 1. Comparison of test results of undisturbed clay samples under two loading schemes.

Sampling Tube No.	Depth (m)	Coefficient of consolidation				C_{v2}/C_{v1}
		Scheme A		Scheme B		
		No. of Tested samples	C_{v1} ($\times 10^3$) (cm^2/sec.)	No. of tested samples	C_{v2} ($\times 10^3$) (cm^2/sec.)	
1	7.6	2	3.146	2	7.543	2.4
2	15.4	2	1.403	2	5.088	1.8
3	19.6	2	2.122	2	16.43	7.7
4	24.7	4	6.124	1	27.09	4.4
5	28.1	2	13.63	-	-	-

(2) to calculate the settlement Δp of clay layer prior to the application of over surcharge under the load of surcharge;

(3) to calculate the settlement Δs of clay layer due to over surcharge after the removal of over surcharge;

(4) to compare Δf and ($\Delta p+\Delta s$). if ($\Delta p+\Delta s$) is equal or large than Δf , the primary settlement of clay is considered completed and both of equivalent time in primary and secondary stage are to be assessed;

(5) to calculate the equivalent time according to the additional settlement of clay caused by over surcharge. The equivalent time in primary stage can be estimated by assuming time factor Tv=0.9 if ($\Delta p+\Delta s$) is larger than Δf.

The method discussed above was already used in a project in Singapore. In the said project, about 8m to 13m depth of very soft clay was to be improved by using vertical drains. According to the contract requirements, this soil improvement work must be completed within 40 weeks. In addition, the settlement rate of clay after improvement was required to be less than 20mm per year.

Vertical drains with over surcharge were used for this project. To ensure that the results of soil improvement can meet the contract requirements, the loading procedure of over surcharge was checked based on the method of equivalent time. The settlement rate of soft clay after removal of surcharge was estimated according to the values of additional settlement and equivalent time of consolidation. The actual loading procedure of applying-removing backfilling was adjusted depending on the calculation and monitoring results. The results of application are proven by the monitoring results to be satisfactory. In addition, it was also noticed that the proposed method is reliable and quite easy to use.

CONCLUSIONS

The effects of over surcharge on soft clay could be assessed in a term of "equivalent time" corresponding to the additional deformation in effective stress – strain –time space. The deformation of clay caused by the process of applying-and-removing of over surcharge is considered as equal effect as additional elapsed time of clay under the load without over surcharge. The results of laboratory tests of undisturbed samples by using two loading schemes with and without preloading (preloading to cause additional deformation of clay) indicate that the process of applying-and-removing of loads apparently reduces the consolidation time of clay, which provides evidence of the concept of equivalent time. According to the theoretical analysis and laboratory tests, a simplified method to assess the effect of over surcharge on soft clay with vertical drains was proposed and successfully applied in relative projects

REFERENCES

Bjerrum, L.(1967), "Engineering geology of Norwegian normally consolidated marine clays as related to settlement of buildings", *Geotechnique (Seventh Rankine Lecture)*, Vol.17, No.2 81-118.

Lowe III, J. (1974). "New concept on the consolidation and settlement analysis", *Journal Of Geotech. Eng. Div.*, Asce GT6, 574-612.

Leroueil, S., Kabbaj, M., Travenas, F. & Bouchard, R. (1985). "Stress-strain-strain rate relation for the compressibility of sensitive natural clay", *Geotechnique,* Vol. 35 No.2 ,159-180.

International Conference on Geotechnical Engineering for Disaster Mitigation & Rehabilitation
Chu, Phoon & Yong (eds) © 2005 World Scientific Publishing Company ISBN 981-256-469-1

POTENTIAL USE OF GEOSYNTHETIC CLAY LINERS (GCL) IN WASTE MANAGEMENT

Z.MOSES RAVINDRANATH
M.Tech (GTE), Department of Civil Engineering,
National Institute of Technology, Surathkal. karnataka, India.

Dr.S.SHRIHARI,
Asst.Professor, Department of Civil Engineering,
National Institute of Technology, Surathkal, Karnataka ,India.

Leachate in Municipal Solid Waste (MSW) landfills are contaminated liquids that contain a number of dissolved or suspended materials. The liner in barrier systems for landfills are to be properly designed, otherwise there may be a danger of contamination of the surrounding soil and ground water. The available soil at the site generally does not have the desired geotechnical properties to be used as liner materials. Use of Geosynthetic products as liner material can isolate the MSW and prevent contamination. In present study an experimental program has been carried out to assess the permeability characteristics of soil-GCL, soil-HDPE combinations as a liner material with leachate of different pH concentrations. Alternate wetting and drying may occur in a GCL in final cover systems of landfill and hence the effect of wetting and drying on the hydraulic conductivity of GCL was monitored. It is observed that the self-sealing capacity of GCL makes them a potentially viable hydraulic barrier to prevent contamination of the environment.

INTRODUCTION

The increasing level of solid waste is, now a days, a serious problem in the urban areas of the world. A high level of growth of population and increasing per capita income have resulted in the generation of enormous solid waste posing a serious threat to environment quality and human health. This is more so in the case of developing countries where large quantities of solid wastes are dumped haphazardly, there by, putting pressure on scarce land and water resources and at the same time adversely affecting the health of human beings.

Municipal Solid Waste (MSW) consists of household waste, construction and demolition debris, sanitation residue and waste from streets. With rising urbanization and change in lifestyle and food habits, the amount of municipal solid waste has been increasing rapidly. In 1947 cities and towns in India generated an estimated 6 million tonnes of solid waste and in 1997 it was about 48 million tonnes. The existing landfills in India are neither well equipped nor well managed and are not lined properly to protect against contamination of soil and groundwater.

At present the disposal of MSW become a serious problem as it causes threat to the environment and leads to health hazards of mankind. Ministry of Environment and Forests, Government of India has imposed rules to every municipal authority responsible for collection, segregation, storage, transportation, processing and disposal of MSW.

The pollution preventions from the landfill operations given in the Gazette of India, Part II, Section 3 specifies the use of composite barrier having 1.5mm HDPE geomembrane overlaying 0.90m of soil (clay or amended soil) having permeability coefficient not greater than $1x10^{-9}$ m/sec. But the use of Composite Clay Liners (CCLs) in all parts of India is not feasible because of unavailability of clay material. In this investigation an attempt has been made to substitute the CCLs with GCLs as landfill barrier materials. Major design concern was the hydraulic conductivity of different liner configurations with different permeating liquids.

MATERIALS AND THEIR PROPERTIES

Physical and chemical characteristics of materials used to build different liner configurations are presented. Some of the characteristics were obtained in the laboratory while the remaining properties have been provided by the manufacturer.

Clay

Clay is used as a primary or secondary liner material. The geotechnical properties of clay used in liner configurations were determined in the laboratory. It was observed that the permeability of clay is less than $1.0x10^{-9}$ m/sec. Specific gravity is 2.7, Liquid limit is 64% and plasticity index is 34%. The optimum moisture content is 20.9 % and maximum dry density is 15.89 kN/m^3.

Geosynthetic Clay Liner (GCL)

In this experimental study Bentofix® NSP 4900-4, a needle punched GCL, was used in the configurations with clay. It contains a continuous layer of sodium bentonite, sandwiched between a needle punched poly-propylene (PP) geotextile as a cover layer and a slit film PP woven geotextile as the carrier layer. The GCL consisted of powdered sodium bentonite and was placed at a mass per unit area of 4930 g/m^2. The lower geotextile was a non-woven, needle punched with a mass per unit area of 220g/m^2, while upper geotextile was a woven geotextile with a mass per unit area of 110g/m^2. The components shall be needle punched uniformly together across the entire GCL and then thermally treated on the woven side.

Wet-Dry response of GCL

Alternate wetting and drying may occur in a GCL in final cover systems of landfills. Wet dry GCL sample was prepared in the laboratory.

GCL of size 190mmx190mm was buried under 600mm height of gravel of 20mm size and permeated with water for 3days. Then the water was removed, the gravel was taken out and the wet GCL was kept in oven at 40 degrees centigrade for 24 hrs. The GCL was subjected to three wet dry cycles. Severe drying and cracking occurred in the bentonite component of GCL. It is observed that after three wet dry cycles the thickness of GCL was increased and remained 10-12mm even after drying.

High Density Polyethylene (HDPE) Geomembrane

HDPE geomembranes are manufactured using approximately 97% high molecular weight polyethylene, 2 to 3% carbon black, and 0.5 to 1.0% stabilizers and antioxidants. Some salient features of HDPE geomembrane include chemical resistance, low permeability and ultraviolet resistance. The thickness of this geomembrane is 2 mm.

Synthetic Leachate

The synthetic leachate is prepared with the following chemical composition (Table 1).

Table1. Chemical composition of Synthetic leachate

Parameters	Leachate I	Leachate II	Control water
COD (mg/litre)	300-350	300-350	300-350
BOD (mg/litre)	300-320	300-320	300-320
pH	8.02-8.1	5.14-5.6	7.1-7.5
Hardness	150-170	150-170	150-170
NH_4 (mg/litre)	20-30	20-30	20-30
Cl(mg/litre)	20-30	20-30	20-30
K (mg/litre)	10-15	10-15	10-15
NO_3 (mg/litre)	10-15	10-15	10-15
Fe (mg/litre)	2.5-3.0	2.5-3.0	2.5-3.0
Ca (mg/litre)	700-710	700-710	700-710

EXPERIMENTAL METHODOLOGY

For determination of hydraulic conductivity of soil and liner material configurations, odometer test was conducted as per IS: 2720 (Part 15).

Preparation of odometer sample

About 1.6 kg of clay sample passing through 425 micron was collected and 20.9% of water (OMC) is added and the soil is compacted in compaction test mould in 3layers with 25blows per layer, so as to get the maximum dry density. The extracted sample was trimmed so as to accommodate the liner materials at the top of the each odometer ring. The arrangement of each liner material is shown as schematic diagram in fig 1.

Then the consolidation characteristics of the above shown samples viz., coefficient of consolidation (C_V), coefficient of compressibility (a_v), coefficient of volume change (m_v) have been obtained.

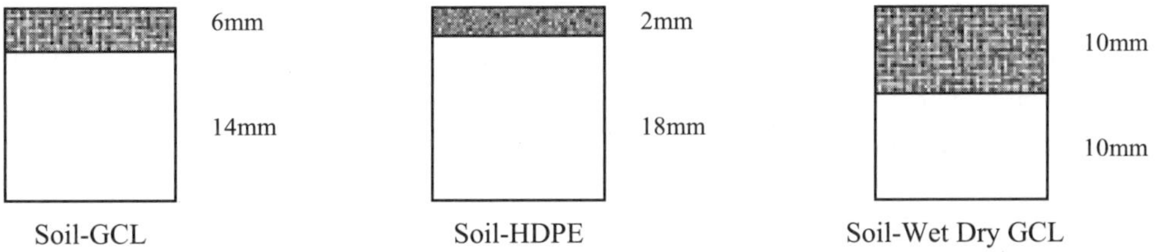

Fig. 1. Soil- Liner material configurations

RESULTS AND DISCUSSIONS

Each configuration was tested for hydraulic conductivity with permeating liquids having different values of pH over a period of 7days and the results are presented.

Soil-GCL configuration

It is observed that the hydraulic conductivity of this configuration increases at initial load increments upto 175 kPa. The hydraulic conductivity then decreases rapidly and this variation is intense when soil-GCL configuration subjected to control water with pH 7.2 as permeating liquid. This is because the GCL is initially placed in dry condition and absorbs more liquid and is not misinterpreted as liner leakage. Subsequently when GCL got hydrated and as the consolidation load increases the hydraulic conductivity for all the soil-GCL configurations irrespective of leachate concentrations stabilizes and shows low hydraulic conductivity at the end of the test.

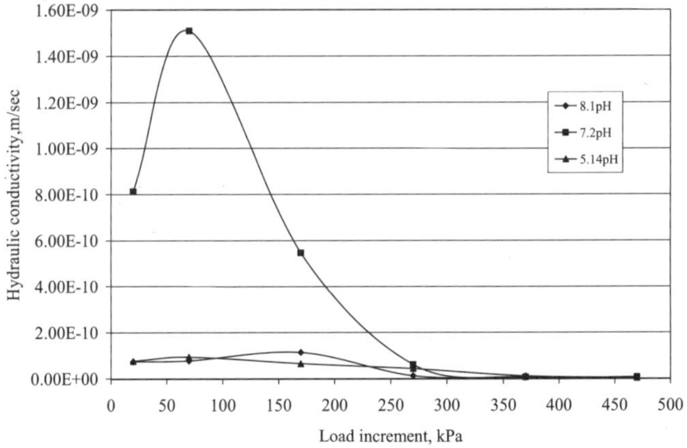

Fig. 2. Hydraulic conductivity Vs Load increment for Soil-GCL configuration

Soil-HDPE Geomembrane configuration

The hydraulic conductivity of this configuration increases at initial load increments for all configurations and this trend is observed till the pressure reaches about 175 kPa (fig3).The hydraulic conductivity for all soil-HDPE configurations, irrespective of pH values of leachate, stabilizes at the end of the test. The permeability values of soil-HDPE configuration are less than the values of soil-GCL configuration. But soil-HDPE configuration works differently when the HDPE geomembrane is having defects, and will show high permeable values (Giroud and Badu-Twenoboah, 1992).

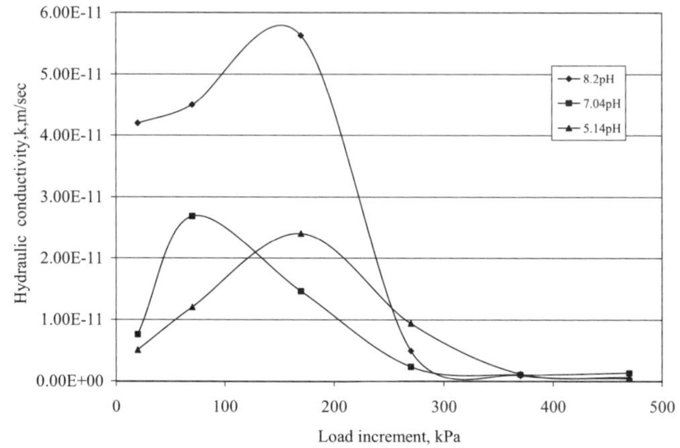

Fig. 3. Hydraulic conductivity Vs Load increment for Soil-HDPE configuration

Soil- wet dry GCL configuration

In this test, the GCL which was subjected to wet and dry cycles has been used. The hydraulic conductivity of this configuration is relatively high in comparison with other configurations at the initial stages. This is because of the cracks developed in GCL due to the wet and dry cycles (fig 4). Initially the GCL is placed in dry condition and hence the hydraulic conductivity is high. As GCL gets rehydrated, it gets self-sealed and the hydraulic conductivity reduces for all the three configurations.

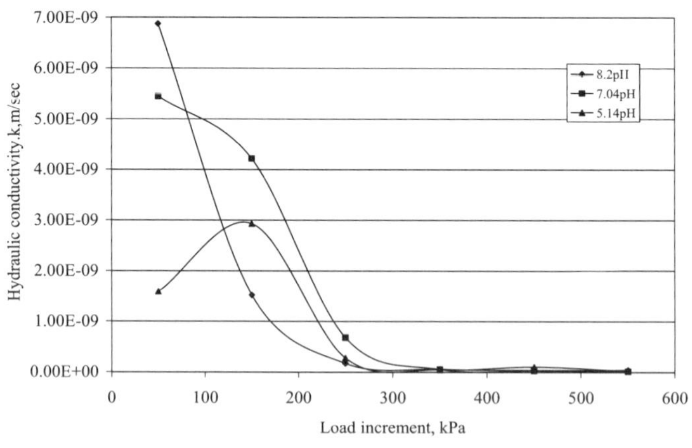

Fig. 4. Hydraulic conductivity Vs Load increment for Soil-WD configuration

EFFECT OF pH ON LINER MATERIAL CONFIGURATION

The fig 5-fig 7 shows the plots of hydraulic conductivity against load increment for the permeating liquid having different values of pH. HDPE geomembrane has no effect on its hydraulic conductivity and dry GCL also follows the same trend for the leachate having pH 5.14. But WD GCL has more cracks due to the alternate wet and dry cycles and because of its initial cracks has more hydraulic conductivity and subsequently when it becomes rehydrated it shows low permeability(fig 5).

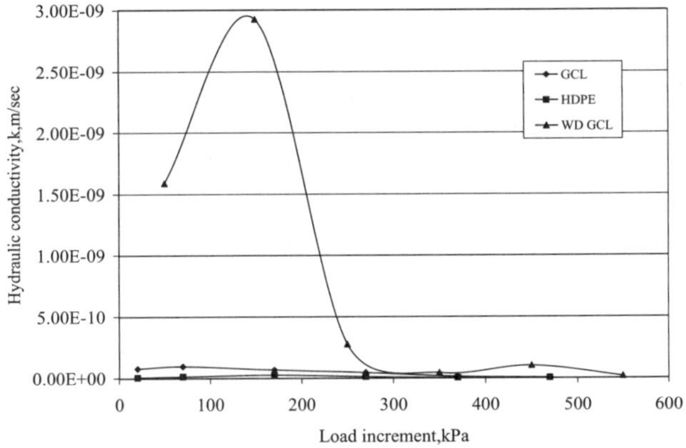

Fig. 5. Effect of Leachate having pH 5.14 on Liner configuration

The soil-WD GCL configuration shows initially a high value of hydraulic conductivity and the rest of the two configurations were unaffected with the leachate having pH 8.2. But all the three configurations sustain almost constant hydraulic conductivity at the end of the test (fig 6).

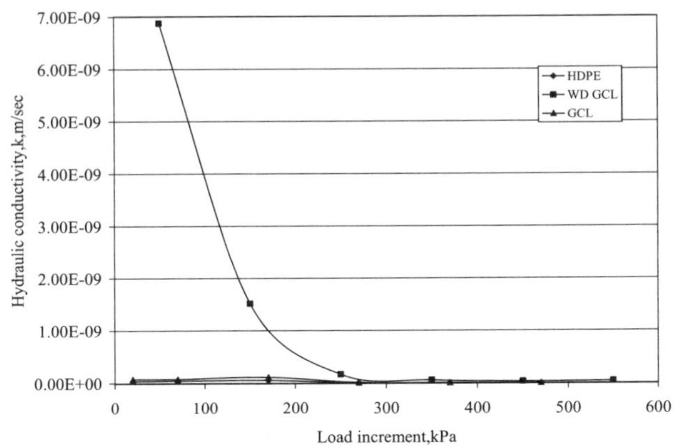

Fig. 6. Effect of Leachate having pH 8.1 on Liner configuration

WD GCL and GCL configurations shows variations in hydraulic conductivity at the initial stages of load increment, where as HDPE has no effect on its hydraulic conductivity, when the permeating liquid is control water having pH 7.04. At the end of the test all the three configurations show constant low hydraulic conductivity (fig 7).

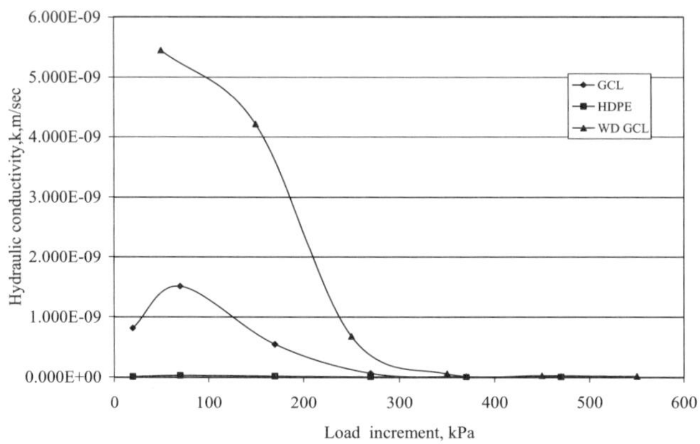

Fig. 7. Effect of Leachate having pH 7.04 on Liner configuration

CONCLUSIONS

The purpose of this experimental study was to propose and evaluate GCL as the ideal barrier in landfill systems. This objective was approached by carrying out hydraulic conductivity tests on various soil-liner material configurations. The following are the important conclusions drawn from the results of this investigation:

1. The hydraulic conductivity of soil-GCL configuration when subjected to control water, having pH 7.04, is high at initial stages of load increment and is 6.659 x 10^{-12} m/sec when the load was 370 kPa. The condition of hydration was found to be very important and much lower hydraulic conductivity generally resulted when the first wetting liquid is water rather than leachate.
2. Initially high value of hydraulic conductivity of the desiccated GCL is observed but this may not be representative of the true field conditions because the overlying cover soil would likely adsorb some of the incoming rainfall and cause a more gradual wetting of the GCL.
3. The hydraulic conductivity values of soil-WD GCL are less and the results are encouraging and suggest that GCLs may be an attractive option to use when some degree of cycling in water content is anticipated within the hydraulic barrier and proves is most effective as barrier in landfills.
4. HDPE Geomembrane is an excellent liner material having low permeability of 4.22 x 10^{-13} m/sec. However, if there is a defect in the geomembrane, the liquid leaks through the geomembrane defect, which in turn will affect the intimate contact of the liner materials.

5. The effect of pH of permeating liquid is not significant as the period of testing is less. The authentic effect of pH on liner material configuration can be studied precisely by long term testing.

ACKNOWLEDGMENTS

I am thankful to Radhika for her support, encouragement and guidance all the way through my work. I thank Garware-Wall Ropes Ltd., Chennai for supplying of GCL.

REFERENCE

Aarne P.Versilind, William A. Worrel (2002) "*Solid Waste Engineering*" Thomson Brooks/cole

Narejo,D.B and Memon,G.Q (1995), "*Compatibility of Geosynthetic clay liners with three Pennsylvania municipal solid waste leachates*", Geosynthetics International, vol2, No.5, pp. 889-892.

Giroud J.P., Badu-Tweneboah.,K (1992) "*Rate of Leakage through a composite liner due to Geomembrane Defects*" Geotextiles and Geomembranes, Vol.11 No.1, Elsevier Science Publishers Ltd, England.

Quigley M Robert (1994) "*Municipal solid waste landfilling*", Proceedings of the First International Congress on Environmental Geotechnics, Alberta Canada pp 589 - 597

SP 36 Part1(1987), compendium of Indian standards on soil engineering- part 1, "*Laboratory Testing Of Soils For Civil Engineering Purposes*", Bureau of Indian standards 1987.

The Gazette of India (2000) Extraordinary-Part II section 3. New Delhi, 2000.

Quian Xuede, Robert M.Koerner, Donald H.Gray (2002) "*Geotechnical Aspects of Landfill Design and Construction*", Prentice Hall, New Jersey.

RISK ASSESSMENT & MANAGEMENT

International Conference on Geotechnical Engineering for Disaster Mitigation & Rehabilitation
Chu, Phoon & Yong (eds) © 2005 World Scientific Publishing Company ISBN 981-256-469-1

WIRELESS REMOTE MONITORING OF GEOTECHNICAL SYSTEMS

TAREK ABDOUN
Department of Civil and Environmental Engineering, Rensselaer Polytechnic Institute, 110 8th Street, Troy, NY 12180, USA

LEE DANISCH
Measurand Inc., 2111 Hanwell Road, Fredericton, New Brunswick, Canada

VICTORIA BENNETT
Department of Civil and Environmental Engineering, Rensselaer Polytechnic Institute, 110 8th Street, Troy, NY 12180, USA

Landslides, lateral spreading and other similar forms of ground failures due to natural disasters are often catastrophic primarily because they occur without much warning. As our society becomes ever more complex and urban areas continue to spread, the economic and societal costs of landslides and other ground failures will continue to rise. This state of affairs stems from the limitations of existing real-time sensing and monitoring tools as well as inadequate predictive capabilities of current computational models. The work presented in this paper constitutes a major step in the direction of establishing a low cost wireless remote monitoring system for active ground. A ShapeAccelArray sensor is being developed, taking advantage of promising new advances in the fiber optic and micro-machined electromechanical sensor (MEMS) technologies. This sensor array is capable of simultaneously measuring 2D soil acceleration and 3D permanent ground deformation down to tens of meters of depth. This paper presents the preliminary design of the new sensor array as well as preliminary results from tests aimed at validating and calibrating the accelerations and displacements measured with this sensor array.

INTRODUCTION

Over the past 20 years, each instance of natural disasters resulted in great human suffering and many billions of dollars in losses. Much of the destruction is associated with ground failures and other geotechnical phenomena, and the loss exposure continues to rise due to expansion of urban areas and increasing complexity of our society. Landslides during the rainy season occur in essentially every state. The US Geologic Survey (USGS) estimates that landslides cause in excess of $2 billion in damages and about 25 to 50 deaths each year in the U.S. (Fig. 1), and worldwide they are responsible for thousands of fatalities and hundreds of billions of dollars lost. Failures of dykes, levees and dams during floods are also very dangerous and expensive. FEMA and the US

Army COE estimate that at least twenty-five of these disastrous floods occurred during the 1988-2002 period with 911 deaths and economic loss of about $140 billion.

Figure 1. Aerial view of the 1995 La Conchita, CA Landslide .

Figure 2. Photo of the San Fernando dam failure during the 1971 San Fernando Earthquake.

Earthquakes are another significant source of geotechnical problems. In addition to triggering landslides and failures of earth embankments (dams, coastal dykes, and highways), the phenomenon of liquefaction of loose, water-saturated sands and other granular soils is another major hazard specifically related to earthquakes. Nearly all major seismic events affecting an urban or industrial area result in liquefaction. This liquefaction leads to ground failure, significant permanent vertical deformations and lateral spreading. These ground failures are usually associated with very costly damage to port facilities, bridges, dams, buried pipes, houses and buildings of all types. For instance, the 1971 San Fernando, California earthquake caused more than five hundred million dollars in damage (National Research Council, 1982), which included a failure of the San Fernando dam (Fig. 2). The 1995 HyogoKen Nanbu earthquake in Kobe, Japan, caused more than one hundred billion dollars in total damage, with about ten billion dollars of that total attributed to soil liquefaction and ground deformation.

In spite of the growing awareness of the importance of accelerometer downhole arrays, especially when combined with pore pressure sensors to capture information on the poorly understood phenomena of liquefaction and lateral spreading, these arrays still remain scarce due to their high cost. The recent advances in sensors and wireless networking technologies provide valuable opportunities for creating new knowledge and technology to assess the impact of natural disasters. The newly developed ShapeAccelArray sensor presented in this paper capitalizes on these recent advances in sensors and wireless networking technologies to develop a new low cost wireless sensor array, Danisch *et al.* (2004). The sensor array consist of several MEMS sensors installed in a flexible water tight casing and is capable of measuring the 3D ground deformation as well as 2D lateral soil acceleration every 0.76 m (2.49 ft) interval down to a soil depth of

100 m (328.1 ft). Each sensor array will be connected to a wireless sensor node, to enable real-time monitoring as well as remote sensor configuration (Fig. 3). The new system is designed to be placed in a traditional inclinometer borehole. Figure 4 presents a sketch rendering this vision of several ShapeAccelArray sensors installed in the ground and on a pile foundation at a bridge abutment site where the soil shakes, liquefies, slides and deforms during an earthquake.

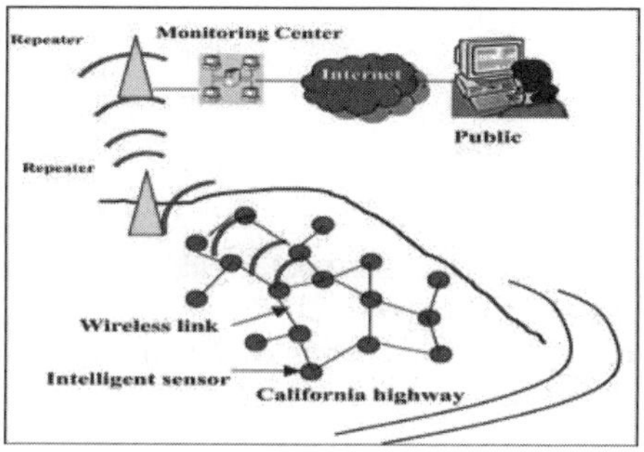

Figure 3. Wireless network for real-time monitoring of geotechnical system.

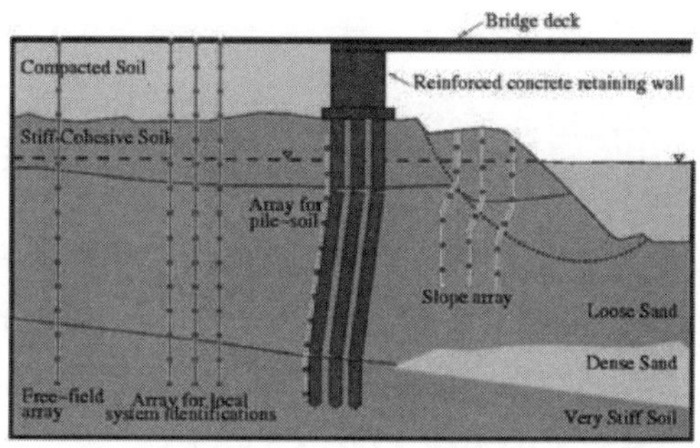

Figure 4. Sketch rendering the vision of Installed ShapeAccelArray sensor in active soil system

SENSOR DESIGN

The ShapeAccelArray sensor is currently being developed in collaboration with Measurand Inc. (www.measurand.com). The ShapeAccelArray sensor is designed to: (1) be low in cost when in volume production, (2) be capable of measuring the 3D soil deformation and 2D accelerations (two lateral directions), (3) provide high resolution measurement in time (sampling rate of 30Hz for dynamic measurement) and space (a sensing element every 0.5 m (1.65 ft)), and (4) contain modular components to enable variable sensor spacing and multiplexing.

MEMS are good candidates for long-term stable measurements of curvature along a substrate. In the future, bipolar fiber optic sensors can be added to resolve torsional movements and near-pole conditions (Danisch *et al.* (1998), Danisch *et al.* (2000), and Danisch *et al.* (2003)). The MEMS accelerometers can be applied to a flexible substrate at intervals, chosen to include at least two sensors within any expected monotonic curve that can be modeled as a circular arc (Fig. 5). The MEMS devices have sufficient resolution and accuracy for many applications, and it is possible to multiplex a large number of them using conventional CANbus or similar high-speed multiplexing technology (so that electrical interference can be minimized by converting to digital signals locally along an array). They also offer the possibility of measuring both differential tilt (bend) at low sampling rate, and, when needed, to measure acceleration at a higher sampling rate (100Hz or higher).

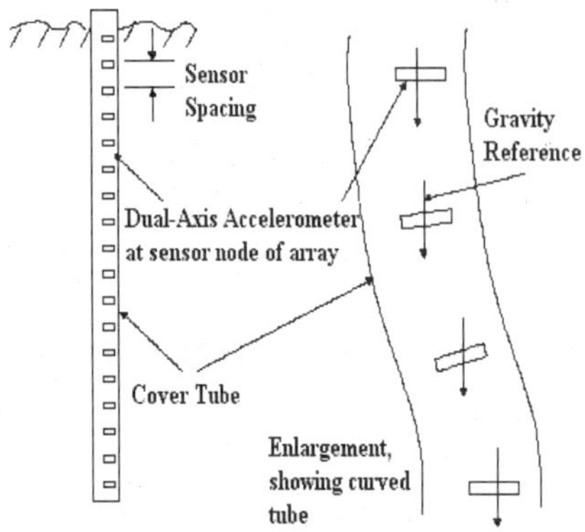

Figure 5. Design of ShapeAccelArray

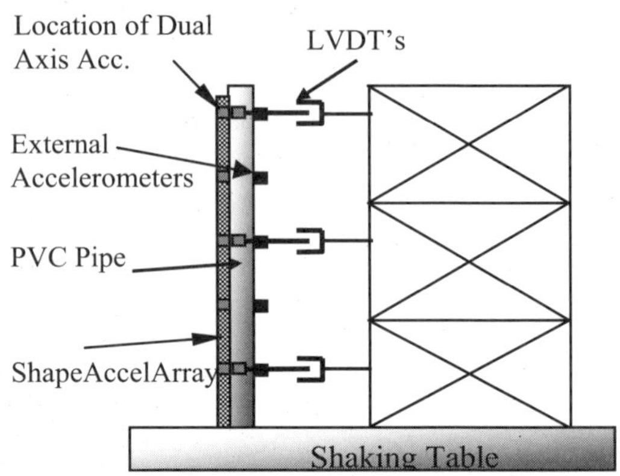

Figure 6. Setup for Testing and calibration

SENSOR CALIBRATION TESTS

A series of calibration tests using the shaking table facility at Rensselaer Polytechnic Institute (RPI) were conducted to evaluate the response of the ShapeAccelArray sensor prototype. Figure 6 presents a sketch of the test setup. A 1.5 m long sensor array, which includes 6 sensing elements placed at 0.25 m intervals was attached to a flexible pipe and placed on RPI's 1g shaking table. The response of the flexible pipe was also monitored using traditional accelerometers and displacement sensors (LVDT) as shown in Figure 6. The test setup was excited by different shaking amplitudes and frequencies.

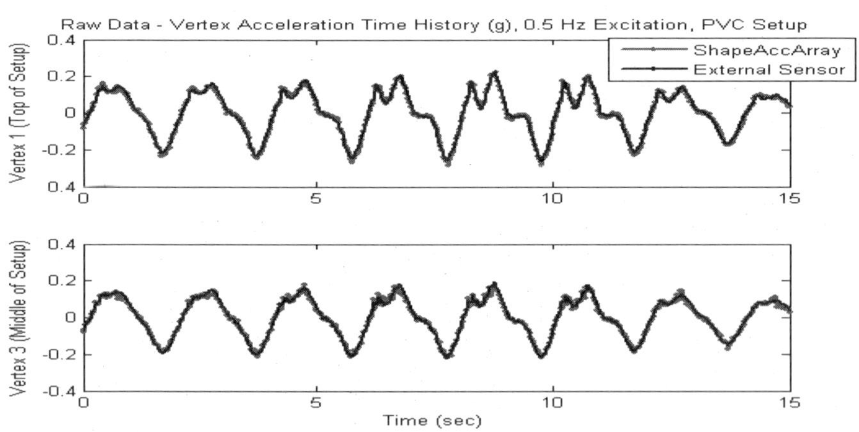

Figure 7. Comparison between acceleration (g) measured using traditional accelerometers and ShapeAccelArray sensor.

A representative sample of test results is included here. Figure 7, presents a comparison between accelerations time histories recorded using the ShapeAccelArray sensor and external accelerometers located at vertex 1(top of the flexible pipe) and vertex 3 (mid height of the flexible pipe). Figure 8, presents a comparison between the lateral deformation time histories measured using the sensor array and an external displacement

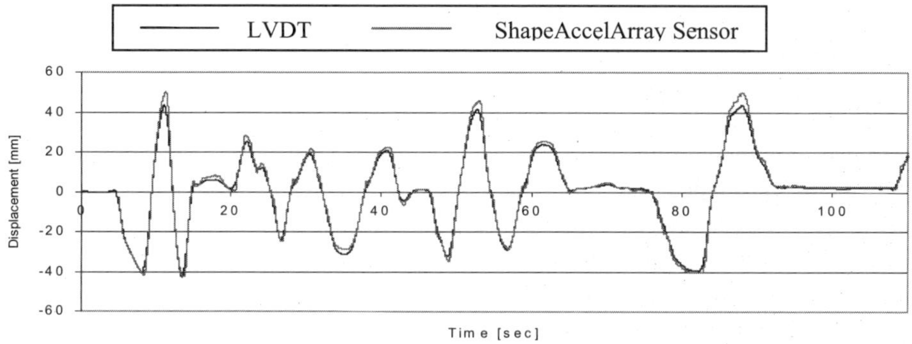

Figure 8. Comparison between lateral displacements (mm) measured using traditional displacement sensors (LVDT) and ShapeAccelArray sensor.

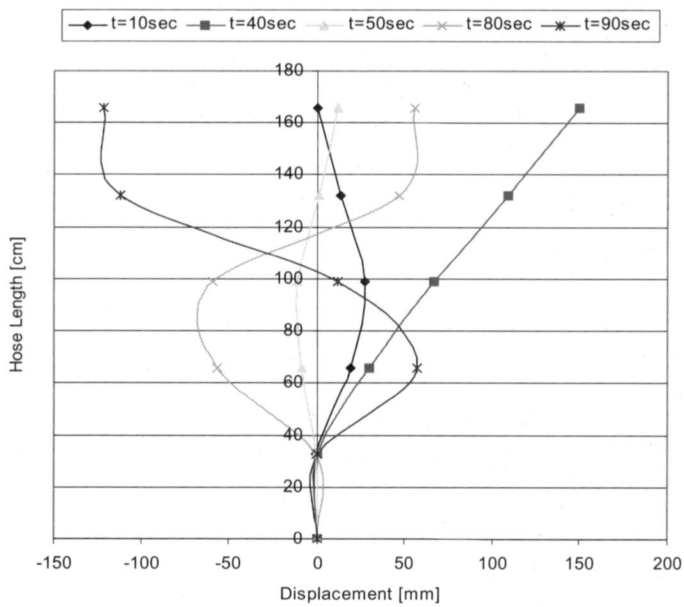

Figure 9. Measured lateral displacement (mm) profiles.

sensor, located at 0.7 m (2.3 ft) from the shaking table surface. Figures 7 and 8 show a very good comparison between the acceleration and lateral displacement measured by the traditional sensors and the sensor array. Figure 9, presents the profiles of the measured lateral displacement using the ShapeAccelArray sensor at different times during shaking. The lateral displacement profiles presented in Figure 9 demonstrate the ShapeAccelArray sensor capability of measuring internal soil static deformation.

CONCLUSIONS

The work presented in this paper constitutes a major step in the direction of establishing a low cost real-time monitoring system for active ground. This ShapeAccelArray sensor is currently being developed for in situ and research applications, taking advantage of promising advances in the fiber optic and micro-machined electromechanical sensor (MEMS) technologies. These preliminary test results demonstrate that the newly developed sensor array is capable of simultaneously measuring acceleration and permanent ground deformation down to several meters of depth.

ACKNOWLEDGEMENTS

This research is supported by the NSF sensor program. This support is greatly appreciated. The authors would also like to express their sincere thanks to Prof. M. Symans for his help in conducting the 1g shaking table tests.

REFERENCES

Committee on Earthquake Engineering. “Earthquake Engineering Research - 1982, Overview and Recommendations.” *National Research Council*, National Academy Press, Washington, DC, 1982.

Danisch, L.A., M.S. Lowery-Simpson, and T. Abdoun. “Shape-Acceleration measurement Device and Method”, *Patent Application*, 2004.

Danisch, L.A. "Fiber optic bending and positioning sensor," PCT 0,702,780, *International PCT patent*, 1998.

Danisch, L.A. "Topological and motion measuring tool," *U.S. Paten*t 6,127,672, 2000.

Danisch, L.A., J.F. Danisch, and J. Lutes. “Topological and motion measuring tool II (ShapeRope)”, *US Patent* 6563107, 2003.

International Conference on Geotechnical Engineering for Disaster Mitigation & Rehabilitation
Chu, Phoon & Yong (eds) © 2005 World Scientific Publishing Company ISBN 981-256-469-1

PROBABILISTIC EVALUATION OF LIQUEFACTION POTENTIAL

HADI BARGHAMADI
Soil and Rock Mechanics Group, Mahab Ghodss Consulting Engineer
Tehran, Iran

MANOUCHEHR LATIFI NAMIN
University of Tehran, Faculty of Engineering
Tehran, Iran

Typically, a deterministic method is used to determine the potential for liquefaction. In a deterministic method the uncertainty of many factors that may contribute to the occurrence of liquefaction are not incorporated. As a result, liquefaction potential is evaluated only by the factor of safety. In this paper, probabilistic concepts are used in determination of liquefaction potential. In order to determine the conditional probability of liquefaction, the Taylor's Series reliability method and Monte Carlo simulation method are used. The "simplified procedure", as embodied in the "NCEER report" will be used as the basis for the probabilistic procedure. The methods which are described in this paper are applied to an actual site after Manjil-Rudbar earthquake of June 21, 1990.

INTRODUCTION

Evaluation of liquefaction potential of soils is an important aspect in geotechnical investigations in earthquake-liable region. The most common deterministic procedure around the world for evaluating liquefaction resistance is "simplified procedure" developed originally by Seed and Idriss (1971) using blow counts from the Standard Penetration Test correlated with a parameter called the cyclic stress ratio which represents the cyclic loading on the soil. Since 1971, this procedure has been modified and updated.

In deterministic methods, the values of parameters that may contribute to the occurrence of liquefaction are considered constant. But there is uncertainty in many parameters, so deterministic methods aren't reliable methods. The probabilistic methods differ in incorporating the uncertainty of parameters compared with deterministic methods. Some probabilistic methods include Haldar and Tang (1979) and the Taylor's Series reliability method, Wolf (1994); USACE (1997, 1998); Duncan (2000).

In this paper probabilistic concepts are used in determination of liquefaction potential. In order to determine the conditional probability of liquefaction, the Taylor's Series reliability method and Monte Carlo simulation method are used.

SIMPLIFIED PROCEDURES FOR DETERMINISTIC LIQUEFACTION EVALUATION

The most common deterministic method to evaluate liquefaction potential of a site, is the simplified procedure" originally developed by Seed and Idriss (1971). The method has been modified and improved since it was developed. A review of the current state-of-practice in liquefaction evaluation using the simplified procedure is given in the "1996 NCEER and 1998 NCEER/NSF Workshops on Evaluation of Liquefaction Resistance", Youd et al.(2001), referred as "NCEER Report" in this paper. The current version of the simplified procedure calculates the factor of safety *FS* against liquefaction of a level ground in terms of the cyclic stress ratio CSR (the demand), and the cyclic resistance ratio CRR (the capacity) according to:

$$FS = \frac{CRR_{7.5}}{CSR} MSF.K_{\sigma} \tag{1}$$

Where $CRR_{7.5}$=Cyclic resistance ratio for magnitude 7.5 earthquakes, CSR=cyclic stress ratio, *MSF* = magnitude scaling factor, and K_{σ}= correction for non-linear effects of confining stress. The complete equations are given in NCEER Report.

PROBABILITY OF LIQUEFACTION

The Taylor's Series reliability method

In the Taylor's Series reliability method, Wolff (1994); USACE (1997, 1998); Duncan (2000), conditional probability of liquefaction P_L is determined, assuming a log-normal distribution of the factor of safety. This requires the calculation of a reliability log-normal reliability index β_{LN} and using the formula:

$$p_L = 1 - \Phi\ (\beta_{LN}) \tag{2}$$

Where $\Phi(\ )$= the standard normal cumulative distribution function. To calculate β_{LN}, the moments of the function of the safety factor must be calculated from the moments of the parameters.

Monte Carlo simulation method

Generation of random numbers with predefined probability distribution functions is one of the most important applications of Monte Carlo simulation method. In Monte Carlo simulation, values of the random variables are generated consistent with their probability distribution, and the function of the safety factor is calculated for each generated set of variables. The process is repeated numerous times, typically thousands, and the expected value, standard deviation, and probability distribution of the function of the safety factor are taken to match that of the calculated values. Monte Carlo method permits one to estimate the shape of distribution of the function of the safety factor, permitting more accurate estimation of probability of liquefaction.

LIQUEFACTION POTENTIAL ANALYSIS IN ASTANEH AFTER MANJIL-RUDBAR EARTHQUAKE OF JUNE 21, 1990

The Manjil, Iran earthquake, M_s=7.7, Berberian et al. (1992), occurred on June 21, 1990. Official estimates indicate that more than 35,000 people lost their lives. During the earthquake, the liquefaction of soils played a significant role in the destruction of many houses and commercial and public buildings. There was dramatic evidence of this widespread liquefaction-induced damage as far away as 80 km from the fault.

Astaneh-Ashrafieh region is located to the east of the major city of Rasht. Liquefaction on the east of Rasht occurred almost entirely in the levee deposits along the banks of the Sefid and Heshmat rivers.

To evaluate the in situ properties of the liquefied soils in the town of Astaneh-Ashrafieh, standard penetration tests (SPT) were conducted in 16 boreholes. It is noted that these field borings were made in the summer of 1991 almost a year after the earthquake. In all cases, the SPT tests were performed using a standard 2-inch split spoon sampler without a liner. The borehole AS-3 located in the area that liquefaction was observed after Manjil earthquake. But in the borehole AS-14 liquefaction was not observed, Yegian et al. (1995).

Figure 1 shows SPT, N-values in boreholes AS-3 and AS-14 from Astaneh-Ashrafieh. The sands from this region are characterized as gray poorly graded fine sands with less than 10 % silts. The density of soil is about 19.5 (KN/m^3). At the time of the earthquake the ground water level was about 2 meters below the ground surface.

According to the peak ground acceleration obtained from this earthquake, Naderzadeh (1991), the peak ground acceleration in Astaneh-Ashrafieh was 0.18g.

To evaluate the conditional probability of liquefaction, the average values and estimated coefficient of variation (COV) of the uncertainty involved parameters have been considered according to table 1.

The standard penetration test (SPT) results show that the distribution function of N_{spt} is lognormal. In this paper the distribution function is assumed normal for the other uncertainty involved parameters.

Figure 1 shows the liquefaction safety factor in different depths of boreholes AS-3 and AS-14. The safety factor has been determined according to the deterministic method, Youd et al. (2001).

Figure 2 shows the conditional probability of liquefaction in the different depths of boreholes AS-3 and AS-14. The conditional probability of liquefaction has been determined according to the probabilistic methods, Monte Carlo simulation method and the Taylor's Series Reliability Method, Wolff (1994); USACE (1997, 1998); Duncan (2000). Monte Carlo simulation was carried with 5000 iterations.

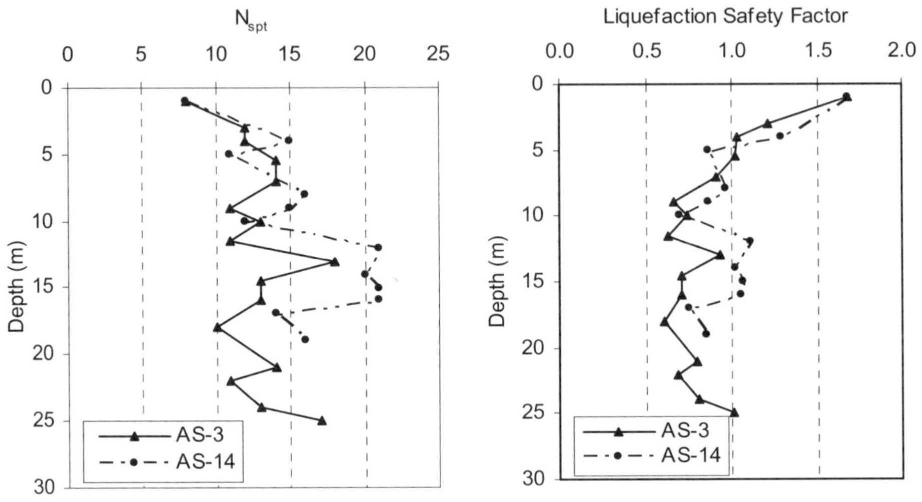

Figure 1. N_{SPT} values, Yegian et al. (1995), and Safety Factor in Boreholes AS-3, AS-14.

Table 1. The average and estimated coefficient of variation (COV).

Parameter	Average value	Estimated COV (%)	Reference
Magnitude scaling factor, MSF	NCEER Report	COV(|MSF-1|)=22	from three-sigma rule
Stress-reduction coefficient , r_d	NCEER Report	COV($1-r_d$)=22	from three-sigma rule
moist unit weights, γ_m	Actual data	COV(γ_m)=9 (SP-SW)	Based on in-place density data from several USBR dams
SPT blow count, N_{spt}	Actual data	In this paper COV=30	COV=15-40% (from Harr (1984), Kulhawy (1992))
Overburden correction factor, C_N	NCEER Report	COV($1-C_N$)=23.1	Based on data from Castro (1995)
Energy correction factor, C_E	0.95	9	Based on data for Safety hammer in NCEER Report
Borehole diameter correction factor, C_B	1.07	1	Based on data for NCEER Report
Rod length correction factor, C_R	0.82	24	Based on data from Skempton (1986)
Sample correction factor, C_S	1.2	3	Based on data for NCEER Report

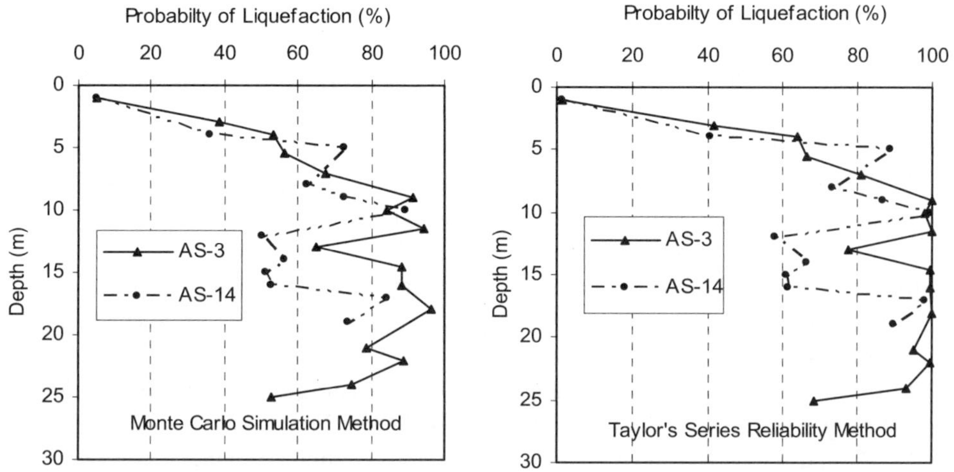

Figure 2. Probability of Liquefaction in Boreholes AS-3, AS-14.

Figure 3 shows the relation between safety factor and probability of liquefaction. According to figure 3, if safety factor is 1.5 and if coefficient of variation of parameters is as indicated in this paper, then the predicted probability of liquefaction is 0.17 in Monte Carlo simulation method and 0.18 in the Taylor's Series reliability method. Also the value of safety factor required to have the probability of liquefaction be less than 0.1 is about 1.6. According to the figure 3 the Taylor's Series reliability method is more conservative than Monte Carlo simulation method.

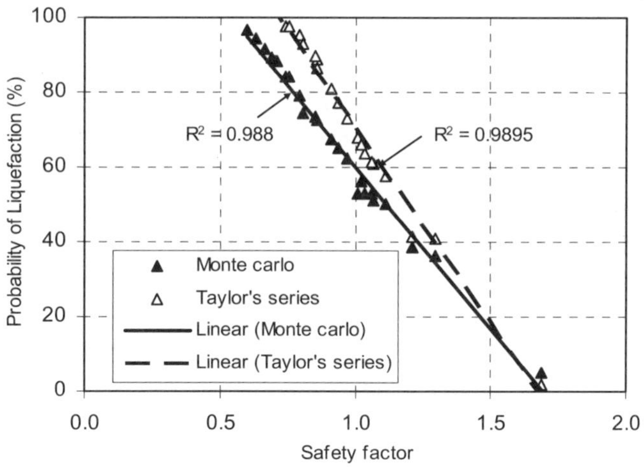

Figure3. Probability of Liquefaction versus Safety Factor.

REFERENCES

Barghamadi, H. (2001). "Liquefaction potential analysis by using Monte Carlo simulation method", MSC thesis, Tehran University, Iran.

Berberian, M., Qorashi, M., Jackson, J.A., Priestley, K., and Wallace, T. (1992). "The Rudbar-Tarom 20 June 1990 earthquake in northwest Pesia: Preliminary field and seismological observations, and its tectonic significance", *Bulletin of the Seismological Society of America*, 82(4), 1726-1755.

Castro, G. (1995). "Empirical methods in liquefaction evaluation", *Primer Ciclo de Conferencias Internationales*, Leonardo Zeevart, Universidad Nacional Autonoma de Mexico, Mexico City.

Duncan, L. M. (2000). "Factors of safety and reliability in geotechnical engineering", *J. Geotech. Envir. Eng.*, ASCE, vol. 126, No. 4, 307-316.

Haldar, A. and Tang, W.H. (1979). "Probabilistic Evaluation of Liquefaction Potential", *Journal of the Geotechnical Engineering Division*, ASCE, Vol. 105, No. GT2, 145-163.

Harr, M.E. (1984). "Reliability-based design in civil engineering", *1984 Henry M. Shaw Lecture*, Dept. of Civil Eng. North Carolina State Univ., Raleigh, NC.

Kulhawy, F.H. (1992). "On the evaluation of soil properties", *J. Geotech. Spec. Publ.*, ASCE, 95-115.

Naderzadeh, A. (1991). "Personal Communications".

Seed, H.B. and Idriss, I.M. (1971). "Simplified procedure for evaluating soil liquefaction potential", *J. Geotech. Engr. Div.*, ASCE, vol. 97, No. 9, 1249-1273.

Skempton, A.W. (1986). "Standard penetration test procedures and the effects in sands of overburden pressure, relative density, particle size, aging and over consolidation", *Geotechnique*, vol. 36, No. 3, 425-447.

USACE (1998). "Risk-based analysis in geotechnical engineering for support of planning studies", Engr. Tech. Ltr. No. 1110-2-554, U.S. Army Corps of Engrs., Dept. of the Army, Washington, D.C. (27 Feb. 1998).

USACE (1997). "Engineering and design introduction to probability and reliability methods for geotechnical engineering", Engr. Tech. Ltr. No. 1110-2-547, U.S. Army Corps of Engrs., Dept. of the Army, Washington, D.C. (30 Sept. 1997).

Wolff, T.F. (1994). "Evaluating the reliability of existing levees", Rep. Res. Reliability of existing levees, prepared for the U.S. Army Engineers Waterways Experiment Station Geotech. Lab., Vicksburg, Miss.

Yegian, M.K., Ghahraman, V.G., Nogole-Sadat, M.A.A., and Daraie, H. (1995). "Liquefaction During 1990 Manjil, Iran, Earthquake, I- Case Histories Data", *Bulletin of the Seismological Society of America*, February.

Youd, T.L. and 20 others (2001). "Liquefaction resistance of soils: Summary report from the 1996 NCEER and 1998 NCEER/NSF Workshop on Evaluation of Liquefaction Resistance of Soils", *J. Geotech. Envir. Eng.*, ASCE, vol. 127, No. 10, 817-833.

International Conference on Geotechnical Engineering for Disaster Mitigation & Rehabilitation
Chu, Phoon & Yong (eds) © 2005 World Scientific Publishing Company ISBN 981-256-469-1

A METHOD OF SAND LIQUEFACTION PROBABILISTIC ESTIMATION BASED ON RBF NEURAL NETWORK MODEL

CHEN GUOXING
Institute of Geotechnical Engineering, Nanjing University of Technology, Nanjing, China

LI FANGMING
Institute of Geotechnical Engineering, Nanjing University of Technology, Nanjing, China

Based on the 344 liquefaction site data during 25 strong earthquakes in the world, through training and testing the neural network model of the Radial Basis Function, the nonlinear relation between corrected blow count of standard penetration test N_1 and cyclic resistance ratio of saturated sand soils CRR is analyzed, and also the empirical equation CRR_{cri} of the liquefaction limit state curve or critical cyclic resistance ratio curve of saturated sand soils is constructed. By the statistic analysis, the probability density functions of liquefaction and non-liquefaction as well as empirical equation between the safety factor and liquefaction probability of saturated sand soils are given, then the empirical equation of the cyclic resistance ratio of saturated sand CRR under the different probability is deduced. The method used in this paper makes the sand liquefaction probabilistic estimation on engineering sites is as easy as the traditional deterministic method of sand liquefaction estimation.

INTRODUCTION

Earthquake-induced saturated sand liquefaction is one of important causes for making ground failure and building damage. Thus, many methods of soil liquefaction estimation are developed. The SPT method is comparatively perfect and universally accepted by the engineering domain in the world. In consistent with the reliability theory of superstructure design, the evaluation of soil liquefaction should also adopt the probability method, which gives the soil liquefaction estimation result under the different probability. Based on the 344 liquefaction site data during 25 strong earthquakes, with Radial Basis Function (RBF) neural network model, the cyclic resistance ratio CRR curve of saturated sand under the different probability is presented.

SAND LIQUEFACTION LIMIT STATE FUNCTION BASED ON THE RBF NEURAL NETWORK MODEL

Design of the Neural Network Model

RBF neural network is a multi-nonlinear dynamic system with good self-adaptive, self-systematical and finer ability of learning, association, compatibility and anti-jamming. It can expediently construct the model for complicated and unknown systems, thereby realizing the self-estimation of sand liquefaction potential under the different influencing factors.

RBF network model comprises three layers, and its construction is shown in Figure 1. A input layer node passing the input signal to a hidden layer is usually a simple linear function while the hidden layer node usually consists of the fundamental function. The fundamental function in hidden layer nodes will be affected in local once the signal is inputted.

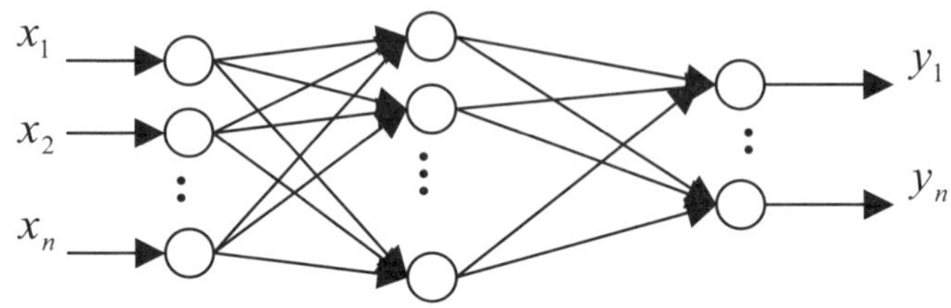

Figure 1. Radial basis function network model

The hidden layer function usually expressed as Gaussian function is

$$u_j = \exp[-\frac{(X-C_j)^T(X-C_j)}{2\sigma_j^2}] \qquad j=1,2,\ldots,N_h \tag{1}$$

Where u_j=the output value of the hidden layer node of No. j, X=input samples, C_j=the central value of the Gaussian function, σ_j=the standardization constant, N_h=the quantity of hidden layer nodes.

The output of RFB network model is a linear combination of the output of hidden layer nodes, and is expressed in the equation (2).

$$y_i = \sum_{j=1}^{N_h} w_{ij}u_j - \theta = W_i^T U \qquad i=1,2,\ldots,m \tag{2}$$

Where $W_i = (w_{i1}, w_{i2}, \ldots, w_{iN_h}, -\theta)^T$, $U = (u_1, u_2, \ldots, u_{N_h}, 1)^T$ .The learning of the RBF network model classifies into two processes. The first one, the central value of the Gaussian function C_j and standardization constant σ_j of each hidden layer node are determined by all the input samples. The second one, after the hidden layer parameters determined, the weights of the output layer W_i can be gained by applying with the least square theory.

Sand Liquefaction Limit State Function

Among the 344 liquefaction site data of 25 strong earthquakes in the world (Xie junfei 1984), there are 206 liquefaction sites and 138 non-liquefaction sites. According to H.B.

Seed's (2001) equation, the cyclic stress ratio of saturated sand layer caused by earthquake ground motion is

$$CSR = 0.65 \frac{\sigma_v}{\sigma_v^{'}} \frac{a_{\max}}{g} r_d MSF^{-1} \quad (3)$$

Where σ_v and $\sigma_v^{'}$ are the total and effective vertical overburden stress, respectively; $a_{\max}$ is the peak value of the horizontal ground motion acceleration; g is the acceleration of gravity; r_d is the stress reduction coefficient; and MSF is the magnitude scaling factor. NCEER suggests using the equation (4) to determine the value of MSF:

$$MSF = (M / 7.5)^{-2.56} \quad (4)$$

The cyclic resistance ratio (CRR) of saturated sand is mainly determined by soil density (adopting SPT-N value representation), vertical overburden stress (σ_v or $\sigma_v^{'}$) , the peak value of horizontal ground motion acceleration, seismic scale and so on. So the liquefaction potential function of saturated sand can be shown as:

$$L = f(N, \sigma_v, \sigma_v^{'}, a, M) \quad (5)$$

In order to calculate conveniently, a reduced acceleration is adopted which considers the influence of horizontal ground motion acceleration and seismic scale defined as 7.5:

$$A_M = \frac{a_{\max}}{g} MSF^{-1} \quad (6)$$

The input layer of the neural network model have 4 neurons, namely $N_1, \sigma_v, \sigma_v^{'}$ and A_M. The determined method of the liquefaction limit state equation of saturated sand proposed by JUANG *et al* (2000) is that any variable determining liquefaction potential such as reduced acceleration A_M, with the trained artificial neural network method, is used to check their state whether from one state convert to another state. Then, through the increase or decrease of the variable, the critical point of converting state is resulted (Fig.2).For example, the liquefaction point A, if the variable A_M is decreased (equals to reduce the seismic stress), becomes non-liquefaction, or non-liquefaction point B, and if the variable A_M is increased, it gets into liquefaction. Thus the critical value of CSR can be found. Then curve fitting of all the critical values of CSR and N_1 can make the liquefaction limit state curve and the critical cyclic resistance ratio obtained:

$$CRR_{cri} = 0.0002 N_1^{2} + 0.005 N_1 + 0.03 \quad (7)$$

The limit state curve of saturated sand of the 344 site data is marked in Fig 3.

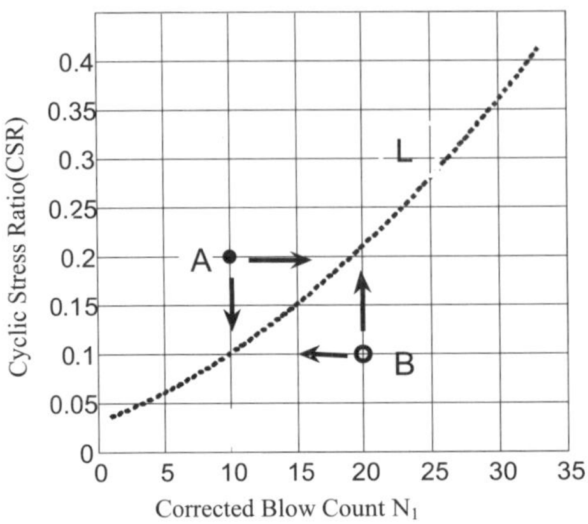

Fig 2. Request limited state curve sketch map

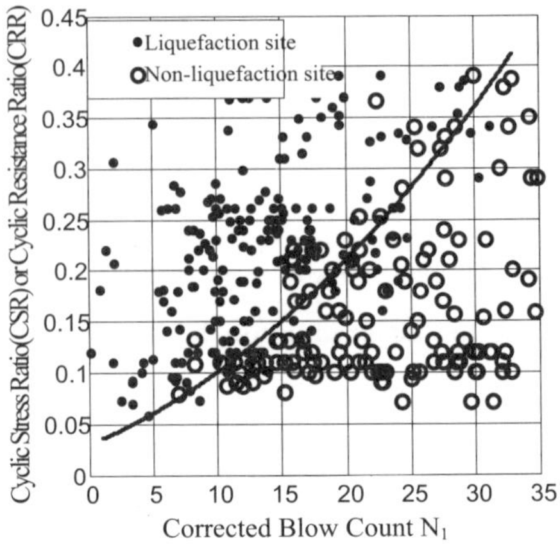

Fig 3. Saturated sand liquefaction limit conditional curve

When the equivalent cyclic stress ratio of soil layers caused by earthquake ground motion is great than CRR_{cri}, the critical cyclic resistance ratio determined in the equation (7), the saturated sand layer will be a liquefaction case, otherwise a non-liquefied case.

PROBABILITY EVALUATION METHOD OF SAND LIQUEFACTION POTENTIAL

Probability Density Function of Sand Liquefaction Potential

The cyclic resistance safety factor of sand liquefaction can be defined as:

$$F_s = CRR_{cri} / CSR \tag{8}$$

Where CSR is a calculated cyclic stress ratio generated by the earthquake ground motion, calculated by the equation (3); CRR_{cri} is a critical cyclic resistance ratio of sand liquefaction, calculated by the equation (7).

With the above 344 site data, the equation (8) is used to calculate the cyclic resistance safety factor of sand liquefaction for every sample. Fig.4 is a histogram describing the safety factor of liquefaction and non-liquefaction cases. By statistic, the probability density function of liquefaction $f_L(F_s)$ and non-liquefaction $f_{NL}(F_s)$ is shown as follows:

$$f_L(F_s) = \frac{1}{F_s\sqrt{2\pi\sigma_L^2}}\exp\left[-\frac{(\ln(F_s)-\mu_L)^2}{2\sigma_L^2}\right] \tag{9a}$$

$$f_{NL}(F_s) = \frac{1}{F_s\sqrt{2\pi\sigma_{NL}^2}}\exp\left[-\frac{(\ln(F_s)-\mu_{NL})^2}{2\sigma_{NL}^2}\right] \tag{9b}$$

where $\mu_L = -0.4627$, $\sigma_L = 0.443$, $\mu_{NL} = 0.4507$, $\sigma_{NL} = 0.4753$.

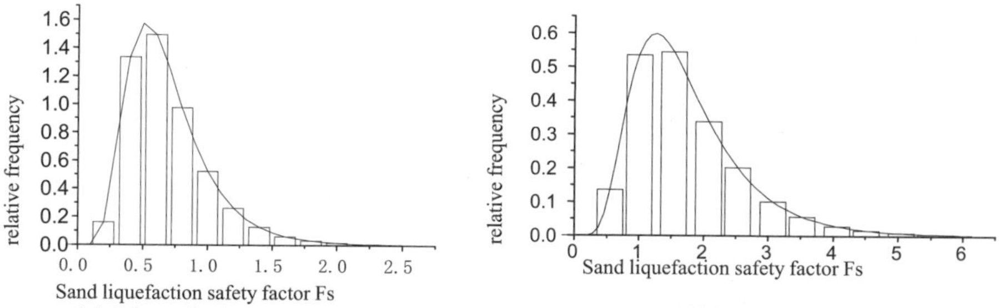

(a) Liquefied Cases (b) Non-liquefied Cases

Fig 4. Probability distribution of the sand liquefaction safety factor

According to the basic concept of probability, if the stylebook is large enough, then

$$P(L/F_s) = f_L(F_s)/[f_L(F_s) + f_{NL}(F_s)] \tag{10}$$

Where $P(L/F_s)$= the liquefaction probability of saturated sand for a given safety factor. With the equation (9) and (10), the liquefaction probability of the 344 site data can be calculated and a Scatters diagram can be drawn accordingly which is shown in Fig.5, and the fitting curve can be presented as:

$$P_L = 1/(1 + F_s^{\,4.297}) \tag{11}$$

If the safety factor equals to 1, the probability of liquefaction or non-liquefaction is both 50%. Juang (2000) pointed out that if the safety factor $F_s = 1$, 30% of the liquefaction probability by H. B. Seed's empirical formula. So it is shown that the cyclic resistance stress curve proposed by H. B. Seed's empirical formula is not the same as the limit state curve of sand liquefaction. By shifting equation (7) and (11), the sand liquefaction resistance stress curve under the different probability can be shown as:

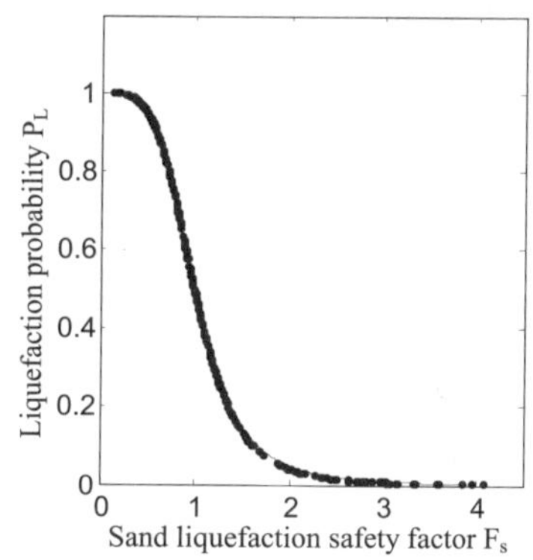

Fig.5 Relation curve of P_L with F_s

$$CRR = [P_L/(1 - P_L)]^{0.233} \cdot CRR_{cri} \tag{12a}$$

$$CRR = [P_L/(1 - P_L)]^{0.233} \cdot (0.0002N_1^{\,2} + 0.005N_1 + 0.03) \tag{12b}$$

The Evaluation Criteria of Sand Liquefaction Potential

In order to be practical and convenient in engineering projects, it suggests that the liquefaction potential of saturated sand is classified into 3 grades under the different liquefaction probability level, and the criteria suggested is shown in table1. According to the importance of engineering projects, an acceptable liquefaction probability level is defined and the estimation criteria of sand liquefaction with different probability can be given by the equation (12). The probability evaluation of sand liquefaction is different to the estimation of sand liquefaction probability. the former is to estimate whether the sand liquefaction will happen under the given probability level. while The latter is given the liquefaction probability of the engineering site so as to make relative decisions,

Table 1. Standard for probability evaluation of saturated sand liquefaction

Liquefaction probability level	Sand liquefaction safety factor	Liquefaction potential evaluation
$0.0 \leq P_L < 0.30$	$F_s \geq 1.2$	non-liquefaction
$0.30 \leq P_L < 0.70$	$0.81 < F_s < 1.2$	possible liquefaction
$0.70 \leq P_L < 1.0$	$F_s \leq 0.81$	liquefaction

CONCLUSION

Based on the RBF neural network model, this paper constructs the empirical equations of the limit state curve and critical cyclic resistance ratio curve of saturated sand as well as the empirical equation between the safety factor and liquefaction probability of saturated sand, then deduces the empirical equation of the cyclic resistance ratio of saturated sand under the different probability. The equation is simple and practical, which makes the sand liquefaction probabilistic estimation on engineering sites as easy and convenient as the traditional deterministic method of sand liquefaction estimation. So it is possible that the method of sand liquefaction probability estimation is applied in the engineering practice and adopted in codes for seismic design.

REFERENCES

Juang,C.H.,Chen J., Tao J., and Andrus, R.D. (2000). "Risk-based liquefaction potential evaluation using standard penetration tests". *J. Can. Geotech*. 37:6, 1195-1208.

Xie Junfei. (1984). "Some Comments on the Formula for Estimation the Liquefaction of Sand in Revised Seismic Design Code". *Earthquake Engineering and Engineering Vibration* 4(2), 95-126.

Youd T.L, and Idriss I. M. et al. (2001). "Liquefaction resistance of soils: summary report from the 1996 NCEER and 1998 NCEER/NSF workshops on evaluation of liquefaction resistance of soils". *Geotechnical and Geoenvironmental Engineering. ASCE,* 127:10, 297-313.

International Conference on Geotechnical Engineering for Disaster Mitigation & Rehabilitation
Chu, Phoon & Yong (eds) © 2005 World Scientific Publishing Company ISBN 981-256-469-1

POTENTIAL EARTHQUAKE RISK OF BUILDINGS IN SRI LANKA

P. B. R. DISSANAYAKE
Department of Civil Engineering, University of Peradeniya
Peradeniya, 20400, Sri Lanka

In this paper, seismic hazard and hazard mitigation in Greater Colombo, Sri Lanka are studied. The objectives of the study are to evaluate the characteristics of local site response against different levels of seismic action and to identify the areas vulnerable to liquefaction damage, and thereby to make a seismic evaluation and hazard level distribution of existing buildings located in the area. The methodology included a series of investigations involving the exploration of the geotechnical structure of the study area, and structural evaluation of existing building stocks against earthquake loading. The analysis demonstrated the possibility of up to twenty times variation in the level of ground shaking at two distinct locations for a particular magnitude of earthquake. A structural evaluation of buildings in the area confirmed that un-reinforced masonry buildings built before 1960 are vulnerable against seismic loading. For the case of moderate to high level of shaking more than 10% of the buildings are exposed to major damage.

INTRODUCTION

Sri Lanka is considered to be in an aseismic zone away from major plate boundaries or any active faults. However, historically there had been many seismic events within Sri Lanka and in the neighbouring areas, which had considerable influence on the island as shown in Abaykoon (1983). Even though some of these are very small in magnitude and some had epicenters far away from the country, the necessity to study the seismic effects on Greater Colombo area has been realized by engineering professionals.

Colombo is the commercial capital of Sri Lanka with having more than 2 million population and further expansion of the city is expected with country's growing economy. Initial seismic risk analysis in Sri Lanka and demonstrated that Colombo is more vulnerable to seismic events than any other part of the country: Abaykoon (1998). Considering the importance of Colombo, a seismic status analysis of Greater Colombo area is contemporary issue.

AIM AND OBJECTIVE

The aim of the paper is to study the seismic hazard and hazard mitigation of Greater Colombo area of Sri Lanka. The main outcome of the study would be an indicator of the seismic status of Greater Colombo and it will provide necessary knowledge to the geotechnical and structural engineers in assessing the local site conditions against seismic risk. It is further expected to help local engineers in introducing appropriate allowances in structural design practice within the area concerned and to adopt preventive or remedial measures to reduce the seismic risk and hazard in critical areas of interest.

The objectives of the study are as follows;

i. Identification of the geotechnical condition,
ii. Identification of locations susceptible to soil liquefaction, and
iii. Seismic evaluation of existing building stocks

METHODOLOGY

Geotechnical condition & Dynamic soil parameters

About 300 boreholes at 200 locations situated within Greater Colombo area have been used to determine the stratigraphy of the study area. This data have been collected from the local consultancy firms and government departments which provide geotechnical investigation services for construction purposes.

The determination of the dynamic soil parameters involves a series of successive calculations for valuating N_{60}, corrected SPT, effective stress, total stress, relative density, untrained shear strength, void ratio, shear modulus and shear wave velocity for each of the soil layers considered. This is a time consuming process involving a series of repetitive calculations with several interpolations. Therefore, it has been decided to formulate all these calculations in a spreadsheet program. This program requires only the entry of soil type, SPT values, and the depth of water table just as it appears in a borehole log sheet. The programme processes all the necessary calculations corresponding to each soil type and displays the soil type, unit weight, shear wave velocity, and thickness along with the depth of ground water table and location details.

Seismic evaluation of existing buildings

There are many types of buildings in Colombo with distinct structural settings. These buildings can be classified upon many aspects such as, the number of stories, construction techniques, materials of construction, period of construction, building configuration and etc. During the current study, many organizations including National Building Research Organization, National Housing Development Authority, Colombo Municipal Council, Department of Census and Statistics were contacted to explore the possibility of getting useful technical information on the building stocks of Colombo.

However, other than a report listing details of 202 historical buildings in Colombo Urban area no data were available as shown by Lewcock (1980). Considering the general construction type of the buildings and the period where there are major changes occur in the construction practices, they can be grouped into two different categories; namely, those constructed after 1960 and prior to 1960.

Most of the buildings in Colombo fall into the first category. Many of these buildings have been constructed using reinforced concrete and therefore, naturally have some shear capacity to tolerate low level of seismic shaking. The buildings in the second category, i.e. those constructed prior to 1960, lack reinforcement. According to Lewcock (1980), a few hundreds of such buildings are found in the Greater Colombo area. Most of these buildings are in poor condition having no capacity to resist any seismic loading. Some of them are situated in Colombo Fort and its neighbouring areas which are thickly populated. Therefore, these buildings are the most vulnerable category in Greater Colombo. It was decided to focus on assessing this type of buildings in the current study.

Forty old buildings (pre 1960 era) identified as weak, and ten buildings constructed after 1960s' were considered for the seismic evaluation. These fifty buildings include twenty seven un-reinforced masonry buildings, twenty two concrete frame buildings and one light metal building.

Evaluation was carried out in the following sequence.

i. Sketch plan showing the location of the building was drawn
ii. Details such as the year of construction, number of stories, usage of building, numbers of occupants, any special identifiers, etc. were obtained.
iii. Evaluation of the building type / facility class
iv. Efforts were made to obtain any other relevant information.

The seismic vulnerability of the structures was determined based on the evaluation procedures recommended by the FEMA of USA (Federal Emergency Management Agency of United States of America). In this procedure each building is given a set of basic and modifying scores to account for various structural conditions affecting the seismic stability. The structural weaknesses were visually inspected from outside the building. The corresponding score sets have been developed based on the expert's opinion obtained to assess the seismic vulnerability of California buildings for different levels of ground shaking and adjusted to evaluate the seismic performance of buildings in the regions out of California where low level of shaking is expected. The seismic hazard score for a particular building is the algebraic sum of the basic and modifying scores corresponding to the level of ground shaking considered. The final score typically range from 0 to 6, with higher scores corresponding to better seismic performance. These hazard scores have been formulated in such a way that the antilogarithm of the negative of the hazard score of a particular building will give the probability against a major structural damage. The analysis was carried out on all the structures against the three levels of shaking namely low, moderate and high level as reported by FEMA (1988). The performance modification factor to account for the effect of local soil profiles is another aspect to be accounted in the assessment procedure. The soil conditions within the study area generally comes into SL1 representing stable deposits of sands, gravels and stiff clays and into SL3 representing the and short to medium stiff clays and sands, exceeding 30 ft in thickness: FEMA (1988). The corresponding modification factors are 0.0 for SL1 and - 0.6 and - 0.8 for SL3 in respect of low rise buildings and 8 to 20 story buildings respectively. In most of the locations considered for the structural assessment there is not enough soil data available in the vicinity of the building assessed. Therefore, it was decided to take up the worst case of the presence of soil profile SL3 in all the locations as recommended by FEMA (1988). The sum of the hazardous scores of each of the buildings indicates the vulnerability of the structure against the level of shaking considered.

OBSERVATION AND DISCUSSION

Soil/rock profiles in the project area and their properties

From the analysis of borehole logs it was found that the depth of the top soil layers is in the range of 5 m to 25 m. Sand profiles were observed in most part of the Colombo Municipal areas and some other adjoining urban areas along the Galle Road up to Moratuwa. Clay deposits were observed as isolated pockets within Colombo Municipality except near Borella where sand as well as clay layers have been observed. Rajagiriya, Malabe, Nugegoda, Nawala, Athurugiriya, Orugodawatta, Negombo, Ja-ela, Etul-Kotte, Boralesgamuwa, Wattala, Kelaniya are also some other locations where sand and clay profiles were observed. The peat deposits were generally observed in the low lying areas of Greater Colombo (some of these areas are now reclaimed) namely Peliyagoda, Orugodawatta, Blomendhal, Yakbedda, Nawala, Battramulla, Maligawatta etc: Senanayake, 1996. In most of the locations the soil profile ends up with the weathered rock.

Seismic evaluation of buildings in Colombo

The final seismic hazard scores corresponding to each of the building assessed are given in Table 1.

Table 1 Structural hazard scores of the buildings assessed in Colombo area

Building ID	Final Score		
	Low Level of Shaking	Moderate Level of Shaking	High Level of Shaking
A1	1.9	1.4	0.4
A2	0.9	0.4	-0.6
A3	1.4	0.9	-0.1
A4	2.4	1.4	0.9
A5	1.4	0.9	-0.1
A6	0.9	0.4	-0.6
A7	0.9	0.4	-0.6
A8	0.4	-0.1	-1.1
A9	1.9	1.4	0.4
A10	0.9	0.4	-0.6

A11	0.9	0.4	-0.6
A12	1.4	0.9	-0.1
A13	1.4	0.4	-0.1
A14	1.9	1.4	0.4
A15	1.9	1.4	0.4
A16	5.9	5.4	4.9
A17	1.9	0.9	0.4
A18	1.4	0.4	0.4
A19	1.4	0.4	0.4
A20	0.9	0.4	-0.1
A21	0.9	0.4	-0.1
A22	1.4	0.9	-0.1
A23	2.4	1.4	0.9
A24	1.9	0.9	0.4
A25	1.9	0.9	0.4
A26	2.4	1.4	0.9
A27	1.9	0.9	0.4
A28	1.4	0.9	-0.1
A29	1.4	0.4	-0.1
A30	0.9	0.4	-0.6
A31	1.9	0.9	0.4
A32	0.9	0.4	-0.6
A33	0.9	0.4	-0.6
A34	1.4	0.4	0.4
A35	0.4	-0.1	-0.6
A36	-0.6	-1.1	-1.6
A37	1.4	0.4	0.4
A38	0.9	0.4	-0.6
A39	1.9	1.4	0.4
A40	1.9	0.4	0.4
A41	-0.1	-0.6	-1.1
A42	0.9	0.4	-0.6
A43	0.9	-0.1	-0.1
A44	2.4	1.4	0.9
A45	1.9	0.9	0.4
A46	2.4	1.4	0.9
A47	1.9	1.4	0.4
A48	1.9	0.9	0.4

A49	1.9	0.9	0.4
A50	1.9	0.9	0.4

CONCLUSIONS

For the case of high level of shaking, all the buildings considered except the light metal structure, registered with poor seismic performance with more than 10% of the buildings exposed to major damage. About 75% of the buildings considered recorded less than 1.0 structural hazard score for moderate level of shaking. All the 27 un-reinforced masonry buildings have been registered with critical hazard scores for all the three levels of shaking. Out of the 22 cases corresponding to concrete framed buildings, 15 satisfied the cut-off score (2.0) for the low level of shaking. The light metal building exhibited best performance with high structural hazard scores. However, as a global measure the basic structural hazard scores can be used to get the overall picture assuming that the average building has no adverse effect due to building configuration or any supplementary strength due to the adoption of any seismic provisions.

It is possible to predict the vulnerability of a building against a given magnitude of earthquake. It is clear that an earthquake of magnitude 6 in Richter scale and epicentral distance grater than 160 km may cause no major damage in any type of the structures present within the study area.

ACKNOWLEDGEMENT

This work was supported by National Science Foundation of Sri Lanka under Research Grants RG/2001/E/03 and RG/2002/E/01, and these grants are gratefully acknowledged.

REFERENCES

Abaykoon S.B.S (1983) "*Seismic Risk Analysis of Sri Lanka,*" Journal of the Geological Society of Sri Lanka, Vol. 6, pp 65-72.

Abaykoon S.B.S. (1998), "Seismic Response of Low Lying Areas in Colombo Sri Lanka," Engineer, Journal of Institution of Engineers, Sri Lanka, Vol. xxviii, Journal No-02, 29-36.

Federal Emergency Management Agency, (1988), "Rapid Visual Screening of buildings for Potential Seismic Hazards; A Handbook", FEMA 154, July 1988, Earthquake Hazard Reduction Series 41, pp 185.

Lewcock, R. (1980) "Report on the listing and presentation of historic buildings and zones in the Colombo Urban Area", United Nations Centre for Human Settlements (Habitat), August-September.

Senanayake, K.S., (1996) "Geotechnical Mapping of Low-lying Areas in and around Colombo City", Asian regional Symposium on Geotechnical Problems and Practices in Foundation Engineering, Colombo, Sri Lanka.

International Conference on Geotechnical Engineering for Disaster Mitigation & Rehabilitation
Chu, Phoon & Yong (eds) © 2005 World Scientific Publishing Company ISBN 981-256-469-1

IRAN'S STRATEGY ON NATURAL DISASTER RISK MANAGEMENT

S.J.GHANEFAR
Phd student, Facualty Engineering, Tehran University
Tehran, Iran

A. TABAR
Deputy General Director, Technical, Criteria Codification and Earthquake Risk Reduction Affairs Bureau MPO
Tehran, Iran

Having confronted with more than 34 out of 41 types of known natural disasters, Iran has one of the longest histories in challenging the natural disasters. In recent years there have been one earthquake with extremely high human and property losses, every five years on the average. Among these earthquakes of the recent years, the 2003 Bam, the 2004 Baladeh, and the 2005 Zarand earthquakes are the most manifesting ones. With respect to the number of natural disaster and the socio-economic impacts, extensive planning have been implemented in the course of the recent two decades in order to mitigate natural disaster effects that in this study some of them will be discussed.

INTRODUCTION

With respect to rise in the number of natural disasters and the following mortality and socio-economic impacts, extensive planning have been implemented in the course of the recent two decades in order to mitigate natural disasters effects and consequences to an acceptable level, some of which will be discussed in the following:

Mainstreaming Disaster Risk Management into Development Policy Planning and Implementation in Iran

A. Revision of the Natural Disaster Emergency Management Structure of the Country

In the course of the recent years, especially after occurrence of the Bam earthquake, a modification to the emergency management system structure has been adopted by taking a new approach regarding more importance and attention to pre planning (public training, promotion of safety precautions, early warning, awareness and preparedness).
Considering the great importance of natural disaster management the First Vise president of the country was nominated by the High Council of natural disaster prevention and management, to observe the trend of related operations.

B. Natural Disaster Management in Policies and Development Plans of the Country

With new approach, due to the great importance of NDM, the necessity of sustainable macro planning for this issue is observed in the 4th Cultural, Social and Economic Plan of the Country. The following pivots have been confirmed in macro policies of the country :

1. Prevention and disaster risk reduction in urban & rural regions and enhancement of safety factor in new constructions.

2. Earthquake risk reduction at present status of urban and rural sectors with human life protection based policies.

3. Establishment of integrated information management with contribution of relevant scientific research institutes and key executive organizations for prompt and precise information dissemination during & after disaster – particularly floods and earthquake.

4. Early warning and prompt information dissemination regarding regions vulnerable and prone to natural disaster, by utilization of early warning systems technology.

5. Codification and implementation of integrated research and scientific programs for social and psychological rehabilitation of victims and comprehensive renovation of damaged urban and rural regions.

6. Expansion, support and regulation of multilateral scientific research on earthquake through development and completion of seismic & strong motion networks, mobilization of earthquake laboratories network in the Country, completion of N.D. early warning systems, N.D. risk zoning maps, and seismic micro zoning of the country.

7. Training and encouragement of authorities and general public to become more acquainted with earthquake prevention and develop adequate judgment along with the knowledge to take effective action during a disaster.

8. Establishing an assessment system for N.D. risk reduction management approach in the country in order to have an efficient disaster management system.

Performed Measures

In recent two decades with knowledge enhancement of lecturers in universities and research centers of the Country and particular attention of authorities to natural disasters noticeable results have been achieved.

Parts of these achievements are as follows:

1. Research

In addition to a noticeable increase in number of specialists, university lecturers and research centers involved in natural disasters and disaster mitigation, enjoyment of modern technologies in development activities, required applied research in preventive measures

2. Preparation and codification of Technical Managerial and Safety Codes, Criteria and Guidelines

Iranian Code of Practice for Seismic Resistant Design of Buildings (Standard no. 2800) and retrofitting guideline for existing structures are among codified and notified codes in the Country.

3. Retrofitting Plan for Public Buildings, Infrastructure Facilities, and Lifelines as a Preventive Measure

Allocating funds, guiding and monitoring retrofitting research studies and performing projects including:
- Major hospitals, rescue centers & fire stations
- Strategic buildings, military and police headquarters
- Universities and schools
- Tehran major bridges & important railroad bridges
- Key fuel facilities and main refineries of the Country
- Major telecommunications centers, radios and TV stations
- Power plants, transmitting & distributing networks, and urban water networks

4. Rural Housing Integrated Plan as a Preventive Measure

5. Natural Disasters Information Networks

Currently there are variety of information networks in the Country, and particularly in geological data and meteorological information analysis some great goals have been accomplished.

Among the aforesaid networks are:
- Iranian Space Agency
- Iran Strong Motion Network
- National Seismic Network of IIEES
- National Geosciences Database of Iran
- Seismic Geodynamics Database of NCC
- IR of Iran Meteorological Organization
- Institute of Geophysics Seismograph Networks

6. Early Warning and Forecasting Systems

In this regard a large number of research have been initiated since Bam earthquake, of which extensive activities have been planned and implemented at Natural Disaster Research Institute of Iran. The implemented research and studies have led to a series of satisfactory out put for forecasting the time, location, and the intensity of the earthquake within a satisfactory range. These designed equipment has been installed as a pilot project in provinces of the country and based on the resulted data, it will be available internationally in near future.

7. International and Regional Cooperation & Exchange of Views

Iran has experienced the establishment of regional centers with regional and international communities, including:

- Regional Center on Urban Water Management (RCUWM) – Tehran
- Regional center on Drought Management at IRIMO
- Natural Disaster Risk Management as joint projects with international organizations :
  - Improving Access to Information on Disaster Risk Management by Establishing Information System
  - Urban Earthquake Risk Reduction Program with Focus on Local & Intermediate Levels in two Selected Cities
  - Facilitating Knowledge Networking for Disaster Risk Management at the Sub-regional Level
- National Disaster Risk Management System as a joint project

CONCLUSION

The Country's main approach towards natural disaster was focused on post-disaster affairs management, but in recent years the preventive measures and comprehensive approaches of disaster risk reduction as, public training and promotion of the culture of prevention, enhancement of applied research, codification of required guidelines, expansion of natural disaster information networks, ... etc. were highly considered by the Government. And the results of the aforementioned policies & activities were clearly observed in Zarand earthquake disaster management in 2004.

REFRENCES

Low of the 4th Economic, Social & Cultural Development Plan of the I.R. of Iran (2005-2009), 2004, MPO

Iran National Report on Natural Disaster Management in WCDR, June 2005, Hyogo, Japan

BHRC (2005): Iranian Code of Practice for Seismic Resistant Design of Buildings, Standard No. 2800, BHRC- PN S 253 3rd ed, Building and Housing Research Center Press.

Quryshi, M and Tehranizadeh, M. (1997): National Earthquake Research Report, Scientific Research Council of Iran, Earthquake Committee Press.

Tehranizadeh, M. (1996): Main Issues of National Earthquake Research Plan of Country, Scientific Research Council of Iran, Civil and Earthquake Commission Press.

International Conference on Geotechnical Engineering for Disaster Mitigation & Rehabilitation
Chu, Phoon & Yong (eds) © 2005 World Scientific Publishing Company ISBN 981-256-469-1

MODELING OF RISK ACCEPTANCE CRITERIA FOR TUNNEL AND UNDERGROUND ENGINEERING

HU QUNFANG
Department of Geotechnical Engineering, Tongji University, 1239 siping road Shanghai 200092 China

HUANG HONGWEI
Department of Geotechnical Engineering, Tongji University, 1239 siping road Shanghai 200092 China

The importance of risk acceptance criteria research and the calculation principle for determining them are analyzed in according with the present state of art. And some fundamental risk acceptance criteria, such as the individual risk (IR), societal risk (SR), economical risk and environmental risk, are comprehensively introduced. The paper furthermore addresses and discusses some key points, future application and development of these criteria for tunnel and underground engineering. Finally a number of problems are highlighted and the needs for further research are stressed.

INTRODUCTION

The 21st is the big century of the utility and development for tunnelling and underground works (Song Kezhi et al; 2004). Due to the inherent uncertainties, including ground and groundwater conditions and their variability, it might be significant cost overrun, fatalities and time delays as well as environmental risks during the tunnelling construction. Also, for the complexity and immaturity of tunnelling technique in urban areas there is a risk of damage to a range of third party about property and persons, which will be of particular concern by public where heritage designated buildings are involved. Furthermore, there is a risk to public protests affecting the course of the projects. Finally, there has been an increasing societal concern on sustainable developments focusing on the conservation of the environment, the welfare and safety of the individual during the last decade. As demonstrated by spectacular tunnel collapse and other fatal disasters in the recent past like shanghai (Ministry of Construction, China, 2003), there is a potential for large scale accidents during tunnelling. In order to reduce and mitigation the risk, the research and application of some methods of risk management and risk control for tunnelling and underground works are paid attention in recent years.

As a consequence the methods of risk analysis in civil engineering, mainly developed during the last decades, are increasing gaining importance as a useful tool in tunnelling engineering applications (M.H.Faber; 2003). It is important to realize that risk analysis is very complex theory which is often entwined with not only many technical aspects but also some political, psychological and social processes. A number of issues that risk analysis attempt to resolve include (Hood C.C et al; 1992): What information is required for 'scientific and rational' risk management and how to make a decision? Where is the line to be drawn between that are to be managed by individuals, group or corporations? Who is to bear what level of risk? What actions make what difference to risk outcomes? These matters or problems are not easily resolved, are not for risk analysis to solve alone and are all related to risk acceptance criteria; namely, what risks are acceptable? And some risk acceptance criteria for decision-making should be established qualitatively or quantitatively in advance (Ohtsubo.H et al; 2000). The paper attempts to summarize broadly an overview of some risk acceptance criteria irrespective of the type of applicable risk (individual, social, environment and other risk). It ends with some discussion and conclusion of the need to establishing some risk acceptance criteria for tunnelling and underground engineering.

THE MODEL OF RISK ACCEPTANCE CRITERION

In the study of risk analysis and assessment, the risk control and reduction measures are taken involving some money, techniques and labors. And these risk measures are determined based on the balance of loss (damage) or benefit. It is obvious that risk is in interrelation with investment and benefit. How to draw a line or play game between the risk and investment or benefit for decision-making is complied with a risk acceptance criterion. The development and implementation of risk acceptance criteria involves (M.H.Faber; 2003): (1) Perception of risk, ensure that levels of system risk, such as safety, economy and environment, is acceptance (or tolerable); (2) Formal decision analysis: analytical techniques to balance or compare risks against benefit (e.g. risk cost-benefit analysis, life-cycle cost analysis); (3) Regulatory safety goals: legislative and statutory framework for the development and enforcement of risk acceptance criteria.

As is known to us, the risk acceptance criteria are the foundation of risk analysis for decision-making. They generally are adopted by the U.S Nuclear Regulatory Commission (NRC), U.K Health and Safety Executive (HSE), Dutch Technical Advisory Committee on Water Defenses (TWA) and other regulatory authorities. Mainly, there are two basic principles for defining the risk level used for a qualitative or quantitative risk analysis (Melchers RE et al; 1993, Sharp JV et al; 1993): 'As Low As Reasonably Possible (ALARP)' and 'As Low As Reasonably Achievable (ALARA)'. The definition for such terms as 'Low', 'Reasonably', 'Possible' and 'Attainable' are highly subjective and prone to being interpreted in a conservative manner. Some expressions and models have been made to define these criteria in more tangible limit in terms of quantitative risks.

THE CLASSIFICATION OF RISK ACCEPTANCE CRITERIA

The risk acceptance criteria are a complex attribute that results from three influenced factors, safety (personal risk and social risk) , economy (potential economical loss, construction schedule) and environment, from a technical perspective of the basic classification of engineering risk (Li Dianqing et al; 2003). And they are established based on not only some quantitative indexes like economical loss or time delays, but also many unquantized indexes such as potential environment damage, the monetary valuation of human life and the public influence, etc. Therefore, the establishment of risk acceptance criteria is a complicated, difficult and controversial task (Ditlevsen O et al; 1996). The risk acceptance criteria can be classified into four elementary categories from the recent overview of recent research around the world (Li Dianqing et al; 2003, S. N. Jonkman et al; 2003): (1) Individual Risk; (2) Social Risk; (3) Economical Risk; and (4) Environmental Risk. Every section starts with an overview of the risk mathematical definition, and then the risk acceptance criteria and their application are described as follows.

Individual Risk

The individual risk (IR), as used by the Dutch Ministry of Housing, Spatial Planning and Environment (VROM), is defined as the probability that an average unprotected person, permanently present at a certain location, is killed due to an accident resulting from a hazardous activity (P.H. Bottelberghs et al; 2000).

$$IR = P_f \cdot P_{d|f} \tag{1}$$

Where P_f is the probability of failure and $P_{d|f}$ probability of dying of an individual in the case of failure, assuming the permanent unprotected presence of the individual.

The risk individual is highly subjectively, which is determined by the individual predilection whether or not the individual is actually voluntary or involuntary. Besides the individual risk as mentioned above, four other expressions are described (T. Bedford et al; 2001). (1) The Loss of Life Expectancy shows the decrease of life expectancy due to various caused; (2) The Delta Yearly Probability of Death computes the intensity at which a given activity is to be performed (in suitable units) to increase the yearly probability of death by 10–6; (3) The Activity Specific Hourly Mortality Rate reflects the probability per time unit while engaged in a specified activity. (4) A variant is the Death per Unit Activity, which replaces the time unit by a unit measuring the amount of activity. A different definition is used by the UK's health and safety executive (HSE) (Health and Safety Executive, 1989) that is the risk which a typical user of a development is exposed to a dangerous dose or worse of toxic substance, heat or blast overpressure. To limit the risks, there are many criteria such as ALARA, Risk Matrix, AFR (Annual Fatality Risk), AIR (Average Individual Risk) and AI (Aggregated Indicator) etc.

Social risk

The societal risk is defined as "the relationship between frequency and the number of people suffering from a specified level of harm in a given population from the realization of specified hazards" (Institute of Chemical Engineering; 1985). Where individual risk gives the probability of dying on a certain location, the societal risk gives a number for a whole area, no matter precisely where the harm occurs within that area. To limit the social risks, there are many criteria such as ALARP, Risk Matrix, F-N curve, PLL (Potential Loss of Life), FAR (Fatal Accident Rate), VIIH (Value of Injuries and Ill Health), ICAF (Implied Cost of Averting a Facility) and LQI (Life Quality Index) etc.

$$1-F_N(x)=P(N>x)=\int_x^{\infty} f_N(x)dx \tag{2}$$

Where $f_N(x)$ is the probability density function (PDF) of the number of fatalities per year; $F_N(x)$ is the cumulative probability distribution function (CPDF) of the number of fatalities per year, signifying the probability of less than fatalities per year.

Economic risk

Besides the loss of life due to certain activities, the economic risks play an important role in risk analysis. A FD-curve displays the probability of exceedance as a function of the economic damage (S. N. Jonkman *et al;* 2003). The FD-curve and the expected value of the economic damage can be derived from the PDF of the economic damage ($f_D(x)$).

$$1-F_D(x)=P(D>x)=\int_x^{\infty} f_D(x)dx \tag{3}$$

$$E(D)=\int_0^{\infty} x\cdot f_D(x)dx \tag{4}$$

Where $F_D(x)$ is the cumulative probability distribution function of the economic loss and $E(D)$ is the expected value of the economic damage.

Novgorodsky's study on the economic risks in the Russian region (A.N.Yelokhin et al; 1997) is an example of the use of a FD-curve. Jansen (H.M.A et al; 1998) has tried to obtain a financial economic risk limit in the form of a FD-curve. The expected value of the economic damage is used as part of cost benefit analyses of flood prevention measures in the UK (D.J. Parker et al; 1997) and in The Netherlands (H.M. Zhou et al; 1995). A limit for the expected economic damage per year for dams has been proposed by BC Hydro (Hydropower & Dams; 1998), in which the financial risks for one dam should not exceed: $E(D)<10000$ (US $/per year).

Environmental risk

The Environmental Risk (ER) is different with the individual risk and social risk for engineering construction activities in a certain circumstance. There is a risk of damage to a range of third property and persons and of pollution to environment. The competitive standing of the Norwegian offshore sector (NORSOK) has proposed the probability of exceedance of the time needed by the ecosystem to recover from the damage as a measure for environmental risk (NORSOK; 1998).

$$1-F_T(x)=P(T>x)=\int_x^{\infty} f_T(x)dx \tag{5}$$

$$E(T)=\int_0^{\infty} x\cdot f_T(x)dx \tag{6}$$

Where $F_T(x)$ is the cumulative probability distribution function of the recovery time; $f_T(x)$ is the probability density function of the recovery time of the ecosystem. A Limit for the environmental risks of offshore activities is set by NORSOK. The following limit to determine the acceptable risk for oil platforms is derived from the quoted criterion (NORSOK; 1998).

$$1 - F_T(x) < \frac{0.05}{x} \tag{7}$$

THE MODEL OF RISK ACCEPTANCE CRITERION FOR TUNNEL AND UNDERGROUND WORKS

It is worthwhile to recognize that the criteria and calculations of risk acceptance have a fundamental and special bearing to different industries and engineering. The determination of risk acceptance criteria is a complex task which is entwined with comprehensive several courses and much technique knowledge. And there are many problems on which need to spend more research. At the present time, the study of risk acceptance criteria of risk analysis is a hot research point to which a great number of researchers pay attention in the world (M.H.Faber; 2003, Li Dianqing et al; 2003). Furthermore, it is the beginning stage of the risk analysis that is applied currently in the tunnel and underground engineering. Although it is fortunate that we have started and made some progress of the application of risk management in tunnel and underground engineering, we still have a great deal of work problems to be resolved. As explained in the above of the paper, the model of the risk acceptance criteria for tunnel and underground work is explained below.

The loss consequences of risk in tunnel and underground engineering may be mainly categorized into four types: economic loss, personnel fatalities, work time delays and the damage to environment. The unitive model of risk occurrence mechanism in tunnel and underground engineering is showed as follow (See Figure 1). Therefore, according to the above risk standards and the model, we can build up a calculation model of risk acceptance criterion for unifying the standard.

$$1 - F_R(x) = P(R > x) = \int_x^\infty f_R(x)dx \tag{8}$$

$$E(R) = \int_0^\infty x f_R(x)dx \tag{9}$$

$$1 - F_R(x) < \frac{C}{x^n} \tag{10}$$

Where $f_R(x)$ is the probability density function (PDF) of the risk loss per unite engineering; $F_R(x)$ is the cumulative probability distribution function (CPDF) of t the risk loss per unite engineering, is the expected value of the risk loss, is the steepness of the limit line and the constant that determines the position of the limit line.

In order to establish the unitive model of risk acceptance criteria in detail, we should design a database, which is used for storing some risk events investigation and record data, and do some research work based on the existing risk acceptance criteria (Beacher

GB et al; 1987) all over the world (See Figure 2). At the present time, we all lack the foundation data and need to do much work of investigation and collection. If we have accumulated enough useful engineering risk data and much experience, the parameter value of formula (8), (9) and (10) can be got from doing statistical study. Moreover, the relation of risk acceptance criteria and the development of national economy will be analyzed and established.

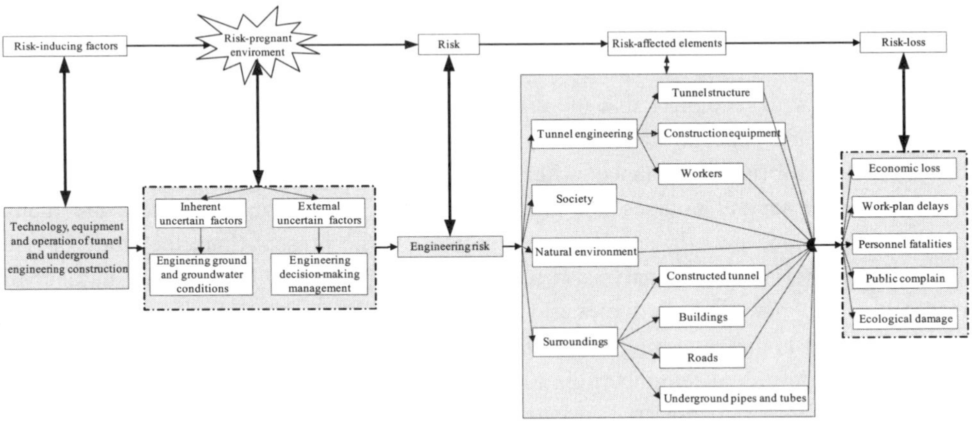

Figure 1 the Model of Risk Occurrence Mechanism in Tunnel and Underground Engineering

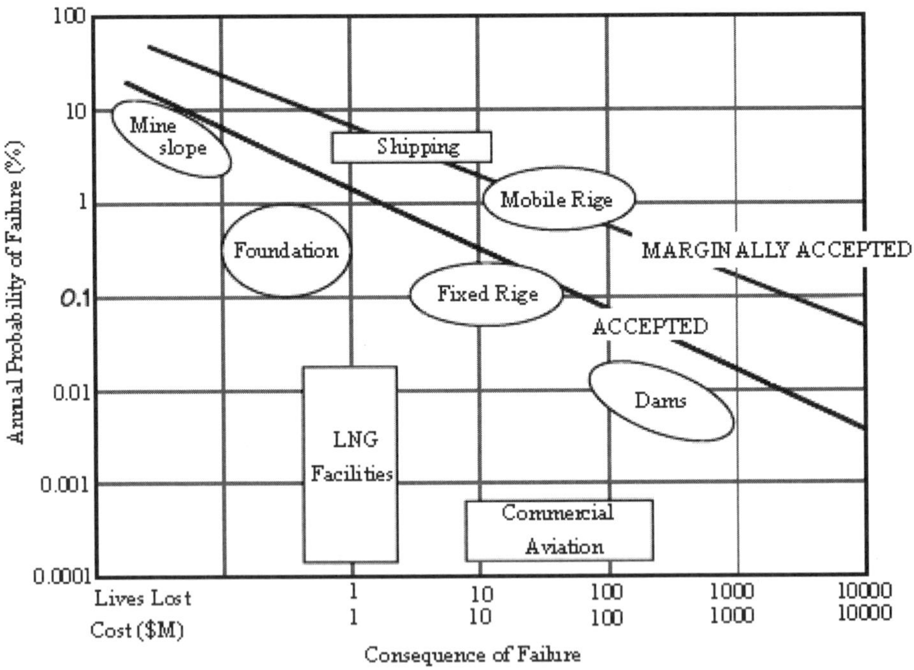

Figure 2 Empirical Rates of Failure for Some Civil Engineering Works

At present time, there is not a technical risk acceptance criterion for the risk analysis in tunnel and underground engineering in China. Furthermore, it is difficult to establish a new one without some basic research in the area. Therefore, in order to carry out the risk management and control and promote the future risk research in tunnel and underground engineering, it is an imperative task for all engineering risk researchers nowadays. And we can get some reference from the reliability-based design theory. The failure rate of tunnel structures can be calculated by the reliability methods considering the variability of design parameters.

Human error is an important source of risk. It is typically found that actual failure rates of engineering construction exceed predicted failure rates through the reliability methods, perhaps as much ad three orders of magnitude. Further research reveals that most of the risk events are the result if human error, for example, some structures not built according to the work plans, materials not meeting the design specification, etc. It is inherently difficult to include human errors involving design, construction, organization and management. Therefore, how to assess the human errors and whether it can be accepted that depend on the context of risk acceptance. But it is a blank research area that needs us to pay attention.

CONCLUSION AND FINAL REMARKS

The application of risk analysis into tunnel and underground engineering in China is a little later than other developed countries. The risk analysis and assessment of Yangtze Channel Tunnel in Shanghai is the first example which is began in 2002 and finished in 2003. Risk analysis is a systematical science which is composed of three main subjects: risk identification and measure, risk assessment and decision-making, and risk management and control. The core of risk in the three subjects is the risk acceptance. Risk analysis involving workers or public or nature resources or economy should be analyzed and assessed based on some risk acceptance criteria, especially when the balance of cost, benefit, risk loss and the public's concerns are to be considered. It can be difficult to perform a comprehensive and quantitative risk analysis because of the many variables involved, insufficient data, distribution of parameters and the lack of a reference base in similar types of structures. So, the application of risk management and control in tunnel and underground engineering is the beginning. And I wish the paper can push and promote the further research of the risk acceptance criteria in tunnel and underground engineering.

REFERENCES

Song Kezhi, Wang Mengshu. 2004) "Shield machine selection lining domestication," *West- China Exploration Engineering*, Vol.5, 99-100 (in Chinese).

Ministry of Construction, China. (2003) *The report of investigation and management of risk accident of Metro No.4 in Shanghai, China.* Vol. 10 (in Chinese).

M.H.Faber, M.G.Stewart. (2003) "Risk assessment for civil engineering facilities: Critical overview and discussion," *Reliability Engineering and System Safety*, Vol. 80, 173-184.

Hood C.C., Jones D.K.C., Pidgeon N.F., and Turner B.A., etc. (1992) *Risk management, risk: analysis, perception and management.* London: the royal society, 92-135.

Ohtsubo.H, Sumi Y. (2000) Committee reports on risk assessment, *Proceedings of the 14th international ship and offshore structures congress*. Nagasaki, Japan, Vol. 2, 1-41.

Melchers RE, Society. Tolerable and the ALARP principle. In: Melchers RE, Stewart MG, Editors. (1993) *Probabilistic risk and hazard assessment.* Netherlands: Balkema, 243-295.

Sharp JV, Kam JC, Birkinshaw M. (1993) Review of criteria for inspection and maintenance and artic engineering (OMAE 1993), Vol.2, New York: *ASME*, 363-371.

Li Dianqing, Tang Wenyong, Zhang Shengkun. (2003) Review of risk acceptance criteria for ocean engineering], *The Ocean engineering*, Vol.21 (5), 96-102 (in Chinese).

Ditlevsen O. (1996) Risk acceptance criteria and design/or decision optimization, *15th IABSE Congress: Structural engineering in consideration of economy, environment and energy*, Copenhagen.

S. N. Jonkman, P.H.A.J.M. van Gelder and J. K. Vrijling. (2003)An overview of quantitative risk measures for loss of life and economic damage. *Journal of Hazardous Materials*, Vol.99 (1-4), 1-30.

A.N.Yelokhin, (1997) Complex risk analysis for Novgorodsky's region's population, in: B.-M. Drottz Sjöberg (Ed.), *Proceedings of the Annual Meeting of the Society for Risk Analysis-Europe*: New Risk Frontiers, Stockholm.

H.M.A. Jansen, (1988) Financieel economische risicolimiet financial economic risk limit, in Dutch), Min. van Volkshuisvesting, *Ruimtelijke Ordening en Milieu, directie stralenbescherming.*

D.J. Parker, C.H. Green, P.M. Thompson, *Urban Flood Protection Benefit*s, Gower Technical Press, 1987.

H.M.Zhou, (1995) *Towards an Operational Risk Assessment in Flood Alleviatio*n, Delft University Press, ISBN 90-407-1146-1.

Hydropower & Dams, (1998) Risk Based Dam Safety Evaluations, *Conference Report: Part 2, Hydropower Dams.*

NORSOK, Risk and emergency preparedness, (Z-013 rev.1), Annex C: Methodology for establishment and use of environmental risk acceptance criteria, (http://www.nts.no/norsok/) 1998.

Beacher, GB, (1987) Geotechnical Risk Analysis User's Guide, *FHWA/*RD-87-011, Fed. Hwy. Admin.

International Conference on Geotechnical Engineering for Disaster Mitigation & Rehabilitation
Chu, Phoon & Yong (eds) © 2005 World Scientific Publishing Company ISBN 981-256-469-1

MINIMIZING KARST RELATED RISKS IN HIGHWAY CONSTRUCTION

M. ZIA ISLAM, P.E.
Michael Baker, Jr., Inc.
201 Gibraltar Road, Suite 120
Horsham, PA 19044, USA

A significant part of South-east Asia constitutes karst geology. Geological hazard related to karst is well-known. Among the various causes of karst collapse or sinkholes, construction activities are the major source. With the rapid growth of infrastructures in south-asian countries in recent times, careful planning is required to minimize the risk of karst collapse. The consequences of karst collapse on highways could be severe and expensive. During 2000-01, a highway agency in the eastern Pennsylvania, USA spent about US$25 million to repair sinkholes during the construction of a highway ramp. This paper discusses the authors' experience with construction induced sinkholes and recommendations for risk minimization during highway design and construction.

INTRODUCTION

Karst collapse or sinkholes occur when water erodes carbonate rocks, such as, limestone and dolomite. Under natural conditions, the formation of new sinkholes during a man's lifetime is relatively rare. In contrast, sinkholes induced by man's activities are comparatively abundant (Newton, 1986). Highway construction activities typically involve lowering of groundwater, diversion of drainage, and large-scale excavations. In karst geology, all of these activities increase the risks of development of sinkholes. Repairing sinkholes at highways has been a common phenomenon in the karstic regions of the United States.

In continental China, the Yunnan-Ghizhou plateau and South China hill-basin karst province have been the areas most severely damaged by karst collapse (Kato et. al. 2003). In Thailand, karst areas are concentrated in the southernmost part where Permian limestone predominates, and middle of the country where Ordovician limestone is the dominant formation. In Malaysia, limestone is present in Kuala Lumpur, Ipoh, and Bau.

Realignment of a highway to a less sinkhole prone area is rarely possible. Therefore, steps need to be taken to assess and minimize the risks related to karst collapses.

CAUSES OF SINKHOLE DEVELOPMENT

Sinkholes induced by construction basically result from the downward migration of soil and other unconsolidated deposits into fractures/openings near the bedrock surface. Infiltration of water in the soft soil zone near the rock crevice increases the unit weight of the soil and induces seepage forces. In addition, the rise in saturation within the soft zone reduces existing suction in the surrounding soil, which decreases the effective strength in the soil. If the strength of soft soil at the rock crevice cannot support the weight of soil above, a 'local failure' zone is developed and the seepage water in the crevice washes these softened soils away. Finally, a small void is created above the rock crevice and gradually enlarges when the hydrological condition is favorable. The stability of this soil-void system is maintained by soils above the void through a mechanism called "arch effect". Cohesion and the thickness of soil above the void are two important factors for the stability of this soil-void system. When the soil thickness above the void reduces due to construction or soil erosion, or infiltration of water reduces the strength of soil, the 'arch-effect' in the soil-void system ceases.

IDENTIFYING THE HIGH RISK AREAS

Identification of areas prone to sinkhole development involves various tasks. Available geologic maps, aerial photographs, subsurface investigation reports from adjacent projects and reports of sinkhole activities could provide valuable information on assessing the risks of sinkhole development in the project area. Once research on historical data and reconnaissance of the project area is performed, a subsurface investigation program needs to be planned. Typical subsurface investigation program includes geophysical testing and test borings. Geophysical testing, such as, Electrical Resistivity Imaging (ERI) and Electromagnetic testing could provide a 'quick' overall map of the subsurface condition, specifically, depth to rock and presence of voids (Photo 1). The results of geophysical testing could be confirmed by test borings. Standard Penetration Test borings provide samples and data on consistency of soil, whereas, air-track drilling is a faster operation when the objective is to determine the top of rock and limits of known voids.

Presence of the following features at a site elevates the risk of sinkhole development in karst geology. The more the number of these features observed at the site, and the more the extent of each of these features, the greater the risk is.

- Subsidence history – presence of existing sinkholes nearby
- Rock quality – fractured rock, soil filled voids within rock
- Geologic features – contacts between formations
- Depth to rock – shallow depth to rock
- Rock pinnacles - significant variation in rock depth within short distance
- Soil consistency – presence of soft soil at soil/rock interface
- Groundwater – groundwater below top of rock, fluctuations in groundwater level

The position of water table in relation to the top of bedrock helps in determining whether a triggering mechanism will be activated by a water-level decline or whether piping can result from diversion of drainage. In areas with a high water table above bedrock, probability of piping is low but sinkholes may result from lowering of the water table (Newton, 1986). Where the water table is below the soil-rock contact, formation of sinkholes during and after rain is common.

Identifying soil zones with low blow counts (N < 5 blows/0.3 m) are critical in karst geology since a soft zone could be an indication of incipient or collapsed soil void. As water flows through the overburden soil, it washes away the finer material into the bedrock fractures/crevices creating a soaked zone of low strength.

Experience indicates that induced sinkholes are common near drainage pipes, basins and excavated ('cut') areas. Examples are presented in the following paragraphs.

Underground drainage pipes often leak and exacerbate sinkhole development. Mellett and Maccarillo (1989) reported that most of the sinkholes developed along the I-78 highway in western New Jersey, USA occurred after periods of heavy rain, and have been associated with ruptures at joints in drainpipes.

Moore (1988) conducted a study involving 72 karst related subsidence problems experienced along highways in East Tennessee, USA over a period of ten years. It was found that 74% of the karst problems occurred in highway ditchlines. Most (93%) of the highway ditchlines were constructed without any impervious material. Water from the untreated ditchlines percolated into the underlying soil/rock interface, and developed sinkholes. Water basins without impervious linings could also accelerate sinkhole development (Photo 2).

Yang et. al (2003) indicated that most of the ground depressions and sinkholes observed along a 10 km long section of a highway in eastern Pennsylvania, USA were found to be concentrated in the areas where the original highway was constructed in cuts.

Decline of groundwater level by pumping could also accelerate sinkhole development. Sites where the water table is shallow, overburden is saturated and thin, and clay-filled openings are at or near the top of bedrock, have higher risk for sinkhole development due to pumping.

RISK MINIMIZATION

Some level of risk for sinkhole development due to construction activities can not be avoided in a karstic region. But, it may be possible to select another site, minimize hazard potential or accept some level of risk.

Alternatives, such as, grouting openings in bedrock, diverting surface run-off through impervious drains, providing impermeable liners for basins, regulating groundwater withdrawal rate from wells, minimizing cuts or excavations are typical engineering practices. Selection of appropriate type of retaining walls (Islam and Ramakrishna, 2005) to reduce amount of excavation could reduce the risk of sinkhole development.

Protection of excavations and trenches during construction should be an important step. Measures are needed to cover excavations during rainy seasons. Foundation excavations and utility trenches should be backfilled properly as soon as possible. Use of flowable backfill should be considered when practical.

Drainage pipe leakage can be a prime source of post-construction sinkhole development. Leakproof piping with sealed joints should be considered.

CONCLUSIONS

- Construction activities is a major source for sinkhole development in karst geology
- Majority of the induced sinkholes are caused by water activities, such as, change in groundwater level, ponding of water in excavations and change in natural drainage
- Assessment for sinkhole potential due to construction requires identification of karstic features, knowledge of site history and understanding of triggering mechanisms.
- Minimizing the risk for sinkhole development requires early identification of potential problem areas so that the proposed design can be modified.

ACKNOWLEDGEMENTS

I am grateful to my employer Michael Baker, Jr., Inc. for its support in preparing this manuscript. I would also like to thank Mr. Bruce Shelly of Pennsylvania Department of Transportation for providing photo no. 2.

REFERENCES

Islam, M.Z. and Ramakrishna (2005). "Highway Widening in Karst", *Proceedings of the 10th Multidisciplinary Conference on Sinkholes and the Engineering and Environmental Impact of Karst,* San Antonio, Texas.

Kato, H., Wakita, K. and Bandibas, J.C. (2003). "Eastern Asia Geological Hazards Map: Paper and Digital Versions", *Journal of Natural Disaster Science,* Volume 25, Number 2, pp. 65 – 74.

Mellett, J.S. and Maccarillo, B.J. (1989). "Highway Engineering Aspects of Karst Terrane near Alpha, New Jersey", *Proceedings of the 3rd Multidisciplinary Conference on Sinkholes,* St. Petersburg, Florida.

Moore, H. (1988). "Treatment of Karst Along Tennessee Highways", *Proceedings Symposium on Geotechnical Aspects of Karst Terrains,* Nashville, Tennessee.

Newton, J.G. (1986). "Development of Sinkholes Resulting from Man's Activities in the Eastern United States", *U.S. Geological Survey Circular,* 968, pp. 1 – 41.

Yang, M.Z., Shelly, B. and Islam, M.Z. (2002). "Geotechnical Investigation for Highway Structures in Karst Area", *Proceedings of the 55th Canadian Geotechnical Conference,* Niagara Falls, Ontario.

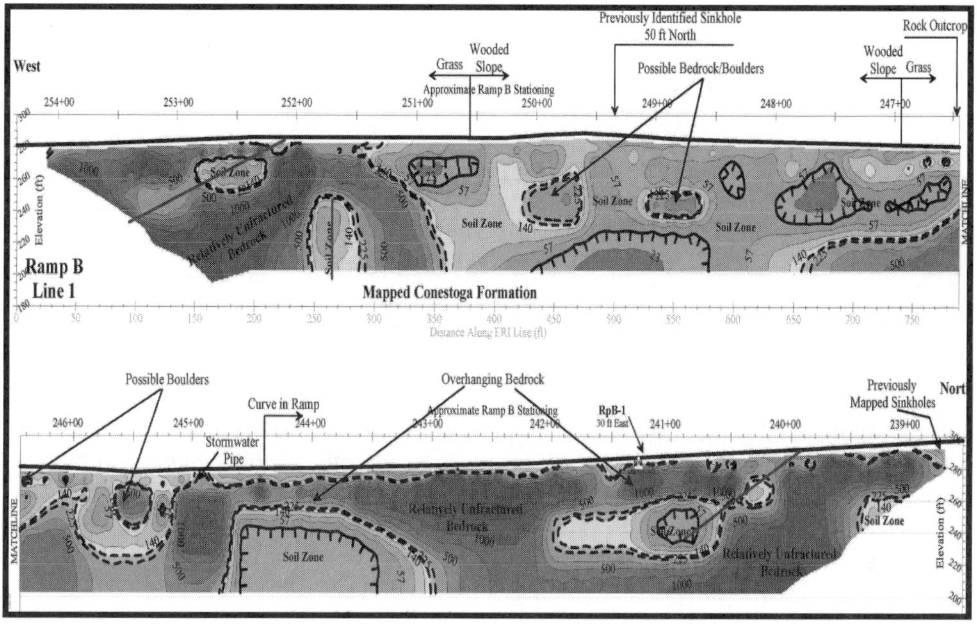

Photo 1. Subsurface mapping by Geophysical testing

Photo 2. Sinkhole development by water infiltration

International Conference on Geotechnical Engineering for Disaster Mitigation & Rehabilitation
Chu, Phoon & Yong (eds) © 2005 World Scientific Publishing Company ISBN 981-256-469-1

NATIONAL CAPITAL OF INDIA: FROM IGNORANCE TOWARDS PREPAREDNESS

EILIA JAFAR
GoI-UNDP Disaster Risk Management Programme
United Nations Development Programme (UNDP)
C- 23, Defence Colony
New Delhi –110024, India

India has witnessed some major disasters in recent memory and the lack of preparedness and inadequacy to face such calamities has been apparently visible. State governments have limited resources and infrastructure facilities to handle a major disaster of the type of Super Cyclone in Orissa (October, 1999) or the earthquake in Gujarat (January, 2001). The Bhuj Earthquake, on 26th January 2001, surprisingly shook the national capital that remained impassive after Koyna and Latur earthquakes. This time Delhi realized, it could also face a similar hazard, but the enormity of destruction would be catastrophic. Since then, a lot has been done in Delhi towards disaster risk reduction. This paper focuses on the dynamic transition from complete ignorance to awareness and preparedness and the role of government in bringing this paradigm shift in approach. The paper further explores the cost effectiveness of different mitigation strategies, adopted or those in the pipeline for Delhi.

BACKGROUND

India has a history of disasters owing to its unique geo-climatic conditions. About 60% land mass is prone to earthquakes of various intensities and a similar percentage is prone to drought. There exists, a long coastline (8000 Km) with two cyclone seasons. According to a report of National Disaster Management Division, Ministry of Home Affairs, "around 30 million people are affected by disasters every year"[1]. Disasters strike in different forms in different regions of the country. Delhi, the national capital is a unique example of City-State combination. Delhi is politically and economically one of the most important states of our country and its ability to mitigate against disasters is of critical significance for the country. As per the latest seismic zones map released by the Bureau of Indian Standards (BIS) in 2000, India is divided into four seismic zones. Zone V is the maximum risk zone and Zone II is the low risk zone. The entire region of Delhi is in Seismic Zone IV, at high risk to earthquakes. Further, on a day-to-day basis, Delhi is at risk to numerous hazards, such as fire, floods, bomb-blasts, other acts of terrorism, industrial and chemical hazards, building collapses, road accidents, water logging, etc.

Earthquake Risk In Delhi

In a recent study by R. N. Iyengar and S. Ghosh seismio-tectonic characterization has been attempted with India Gate in New Delhi as the centre, a circular region of 300 km radius has been assumed as the catchment area for Delhi city[2]. Tectonic features around Delhi city discussed by Valdiya, K. S., (Himalayan transverse faults and folds and their parallelism with subsurface structures of north Indian plains. *Tectonophysics*, 1976, 32, 353–386) have been further improved to map all known faults in a radius of 300 km around Delhi city. Twenty faults have been identified, movement on which can cause ground vibration at Delhi.

Seismic history of Delhi, as per the records available, dates back to the earthquake of 15th July, 1720 (intensity IX in Delhi) - well discussed by Kafi Khan. References are also available of the earthquake of 22 March 1825 near Delhi of intensity VII. In the Khurja earthquake (M6.7) of 10 October 1956, many people were injured in Delhi. The earthquake of 27 August 1960, one of the largest instrumented earthquakes affecting Delhi (M6.0) resulted in 50 persons injured. Damage was reported to one of the minarets of Delhi's Jama Masjid, a national heritage during the earthquake (M4.0) on 28 July 1994. During the 1999 Chamoli earthquake (M6.5), 280 km from Delhi, collapse of a few architectural fins and structural damage in a few buildings were reported. One of the largest instrumented earthquakes affecting Delhi occurred on 27 August 1960 – which resulted in structural and cracks in certain areas. [3]

Flood Risk In Delhi

Delhi has also been prone to floods. 'The flood problem in Delhi has become more serious since the river Yamuna has been jacketed by high embankments in its entire reach of about 210 km upstream of Delhi to protect the rural population in Uttar Pradesh and Haryana. This activity has resulted in a decrease in the valley storage of the river and increase in the peak flood discharge levels during floods.'[10]

Besides the frequent flash floods in most of areas due to poor drainage facilities, there are four major floods witnessed by Delhi. During 1977, Najafgarh drain experienced heavy floods due to discharge from the Sahibi River. 14 persons were reported to have died in a boat mishap. Crop damage was estimated at Rs 10 million. Before the government could get over with the relief operations, another major flood occurred in September, 1978. River Yamuna experienced a devastating flood.. The total damage to crops, houses and public utilities was estimated at Rs. 176.1 million. A decade later, another major flood rocked many parts of Delhi. In September, 1988, many villages and localities were flooded by river Yamuna, affecting approximately 8,000 families. The flood of September, 1995 was the last major flood experienced by river Yamuna in which around 15000 families were affected.[4]

Fire Risk In Delhi

Fire accidents are also very frequent in Delhi. Delhi Fire Service responds to 15000 fire/rescue calls per annum on an average according to official sources. A major fire outbreak in one of the picture halls in Delhi (Uphaar Cinema) is fresh in the memories, with the victims still fighting for justice. Delhi Fire Service has been trying to bring down the fire accidents. Delhi Fire Service possesses special appliances for combating any kind of eventuality arising in the cluster of high-rise buildings and other special rescue jobs. Three Hazmat vans are present in Delhi to deal with all kinds of disasters. Specialized units are also positioned at many places in Delhi.[5]

The safety of Delhi is of prime importance as the National Capital, the hub of national government business, multiplying many-fold the implications of a major disaster not only on the population of Delhi and its infrastructure but also on the functioning of the country itself.

While natural hazards can not be controlled, the vulnerability to these hazards can be reduced by planned mitigation and preparedness measures. There needs to be concerted and sustained steps towards reducing the vulnerability of communities to disasters.

The past couple of years have witnessed a paradigm shift in the approach of central and state governments towards disaster management. The new approach proceeds from the conviction that development cannot be sustainable unless disaster mitigation is built into the development process. Another corner stone of the approach is that mitigation has to be multi-disciplinary spanning across all sectors of development.

DISASTER MANAGEMENT FRAMEWORK IN DELHI

The national government is setting up a National Disaster Management Authority, with Prime minister as its chairman[7] . In the meanwhile, the Government of Delhi has established a nodal agency, namely the *Delhi Disaster Management Authority (DDMA)*, to facilitate, coordinate and monitor disaster management activities and promote good disaster management and mitigation practices in the state. The establishment of the DDMA is a key element in the overall disaster management policy of the State Government. With the hon'ble Lieutenant Governer as the head of the Apex committee that governs this authority, Chief Minster, Ministers, Chief Secretary, Principal Secretaries of relevant departments, Police, Fire Service and joint secretaries etc form the members. The Government of Delhi also proposes to introduce legislation in the form of a Delhi Disaster Management Act to provide a legal framework for disaster management in the state.

DISASTER MANAGEMENT INITIATIVES IN DELHI

The Government of India-UNDP Disaster Risk Management Programme is being implemented in Delhi since July 2003. Office of Divisional Commissioner, acting as the state nodal agency for disaster management, has played an active role by focusing on preparedness and mitigation issues. It is a well-known fact that cost of response can be reduced considerably, if part of the investment is directed towards mitigation. The disaster management activities being taken up by the government include awareness generation in educational institutions, government offices, hospitals, market and residential places etc. Training and capacity building of engineers, architects, builders, masons, officials from line departments and the task forces prepared by local communities is also being done.

Great deal of effort is being made to develop a model techno legal and techno financial regime in the capital. Although the building bye laws of Delhi had been revised in 2001 after the Bhuj earthquake, incorporating the six Indian seismic codes for structural safety, the state government has sent the Model Bye laws prepared by the National expert committee to all the Urban Local Bodies inviting comments and suggestions for modification of existing bye laws. Delhi Disaster Management Authority has also drafted State Disaster Management Policy.

Disaster Management planning has been initiated at community level. . Community Based Disaster Preparedness (CBDP) Plan aims to raise awareness on the various hazards faced by people of a given area, and encourages the community to assume a sense of responsibility to protect itself and to support public and institutional efforts geared towards disaster preparedness, management and mitigation. The process involves social mapping, risk mapping, resource mapping etc and formation of specific task forces like 'fire fighting', 'first aid', 'search and rescue', 'warning & information dissemination' and trauma counseling etc. The Government of National Capital Territory of Delhi, with support from United Nations Development Programme has successfully made hundreds of such plans.

Knowledge networking with key stakeholders in disaster management has been done for effective sharing of skills. Delhi government has partnered with a number of volunteer organizations and all line departments to work in partnership to take forward the activities under the Disaster risk management programme.

An online inventory of resources has been developed using the portal, 'India Disaster Resource Network (IDRN)'. All information regarding availability of skilled manpower, equipments, materials and services in a district or neighboring district are now available at the click of a tab.

INSTITUTIONAL FRAMEWORK

In all the nine districts of Delhi, the Revenue departments have been designated as the nodal agencies for disaster management. The Deputy Commissioners, supported by the Additional District Magistrates have been entrusted with the responsibility of carrying forward the disaster management activities at the district level.

Establishment of Emergency Support Functions

As a part of Delhi State Response Plan, twelve emergency support functions have been identified with primary and support roles assigned to different agencies in Delhi e.g. Communication have been assigned to MTNL as nodal agency and other telecom players as support agencies. All these agencies have prepared their "Standard Operating Procedures" (SOPs) detailing each and every step to be taken pre, during and post disaster. These procedures have been developed based of the Incident Command system

(ICS), and the government officials have been provided special training of incident command system, this being a new concept in our country.

Development of Road Map for Disaster Management

A State Level Road Map for Disaster Management in Delhi on the lines of the National Road Map has been prepared and shared with the central government. This road map aims to sustainable reduction of disaster vulnerabilities of the citizens of Delhi, by systematically addressing issues of awareness on vulnerabilities, systems for effective preparedness and response in government and communities, long-term planning for mitigation, capacity building and training to various stakeholders and regular practice drills.[8] All agencies responsible for implementing the activities proposed in road map have also been identified.

Formation of working groups

Two Working Groups for (i) Prevention and mitigation (ii) Preparedness and response, have been constituted and notified. These groups further consist of sub groups to deal with awareness generation, capacity building, planning, techno-legal aspects etc.

Development of Information, Education and Communication (IEC) material

In order to make the awareness generation activities more effective, the Government of Delhi has developed a lot of resource material comprising of leaflets, posters, planning documents, earthquake tips etc for wide dissemination. The IEC material serves as an agent to generate awareness and induce mitigation and preparedness measures for risk reduction. Multi-media CDs on disasters developed by various government agencies have been distributed widely.

Incident Command System

Delhi is perhaps the first state in the country to have used incident command system in an earthquake mock drill held last year. "This system provides for specialist incident command teams with an incident commander and officers trained in different aspects of incident management- logistics, operations, planning, safety media management etc.[9]

Emergency Operation Centres

Delhi government has set up Temporary Emergency Operation Centers at State and District level, with assistance from the budget of Government of India-UNDP Disaster Risk Management programme. Equipments have been obtained for these rooms. Hazard zone wise standard layouts have been sent by the central government to all states. Construction work has already been started for developing Emergency Operation Centres as per drawings in a few districts in Delhi. Once fully functional, these Emergency Operation Centres are expected to function round the clock and serve as control rooms to look after law and order issues as well as disaster management.

DISASTER MITIGATION STRATEGIES FOR DELHI

It has been proven over and over again that the impacts of natural hazards can be lessened and even eliminated by appropriate actions taken well before the hazardous event. The most effective way to ensure that this action takes place is the preparation and implementation of a comprehensive hazard mitigation plan.

Model techno-legal framework

The strategies for mitigation need to address to the safety of two kinds of structures- the upcoming construction and the existing building stock. The state government is planning to develop a model techno legal as well as techno-financial framework for Delhi. The Government of Delhi has trained a large number of Engineers, Architects and masons on disaster safe construction practices. Sensitization programmes have been organized for Builders, Contractors and town planners. A system of special audit of buildings is being put in place with a view to ensuring that the new constructions conform to the building byelaws being modified as per the recommendations for Model Building Bye Laws, developed by the National Core Group for Earthquake Mitigation constituted by the National government consisting of experts in earthquake engineering and administrators. The above strategies being adopted by the government of Delhi ensure long term, sustainable qualitative benefits resulting in disaster risk reduction.

Retrofitting of buildings

With the change in the Indian seismic codes in 2002, many of the buildings in this country that conformed to the codes/ standards of the earlier days are now found seismically deficient as per the current code and many present the risk of life-threatening damage. It is this vulnerability of the built environment in the State that will cause even a moderate earthquake in the region to transform into a disaster in the Delhi region. To influence all new constructions, the Government of Delhi has embarked on capacity building programmes for all the key players in the construction fraternity. But the vulnerability associated with our existing buildings needs to be addressed urgently through retrofitting. Retrofitting is the process of assessing and strengthening existing buildings to perform as per required criteria in earthquakes.

But, it is neither economically feasible to seismically retrofit every building not conforming to the current codes with no/ inadequate seismic provisions, nor is it culturally acceptable to replace them all. Thus, first we need to identify the structures that present a significant risk to life safety and predict the expected performance for such buildings of high importance or those needed for emergency response (lifeline buildings) and then find retrofit solutions for these structures identified as high risk.

Since expertise in retrofitting is limited in the country, the Ministry of Home Affairs, Government of India and the Government of Delhi have embarked on a pilot project for demonstrative retrofitting of five lifeline buildings in the city with support from USAID and Geo Hazards International (GHI). The five lifeline buildings selected by the Govt. of Delhi include the Police headquarters, The Delhi secretariat building, the office of Delhi Disaster Management Authority, an important Hospital and a school. Non Structural Safety assessment has also been carried out and Non Structural Mitigation measures have been initiated in the school and hospital. Though the project envisages only the development of retrofitting plans for these five buildings, the Government of Delhi should carry out the actual retrofitting of these five critical buildings immediately and utilize the opportunity provided by the project to start a movement for retrofitting in the Governmental and non-Governmental setup. Moreover, as this is a pilot project, the five examples set by the Project have to increase in numbers rapidly to more examples spread geographically around the state and involving all agencies carrying out construction in the State viz. Municipal Corporation of Delhi, New Delhi Municipal Corporation, Delhi Development Authority, Central Public Works Department etc.

Insurance Policies

The Insurance sector in India is regulated by Insurance Regulatory and Development Authority. The Tariff Advisory Committee (TAC) lays down tariff rates for some of the general insurance products. Earlier earthquake cover used to be a part of fire insurance policy but since May 2000, the Tariff Advisory Committee effected a reduction of 22 per cent in the fire insurance premium and provided for earthquake risk as an add-on cover. Other add on covers available are STFI (Storm, Tempest, Flood and Inundation) covers. The compensation for damage depends not only on the type of cover opted for but also, on the basis of option like replacement cost basis or market value basis. Many insured people affected by the recent tsunami were deprived of compensation for simple reason that they had either not gone for complete cover or because they had opted for cover policy on a market value basis.[10] A better understanding of the insurance procedures and general awareness is required. The premium cost for complete cover is quite high and given the uncertainty about the time when a disaster will strike a similar investment in other mitigation measures like low level retrofitting may turn out to be more cost effective.

ANALYSIS AND CONCLUSION

Given the seismic vulnerability of the region, an earthquake of a high magnitude is likely to hit Delhi in the near future. In the last few years, Delhi has come a long way from the state of utter ignorance towards disaster vulnerability, to a conscious state. The initiatives to reduce disaster risks taken up by the Government of Delhi are definitely in the right direction. Disaster management in Delhi, however, is yet at its nascent stage. Sustainability is the key word in development process. Whether Delhi will reach the stage of a prepared and safe city, or not, will depend on how long and how seriously the efforts are maintained. While much focus has been laid on awareness generation activities, development of a strategy for disaster mitigation in the state is equally essential. The aim of this strategy should be to reduce losses in the event of a future occurrence of a hazard. Structural mitigation may comprise construction of disaster resistant structures or retrofitting of existing building stock. Land use and site planning may also be done with a disaster mitigation perspective. Prevention, mitigation, preparedness and response are multi disciplinary activities involving a large number of departments. The setting up of DDMA and Emergency Support Function Plan has provided a platform for the officials of these departments to interact with each other. This information, is however mostly limited to the nodal officers nominated for disaster management in their departments. There is a need for further dissemination of disaster prevention, preparedness, mitigation and response related information, so that the efforts do not lack efficiency.

The State government may introduce the concept of disaster insurance as a potentially important mitigation measure. This will not only generate awareness towards vulnerable areas but also motivate people to settle in relatively safer constructions.

In the case of the Structural Mitigation measures, though a few initiatives for reducing the vulnerability of future constructions are being undertaken, there is a need for drawing up an action plan for a systematic reduction in vulnerability of the existing buildings and structures as well as the future constructions in the form of a Delhi Earthquake Loss Reduction Plan.

The efforts of the national government to introduce disaster management as a compulsory paper in all fields of study at graduation level in all states, if and when realized, will ensure sustainable disaster awareness among the youth, leading to disaster risk reduction in due course. The road map developed by the state government is quite comprehensive and a proper execution of the planned activities will ensure a well aware and prepared capital city. The challenge ahead is to implement the road map, unlike many other proposals and schemes of the government which are limited to paper work and contribute to piles of files and folders.

REFERENCES

Disaster Management in India; A status Report, NDM Division, MHA, Government of India.

Microzonation of earthquake hazard in Greater Delhi area, CURRENT SCIENCE, VOL. 87, NO. 9, 10 NOVEMBER 2004

Bill to enact legislation on disaster management, Taja news, March 10, 2005 , see also "Disaster management Authority gets a nod, Financial Express, Feb 25, 2005.

Status Report of Disaster Management in Delhi, Delhi Disaster Management Authority, Office of Divisional Commissioner, Government of Delhi.

Disaster Management in India; A status Report, NDM Division, MHA, Government of India

Teri, Report No. 2000EE65, State of Environment Report for Delhi 2001: Key environmental concerns.

Pushpa Girimaji, The Tribune, Online Edition, February 11, 2001, Chandigarh

www.asc-india.org/seismic/hdc.htm, www.gisdevelopment.net
www.gisdevelopment.net
www.dfs.delhigovt.nic.in

International Conference on Geotechnical Engineering for Disaster Mitigation & Rehabilitation
Chu, Phoon & Yong (eds) © 2005 World Scientific Publishing Company ISBN 981-256-469-1

MAPPING LIQUEFACTION HAZARD IN CHARLESTON QUADRANGLE, SOUTH CAROLINA

DAVID KUN LI
Department of Civil Engineering, Clemson University, Clemson, SC 29634-0911,USA.

C. HSEIN JUANG
Department of Civil Engineering, Clemson University, Clemson, SC 29634-0911,USA.

Using the cone penetration test (CPT) data and the earthquake loading information, which includes the magnitude and peak ground acceleration, collected from the web-based database provided by USGS National Seismic Hazard Mapping Project, the liquefaction potential and the probability of liquefaction-induced ground failure of each site in the Charleston quadrangle can be calculated. With the locations of the CPTs and the probability of ground failure of the sites, the liquefaction hazard map of the Charleston quadrangle can be developed. The methodology proposed in this study provides an applicable way to develop a liquefaction hazard map, which enables the engineer and the policy maker to make an informed decision.

DATABASE

To assess the liquefaction hazard in Charleston quadrangle, South Carolina, cone penetration test (CPT) measurements are used. Sevety-seven CPTs in Charleston quadrangle from the USGS data reports (Fairbanks et al., 2004) are adopted in this study. Among the 77 CPTs, there are 6 different kinds of surface geologies including various deposits of recent man-made fill and Holocene to Pleistocene natural sediments. The deposits aged less than 300 years old are designated as artificial fill (af), which include sands or clayey sands of diverse origin. Holocene-age (<10,000 years or < 10 ka) deposits in this study include tidal-marsh clayey sands and clays (Qht). Early Holocene to late Pleistocene deposits include estuarine silty to sandy clays (Qhec) aged from 6 ka to 85 ka. Pleistocene deposits include beach to barrier-island sands (Qhes) aged from 33 ka to 85 ka. Weems et al. (1997) mapped three facies of the Wando Formation, which are older Pleistocene deposits about 70 ka to 130 ka in age, in the Charleston quadrangle. CPTs adopted in the present study locate in two of the three facies, which are: a clayey sand and clay facies deposited in fluvial to estuarine environments (Qwc) and a barrier sand facies (Qws).

Of the 77 CPTs in study, 40 are located in af, 5 in Qht, 2 in Qhec, 7 in Qhes, 3 in Qwc, and 20 in Qws surficial deposits.

MAJOR METHODOLOGY

Obtaining earthquake loading

The first step to assess the liquefaction hazard is to obtain the earthquake loading, which includes the information of moment magnitude and peak ground acceleration.

Amplification factor

The peak ground acceleration (a_{max}) provided by the USGS National Seismic Hazard Mapping Project web site (http://eqhazmaps.usgs.gov) is for rock outcrop conditions and still need to account for amplification effect of surface soil columns.

The amplification factor of peak ground acceleration (a_{max}) is calculated based on the empirical equations recommended by Stewart et al. (2003). The soils of "Qws", "Qwc", "Qhec", and "Qhes" are categorized as “Qa” (Quaternary alluvium) by Stewart et al. (2003). The soils of "Qht" are categorized as “Hlm” (Holocene lacustrine/marine). Stewart et al. did not provide empirical equations for “artificial fill”. In this study, “artificial fill” with the information of mean shear wave velocity of top 30m soils (V_{S30}) is categorized according to NEHRP categories. The peak ground accelerations of the artificial fill soils with V_{S30} are then calculated using empirical equations for NEHRP categories.

Since most of the natural soils in this area are categorized as “Qa”, a uniform surface geology “Qa” is assumed for all the natural soils and the soils beneath artificial fill in the Charleston area for simplification in further study. It follows that the empirical equation for “Qa” proposed by Stewart et al. (2003) can be used to calculate the surface peak ground accelerations for all 77 CPT sites.

Hazard level

USGS web site lists 6 different hazard levels: ground motion with a 1 % probability of exceedance in 50-yr period (4975-yr return period), 2 % probability of exceedance in 50-yr period (2475-yr return period), 5 % probability of exceedance in 50-yr period (975-yr return period), 10 % probability of exceedance in 50-yr period (475-yr return period), 20 % probability of exceedance in 50-yr period (224-yr return period), and 50 % probability of exceedance in 75-yr period (108-yr return period). In this study, liquefaction hazard analysis is conducted using ground motion with a single hazard level, 10 % probability of exceedance in a 50-yr period, which corresponds to 475-yr return period and is most commonly used in practice.

Probability of ground failure for a site

The procedure for determining factor of safety and probability of liquefaction for a soil element in a CPT profile and the probability of ground failure at a site is described in the following.

The first step is to calculate the cyclic stress ratio (CSR) defined as (Idriss and Boulanger, 2004):

$$CSR = 0.65\left(\frac{\sigma_v}{\sigma_v^{'}}\right)\left(\frac{a_{\max}}{g}\right)(r_d) / MSF / K_\sigma \tag{1}$$

where σ_v = vertical total stress, σ'_v = vertical effective stress, a_{max} = peak horizontal ground surface acceleration, g = acceleration of gravity, r_d = a depth-dependent shear stress reduction factor, MSF = a magnitude scaling factor, and K_σ = an overburden correction factor.

The liquefaction resistance, expressed in terms of cyclic resistance ration (CRR), is calculated by the following equation (Juang et al., 2005):

$$CRR = \exp\,[-2.88 + 0.000309\,(q_{c1N,m})^{1.8}] \tag{2}$$

where $q_{c1N,m}$ = stress-normalized cone tip resistance adjusted for the effect of fines.

Because piezocones were used in the collection of the CPT profiles in Charleston quadrangle, the tip resistance corrected to pore water pressure, q_t, is used in the calculation rather than tip resistance, q_c.

After CSR and CRR are calculated, probability of liquefaction (P_L) or factor of safety (FS) could be used to assess the liquefaction potential of a soil element at a given depth. The factor of safety FS is defined as: FS = CRR/CSR. The probability of liquefaction P_L can be calculated with the following equation (Juang et al., 2005):

$$P_L = \frac{1}{1 + (\frac{FS}{0.81})^{5.45}} \tag{3}$$

The liquefaction potential index, denoted herein as I_L, was proposed by Iwasaki et al. (1982) for predicting the potential of liquefaction to cause foundation damage at a site. They defined the index I_L as follows:

$$I_L = \int_0^{20} Fw(z)dz \tag{4}$$

where the depth weighting factor, $w(z) = 10 - 0.5z$, and z = depth from 0 m to 20 m. The variable F is a function of factor of safety (FS) against liquefaction at a given depth, and is defined as: $F = 1 - FS$ for $FS \leq 1$; and $F = 0$ for $FS > 1$.

In this study, however, a new definition for the variable "F" is adopted based on the probability of liquefaction at a soils element at a given depth:

$$F = P_L - 0.35 \quad \text{if } P_L \geq 0.35,$$
$$F = 0 \quad \text{if } P_L < 0.35. \tag{5}$$

Finally, with the modified liquefaction potential index, the probability of ground failure (P_G) can be calculated using the following empirical equation:

$$P_G = \frac{1}{(1 + e^{4.71-0.71 I_L})} \tag{6}$$

It should be noted that some of the 77 CPT profiles did not reach 20 m. In the local geotechnical reports, these CPTs did reach the Ashley Formation, a Tertiary soil formation, locally known as the Cooper Marl. The Tertiary soils are considered non-liquefiable (Youd and Perkins, 1978). Thus, those CPT profiles are used in this study.

RESULTS

One contour map obtained in this study is based on the ground motion corresponds to 475-yr return period. The amplification factors are calculated using the equations for "Qa" and "Hlm" proposed by Stewart et al. (2003). For the soils of "af" with information of V_{S30}, equations for NEHRP categories are adopted to calculate the amplification factors. 55 CPTs in all are used in this analysis. With the probabilities of ground failure and locations of these 55 sites, the liquefaction hazard contour map can be plotted as shown in Figure 1.

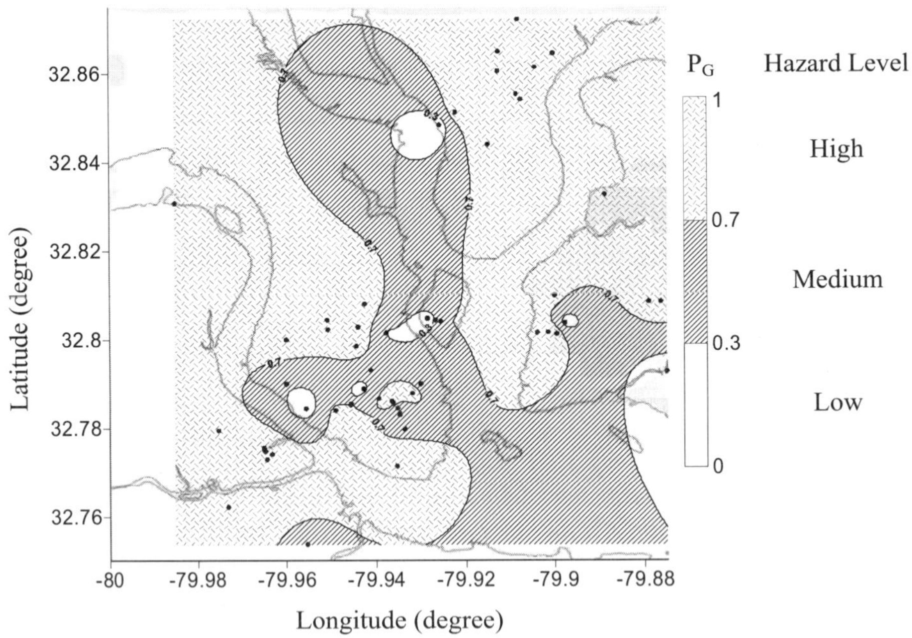

Figure 1. Liquefaction hazard contour map for Charleston quadrangle, South Carolina with 55 CPTs

Another contour map is shown as Figure 2 which was obtained based on the assumption that the natural soils in the study are all "Qa". Then the empirical equation for "Qa" proposed by Stewart et al. (2003) is used to calculate the surface peak ground accelerations. This map is also based on the ground motion corresponds to 475-yr return period in the Charleston quadrangle.

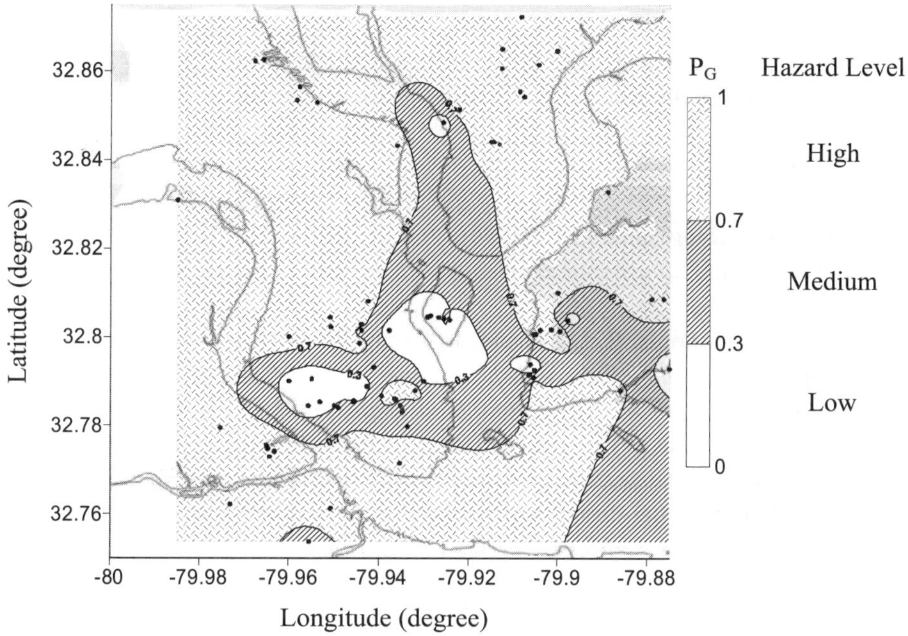

Figure 2. Liquefaction hazard contour map for Charleston quadrangle, South Carolina with 77 CPTs (assuming all soils are "Qa")

Based on the contour, three zones of liquefaction hazard are identified. The liquefaction hazard is high if $P_G > 0.7$; the hazard level is medium if $0.3 \leq P_G \leq 0.7$; and the hazard level is low if $P_G < 0.3$. Elton and Hadj-Hamou (1990) assessed the liquefaction risk of Charleston peninsula and their study shown the area of greatest risk lies along the southern tip of the peninsula and intermediate risk for the central part of the peninsula and the lowest risk for much of the coastline. The result of zonation in this study also shows the highest liquefaction hazard lies along the southern tip. While for the other parts of the peninsula, thc result of this study is more conservative.

SUMMARY AND CONCLUSIONS

Using the methodology developed in this study, along with cone penetration test (CPT) measurements and the earthquake loading information collected from USGS National Seismic Hazard Mapping Project, the liquefaction potential and the probability of liquefaction-induced ground failure of each site in the Charleston quadrangle are evaluated. The liquefaction hazard map of the Charleston quadrangle is prepared in term of contour of the probability of ground failure. Based on the contour, three zones of liquefaction hazard are identified. This paper provides an applicable methodology for developing liquefaction hazard map.

ACKNOWLEDGEMENTS

This study was supported by the National Science Foundation through Grant CMS-0218365. Their financial support is greatly appreciated. The opinions expressed in this paper do not necessarily reflect the views of the National Science Foundation.

REFERENCES

Elton, D.J. and Hadj-Hamou, T. (1990), "Liquefaction potential map for Charleston, South Carolina," *Journal of the Geotechnical Engineering Division*, ASCE, vol. 116, No. 2, 244-265.

Fairbanks, C. D., Andrus, R. D., Zhang, J., Camp, W. M., Casey, T. J., and Cleary T. J. (2004), "*Electronic files of shear-wave velocity and cone penetration test measurements from the Charleston quadrangle, South Carolina*," Data report to the USGS, Department of Civil Engineering, Clemson University, Clemson, SC.

Idriss, I.M. and Boulanger, R.W. (2004), "Semi-empirical procedures for evaluating liquefaction potential during earthquakes," *Proceedings, Joint Conference, the 11th International Conf. on Soil Dynamics and Earthquake Engineering (SDEE), the 3rd International Conf. on Earthquake Geotechnical Engineering (ICEGE)*, Berkeley, CA, 32-56.

Iwasaki, T., Arakawa, T. and Tokida, K. (1982), "Simplified procedures for assessing soil liquefaction during earthquakes," *Proceedings of the Conference on Soil Dynamics and Earthquake Engineering*, Southampton, UK, 925-939.

Juang, C.H., Fang, S. Y., and Li, D. K. (2005), "Reliability analysis of soil liquefaction potential," Geotechnical Special Publication, No. 133, ASCE, Reston, VA, January 2005.

Stewart, J.P., Liu, A.H., and Choi, Y. (2003), "Amplification factors for spectral acceleration in tectonically active regions," *Bulletin of the Seismological Society of America*, Vol. 93, No. 1, 332-352.

Weems, R. E., Lemon, E. M., Jr., and Chirico, P. (1997), "*Digital geology and topography of the Charleston quadrangle, Charleston and Berkeley Counties, South Carolina*," USGS Open-file Report 97-531, U.S. Geological Survey, Reston, VA.

Youd, T.L. and Perkins, M. (1978), "Mapping liquefaction-induced ground failure potential," *Journal of the Geotechnical Engineering Division*, ASCE, Vol. 104, No. 4, 433-446.

International Conference on Geotechnical Engineering for Disaster Mitigation & Rehabilitation
Chu, Phoon & Yong (eds) © 2005 World Scientific Publishing Company ISBN 981-256-469-1

VULNERABILITY ASSESSMENT TO LIQUEFACTION HAZARD INDUCED BY RISING SEA-LEVELS DUE TO GLOBAL WARMING

MURAKAMI, S., YASUHARA, K., SUZUKI, N., NI, WEI AND KOMINE, H.
Department of Urban and Civil Engineering, Ibaraki University, 4-12-1 Hitachi, Ibaraki 316-8511, Japan

This paper describes a procedure for liquefaction hazard mapping that incorporates rising sea levels caused by global warming, along with an application of that mapping procedure to an objective coastal region. Vulnerability to liquefaction hazard in the objective region was assessed by comparison of liquefaction hazard maps before and after sea-level rising caused by global warming.

INTRODUCTION

Ground damage caused by liquefaction is a serious problem. Attention should be given to groundwater level (GWL) changes because liquefaction potential is sensitive to rising GWL. In this century, sea-level rises (SLR) caused by global warming and global climate change have engendered rising GWL in coastal regions. The capabilities of prevention and mitigation against geo-disasters is reduced with rising GWL (Yasuhara *et al.,* 2004). Therefore, it is necessary for sustainable development to consider the influence of global climate change on future situations of GWL. This study shows a procedure for liquefaction hazard mapping considering SRL and an application of the mapping procedure to Yokohama and Kawasaki cities in Japan. Vulnerability to liquefaction hazard in the objective region has been assessed through comparison of liquefaction hazard maps before and after SLR caused by global warming.

LIQUEFACTION HAZARD MAPPING IN CONSIDERATION OF GLOBAL CLIMATE CHANGE

Liquefaction potential depends upon ground properties, seismic magnitude, and GWL. Spatial distribution of GWL in an objective region depends on characteristics of rainfall, groundwater abstraction, and hydraulic boundary conditions. Where the objective area is a coastal zone, one hydraulic boundary condition is the sea level. Global climate change will influence the balance of water circulation and global warming will cause SLR. The liquefaction potential increases with rising GWL. Therefore, for disaster prevention and mitigation of earthquakes, it is important for coastal regions to consider future situations of GWL subjected to SLR. Such measures should take into account the GWL, not only of the present situation, but also that resulting from future trends.

Figure 1 shows a procedure for liquefaction hazard mapping in consideration of

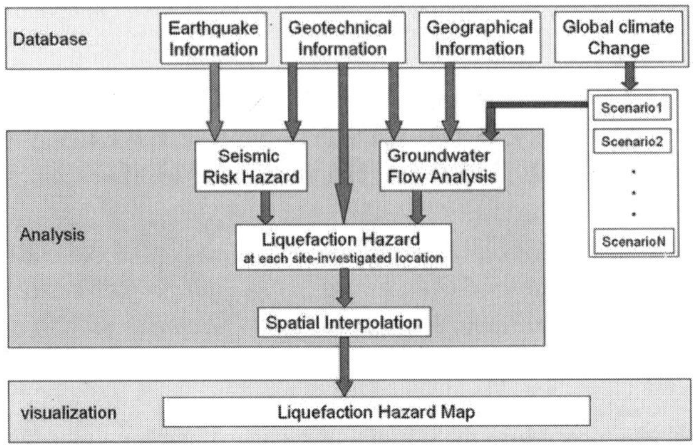

Figure 1. Procedure for liquefaction hazard mapping considering global climate change

global climate change. Calculating the liquefaction hazard at a location in an objective region usually requires information on soil properties through depth, GWL and seismic shear stress. In recent years, geo-information digital databases, including boring logs, N-values, etc. have been made from site investigations in some regions in Japan. The geo-information database is available to predict geo-disasters such as earthquake-induced liquefaction. The database supplies necessary information of soil properties for calculating the liquefaction potential. Seismic shear stress in the ground is predicted considering assumed earthquake events or uses the same value ordained in a design code. Using groundwater flow analyses, the GWL is predicted on the assumption that global climate change will cause a different water circulation scenario from that of the present situation.

The present study calculates the liquefaction hazard at each location using a method proposed in The Japanese Highway Bridge Code (JHBC; so called *"Dorokyo-shihousho"*), which was established in 1996 by the Ministry of Construction (currently the Ministry of Land and Transportation). This judgment procedure is based on the liquefaction resistance factor, F_L, which is defined as

$$F_L = \frac{R}{L}, \tag{1}$$

where R is the dynamic strength ratio determined by cyclic shear strength of soil and the correction coefficient concerning earthquake movement. Cyclic shear strength is estimated using unit weight and fine content of soil, N-value, and GWL. In addition, L is the seismic shear stress ratio generated during earthquakes; L is determined by seismic intensity, which depends on the ground classification and type of earthquake, just as R does. Liquefaction occurs when F_L, given by Eq. (1), is below 1.0.

Evaluation of possible liquefaction through depth in an objective location can be

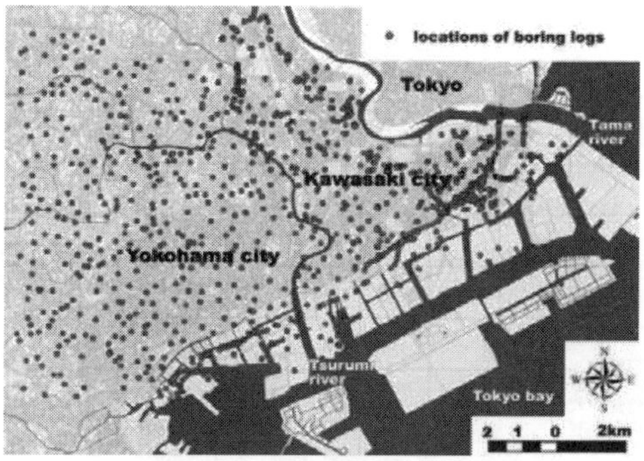

Figure 2. Locations of boring logs in objective area

performed by integration of the liquefaction potential, P_L, through depth using

$$P_L = \int_0^{20} F(z)w(z)dz , \tag{2}$$

where *F(z)* is a function that is $F(z) = 1 - F_L$ when $F_L \leq 1.0$ and $F(z) = 0$ when $F_L > 1.0$. Here, *w(z)* is a weighting parameter defined as $w(z) = 10.0 - 0.5z$ (*z*: GL-m) (Iwasaki *et al.*, 1980).

The GWL is predictable using groundwater flow analyses. The sea level, which is a hydraulic boundary condition, is set as 0.8 m for obtaining simulated results considering SLR resulting from global warming. According to a report of the Intergovernmental Panel on Climate Change (2000), 0.8 m is the worst-case scenario of SLR in 2100.

The objective regions in this study are Yokohama and Kawasaki cities in Japan, as shown in Fig. 2. The figure also shows approximately 700 locations of site investigations in the geo-information database. These data have been used for groundwater flow and liquefaction hazard analyses.

GROUNDWATER LEVEL SIMULATION CONSIDERING RISING SEA-LEVELS CAUSED BY GLOBAL WARMING

Groundwater flow analyses based on finite element method were performed to obtain the GWL for calculating the liquefaction potential. The solving equation of the analysis is a 2D model for unconfined groundwater flow in a steady condition:

$$\frac{\partial}{\partial x}\left\{k(h - z_0)\frac{\partial h}{\partial x}\right\} + \frac{\partial}{\partial y}\left\{k(h - z_0)\frac{\partial h}{\partial y}\right\} = 0 , \tag{3}$$

where h is the GWL, k is the coefficient of permeability of the permeable soil layer, z_0 is the depth of the non-permeable surface and x, y are coordinates of the horizontal plain. Using Eq. (3), distribution of steady GWL subjected to a certain amount of SLR was simulated.

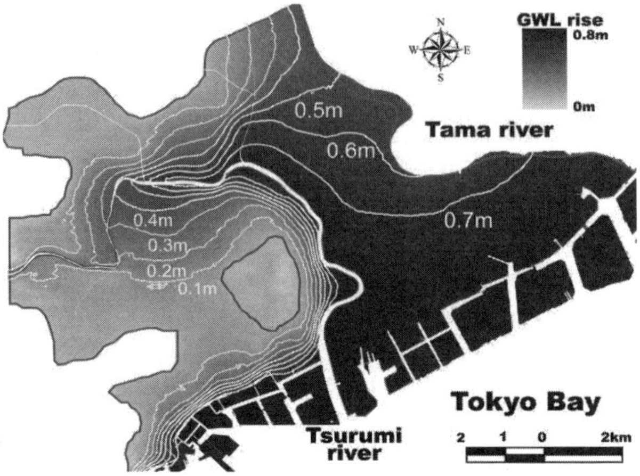

Figure 3. Rise in GWL subjected to 0.8 m SLR

Distributions of the permeable soil layer, impermeable rock surface and ground level in the objective region were estimated using spatial interpolation method with a Geographical Information System (GIS). The coefficient of permeability was calculated using a linearly weighted combination of all soil layers up to the impermeable soil surface in each location. The weighting parameter was determined as the ratio of thickness of the soil layer to the depth of the impermeable rock surface. The coefficient of permeability of soil was assumed by referring to representative values of soils in Japan. Boundary conditions around the analytical area were determined considering not only the sea level, but also the river level affected by SLR.

Groundwater simulations before and after 0.8 m SLR were performed using the numerical method described above. Distributions of the GWL rise subjected to 0.8 m SLR are represented in Fig. 3. The GWL rise in reclaimed islands is regarded as the same value as SLR because the sea surrounds the islands. The rise in GWL is close to 0.8 m when approaching the coastline. However, the GWL rise is not always large near the coastline because the figure shows a larger value of the GWL rise along the Tsurumi River. The GWL rise on each side of the river is different. For that reason and others, it is necessary for investigation of groundwater-related geo-disasters to consider regional characteristics of the ground.

VULNERABILITY ASSESSMENT OF LIQUEFACTION POTENTIAL IN A COASTAL REGION

Distributions of liquefaction hazard in the objective region before and after SLR were compared to investigate vulnerability to liquefaction disaster induced by SLR caused by global warming. Figure 4 shows liquefaction hazard maps before and after 0.8 m SLR. Estimations of liquefaction potential at all site-investigation locations were calculated by Eq. (2). The geo-information database and simulated results of GWL before and after SLR were used as analytical parameters that are soil properties through the depth and GWL at the locations. For determining the seismic shear stress of the ground, seismic movements were assumed to be TYPE II, with earthquake motion corresponding to the Great Hanshin Earthquake in 1995, as specified in the JHB design code. Calculating the liquefaction potential at all locations, distributions in the region were represented as liquefaction hazard maps using spatial interpolation method with GIS. Comparing the map after SLR with others at the present situation of GWL, it is recognized that a severe liquefied area expands slightly around the coastline and riverside.

Using the liquefaction-hazard-map procedure considering SLR caused by global warming, the distributions of liquefaction potential were estimated for cases where the SLR is 0, 0.2, 0.4, 0.6, 0.8, 1.0, 1.2, 1.4, 1.6, 1.8 and 2.0 m. Based on these results, Fig. 5 shows the relationship between SLR and area of PL-values over 5 in the objective regions.

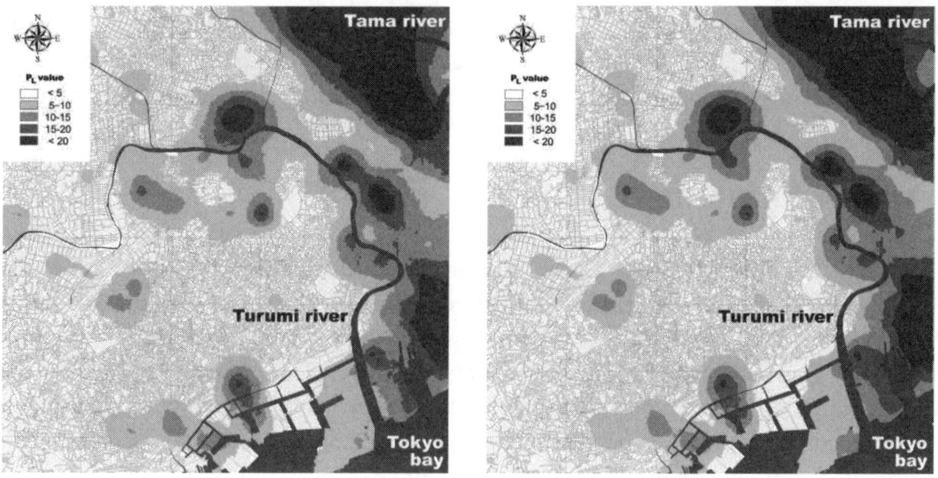

(a) Before SLR (b) After 0.8 m SLR

Figure 4 Liquefaction-hazard maps

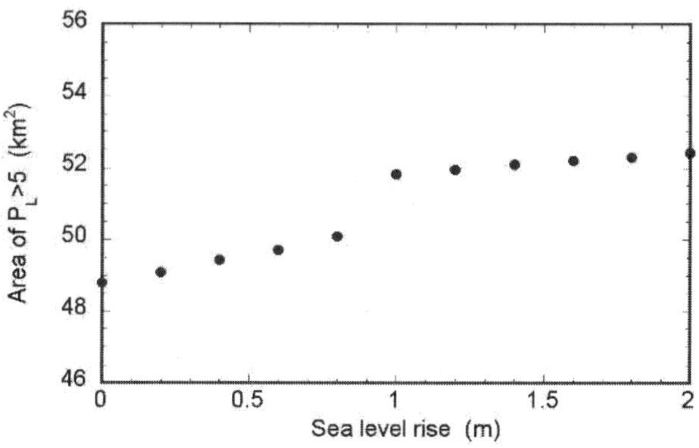

Figure 5 Relationship between the rising sea level and the area of the P_L-value-over-5 region.

The area in which liquefaction potential is high increases with the rising sea level. The increment ratio of the area to SLR increases rapidly from 1.0 m. In regions around the coastline and riverside affected by SLR, disaster prevention against earthquake liquefaction is vulnerable to SLR resulting from global warming.

CONCLUSIONS

A procedure for liquefaction hazard mapping considering global climate change was proposed and applied to a coastal region in this study. Comparison of liquefaction hazard maps before and after SLR caused by global warming illustrates the changing vulnerability to liquefaction hazard. Those results show that the area of high potential liquefaction increases with rising GWL caused by SLR. Regions in which the liquefaction potential is high are not only those around coastlines and riversides affected by SLR. Therefore, for prevention and mitigation of liquefaction by earthquakes, it is necessary to consider regional variations of GWL caused by SLR attributable to global warming.

ACKNOWLEDGMENTS

This study was partly supported by Grants-in-Aid from the Saneyoshi Fellowship Foundation, the Ministry of Education, Culture, Sports, Science and Technology, and the Ministry of Environment of the Japanese Government. We wish to express our sincere thanks for financial support from these organizations. Geotechnical information, and support in Yokohama and Kawasaki cities were provided by the Kanagawa Environmental Science Research Institute in Yokohama and the Department of Environmental Protection, Environmental Directorate, Kawasaki City.

REFERENCES

IPCC WGI (2001): "Technical Summary, Climate Change 2001: The Scientific Basis", *Contribution of Working Group I to the Third Assessment Report of the Intergovernmental Panel on Climate Change,* Cambridge University Press, 83p.

Iwasaki, T., Tatsuoaka, F., Tokita, K. and Yasuda, S. (1980): "Prediction of liquefaction potential during earthquakes", *Tsuchi-to-Kiso*, J. Japanese Geotechnical Society, Vol. 28, No. 4, pp. 23-29 (in Japanese).

Japan Road Association (1996) "Specification for Road Bridge V. Earthquake-resistance design" (in Japanese).

Yasuhara, K., Murakami, S. and Fukuda, T. (2004) "GIS Application for Prediction of Liquefaction Potential Caused by Rising Groundwater Level", *Proc. Int. Symp. Engineering Practice and Performance of Soft Deposits*, pp.433-438.

International Conference on Geotechnical Engineering for Disaster Mitigation & Rehabilitation
Chu, Phoon & Yong (eds) © 2005 World Scientific Publishing Company ISBN 981-256-469-1

ANALYSIS BY USING GIS/RS TECHNOLOGY ON THE GEOHAZARD AUTO MAPPING OF SHOU-SHAN SLIDING AREA IN TAIWAN

KUNG-JUNG TSAI
Department of Civil Engineering
National Pingtung University of Science and Technology, Taiwan
ZHENG-WEI SHI
Department of Civil Engineering
National Pingtung University of Science and Technology, Taiwan

Shou-Shan landslides located at Kaohsiung in southwestern Taiwan have been recognized as serious problems by the public concern. It is of primary importance to realize the situation of sliding areas become unstable and identify the factors trigger the slope movement in these limestone areas. The geological structure of slope, the lithology of rock, the intensity of heavy rainfall and the stage of morphological development around these areas determine susceptibility to Shou-Shan landslides. Only an accurate diagnosis can make it possible to propose effective remedial measures. As a result, both field investigations and laboratory examinations have been conducted since 1995. Also, the great variety of intensive land use in these areas reflects the diversity of factors that may disturb the stabilization of Shou-Shan sliding areas.

This project has been conducted for three year. Two sliding areas in these regions have been selected as experiment districts due to their scale and history of sediment disaster occurrence. Geological investigation, soil mechanical properties determination, groundwater table observation, meteorological data collection and GIS/RS application were required for this study. The critical failure surface of sliding zone can be determined by slope stability analysis and boring test in the field. Also, the potential sliding zones can be estimated roughly by using GPS/GIS/RS integration. All results were presented graphically. It is expected that the result can be used as an indication applied on the stabilization for Shou-Shan sliding areas at Kaohsiung city in southwestern Taiwan.

INTRODUCTION

Large scale landslides have attracted the attention of people in the same way as other uncontrollable natural disasters, such as earthquake, volcanism, floods and debris flow, which threaten human life and their properties. It is a major problem not only today but will be an even serious problem in the future. The problem of landslides is strong related to human activities such as deforestations, hill slope development and surface mining operation. Because of the great damage caused by landslides to forest stands, farmland, buildings, reservoirs, and other engineering structures, they also can be recognized as a serious economic problem.

Shou-Shan sliding areas, are one of the main regions for the hill slope development in Taiwan island. The effect of these sliding areas on the growth of city is well demonstrated in the development of Kaohsiung metropolitan. Actually, Large-scale movement of Shou-Shan sliding area has a detrimental effect on the natural environment and public facilities around the Kaohsiung city. Factors governing Shou-Shan landslides are complex and complicate. Various studies have shown that most sliding area in these regions are human activity related. The great diversity of factors and complexity of interrelationships, as well as the practical relevance of Shou-Shan landslide, can be recognized only by systematic field studies. Therefore, the possible approach used to prevent these areas from hazard attack was to intensively perform the field investigations by using GPS/RS technology (Global Positioning System/Remote Sensing) and conduct the laboratory experiments based on the soil mechanical properties determination. Database established by GIS/RS integration. was also critical in this study. The best result can be achieved by the integration of GPS/GIS/RS. Since the field investigations can provide a clear picture of the origin, cause, and nature of landslide processes by checking their information against the result of static analysis and soil mechanical tests. From what has been said above, a sound treatment for landslides is based on careful observations for possible site of soil mass movement before, during and even after carrying out stabilization measures. Indispensable geological survey, hydrological analysis, geomorphologic comparison, soil mechanical parameter investigations and GPS/GIS/RS integration shall serve the propose of this research.

The propose of this study is to provide an effective method and optimized decision for the monitoring and designing of Shou-Shan landside control project. On the other hand, this study can give a reference on the futural planning of hill slope development at Kaohsiung city in southwestern Taiwan.

METHODS & PROCEDURES

Site Characterization

The study site has total area of 1200 ha (Fig.1), of which about 90% is classified as slope land. It is about 5.5 kilometers long running from north to south and has a maximum width of 2.5 kilometers between east and west bond. One significant high range is distinguished on the central part of the region. The highest peak is the Kaohsiung Mountain with the elevation of 350 meters above sea level.

Shou-Shan sliding area is located within the western foothills geologic province of Neocene elastic sediments. The predominant underlying rock is Kutingteng formation composed of sandstone, siltstone, mudstone, shale and organic limestone reefs in a monotonous alternating sequence. Organic limestone reefs are the important rocks that are distributed sparsely in the Pliocene rocks of southwestern Taiwan. The limestone reef occurs either as isothermal or as bios Rome bodied emplaced at different zones in the Pliocene strata. The lentil often thins out and tapers off at a short distance and grades into the adjacent better-stratified muddy sediments. These limestone reefs are whitish gray, massive and porous with litter or no evidence of stratifications.

Soils of Shou-Shan sliding areas lie in the transition zone of Entisols and Inceptsols. Properties observed in these soils resulted from the interaction of five soil forming factors, such as climate, organism, parent material and relief, combine to create a set of processes which modify soils through time (Morgan & Rickson 1995). The soils in these areas are weakly developed and usually silt clay loam or clayey loam in texture. Rapid penetration of water into mudstones is due to the formation of many coarse-textured soils coming from the weathering of limestone reefs. This subsurface flow not only aggravates soil erosion but also induces slope movement in the area. Therefore, hydrological factors play an important part on controlling landslide of the region. Most of soils have a thermal temperature regime. Entisols and Inceptisols are locally common due to steep slope. They are also calcareous, and infertility in soil profile.

Shou-Shan sliding areas are humid and tropical with annual precipitation ranges from 90 to 1800mm. High rainfall amount are intensively concentrated on storm season during summer time. They trigger the soil erosion and induce large-scale slope movement within the region. The extreme temperature range fluctuated from 12°C to 34.3 °C which is characterized the tropical climate of the region. The warmest month is June with mean monthly temperature of 27 °C, but the coldest month occurred at December with temperature of 12 °C. The mean relative humidity of whole year is 77.5%. Average annual precipitation for the period 1960~2003 is 1673mm; about 1199.5mm of this occurs as heavy rainfall during the period of May to August （Table 1）. Summer precipitation comes as showers, typhoon and high intensity thunderstorms. Approximately 63% of the annual precipitation falls in the months June to August. In general, the temperature is warm during the summer season but is cold in the winter time. The primary land use in Shou-Shan area is native forestry and rangeland. A limited amount of native rangeland has been converted to agricultural land uses including orchard planting and economic reforestation. An increasing amount of land is being used for urban development including multi-family doweling, schools, and temples, surface mining sites, parks, roads and commercial activities (Fig.2).

Field Investigations

Several sliding and slumping slopes, located at Shou-Shan area, were investigated by using GPS in the research. Two of them have been selected as experiment districts. There were chosen as study sites due to their recently potential slope failures and their relatively continuous sliding history. One of the experiment sites, called New Zoo area, is located at the eastern facing of Shou-Shan range. Another one lies in the western part of the range and gives a name of Chung-Shan University （Tsai 2004）.

The purpose of field investigations is to determine the natures, cause, and extent of each type of material that may have a negative effect on the stability of these sliding areas (Das 2005). Detail knowledge of the landslide from toe to crest is essential. The scope of field investigations should include topography, groundwater, erosion, vegetation, geomorphology and history of sliding slope changes (Zaruba 1982). The topography of landscape is an overt clue to past landside activity and potential instability. A detail survey of the sliding area and the preparation of contour map are required for this project. Geologic structure is frequently a major factor in governing landslides. Although this topic includes major large-scale structural factures such as folds and faults, the minor structural details including joints, small faults, and local shear zones may be even more important. Hence, the sliding area should be mapped geologically in detail by using GIS technology （Fig.3-4）.

Table.1. The metrological data of Shou-Shan area (1960~2003).

Item \ Month	1	2	3	4	5	6	7	8	9	10	11	12
Average Temp.(°C)	22.4	23.1	24.9	27.3	29.6	30.4	31.0	30.3	29.7	28.4	26.3	23.4
Maximum Temp.(°C)	28.0	28.7	30.2	32.0	33.4	34.2	34.3	33.9	33.6	32.1	30.6	28.3
Minimum Temp.(°C)	12.2	12.5	14.9	17.7	21.7	22.4	24.0	23.9	23.1	20.5	16.4	12.0
Relative Humity(%)	74.0	74.3	75.3	77.6	78.3	81.7	80.9	82.0	80.4	77.1	74.5	74.5
Average ppt.(mm)	22.3	22.7	36.6	67.6	185.4	388.4	340.5	364.8	222.9	82.0	16.4	13.1
Maximum ppt.(mm)	83.9	87.4	181.2	262.2	727.0	1082.4	927.1	1375.9	781.9	488.9	94.9	110.0
Minimum ppt.(mm)	0.0	0.0	0.0	0.0	0.0	30.5	23.0	19.0	19.2	0.0	0.0	0.0

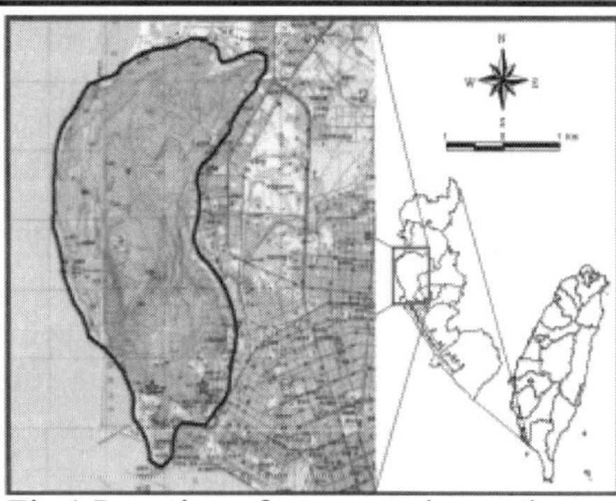

Fig.1.Location of two experimental sites in study area

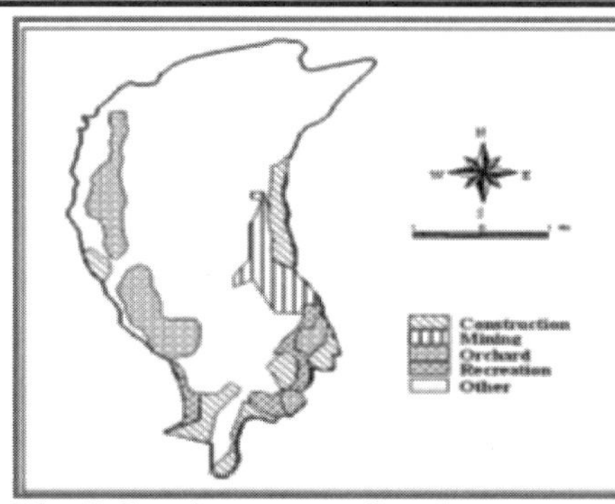

Fig. 2. Land use map of study area.

Fig.3.Topographic & Geologic map of Shou-Shan area.

Fig.4 .The satellite image、DEM、contours of Shou-Shan Community

Laboratory Experiments

Laboratory tests complement field investigations are to give more complete picture of soil materials within the slope and their engineering properties. Furthermore, it is possible in the laboratory to establish the changes in soil behavior due to the changes in environment. The major laboratory tests (McCarty, 2002) for determining the shear strength of soils include the direct shear test, the triaxial compression test, the unconfined compression test, mechanical analysis, and soil physical properties determinations. The direct shear strength is usually in terms of the soil stress. The triaxial compression test can be used for determining either the total strength or the effective strength. Mechanical analysis is used to determine the soil texture. Soil physical properties can provide the information to evaluate the causes of landslides. These tests were conducted in the laboratory during the period of this study (Table 2).

Shou-Shan sliding areas are greatly affected by the non-homogeneity and anisotropy of soil mass within the sliding zone as well as by the conditions in the surrounding environment (Lee, et al 2002). Consequently, the validity of result analyses depends on the ability of investigator who can give a true picture of the geological properties at sliding site in relation to the mechanical properties of rocks and soils weathering from limestones and mudstones. The initial stresses present in the ground are also important, as well as the hydro geological condition, the fluctuation of groundwater or other climatic factors that attack the landslide in Shou-Shan areas. However, each of these factors is a function of complex interrelations between each other.

Tables 1 through 2 and figures 4 through 8 illustrate the typical results of this research. These accompanying tables, satellite images, and GIS map layers show how various physical parameters affect the results. Table 2 shows the soils collected from sliding sites are weakly developed and usually silty clay loam or clayey loam in texture. Much more important is the influence of high pH-values and organic matter content. Unfavorable effect of high pH-values (pH=7.0-8.2) is a shift of electric soil potential towards negative values that may generate an additional water inflow by eletroosmosism. This caused a further decrease of soil shear strength within the slope. Quite general, any significant deviation from normal soil pH has an unfavorable effect on shear strength. Hence, slightly alkaline soil solution in these areas has a detrimental effect on the stability of sliding zone (Hilled, 1982).

Slope stability analysis by using Simplied Bishop and Janbu's modified methods (Lee, 2002) that have been computerized by the PC program at National Pingtung University of Science and Technology. This result can define the location of all possible critical shear failure surfaces. At this point, an automatic search program has been developed to further refine the location of the most critical shear failure surface in this research. Therefore, the prediction of landslide can be effectively controlled by the database developed from GIS technology combined with Visual Basic program. From what approaches we have done can give a general information related to the geohazard automatically mapped by using GPS/GIS/RS integration.

Table .2. The mechanical characteristics of soils at Shou-Shan area.

No	Sample	slieve analysis (%)			W(%)	e	r_t (t/m^3)	G	LL (%)	PL (%)	PI (%)	PH	Tot.N (%)	O.M (%)	Ex.K (ppm)	Ava.P (ppm)	CEC (me/100g)	K (10^{-8} cm/sec)	Direct Shear test				Qu (kg/cm^2)	Cc	soil class
		G	S	CL															C_d (kg/cm^2)	ϕ_d (°)	Cu (kg/cm^2)	ϕu (°)			
BH-1	S-1	0	12	88	26.54	0.83	1.88	2.72	35.6	16	20	7.8	0.18	2.4	136.0	18.55	26.80	1.571	~	~	~	~	~	0.11	CL
BH-1	S-2	0	19	81	27.49	0.72	2.00	2.70	41.4	20	22	~	~	~	~	~	~	~	0	26	0	14	2.87	~	CL
BH-1	S-3	0	8	92	24.07	0.71	1.95	2.70	39.5	18	21	7.6	0.15	~	~	~	~	~	~	~	~	~	~	~	CL
BH-2	S-1	0	10	90	31.16	0.75	2.03	2.71	45.4	21	24	7.9	0.20	2.7	161.0	16.39	27.84	1.069	~	~	~	~	~	0.14	CL
BH-2	S-2	0	19	81	19.79	0.48	2.17	2.70	28.2	8.9	19	~	~	~	~	~	~	~	0	27	0	14	4.27	~	CL
BH-2	S-3	0	5	95	21.00	0.53	2.15	2.72	23.1	4.7	18	8.1	0.21	~	~	~	~	~	~	~	~	~	~	~	CL
BH-3	S-1	0	9	91	25.60	0.77	1.90	2.69	36.9	22	15	8.0	0.21	3.2	175.0	14.23	29.46	1.124	~	~	~	~	~	0.16	CL
BH-3	S-2	0	21	79	23.80	0.65	2.03	2.71	30.4	12	12	~	~	~	~	~	~	~	0	25	0	15	3.65	~	CL
BH-3	S-3	0	13	87	22.67	0.62	2.04	2.70	26.7	11	15	8.2	0.23	~	~	~	~	~	~	~	~	~	~	~	CL
BH-4	S-1	15	42	43	29.85	0.76	1.97	2.68	~	NP	~	8.0	0.20	3.4	185.0	17.36	17.82	51.26	~	~	~	~	~	0.12	GM
BH-4	S-2	0	19	81	25.02	0.68	2.01	2.71	30.4	19	12	~	~	~	~	~	~	~	0	26	0	18	4.02	~	CL
BH-5	S-1	16	48	36	28.85	0.56	1.67	2.67	~	NP	~	7.8	0.29	2.5	160.0	11.25	13.90	60.77	~	~	~	~	~	0.15	GM
BH-5	S-2	35	36	29	20.41	0.58	2.05	2.69	~	NP	~	8.1	0.21	~	~	~	~	~	0	28	0	19	3.91	~	GM
BH-5	S-3	21	37	42	21.61	0.61	2.03	2.69	~	NP	~	7.7	~	~	~	~	~	~	~	~	~	~	~	~	GM
BH-6	S-1	43	31	26	12.60	0.50	1.99	2.66	~	NP	~	8.2	0.23	2.3	196.0	14.39	10.16	79.97	0	26	0	19	3.28	0.16	GM

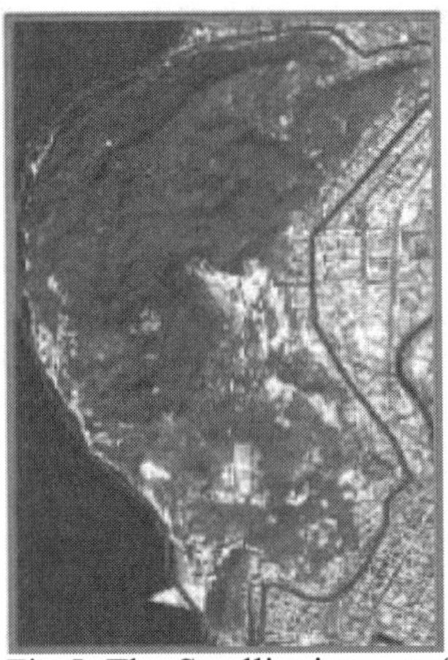

Fig.5. The Satellite image of Shou-Shan area

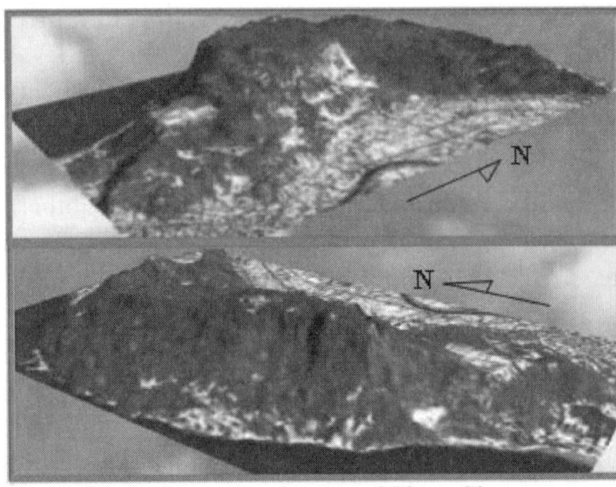

Fig.6. The 3D satellite image of Shou-Shan area

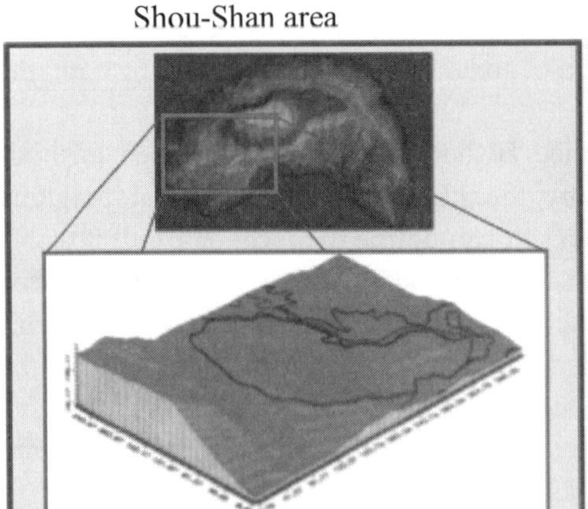

Fig 7. 3D grid map of potential sliding area at Shou-Shan, Kaohsioung

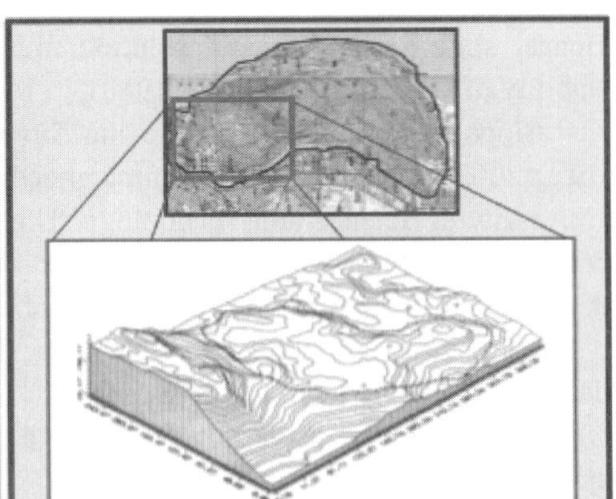

Fig 8 . 3D contour map of potential sliding area at Shou-Shan, Kaohsioung

CONCLUSION AND RECOMMENDATION

1) The major factors contributing to Shou-Shan landslides are predominated by weakly geological structure, of lime stones and mudstones, high rainfall intensity due to thunderstorm during typhoon season, great variety of ground water table fluctuation, poorly developed soil formation, sail strength decrease within the sliding zone and intensive human activities on the hill slope of Shou-Shan area.
2) High pH-value within the soil profile has an unfavorable effect on the slope stability in this region. Alkaline solutions effect increases the swelling capacity of clay and decreases the shear strength of soil formation in the sliding zone.
3) The distribution of potential sliding zones is presented graphically as Fig 7-8 that can be used to identify the possible extent of sliding zone in the region.
4) Further research should make it possible to develop accurate instruments that can be set up in the field for monitoring the stability of the whole Shou-Shan sliding area.
5) GPS/GIS/RS integrated technology should be widely used to investigate the potential sliding area around Shou-Shan region in the near future.

REFERENCES

Das, B.M (2005), "*Fundamentals of Geotechnical Engineering*", 2nd.ed., Thomson Canada Limited, CA. pp.240-290.

Hillel. D. (1982), "*Introduction to Soil Physics*", John Wiley & Sons Inc. P:21-37.57-104.

Lee W.A., T.S. Lee, Sharma, G. E. Boyce (2002), "*Slope Stability and Stabilization Methods*", 2nd ed. John Wiley & Sons Lnc. pp: 329~598.

McCarthy D. F. (2002), "*Essentials of Soil Mechanics and Foundations*", 6th ed. Prentice Hall, NJ. pp: 396~440.

Morgan R.P.C. and R.J. Rickson (1995), "*Slope Stabilization and Erosion Control, A Bioengineering Approach*", E & FN Spon London UK. pp: 221~248.

Zaruba,O. & Mencl,V (1982), "*Landslides and Their Control*", 2nd ed, Czechoslovak Academy of Science, P:187-201

Tsai Kuang-Jung , Yu Sheng-Horng, Lu Chih-Kaung. (2004). "Investigations on the Stability Analysis of Shou-Shan Landslides at Kaohsiung City in Southern Taiwan", *Proceedings of International Symposium on Landslide and Debris Flow Hazard Assessment*, Taipei of Taiwan, Vol. 4,1-12.

OTHERS

International Conference on Geotechnical Engineering for Disaster Mitigation & Rehabilitation
Chu, Phoon & Yong (eds) © 2005 World Scientific Publishing Company ISBN 981-256-469-1

UNDERWATER DETECTION OF SCOUR OF ABUTMENT FOUNDATION AND ITS INFLUENCE ON STABILITY OF BRIDGE PIERS

FANG LI-GANG
School of Civil Engineering and Architecture, Central South University, Changsha, Hunan, China

DUAN LIANG-LIANG
School of Civil Engineering and Architecture, Central South University, Changsha, Hunan, China

LIANG KAI
School of Civil Engineering and Architecture, Central South University, Changsha, Hunan, China

Scour is one of the major causes for bridge failure. To prevent this type of failure, it is necessary to detect the scour of abutment foundation. As a dependable method to measure scour, the underwater detection technology has been used to determine the scour degree of abutment foundation of the Zishui bridge of Xiangqian railway in China, and the transverse and vertical displacements of the bridge piers were analyzed resulted from different scour degrees of the abutment foundation. The results showed that the maximum transverse and vertical displacement of the bridge piers depended mainly on the ultimate scoured area of foundation base.

INTRODUCTION

At the bridge site of a river, the scour around piers and abutments is a common occurrence and poses a challenging problem to the engineers due to its detrimental effect to the foundation of piers and abutments. The scour of foundations of bridges has caused unexpected problems, e.g. catastrophic. Scouring will often be most harmful around the upstream face of a pier, and may eventually cause the pier foundation to become undetermined. Abutments, located at either end of the bridge, help to transmit the weight of the bridge including traffic to the foundation bed. On the other hand, piers are located within the bridge span. Based on the supply of sediment by the approaching flow, localized scour can be classified in two ways: clear water scour and live bed scour. Clear water scour is the situation where no sediment is supplied by the approaching flow to the scour zone. Live bed scour, on the other hand, occurs where there is a supply of sediment by the approaching flow to the scour zone. The flow field around piers has been well researched. Therefore, it is very important to detect underwater conditions and study the influence of scour on stability of piers.

The Zishui Bridge located in Lengshuijiang to Xinhua section of Xiangqian railway was designed and constructed with 4×60m through steel truss beam; all the piers and bridge abutments were built in 1937 and still in use now. The spread footings were used for the bridge. It was a period for resuming in construction time of the new and old line of Hunan province from 1958 to 1960, during this period, the bridge was resumed and constructed, and installed No 1 and No 5 piers. Due to the desire of navigable fairway, the top of existed pier have been heightened for 2.45m. The crack of wing walls of two abutments was serious, and the crack also existed in abutment body, therefore, it was needed for removing and reconstruction according to design requirement.

Because of the impact of technology and economy at that time, detailed engineering investigation has not been done, and lower rock stratum and karst development condition of the foundation was not in detail. In order to ensure safety of Zishui Bridge at stake K198+475 of Xiangqian railway and safe traffic of the bridge, underwater detection and geologic examination were used for abutment foundation and sub-grade condition of the piers, scour and damage condition of sub-grade and foundation were known, and finite element calculation model and manual computation method both can be used for the calculation and analysis of compression stress of foundation base and displacement of the pier top under the influence of different scour degree, to determine the impact of scour on stability of the piers.

ENGINEERING GEOLOGIC CONDITIONS OF THE BRIDGE SITE

The site exploration was carried out using coring to determinate the locations and thickness of the rock strata below the piers. The exploration revealed the main revealed stratum of the working area includes Quaternary residual diluvial horizon and Liujiatan section and Zhongya section of Carboniferous system Yanguan Group, it is divided from new to old as follows now: Quaternary residual diluvial horizon (Qh): consist of clay, residual clay, sand and gravel stone, mainly distribute on both sides along the river and low-lying point of gully. Liujiatan section and Zhongya section of carboniferous system Yanguan Group (Cly): greyblack middle thick silllike argillaceous limestone with greyblack flaggy siliceous rock or firestone band. The thickness is 65.6m, this set of stratum are distributed in the range of the whole working area. There is no obvious big regional structure existing in the working area.

UNDERWATER DETECTION

Because of the use of this bridge for a long time, and the design and construction information about this bridge construction are incomplete, its abutment foundation are required to be detected to verify some concrete data and information about this bridge: the form and size of the foundation, actual state of scour for the foundation etc. In order to detect the foundation and its concrete conditions of scour, the method of underwater detection must be used. Therefore, the method that the divers perform underwater tracing and picture photography should be used.

The Canon digital camera S30 with a waterproof case was used for underwater detection.

The maximum water depth is in the No.4 pier site with about 15m, all its foundation was flooded by water and the pier body was also flooded partly, this site is main navigation channel of this period now. And, there is rapidest current of water, the muddiest water quality, and relatively deep water depth in the No.4 pier site by observation, so its scour situation should be the most serious. In order to verify its foundation form and to measure and grasp scour situation, we should carry on careful underwater detection.

From the underwater photo, we could see that the foundation concrete on water face was scoured and destroyed, and the water face existed defect place, the foundation base was scoured seriously, and existed pore space.

For the condition of water return area in downstream of bridge foundation, the foundation concrete was kept well, and the foundation base was covered by silt.

The result of the detection is: though there was still 1.90m buried depth in some place of the No.4 pier foundation, all the upstream face and some downstream face surface have serious scour vugh, and the concrete of the foundation itself has presented the defect too.

THE INFLUENCE OF SCOUR ON STABILITY OF THE PIERS

The influence of scour on horizontal stability

When partial scour condition appear on horizontal bottom surface of the foundation resulted from scour (the upstream), as Fig. 1 shows, set horizontal size of the foundation as b, the size of scoured part is x, the bending moment of foundation base caused by horizontal wind force is M_1, According to *Fundamental code for design of railway bridge and culvert of P. R China,* $K_0 \geq 1.5$, and then

$$K_0 = \frac{\frac{b-x}{2}}{\frac{x}{2}+\frac{M_1}{N}} \geq 1.5 \tag{1}$$

There:

$$x \leq 0.4\left(b-\frac{2M_1}{N}\right) \tag{2}$$

It is obvious that when the scoured range of foundation base exceeds above-mentioned value, the pier foundation may be overturned. Even if no consideration of the influence of wind-force, scoured range of foundation base cannot exceed 40% of original size, and cannot exceed 20% of original of original size with consideration of wind load condition.

The influence of scour on the foundation base stress and eccentric examination

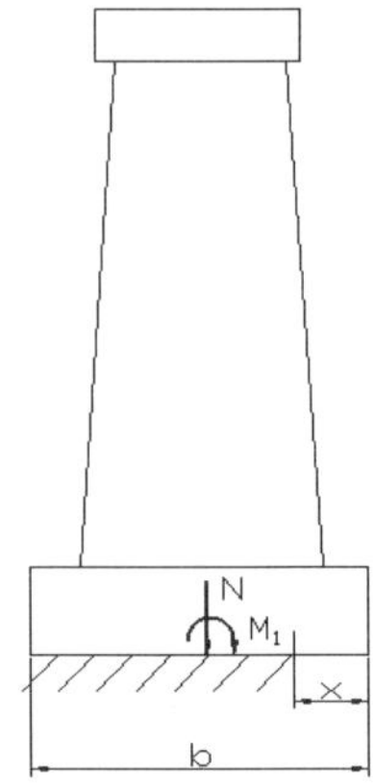

Figure 1. Simplified Section Figure of NO.4 Pier

Because scour holes of piers are mostly formed at upriver, after scour of foundation base, the original foundation under unidirectional stress (straight line bridge) has became the foundation under bidirectional stress, and stressed area of foundation base has decreased, all above would cause the condition under stress of the foundation to be worsened which extent changes with scoured position and range of foundation base, so we should pay enough attention to these. In addition, scour caused the great mass of cobble used for protecting bridge foundation to be washed away, existing buried depth can not satisfied with the minimum request of specification.

THE INFLUENCE OF SCOUR ON STABILITY OF THE PIERS BY THE FINITE ELEMENT ANALYSIS

In order to analyze the impact of scour on stability of the piers, the finite element analysis is used, to calculate and analyze compression stress of foundation base of the piers and displacement of the pier top under the influence of all kinds of degree of scour separately. The scoured method of foundation base simulated by this model is changing support pattern of foundation base. If the scoured area is 10% of the area of foundation base along left part, the 10% of foundation base part will be dealt with hangup without any constraint form, but will only remain 90% of the area to restrain; displacement constraint will be adopted for the constraint form.

In order to analysis and calculation of the maximum transverse displacement of the top of pier caused by overturn of the pier body which resulted from scour for foundation base, calculation method under the condition of worst loading combinations will be adopted for this finite element method. At this moment, only the vertical and horizontal load combination will be considered, and for reaching most unfavorable combination, transverse wind load should be perpendicular to horizontal structure of the bridge, and contrary to the direction of current scour.

For scour condition of the pier with different degrees, in the course of this finite element analysis, the other analysis conditions (such as load, the pier entity's form) are all the same except support ways of the foundation bottom. This analysis has separately calculated compressive stress of foundation base and displacement of the top of pier under different scour degrees which scoured area of the pier are 0% (i.e. without scour), 10%, 20%, 30%, 40% and 50%.

The result and summarize of finite element structural analysis of No.4 pier are shown in table 1.

Table 1. Maximum compressive stress of foundation base and displacement of the top of pier of No.4 pier.

Items / scour	Maximum compressive stress of foundation base σ_{MAX} (kPa)	Maximum transverse displacement Δx_{MAX} (10^{-3}m)	Maximum longitudinal displacement Δy_{MAX} (10^{-3}m)	Maximum vertical displacement Δz_{MAX} (10^{-3}m)
0%	608.26	0.341	0.959	3.202
10%	584.77	0.656	0.955	3.281
20%	555.91	1.600	0.943	3.591
30%	671.19	4.657	1.256	4.483
40%	864.19	10.029	1.612	6.075
50%	1369.70	20.875	2.227	10.804

According to calculation and analysis hereinbefore, the maximum compressive stress of foundation base at initial stage reduces slightly with the increase of scoured area with a little variety. When the scoured area is up to 20%, the maximum compressive stress of foundation base increases with the increase of the scoured area, and increment is relatively large. When the scour area reaches 50%, σ_{MAX} =1369.70kPa, has already exceed its allowable stress $[\sigma]$ =1080kPa. It can be calculated by linear interpolation that when scoured area is up to 44.5%, $\sigma_{MAX}=[\sigma]$ =1080kPa, i.e. the ultimate scoured area should be 44.5%.

Being similar with the analysis result of the maximum compressive stress of foundation base, the analysis curve of the maximum transverse, longitudinal and vertical displacement of the top of No.4 pier have similar condition. When the scoured area reaches 20%, displacement value increases rapidly.

THE INFLUENCE OF SCOUR ON STABILITY OF THE PIERS BY THEORETICAL CALCULATION

The horizontal displacement of the top of the pier includes elastic displacement of the pier itself and the displacement of inclination of the pier caused by differential settlement of the foundation which can calculated from the angle of inclination.

The theoretical calculation result of transverse displacement of the top of the NO.4 pier resulted from inclination of the pier is shown in Table 2.

Based on Fundamental code for design of railway bridge and culvert and Eq. (1), transverse dimension of the foundation is 13.53 m, bending moment of foundation base caused by transverse wind force is 7268.65 kN·m, vertical load N is 24394 kN, by calculation, under the action of eccentric moment of centroid of area and wind load, limiting state of equilibrium appears when the dimension of scoured part is 5.17m, it will be calculated from this that the pier will be overturned when scoured area of foundation base is up to 36.8%.

Table 2. Theoretical calculation result of transverse displacement resulted from inclination of the pier.

Scour	10%	20%	30%	40%	50%
Maximum transverse displacement Δx_{MAX} (10^{-3}m)	2.22	3.52	5.76	10.26	18.12

CONCLUSION

When situation of scour is that the scoured area is 0 %,(i.e. the condition without scour), the calculation result according to theory of elasticity is similar to the result of the finite element analysis, and the result of the finite element analysis is greater than theoretical calculation result. The different conditions that the scoured area increases by 10% will be calculated separately by the same method till the scoured area reached 50%, at this moment, the maximum compressive stress of foundation base has already been greater than allowable compressive stress of foundation base at this moment, i.e., $\sigma_{MAX} > [\sigma]$. It indicates that bearing capacity of sub-grade at this moment is not enough, and sub-grade would be destroyed. When the scoured area is up to 44.5%, $\sigma_{MAX} = [\sigma]$ =1080kPa, i.e. the ultimate scoured area should be 44. 5%.

When scoured area of foundation base is 20%, various kinds of deformation and stress of the foundation change violently, it can be concluded that the foundation will be failure.

REFERENCES

Li Ke-chuan, Luo Xue-shu (2000). "*Foundation Engineering*", 2nd edition, Beijing: Railway Press

Liu Cheng-yu (2000). "*Soil Mechanics*", 2nd edition, Beijing: Railway PressQiu Bo-yong, Qiao Jiang-dong, Sheng Xin-wang, Wen Yu-song (2000). "*Bridge Engineering*". Beijing: Railway Press

International Conference on Geotechnical Engineering for Disaster Mitigation & Rehabilitation
Chu, Phoon & Yong (eds) © 2005 World Scientific Publishing Company ISBN 981-256-469-1

ANALYSIS OF LONGYANGXIA DAM DEFORMATION BASED ON SEEPAGE AND CREEP COUPLING METHOD

GUO HAIQING
Geotechnical Institute, Hohai University, 1, Xikang Road
Nanjing, Jiangsu, 210098, China

GU CHANGCUN, XU WEIYA
Geotechnical Institute, Hohai University, 1, Xikang Road
Nanjing, Jiangsu, 210098, China

Dam failures are mainly caused by cracks or failures of their foundation rocks, which are directly related to water seepage in the rock. This kind of fluid-rock interaction has an important influence on deformation and stress characters of the dam-rock system. By use of visco-elastic constitutive models and finite element solution method, the stress and seepage fields of foundation rocks are studied as a coupled system in this paper. Using this coupled models, the deformation doubts of the continuous displacement of the 13th dam section of the Longyangxia dam are analyzed and explained reasonably.

INTRODUCTION

The Longyangxia hydropower gravity arch is the main part of the major hydropower generation and water resources utilization systems built along the Yellow River in China. However, continuous displacement towards the left bank at the 13th dam section puzzled both the dam safety administration and engineers. Using the seepage and creep coupling theory and finite element method, this paper represents the results of a research effort devoted to investigate its causes.

The theoretical foundations of the theory and the FEM formulations presented in this paper are based on works in Oda (1986), Ohnishi and Kobayashi (1993), Shen et al. (2000) and Wu et al. (2001).

BASIC EQUATIONS

The balance differential equations expressed by displacements and general water heads can be given as

$$\left.\begin{aligned}
&G\nabla^2\delta_x + \frac{G}{1-2v}\cdot\frac{\partial\varepsilon_v}{\partial x} - r_w\frac{\partial h}{\partial x} + X_0 = 0\\
&G\nabla^2\delta_y + \frac{G}{1-2v}\cdot\frac{\partial\varepsilon_v}{\partial y} - r_w\frac{\partial h}{\partial y} + Y_0 = 0\\
&G\nabla^2\delta_z + \frac{G}{1-2v}\cdot\frac{\partial\varepsilon_v}{\partial z} - r_w\frac{\partial h}{\partial z} + Z_0 + r_w - r_c = 0
\end{aligned}\right\} \tag{1}$$

where, G is the shear module, E is the Young's modulus, ν is the Poisson's ratio, r_w is the density of water, r_c is the saturated density of the concrete or foundation rock, h is the water head. In addition, ∇^2 is the Laplace operator, ε_v is the volume strain, $\delta_i(i = x, y, z)$ are displacement components and X_0, Y_0, Z_0 are equivalent body force components caused by initial strain $\{\varepsilon_0\}$.

To get equation q, we have four hypotheses: a) the dam and its rock base are isotropic continuum media in different areas. b) The seepage follows the Darcy law. c) The grain skeletons' deformations of dam concrete and foundation rock are ignored. d) The deformations of dam concrete and its foundation rock are mainly caused by deformations of void spaces and cracks between grain skeletons mentioned above.

According to the law of mass conservation, the continuity equation of the water is derived as given in equation (2). For simplification, the equation is expressed along the main seepage directions denoted as coordinate axes x, y and z.

$$\frac{\partial}{\partial x}(k_x \frac{\partial h}{\partial x}) + \frac{\partial}{\partial y}(k_y \frac{\partial h}{\partial y}) + \frac{\partial}{\partial z}(k_z \frac{\partial h}{\partial z}) = \frac{\partial \varepsilon_v}{\partial t} + r_w n \beta \frac{\partial h}{\partial t} \tag{2}$$

where $[k]$ is the permeability tensor, whose main components are k_x, k_y, and k_z; n is the void ratio, and β is the compression parameter of the water.

The basic equations for the coupling analysis of stress and seepage fields are composed of equations (1) and (2), whose boundary conditions include (a) displacement boundary condition $\{\delta\} = \{\delta_0\}$, (b) stress boundary condition $\sigma_{kl} m_l = \sigma_{k_0}$ (where $k, l = 1,2,3$), (c) water head boundary condition $h = h_0$, (d) seepage boundary condition $-k_n \frac{\partial h}{\partial n} = q_0$, which should be specified according to the site conditions.

For studying the viscous deformation caused by the creep of rock foundation upon time-dependent loading, different visco-elastic constitutive models are developed to identify the most suitable models and parameters for more accurate simulation of the time-dependent deformation of the dam-foundation system.

The time-dependent deformation of the foundation rock, caused by loadings, is described by a Burgers model, which is composed of a Kelvin model and Maxwell model in series. The partial strain expression of Burgers model is

$$\begin{aligned} e_{ij,K} = & \left\{ e_{ij,0K} + \frac{1}{2\eta_K} \int_{t_0}^{t} s_{ij} \exp[\frac{E_K}{\eta_K}(t - t_0)] dt \right\} \\ & \exp\left[-\frac{E_K}{\eta_K}(t - t_0) \right] + e_{ij,0M} + \frac{1}{2E_M} \int_{t_0}^{t} \dot{s}_{ij} dt \\ & + \frac{1}{2\eta_M} \int_{t_0}^{t} s_{ij} dt \end{aligned} \tag{3}$$

Specially, when $s_{ij} = s_{ij0}$ is a constant value, and $\dot{s}_{ij} = 0$, the equation (3) becomes

$$e_{ij} = e_{ij,0K} \exp[-\frac{E_K}{\eta_K}(t - t_0)] + \frac{s_{ij0}}{2E_K}$$
$$\left\{1 - \exp[-\frac{E_K}{\eta_K}(t - t_0)]\right\} + e_{ij,0M} + \frac{s_{ij0}}{2\eta_M}(t - t_0) \tag{4}$$

The visco-elastic strain of the Maxwell component will change into $\{\varepsilon_{v,M}\}_{t_0}$ at time t_0. If $t = t_0 + \Delta t$, and the stress remain the same during the increment of Δt, the visco-elastic strain increment of Burgers model can be derived from equation (4) as given by

$$\{\Delta\varepsilon_v\}_t = \left[1 - \exp(-\frac{E_K}{\eta_K}\Delta t)\right]\left(\frac{1}{E_K}[C]\{\sigma\} - \{\varepsilon_{v,K}\}_{t_0}\right)$$
$$+ \frac{\Delta t}{\eta_M}[C]\{\sigma\} \tag{5}$$

where E_K, η_K, η_M is the stretch (compression or shear) modulus and viscous parameters and $[C]$ is the Poisson's ratio matrix, respectively.

For the basic equations of coupled stress-flow analysis mentioned above, the finite element method is used to solve the coupled partial differential equations in this paper.

The FEM solution scheme for the coupled equations of displacement and seepage fields is

$$\left.\begin{aligned} &[\overline{K}]^T\{\Delta\delta\} - [K']\{h\}_{m+1} = \{\Delta F_1\} - [K']\{h\}_m \\ &[K']^T\{\Delta\delta\} + ([S] + \theta\Delta t[\widetilde{K}])\{h\}_{m+1} = \Delta t(\theta\{\Delta F_2\} + \{F_2\}_m) \\ &\qquad + ([S] - (1-\theta)\Delta t[\widetilde{K}])\{h\}_m \end{aligned}\right\} \tag{6}$$

where θ is the integration parameter, $[\overline{K}]$ is the general stiffness matrix, $[K']$ is the general coupling matrix, $[S]$ is the general compression matrix, $[\widetilde{K}]$ is the general seepage matrix, $\{F_0\}$ is the equivalent nodal load vector by the initial strain, $\{F_1\}$ is the equivalent nodal load vector, $\{F_2\}$ is the equivalent nodal discharge vector, and $\{h\}_{m+1}$ is the general water head vector at time t_{m+1}, respectively.

If displacement increment $\{\Delta\delta\}$ and super-static water pressure $\{\Delta p\}$ are taken as the unknown quantities, and the full Hermit differences are adopted with $\theta = 1$, the equation (6) can be re-written as

$$\left.\begin{aligned} &[\overline{K}_t]\{\Delta\delta\} - [K']\{\Delta p\} = \{R\} - \{R_t\} \\ &[K']^T\{\Delta\delta\} - ([S] + \Delta t[\widetilde{K}])\{\Delta p\} = 0 \end{aligned}\right\} \tag{7}$$

where $\{R_t\}$ refers to the part of loads balanced by the stresses related to the displacement happened before the time t_m, and are calculated by the following equation

$$\{R_t\} = \sum_{l=0}^{t_m} [\overline{K_t}]_l \{\Delta\delta\}_l \tag{8}$$

where l is the number of calculating periods of time before the time t_m.
According to the incremental initial-strain method used in equation (7), the initial stress field and seepage field of dam-foundation system need to be created by iterations starting from the beginning of the time-marching process. The element stress remains unchanged during the subsequent time step Δt, so that the seepage coefficients, which are decided by the state of stress at the beginning of the time step, remain the same value. They change step by step with loads increasing gradually at each time step. The water heads at the end of every time step are the initial ones of the next time step.

ANALYSIS OF LONGYANGXIA DAM'S DEFORMATION

The main part of Longyangxia dam is a gravity arch dam of 396 meters in length, 29.2/80 meters in width (at the top/base), 2610 meters for DCL and 178 meters in height. The dimension of the FEM model is 540 m in both length and width and 360 m in height. When dividing the FEM mesh, the geological structures in the foundation rock and their engineering treatment measures were taken into account. While laying out the element nodes, the locations of the in situ measuring points of displacement, temperature and stresses are considered. The FEM mesh, which is consisted of 21,189 eight-node-hexahedron elements with 24,873 nodes, is illustrated in Figure 1. The material parameters are listed in Table 1.
From April 16, 1990 to May 1992, the water level of Longyangxia reservoir dropped from 2575.04 m to 2533.15 m. From May 1992 to December 1994, the water level rose from 2533.15 m to 2577.58 m. From January 1995 to July 1998, the water level dropped from 2577.58 m to 2533.54 m again. After July 1998, the water level rose from 2533.54 m to 2581.08 m. These water level variations were used for deriving the water head loading conditions.
The tangential displacements of Longyangxia dam's typical dam section on April 16, 1989, April 16, 1990, and April 16, 1996, and December 31, 1999 were calculated and compared with measured data. The results are listed in Table 2 and illustrated in Figure 2. It can be seen that: the calculated values are close to the measured ones, and he calculated tangential displacements above the 2500m level on April 16, 1996 are as follows: -3.04mm, -6.91mm, -6.33mm, -5.77mm, and -4.68mm, respectively. These data show that the 13th dam section moved towards the left bank. This was caused by the continuing high water level, about 100m ~ 110m higher than adjacent uplift pressure, and the creep displacements of dam body and rock foundation.
The calculated results, which agreed well with the measured ones, clearly indicated that the main reason for the continuous displacement of Longyangxia dam towards the left bank after July 1989, was caused by the influence of the seepage-stress combined operations on its rock foundation.

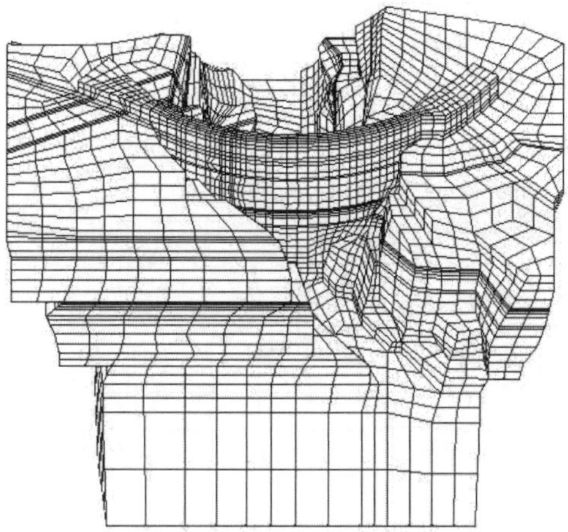

Figure 1. The FEM mesh of the Longyangxia dam model

Table 1. Parameters of Longyangxia dam and foundation materials

		Density kg/m^3	Elastic constants		Viscous constants				Seepage parameters	
			Modulus (GPa)	Poisson's ratio μ	E_K GPa	η_K $GPa \cdot S$	E_M GPa	η_M $GPa \cdot S$	cm/s	
Dam concrete		2400	20	0.18	250	2.3×10^5	66.7	1.15×10^9	Dam	1.0×10^{-7}
Rock foundation	2580 m ~ 2560 m	2650	8	0.25	50	4.0×10^4	15.0	8.5×10^9	Rock	1.0×10^{-9}
	2560 m ~ 2540 m	2700	12	0.23	200	3.6×10^4	20.0	5.4×10^9		
	2540 m ~ 2500 m	2750	16	0.22	200	2.8×10^4	40.0	5.0×10^9	Crack	1.0×10^{-4}
	Below 2400 m	2755	22	0.22	300	2.0×10^5	51.0	1.8×10^9		
	G_4*	2600	3	0.25	30	5.4×10^4	10.0	9.5×10^9	Curtain	1.0×10^{-8}
	F_{18}	2600	4.5	0.25	40	4.0×10^4	15.0	9.0×10^9		
	F_{71}, F_{73}, F_{32}, F_{67}	2600	3.2	0.25	30	5.4×10^4	10.0	9.5×10^9	Drainage	1.0×10^{-2}
	A_2+F_{120}	2600	5.4	0.25	40	4.0×10^4	15.0	9.0×10^9		

*G_4, F_{18}, F_{71}, F_{73}, F_{32}, F_{67}, and A_2+F_{120} are geological structures and cracks in Longyangxia dam foundation.

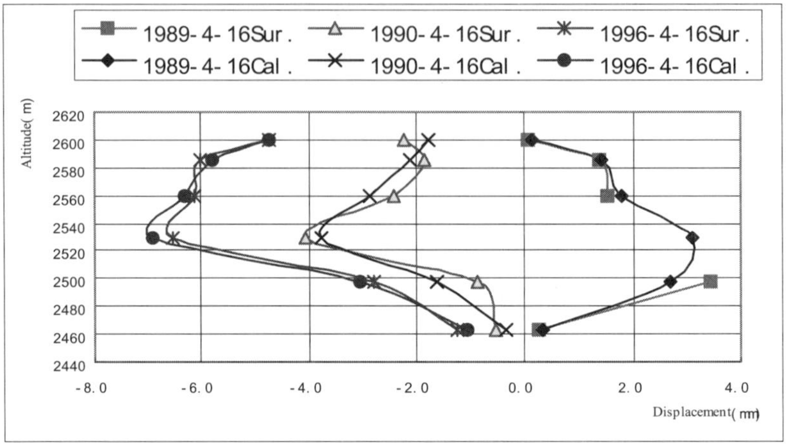

Figure 2. Survey values and calculation values of the 13th dam section's displacements

Table 2. Calculated values of time-dependent viscous displacements (mm)

Elevation \ Date	1989-4-16	1990-4-16	1996-4-16	1999-12-31
2463.3 m	0.12	-0.28	-1.01	-1.75
2497 m	1.05	-0.44	-2.85	-3.82
2530 m	1.24	-0.79	-4.44	-5.97
2560 m	0.89	-0.61	-3.69	-5.48
2585 m	0.57	-0.36	-1.84	-2.26
2600 m	0.06	-0.31	-4.14	-5.49

REFERENCES

Oda, M. 1986. An equivalent continuum model for coupled stress and fluid flow analysis in jointed rock massed. Water Resource Research 22(13): 1845 - 1856

Ohnishi, Y. & Kabayashi, A. 1993. Thermal-hydraulic-mechanical coupling analysis of rock mass. In Hudson J. A. (ed), Comprehensive Rock Engineering, Pergamon Press: 191 - 208

Shen Zhenzhong. Xu Zhiying. & Luo Cui. 2000. Coupled analysis of viscoelasticity stress field and seepage field for the Three Gorges dam's foundation. Engineering Mechanics. 17(1): 105 - 113.

Wu Zhongru. Gu Chongshi. & Wu Xianghao. 2001. Theory and its applications of safety monitoring of roller concrete dam. Science Press.

International Conference on Geotechnical Engineering for Disaster Mitigation & Rehabilitation
Chu, Phoon & Yong (eds) © 2005 World Scientific Publishing Company ISBN 981-256-469-1

BIOLOGICAL PERSPECTIVE OF TSUNAMI: A RELIEF STRATEGY - BIOPESTICIDE SPRAY OPERATION FOR THE MOSQUITOES, FLIES AND DISEASE CONTROL AT THE TSUNAMI AFFECTED AREAS OF TAMIL NADU AND PONDICHERRY, INDIA

MURUGAN, K., VANITHAKUMARI, G. AND NOORTHEEN, A.
Department of Zoology, Bharathiar University, Coimbatore - 641 046, India.

Biopesticides spray operations have been done at the Tsunami affected areas of Tamil Nadu and Pondicherry for the control of mosquito vectors and flies. Biopesticides (plant extracts and bacterial toxins) spray operation at the Tsunami affected areas of Tamil Nadu such as Velankanni, Nagapattinam and Cuddalore, and Pondicherry, India for the control of mosquito, flies, and other diseases causing microorganisms. Our bioformulations from Department of Zoology, Bharathiar University, Coimbatore, India possesses mosquito larvicidal effect, antibiotic, antibacterial, antiviral properties and it has right combination of herbal and bacterial toxins.

INTRODUCTION

Tsunami occurred on 26th December 2004 and in India more than 8000 people were killed due to the massive attack. The worst affected districts in the Tamil Nadu were Viz., Vellankanni, Nagapattinam d Cuddalore, Tamil Nadu, Pondicherry, India. Official sources said that due to massive death of animals and human beings, the public health has been spoiled and it will cause further outbreak of infectious diseases like diarrhea, cholera and mosquito transmitting diseases. After a week there was a report of incidents of mosquito transmitting diseases such malaria, and dengue. Tsunami made breeding ground for mosquito vector and thereby invasion of mosquito transmitting diseases at the Tsunami affected areas such Vellankanni, Nagapattinam and Cuddalore Tamil Nadu, India

Team has also sprayed biocides with suitable herbal combinations and had sprayed on the streets and human dwelling areas at the Thevanampatinam and Thalanguda villages of Cuddalore district, Tamil Nadu, India. Due to evasion of chicken pox due to the dumping of old clothes donated by people, our operations had a timely impact on further outbreak of viral fever in populations at the villages. District Collector, Thiru. Gangan Deep Singh Bedi I.A.S., Cuddalore had visited the Operational spot and appreciated the Team's work. Hence, the above said spray operation not only helped in control of mosquito larvae and there by solving vector borne diseases at helped in indirectly killing microorganisms to provide sustainable protection against infectious diseases like chicken pox and dermal diseases.

In the present study, our biopesticides application was effective against mosquito population and flies. Our biopesticides applications were a right combination of botanicals really had good impact on the other viral and bacterial diseases prevalent in the Tsunami affected areas of Tamil Nadu. For example, at Cuddalore and Nagapattinam District, Tamil Nadu, India Chicken pox has been noted during the 3rd and 4th week of January, 2005. It may be due to viral invasion at the Tsunami affected area and it might have been due to the dumping of old cloth donated by public and it has been found that the dormant viral bodies had got virulent after soaking in to the water. Hence, our herbal extracts were capable of destroying viruses and had inhibitory effect on the growth of bacteria and in total it had an ability to destroy growth of disease-causing organisms (Antimicrobial). Therefore, our biopesticides operation was really successful in spraying in the infected area (death of animals and human being). Some of the active chemicals present in the neem and other herbals offer a non-toxic alternative to synthetic drugs and disinfectants.

People of Tsunami affected areas have already had a severe shock, had lost all the possessions in the form of materials and relations. Due to the invasion of newer disease causing organisms producing chicken pox, dermal diseases and mosquito transmitting diseases like malaria, dengue and Japanese encephalitis etc., as well as the spread of infectious diseases from the burial bodies due to the pollution that has emerged from the dense population at concentrated area, this has created, in turn will an enhancement of stress and trauma already faced by the Tsunami affected people. Hence, eradication of these kinds of situational diseases by using synthetic chemicals will pose newer intolerable problems to the Tsunami affected people.

Occurrence of dermal disease was also a major problem in the Tsunami affected areas of Tamil Nadu. Our biopesticides operation at the streets, houses and temporary shelters had definite protective effect on the people, because our herbal formulations have nutritive fatty acids and contain all these viral nutrients in significant quantities. Our herbal ingredients had already been reported to be used for treatment of many skin ailments. It is important to note that neem and turmeric have rejuvenating qualities and are absorbed quickly by the skin.

MATERIALS AND METHODS

Biopesticides spray operations have been carried out at the various villages of Velankanni (Kallar, Sebastian Street and Seruthur), Nagapattinam (Akkaraipettai) and Cuddalore (Thevanampattinam and Thalanguda). Team consists of 10 members of the Zoology Department, Bharathiar University, Coimbatore 641 046, and India. Spray operation made at the various mosquito-breeding places of Tsunami affected areas in the above said areas. Selection of Tsunami affected areas was done with help of Official Co-ordinator of Tsunami Task Force, Tamil Nadu. Spray operations done in and around the houses, streets, and water stagnant areas such as cesspool, ditches, and other open house hold containers. Biopesticides spray also done at the temporary shelters of the Tsunami affected people, temples and schools etc., Biopesticides have been sprayed uniformly on the surface of the breeding habitat such as ditches, cesspool and other temporary water bodies at Tsunami affected areas of Tamil Nadu. Dipper sampling and counting of the larvae monitored and larval density before 24 hrs, 48 hrs and 72 hrs after the treatment. A separate sample was taken to determine the special composition of each larval habitat.

RESULTS AND DISCUSSION

It is important to note that the there was higher incidence of flies, which have emerged from the buried bodies and other dead animals at the Tsunami affected areas. After Tsunami incidence the dead bodies have been buried in a single place and it was a conducive right atmosphere for the flies complete fit to their complete their life cycle. Hence, spray operation was also done at the mass burial places of the Tsunami affected areas for the control of flies, which cause further infectious diseases. The biopesticides have greatly affected flies larvae (maggots), which could not molt into pupa for further emergence. Since, plant derived compounds distinctly affect the developmental stages of insects, it has distinct advantages over synthetic pesticides.

Lethal concentration were also worked out to find out the efficacy of our sprayed biopesticides against mosquito larvae populations of Tsunami affected areas and mosquito larvae collected from Coimbatore (Table 1). The result of the lethal dose has shown interesting trend. Higher doses were needed to kill the mosquito larvae collected from the Tsunami affected areas of Tamil Nadu. The percentage difference between the lethal value were also worked out and it was interesting to note that the percentage difference of LC_{50} and LC_{90} values were higher for the mosquito population from Tsunami affected areas larval population than the mosquito population from Coimbatore, Tamil Nadu, India.

The team has also inspected and analyzed the efficacy of biopesticides at already sprayed areas of mosquito and flies infected areas after 24 hrs, 48 hrs and 72 hrs of durations (Table 2). Really, spray operations were very successful and it has been noted that, biopesticides greatly affected the egg and larval populations of mosquito vectors and had eradicated them completely.

Synthetic organic insecticides, although has higher efficacious against target species such as mosquito and flies, it can be detrimental to a variety of animal life including man. In addition to adverse environmental ill effects, they affected non-target organisms also moreover most of mosquito species have become physiologically resistant to these synthetic organic insecticides. The present biopesticides spray operation had significant impact on mosquito and flies and the oily nature of organic extracts of plants and bacterial toxins have interfered with the larval respiration as well as had induced toxicity to the gut system and brought out mortality of mosquito larvae.

The team has done good work for the cause of public and this operation is the need of the hour to control the infection from mosquito, flies and thereby prevent the health hazards and relieve the people from further stress and strain in the present situations. We need long-term project to control vectors and other infectious diseases like chicken pox and other dermal diseases.

Management plans and awareness programme should be adapted for the control and prevention of mosquito, flies and other disease causing organisms at the Tsunami affected areas by following these strategies.

i) Avoid stagnant water for the prevention of mosquito breeding
ii) Need good environment (free from garbage and plant animal debris) for control of flies.
iii) Public toiletry facilities to be improved.
iv) Stop usage of old cloths, from others.
v) Good drainage facilities.

Training and awareness programme to be include:

i) To give demonstration about the life history and distribution and seasonal abundance of mosquito vectors and flies to the people of Tsunami affected villages.
ii) To show video presentation about the pathogenic organisms of mosquito transmitted diseases such as malaria, filariasis, dengue and other disease (chicken pox, dermal and other diseases).
iii) To teach the necessary safeguard measure against future natural disasters like Tsunami, flood, fire etc.,

The killer waves have not only washed boats, other implements and other household articles, but also had filled water in the opened materials that also had given enough moisture in about 2 to 3 km radius for the mosquito breeding. It is also of doubt that the minerals of volcano particles from deep sea might have contaminated the soil for the early proliferation of the microorganisms and mosquito breeding.

Our spray operation had good impact on the control measures as

i) Mosquito larval control at the breeding habitats of Tsunami affected areas was effective for the further prevention of mosquito-transmitted diseases such malaria, filarial and dengue fever.
ii) Effective flies control was evident at the breeding habitats of Tsunami affected areas (plant and debris of animal and human beings).
iii) Prevention and control of spread of further infectious disease causing microorganism, which cause dermal, chicken pox and diarrhea (virus, bacteria and fungi).
iv) Suggestions to the locals for planting of medical and aromatic plants such as neem, pongamia and nochi at the Tsunami affected areas of Tamil Nadu, India and also along the entire coastal system explaining their role as physical barriers, as well as readymade pharmaceutical uses for the emergency medical purposes. Suggestions for use of appropriate animal system (Rats, Dogs etc.,) could also be employed for the forecasting for the forecasting disaster, as they indicate through their changed behaviors.

Table 1. Toxicity evaluation of Biopesticides on the larvae 4th instar larvae of *Culex quinquefasciatus* Liston at Tsunami affected area, Akkaraipettai, Nagapattinam District and Thevanampattinam, Cuddalore District, Tamil Nadu., India.

Concen- tration (ppm)	Number of Larvae	Observed Mortality in Akkarai- pettai	Probit Analysis of Akkarai- pettai	Observed Mortality in Thevanam- pattinam	Probit Analysis of Thevanm- pattinam
2.00	100	24e	LC_{50}	29d	LC_{50}
4.00	100	32d	14.57	37cd	10.80
8.00	100	49c		52c	
16.00	100	60b	LC_{90}	67b	LC_{90}
32.00	100	69a	49.38	75a	43.33

Within a column means followed by the same letter(s) are not significantly different at 5% level by DMRT.

Table 2. Field trial of Biopesticides at Tsunami affected area, Nagapattinam, Tamil Nadu.

Sl. No.	Larval Density			
	Before treatment	After Treatment		
		24h	48h	72h
1.	20	4	-	--
2.	25	2	-	--
3.	16	5	2	--
4.	14	5	1	--
5.	22	3	1	--
6.	18	1	-	--
Total	115	20	4	-
Average	19.16	3.33	0.66	-
% Reduction		82.6%	96.5%	100%

1. Place : Akkaraipettai, Nagapattinam District, Tamil Nadu, India.
2. Habitat : Drainage (Tsunami contaminated water)
3. Size : 3mx 2m
4. Depth : 1cm
5. Species : *Culex quinquefasciatus*
6. Stage : Larvae
7. Calculation : 3x 2= 6 x 1= 6 liters.
8. Required concentration: 296.28 ppm

CONCLUSION

People of Tsunami affected areas have already had a severe shock, had lost all the possessions in the form of materials and relations. Due to the invasion of newer disease causing organisms producing chicken pox, dermal diseases and mosquito transmitting diseases like malaria, dengue and Japanese encephalitis etc., as well as the spread of infectious diseases from the burial bodies due to the pollution that has emerged from the dense population at concentrated area, this has created, in turn will an enhancement of stress and trauma already faced by the Tsunami affected people. Hence, eradication of these kinds of situational diseases by using synthetic chemicals will pose newer intolerable problems to the Tsunami affected people. This is just a small effort of ours. However, exploration of entire coastal system needs similar treatment in a big way. Authorities should concentrate on such management strategies to prevent the outbreak of epidemics and encourage researchers involved in such activities. The ad "prevention is better than cure" is very apt in this situation. The only plant-based insecticides will be effective and long lasting as compared with synthetic pesticides. Moreover, biopesticides of plant origin need to be tested to provide a cheap and environmentally safe solution to the present problems.

International Conference on Geotechnical Engineering for Disaster Mitigation & Rehabilitation
Chu, Phoon & Yong (eds) © 2005 World Scientific Publishing Company ISBN 981-256-469-1

APPLICATION OF FREEZING METHOD TO RECOVER TUNNEL ACCIDENT IN COMPLEX STRATUM OF NANJING SUBWAY

YANG PING & ZHU FENG-BIN

Faculty of Civil Engineering, Nanjing Forestry University,Longpan road 159#, Nanjing Nanjing,Jiangsu, China

By the application of the freezing method to handle several engineering accident in Nanjing subway's complicated stratum, this paper presented the freezing design method, and analyzed the monitoring result and pattern of freezing-induced ground surface deformation in the sites. It is suggested that bore-unloading could counteract a portion of frost heave deformation and the grouting was to be the key measure to minimize the thaw settlement. Guidelines are provided for the application of the freezing method to deal with related engineering accident and construct subway in complicated stratum.

INTRODUCTION

The main stratum which the shield tunnel of Nanjing subway line #1 traversed are peaty clay, silty soil and fine sand. The peaty clay possesses high compressibility and easily fluid. Silty soil and fine sand with high water content and strong penetrability easily bring about spurting water and gushing sand. Because of the soft and weak stratum, several methods of improving soil have been used at the enreance of Zhang Fuyuan southern shield tunnel (Yang et al., 2003) and in five out of six corridors of line #1, but all was failing, only the freezing method succeeds to reconstruct and guarantee the safety.

HANDLING THE ACCIDENT AT SHIELD ENTRANCE

Permeability grouting and soil mixing methods for soil improvement were used at the entrance of Zhang Fuyuan southern shield tunnel .When punching concrete wall, a great deal of flow-sand twice gushes from both sides around the cave for shield, the quantity of which estimated to be about 110m^3. The ground surface's settlement is about 1.5m in the 20m^2 area of the east, and about 1m in the 1.5m^2 zone of the south west. The settlement of the whole region near shield entrance is larger (Fig.1). In order to insure the pipelines and the transportation being in normal usage and safety, and to guarantee the shield machine safely passing doorway,the freezing method was applied to improve the stratum.

Frozen wall and freezing parameter

By experiment, obtained the compression strength of frozen silty soil and fine sand is 9.95MPa and 9.57MPa at -10°C, and the tension strength is 1.81MPa and 1.72Mpa, respectively. The thickness of frozen wall can be calculated by empirical formula:

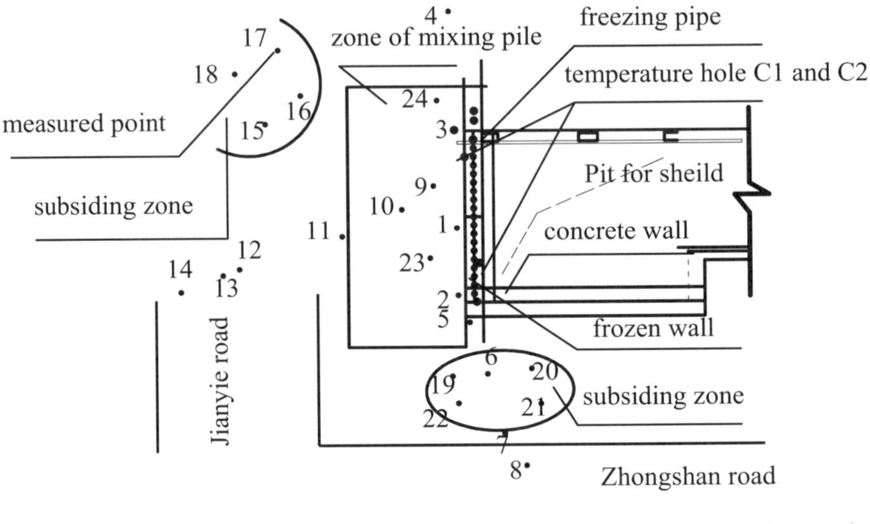

Figure 1. Layout of freezing pipe and displacement point near the entrance

$$e = [\frac{kB \cdot D \cdot p}{2\sigma_c}]^{1/2} \quad (1)$$

where e is frozen wall's thickness; k is safety coefficient, adopting 3; B is empirical coefficient, adopting 1.2; D is excavation diameter, 6.5m; p is the total water and earth pressure, 0.217Mpa; σ_c is the compression strength of frozen soil.

The thickness of frozen wall is calculated to be 0.5m. The overlapping length between the frozen wall and concrete wall which is around the cave for shield is usually 1-2m according to experience, and taking the larger water-earth pressure into account, the width and depth of frozen wall are designed to be 9.5m and 18.5m, respectively. The freezing depth above the cave is 3m. The zone from 3m above the cave to the ground surface is not frozen so as to reduce freezing-induced influence on ground surface.

Freezing pipes are set up perpendicular to ground surface. Because there have been soil mixing piles in the zone out of frozen wall ,which can bear lateral earth pressure, the design of the frozen wall should be calculated according to the demand of sealing water and according to the distance between soil-cement mixing piles and concrete wall. The thickness of 0.5m is deemed enough. It is more difficult to freeze soils near the concrete wall due to the large thermal conductivity of concrete. Freezing pipes should be lay near the concrete wall, And should take measures to preserve heat of the wall.

The layout of freezing pipes and frozen wall are shown in Fig.1. the 21 freezing pipes with diameter of 108mm and 2 temperature holes were laid out with the bottom at a depth of 18.5m. The distance between two pipes is 450mm and the distance between the freezing pipes and concrete wall is 200mm. The brine temperature should come to -22°C after freezing for 7days and start excavation of the cave of shield after freezing for 15 day. The lowest temperature of recycling brine is -24°C ~-28°C and the average temperature of the frozen wall is no more than -10°C. When excavating the cave for shield, the temperature at interface between the frozen wall and concrete wall is less than –3°C.

Monitoring and analysis of the temperature and deformation

In order to monitor the temperature and displacement during the freezing, measured temperature holes C1and C2 have been set up in frozen wall. And there are totally 24 points of monitoring displacement on the ground surface. Fig.2 shows the temperature change of the temperature holes with freezing time. C1 is near the concrete wall, where there is direct exchange of heat with hot atmosphere(20°C).So its temperature descends slowly, but is basically uniform at various depth. The temperature come to -7°C after freezing for 15 days. Whereas, the C2's temperature descends quickly to a lowest value

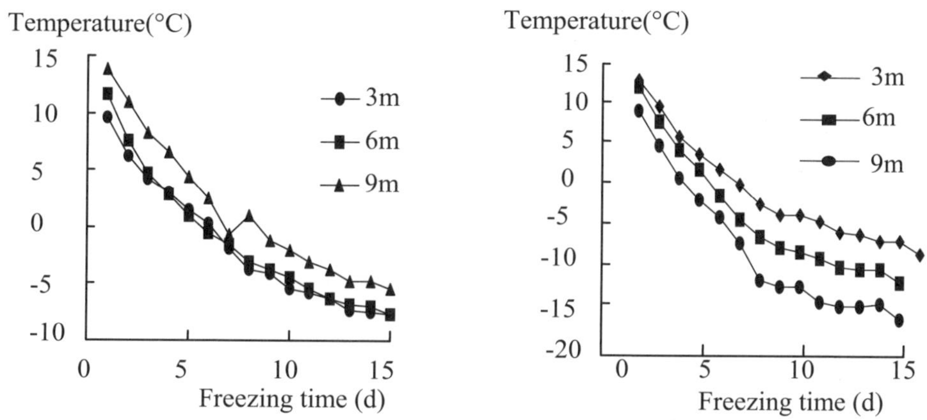

(a) Temperature hole C1 (b) Temperature hole C2

Figure 2. Curves of time and temperature in different depth

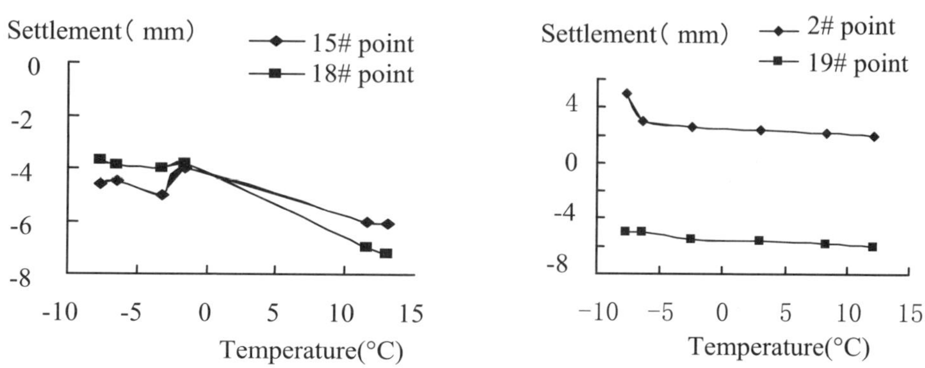

(a) Near subsiding area of north east accident (b) Near frozen wall

Figure 3. Curves of ground settlement and temperature in holes

of -16.5°C as it is adjacent to the side of mixing piles. Moreover, because the hole leans to freezing pipes with the increase of the depth, the temperature discrepancy at different depth is very large. Such as ,the temperature at the depth of 3m is only -8°C, while it is -16.5°C at 9m. The figure also shows it is read for excavation after freezing for 15 days

Fig.3 shows that the measured points 15# and 18#, which lie in the flow sand area east to the frozen wall(Fig.1). It appears that quite large initial settlements before freezing, and there is frost heave deformation of 4mm when the temperature descends, but the total settlement is 3-4mm at last. The settlement at the point 2# and 19# are not too large, with a maximum value of 5mm at 2# and only 1mm at 19# . This is because point 19# lies in the accident-induced subsiding zone. According to the results measured at 24 points, it is found that the maximal frost heave deformation is no more than 5mm.

THE APPLICATION OF FREEZING METHOD TO TUNNEL CORRIDOR

Corridors of tunnels of subway line #1in Nanjing are almost located at depth of 14m to 17m below ground surface. The stratum consists of silt clay layer, silty soil and fine sand layer, which have the properties of high moisture content, high sensitivity, low strength, large voids ratio and big penetrability, They belong to unstable stratum. The construction of five corridors once used several methods , but all was failing, only the freezing method succeed to reconstruct them. The freezing method is very effective on improving soil in weak layer with rich water. Sometimes ,it may be unique method.

Taking only San-Zhang corridor for example, the layout of freezing pipes, freezing parameter and monitoring frost heave and thaw settlement in the site are presented below. The corridor with a diameter of 3.2m is at depth of 17.13m located in liquefiable fine sand layer. The corridor and pumping-pit are constructed in one operation (Fig.4).

Layout of freezing pipes and determining freezing parameters

The freezing pipes were in line with rectangle, and two upper angles of wall were changed into arc(Fig.5). Excavation outline of the corridor is a height of 4.24m and breadth of 4.16m. Frozen wall was designed according to frozen powder sand strength for the corridor located in powder sand layer ,and a design thickness of 1.6m and average temperature of -10°C were obtained. The horizontal freezing pipes with 108mm diameter were installed in the direction of parallel to corridor axis, as shown in Fig.4 and fig.5. The detail parameters about freezing method applied to the corridor are shown in table 1.

Monitoring and analysis of frost heave and thaw displacement

Twelve monitoring points(1 to 12) of displacement were positioned on the ground surface in the direction perpendicular to the corridor axis(Fig.5).The frost heave and thaw displacement were obtained during different construction stages and shown in Fig.6.

Fig. 6 shows that the displacement above corridor axis is the biggest and it decreases with the increase of the distance from centerline. The deformation is divided into four stages. The first is drilling holes and installing pipes stage(25days), during which the

Table 1. Freezing parameters

Parameter name	Value	Parameter name	Value
Design thickness of frozen wall	1.6m	Design space of freezing pipes	0.6~0.84m
Recycling brine temperature	-22~-25°C	Bottom space of freezing pipes	1.0m
Average temperature of frozen wall	-8~-10°C	Time of forming sealed wall	25day
Amount of measure temperature holes	10	Positive freezing time	32day
Freezing pipes Amount	55	Maintainance freezing time	35day
Unload holes	4	Time of whole construction	95day

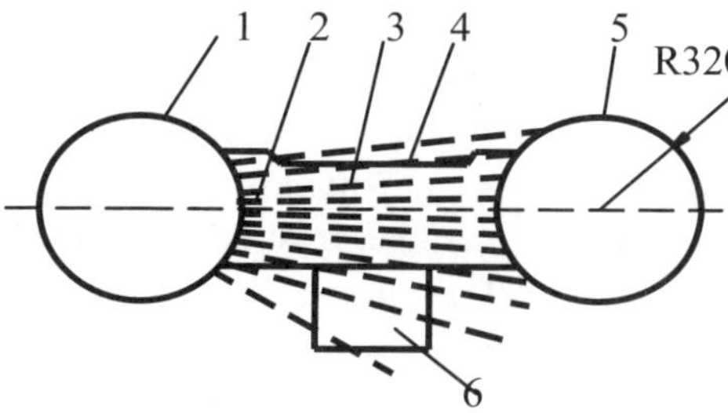

Figure 4. Freezing corridor of tunnels

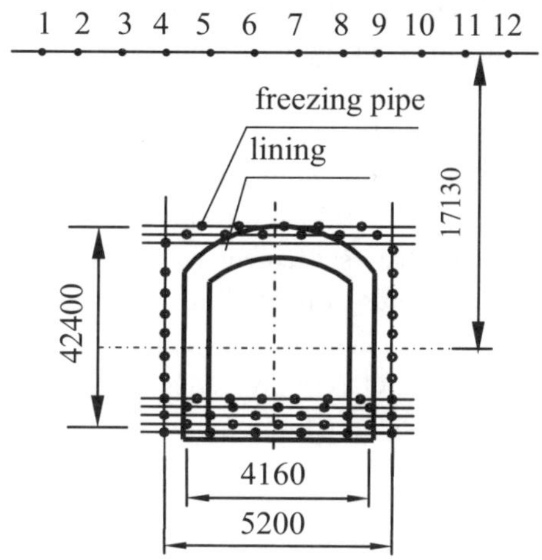

Figure 5. Layout of freezing pipes and point measured (mm)

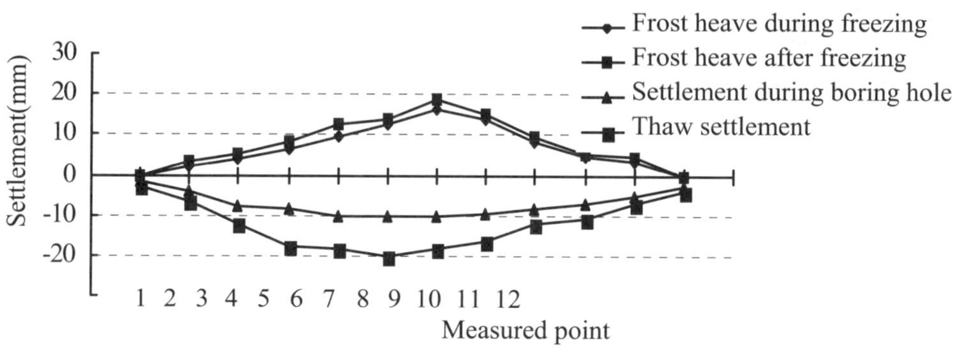

Figure 6. The displacement distribution during different construction stages

settlement of ground surface is not big with a maximum value of 10.3mm. This is due to partial water and soils running off holes and the ground volume is lost a little. But grouting into holes is adopted for reducing the settlement, which is very effective. The second is freezing stage after installation of pipes (32 days), during which the frost heave deformation is small at the beginning of freezing for the temperature of most soils being above 0°C, but the frost heave deformation increases greatly when soil temperature falls below zero degree. The reason is that original water and water migrating to the freezing front at 0°C is changed into ice and the soil swells rapidly. The figure shows the maximum heave deformation is 16.5mm after taking the measure of drilling four unload holes. The third is maintained freezing stage, i.e. excavation and installing lining stage (35days), during which the heave displacement rate reduces with an increment of only 2.6mm.The final maximum frost heave is 19.1mm. The fourth is unfreeze stage, during which thaw zone is grouted with injecting-pipes placed in advance to prevent the large ground surface subsidence, As such, the maximum thaw settlement is 20.4mm. The deformations during every stage are satisfied with allowance deformation of structure.

The soil will be disturbed during improving stratum with freezing method, especially, when the unfreezing time lasts longer. The disturbance may influence and even damage tunnel lining and buildings and pipelines around the tunnel(Brown et al.,2003). Some research results show that freezing-induced deformation may be bigger. Moreover, thaw settlement was about two to three times larger than frost heave without taking measures of unloading and grouting (Konrad J.M.,2002). But the examples in this paper proved that drilling unload holes and grouting with injecting-pipes placed in advance can effectively reduce frost heave and thaw settlement in order to protect buildings and tunnel lining.

CONCLUSION

Freezing method has been effectively applied to recover tunnel accidents in complex stratum of Nanjing subway. The examples proved that the frost heave and thaw settlement in the technique can be completely controlled within allowance value by means of taking measures during different stages. For example, partial frost heave can be compensated by drilling unload holes and thaw settlement can be reduced by grouting. So freezing method should be adopted prior to other methods when recover accidents and improves soil of shield entrance ,corridors and pumping-pit in weak stratum with rich water.

REFERENCES

Brown, D.A. (2003), “Hull wastewater flow transfer tunnel: recovery of tunnel collapse by ground freezing”, Geotechnical engineering, Vol.157,No.2,77~83

Konrad, J.M.(2002). “Prediction of freezing-induced movements for an underground construction project in Japan”, Can.Geotech.J.,Vol.39,No.6,1231-1242

Yang, P. and She, C.G.(2003), “Application of artificial freezing method in Zhang Fuyuan station of Nanjing subway”, Rock and Soil Mechanics ,Vol.24,No.2,365-368

Author Index